Methods in Enzymology

Volume XXIV
PHOTOSYNTHESIS AND NITROGEN FIXATION
Part B

METHODS IN ENZYMOLOGY

EDITORS-IN-CHIEF

Sidney P. Colowick   **Nathan O. Kaplan**

Methods in Enzymology

Volume XXIV

Photosynthesis and Nitrogen Fixation

Part B

EDITED BY

Anthony San Pietro

DEPARTMENT OF PLANT SCIENCES
INDIANA UNIVERSITY
BLOOMINGTON, INDIANA

1972

ACADEMIC PRESS   New York and London

ACADEMIC PRESS, INC.
111 Fifth Avenue, New York, New York 10003

United Kingdom Edition published by
ACADEMIC PRESS, INC. (LONDON) LTD.
24/28 Oval Road, London NW1

LIBRARY OF CONGRESS CATALOG CARD NUMBER: 54-9110

PRINTED IN THE UNITED STATES OF AMERICA

Table of Contents

Section I. Methodology

Section II. Inhibitors

Section III. Synthesizing Capabilities of the Photosynthetic Apparatus

Section IV. Nitrogen Fixation

Contributors to Volume XXIV

Article numbers are shown in parentheses following the names of contributors.
Affiliations listed are current.

M. Avron (7, 24), *Department of Biochemistry, The Weizmann Institute of Science, Rehovoth, Israel*

G. Ben-Hayyim, *Department of Biochemistry, Tel Aviv University, Tel Aviv, Israel*

William A. Bulen (40), *Charles F. Kettering Research Laboratory, Yellow Springs, Ohio*

R. C. Burns (43), *Central Research Department, E. I. DuPont de Nemours and Co., Wilmington, Delaware*

R. H. Burris (37), *Department of Biochemistry, University of Wisconsin, Madison, Wisconsin*

Warren L. Butler (1), *Department of Biology, University of California at San Diego, La Jolla, California*

David T. Canvin (21), *Department of Biology, Queen's University, Kingston, Ontario, Canada*

C. Carmeli (7), *Department of Biochemistry, Tel Aviv University, Tel Aviv, Israel*

Britton Chance (29, 30), *Department of Biophysics and Physical Biochemistry, Johnson Research Foundation, University of Pennsylvania School of Medicine, Philadelphia, Pennsylvania*

Roderick K. Clayton (12), *Division of Biological Sciences and Department of Applied Physics, Cornell University, Ithaca, New York*

Richard A. Dilley (4), *Department of Biological Sciences, Purdue University, Lafayette, Indiana*

Harold J. Evans (41), *Botany Department, Oregon State University, Corvallis, Oregon*

H. Fock (21), *Botanisches Institut, Johann Wolfgang Goethe-Universität, Frankfurt, West Germany*

David C. Fork (10), *Department of Plant Biology, Carnegie Institution of Washington, Stanford, California*

David Geller (6), *Department of Pharmacology, Washington University School of Medicine, St. Louis, Missouri*

Martin Gibbs (22), *Department of Biology, Brandeis University, Waltham, Massachusetts*

N. E. Good (3, 32), *Department of Botany and Plant Pathology, Michigan State University, East Lansing, Michigan*

T. W. Goodwin (35), *Department of Biochemistry, Johnson Laboratories, The University of Liverpool, Liverpool, England*

Thomas P. Hatch (36), *Department of Microbiology, University of Chicago, Chicago, Illinois*

G. E. Hoch (15, 25), *Department of Biology, University of Rochester, Rochester, New York*

R. W. F. Hardy (43, 44), *Central Research Department, E. I. DuPont de Nemours and Co., Wilmington, Delaware*

R. D. Holsten (44), *Central Research Department, E. I. DuPont de Nemours and Co., Wilmington, Delaware*

T. Horio (8), *Division of Enzymology, Institute for Protein Research, Osaka University, Osaka, Japan*

Y. Horiuti (8), *Research Laboratories, Toyo Jozo Co., Ltd., Shizuoka, Japan*

S. Izawa (3, 32), *Department of Botany and Plant Pathology, Michigan State*

University, East Lansing, Michigan

ANDRE T. JAGENDORF (9), *Section of Genetics, Development and Physiology, Division of Biological Sciences, Cornell University, Ithaca, New York*

PIERRE JOLIOT (11), *Institut de Biologie Physico-Chimique, Fondation Edmond de Rothschild, Paris, France*

BACON KE (2), *Charles F. Kettering Research Laboratory, Yellow Springs, Ohio*

ROBERT KLUCAS (41), *Department of Biochemistry and Nutrition, The University of Nebraska, Lincoln, Nebraska*

BURTON KOCH (41), *The Department of Soil Science, The University of Hawaii, Honolulu, Hawaii*

B. KOK (19), *Research Institute for Advanced Studies, Baltimore, Maryland*

JUNE LASCELLES (36), *Bacteriology Department, University of California, Los Angeles, California*

ERWIN LATZKO (22), *Chemisches Institut, Technische Hochschule, Munich, West Germany*

JACK R. LECOMTE (40), *Charles F. Kettering Research Laboratory, Yellow Springs, Ohio*

WALTER LOVENBERG (42), *National Heart and Lung Institute, National Institutes of Health, Bethesda, Maryland*

E. I. MERCER (35), *Dept. of Biochemistry and Agricultural Biochemistry, University College of Wales, Aberystwyth, Wales*

LEONARD E. MORTENSON (39), *Department of Biological Sciences, Purdue University, Lafayette, Indiana*

SATORU MURAKAMI (17), *Department of Biology, College of General Education, University of Tokyo, Tokyo, Japan*

K. NISHIKAWA (8), *Division of Enzymology, Institute for Protein Research, Osaka University, Osaka, Japan*

LESTER PACKER (17), *Department of Physiology — Anatomy, University of California, Berkeley, California*

W. D. PHILLIPS (26), *Central Research Department, E. I. DuPont de Nemours and Co., Wilmington, Delaware*

MARTIN POE (26), *Merck Institute for Therapeutic Research, Rahway, New Jersey*

JESSE RABINOWITZ (38), *Department of Biochemistry, University of California, Berkeley, California*

KENNETH SAUER (18), *Department of Chemistry and Laboratory of Chemical Biodynamics, University of California, Berkeley, California*

ANTONIO SCARPA (30, 31), *Department of Biophysics and Physical Biochemistry, Johnson Research Foundation, University of Pennsylvania School of Medicine, Philadelphia, Pennsylvania*

MARTIN SCHWARTZ (13), *Division of Science, University of Maryland, Baltimore County, Baltimore, Maryland*

JEROME A. SCHIFF (28), *Department of Biology, Brandeis University, Waltham, Massachusetts*

G. R. SEELY (20), *Charles F. Kettering Research Laboratory, Yellow Springs, Ohio*

NOUN SHAVIT (27) *Negev Institute for Arid Zone Research, Beer Sheva, Israel*

KAZUO SHIBATA (16), *Laboratory of Plant Physiology, The Laboratory of Physical and Chemical Research, Rikagaku Kenkyusho, Wakohshi, Japan*

ROBERT M. SMILLIE (33), *Plant Physiology Unit, Botany Department, University of Sydney, Sydney, N.S.W. Australia*

P. K. STUMPF (34), *Department of Biochemistry and Biophysics, University of California, Davis, California*

A. TREBST (14), *Department of Biology, Ruhr University, Bochum, Germany*

RICHARD W. TREHARNE (23), *Charles F. Kettering Research Laboratory, Yellow Springs, Ohio*

ELLEN C. WEAVER (5), *Department of Biological Sciences, San Jose State College, San Jose, California.*

HARRY E. WEAVER, *Hewlett-Packard Company, Palo Alto, California*

N. YAMAMOTO (8), *Faculty of Anatomy,*

School of Medicine, Kitasato University, Asamizodai, Sagamihara-shi, Kanagawa, Japan

K. L. ZANKEL (19), *Research Institute for Advanced Studies, Baltimore, Maryland*

Preface

Almost a decade ago, Martin D. Kamen ("Primary Processes in Photosynthesis," Academic Press, 1963) presented a most interesting temporal analysis of photosynthesis and correlated it with the "level of ignorance" about the nature of the sequential processes which occur in the successive "eras" of photosynthesis. Ignorance was great in the time span from pt_s 9 to 4 (the pt_s terminology denotes "logarithm of the reciprocal of time, expressed in seconds"). An additional correlation was that the "level of ignorance" was generally related to the somewhat limited number and sophistication of approaches then available to investigate the various "eras" of photosynthesis. Clearly, the widespread availability during the past decade of new, more sophisticated and powerful investigative techniques has resulted in a decrease in both the time span (perhaps presently pt_s 9 to 6) and "level of ignorance."

The presentations in the first three sections of this volume provide an up-to-date analysis and descriptions of the chemical and physical investigative techniques used to probe the mysteries of photosynthesis; a description of the widely used buffers and inhibitors; and discussions of the synthesizing capabilities of the photosynthetic apparatus. The last section is concerned with the analytical techniques applicable to the study of biological nitrogen fixation as well as with the isolation of components involved therein.

My deepest gratitude is extended to all the authors for their excellent contributions. The boundless patience and cooperation of the staff of Academic Press are gratefully acknowledged. Excellent secretarial assistance was provided by Mrs. Virginia Flack and Mrs. Cheryl Fisher.

ANTHONY SAN PIETRO

METHODS IN ENZYMOLOGY

EDITED BY

Sidney P. Colowick   Nathan O. Kaplan

VANDERBILT UNIVERSITY
SCHOOL OF MEDICINE
NASHVILLE, TENNESSEE

DEPARTMENT OF CHEMISTRY
UNIVERSITY OF CALIFORNIA
AT SAN DIEGO
LA JOLLA, CALIFORNIA

METHODS IN ENZYMOLOGY

EDITORS-IN-CHIEF

Sidney P. Colowick and Nathan O. Kaplan

Section I

Methodology

[1] Absorption Spectroscopy of Biological Materials

By WARREN L. BUTLER

The purpose of this chapter is to promote spectroscopy on biological materials. Pigments play central roles in many basic biological processes. The discovery and study of these pigments by spectrophotometric measurements affords the biochemist one of his most powerful techniques, especially when measurements can be made on intact tissue or cellular homogenates. It has previously been demonstrated that dense, light-scattering samples, such as frozen homogenates, tissue slices, intact peanuts, or even a block of wood, are amenable to spectroscopic examination.[1-3]

Emphasis will be on low-temperature absorption and derivative spectroscopy, because these techniques greatly enhance the power of the method. Instrumentation for making spectrophotometric measurements on dense, highly scattering samples and theoretical and practical aspects of the measurements will constitute the major theme of the chapter.

Instrumentation

The discussion of instrumentation will describe the construction and use of a single-beam recording spectrophotometer. Such an instrument lacks the convenience of having a reference beam for difference spectroscopy or baseline compensation but has a number of advantages for making spectroscopic measurements on dense, light-scattering materials which may cover a wide range of optical densities over the spectral region of interest. The photometric range of a split-beam instrument is limited by cross talk between the two beams, whereas the photometer of a single-beam instrument should be capable of covering at least seven to eight decades of light intensity. The dynamic range of a single-beam spectrophotometer is generally limited by a component other than the photometer. Commercial split-beam spectrophotometers use a horizontal light path through the sample, whereas a vertical light path has advantages for light-scattering materials. Construction of a single-beam

[1] W. L. Butler and K. H. Norris, *Arch. Biochem. Biophys.* **87**, 31 (1960).

[2] K. H. Norris and W. L. Butler, *IRE Transactions on Bio-Medical Electronics*, Vol. BME-8, No. 3: 154 (1961).

[3] W. L. Butler and K. H. Norris, *in* "*Modern Methods of Plant Analysis*," pp. 51–72. Springer-Verlag, Berlin and New York, 1962.

spectrophotometer largely from commercially available modules is a relatively simple task and can be accomplished for less than 5000 dollars.

Results and experience from two single-beam recording spectrophotometers will be discussed. One instrument employs a Bausch and Lomb 500-mm Monochromator, an electrical baseline compensator, and a direct readout from the photometer to an $X–Y$ recorder. The other employs a Cary Model 14 Monochromator with the photometer on line with a small computer (Digital Equipment Corp. PDP 8-I). Spectra are stored in the memory of the computer for later manipulation and display as absolute spectra, difference spectra, or derivative spectra on an oscilloscope or $X–Y$ recorder.

Even though the computerized spectrophotometer has many advantages, particularly for difference spectroscopy, the main emphasis will be on the simpler and less expensive instrument. Much of the spectral information can be obtained more conveniently with the simple instrument.

Monochromator and Lamp. Spectral purity, spectral resolution, and the spectral energy distribution should be considered in selecting a monochromator. Spectral resolution of 1 nm or less is required for low-temperature spectroscopy of cytochromes, which limits the usefulness of some inexpensive monochromators. Prism monochromators have the advantage of having a smooth spectral distribution of energy with the wavelength, but the resolution is poor at longer wavelengths, and the wavelength display is not linear. Grating monochromators have the advantage of a linear dispersion capable of high resolution at all wavelengths, but the spectral distribution may have "flares" in certain wavelength regions. Examples of such "flares" and the problems of baseline compensation will be discussed in the section on *Spectral Characteristics*.

The spectral purity of the monochromator may set the upper limit of the photometric range. Examples of how stray light limits the photometric range and accuracy and a comparison of the stray light from a single and a double monochromator will be presented in the section on *Stray Light*.

A tungsten–iodine lamp with a well-regulated dc power supply provides a good source for measurements in the visible and near-infrared regions. The color temperature of these lamps is somewhat higher than that of regular tungsten lamps so that more energy is available, particularly at shorter wavelengths. At high sensitivity, it is possible to observe the absorption line structure of the iodine vapor as a series of absorption bands in the 550–650-nm region, but in most applications the iodine absorption bands are not strong enough to be a serious problem.

Sample Compartment. A vertical light path is more convenient than a horizontal path for spectrophotometric measurements on dense, scattering samples [Fig. 1(a)]. Cuvettes for liquid samples can be made from metal cylinders with clear plastic windows pressed into the bottom. It is important that the walls of the cuvettes be opaque to avoid light being transmitted around the sample to the phototube. When measuring high optical densities, all possible precautions should be taken to avoid stray light. The vertical light path has the advantage with suspensions that the settling particles remain in the measuring beam. An open-topped Dewar with optical windows on the bottom [Fig. 1(b)] provides a convenient low-temperature system for a vertical optical path.[2] The liquid nitrogen surrounds the metal-walled cuvette and is not in the optical path. The opaque top and metal walls of the cuvette ensure that only light transmitted through the frozen sample is measured.

Photometer. The development of operational amplifiers (OA) has greatly simplified the design and construction of photometers. The single-beam instrument requires a wide-range photometer to cover the range of phototube currents from 10^{-4}–10^{-11} A. A split-beam instrument generally uses the signal from the reference beam to adjust the gain of the phototube so that the photometer handles a spectrophotometric range of about two decades. The simple single-beam instrument uses a fixed phototube gain (set by the phototube high-voltage power supply) so that a range of 7 decades is required. It is generally desirable to have a logarithmic photometer to cover such a wide photometric range.

The OA acts as a negative feedback amplifier to oppose the voltage difference across its input terminals. The anode of the phototube is con-

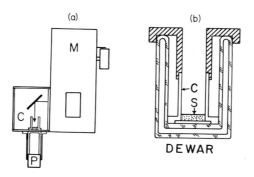

Fig. 1.   (a) Diagram of monochromator (M) with sample compartment (C) and phototube (P). (b) Diagram of Dewar with cuvette (C) and sample (S).

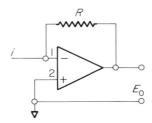

nected to the input terminal 1, called the summing point, and terminal 2 is connected to the ground. With a resistor in the feedback loop from the output to the input, the amplifier generates a voltage E_0 which produces a current through the feedback resistor (R) to nullify the input current (i)

$$i + E_0/R = 0$$

or

$$E_0 = -Ri$$

The input terminal is called the summing point because the current may be from a variety of sources which are summed together and which together determine the output voltage. Since the input current is nullified, no voltage appears between the input terminals, and terminal 1 is essentially at ground potential. A linear amplifier of wide dynamic range can be achieved by using different feedback resistors. For example, with $R = 10^9\ \Omega$, $E_0 = 1$ V for $i = 10^{-9}$ A or for $R = 10^4\ \Omega$, $E_0 = 1$ V for $i = 10^{-4}$ A. The quality of the OA sets the lower limit on current measurement and instruments capable of measuring 10^{-14} A are readily available.

A logarithmic response is obtained if a transistor is used as the feedback element. In that case, E_0 is equal to $0.059 \log i$. The transistor feedback element provides a logarithmic response over the range 10^{-11}–10^{-4} A. The change of voltage for a change of current is

$$\Delta E = 0.059 \log i_1/i_2\ \text{V}$$

or ΔE is about 60 mV for every tenfold change of current (i.e., 60 mV/OD unit).

The main problem of the simple transistor feedback log circuit is one of temperature stability. The voltage–current characteristics of the transistor are sensitive to temperature so that the feedback properties change with temperature. However, two identical transistors with two OA can be used in such a way that the effects of a temperature change are canceled. Detailed circuits for temperature stabilization as well as matched transistors and packaged temperature-stabilized log modules are available from manufacturers of OA. The log circuit is also sensitive to temperature because the 0.059 figure (which is $2.3\ RT/F$ where R is the gas constant, T is the absolute temperature, and F is Faraday's constant) changes about 0.3% per degree centigrade. This source of temperature dependence can also be compensated by incorporating a thermistor into the photometer circuit.

A second OA of low quality can be used in the linear mode after the logarithmic amplifier to give some convenient photometric scale such

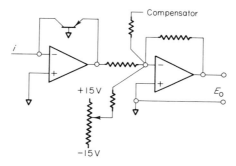

as 1 V/OD (optical density) unit. The summing point on the second OA is also used for a zero control and the baseline compensator. A potentiometer connected to the ± 15 V of the OA power supply will provide a zero adjust. The signal from the potentiometer used for the baseline compensator can also be summed to the photosignal here.

The technology and prices of operational amplifiers has changed rapidly over the last few years. At present, however, one can purchase for less than 40 dollars a temperature-stabilized logarithmic module which covers a range of 10^{-11}–10^{-3} A with a calibrated output of 1 V/decade. This unit would require a ± 15 V stabilized power supply, and at least one additional OA would be needed to provide a summing point for zero adjust and baseline compensation.

The signal from the second OA can be differentiated with a third OA, also of low quality, by connecting the two through a condenser with a feedback resistor on the third OA. The signal is differentiated against time by the RC network, $E_0 = RC(dE_i/dt)$. A condenser is generally

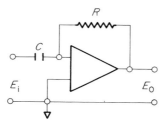

needed in parallel with the feedback resistor to prevent the amplifier from oscillating. Here, also, manufacturers' literature should be consulted for detailed circuitry.

The derivative circuit, however, needs a constant speed for the monochromator drive because the differential is taken with respect to time. The magnitude of the derivative bands is directly proportional to the speed of the scan. The selection of component values for the derivative circuit requires some compromise between speed of response and magnitude of output.

The phototube can be connected so that the current from the anode goes directly to the first OA without passing through a load resistor. The

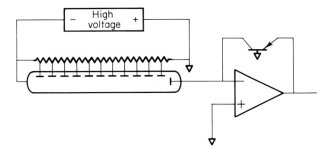

cathode is at the negative high voltage, and the anode is at ground potential because the input terminal of the OA is effectively at ground.

Recorder. If an X–Y recorder is used, the X signal can be taken from a simple potentiometer geared to the wavelength drive. Alternatively the wavelength drive of the monochromator can be mechanically coupled to the paper drive of some strip chart recorders. The Y axis should have range of voltage sensitivities to give a range of photometric sensitivities.

Baseline Compensation. The simplest form of baseline compensation is a simple potentiometer geared to the wavelength drive that supplies a current which changes linearly with wavelength to the summing point of the second OA in the photometer. The voltage across the potentiometer (from a regulated dc power supply and a second auxiliary potentiometer) determines the slope of the current-*vs*-wavelength ramp. The

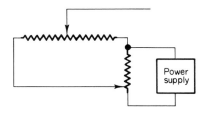

sign and slope of the signal from the scanning potentiometer can be adjusted to make the absorption spectrum horizontal over a limited spectral range.

A multitapped potentiometer (connections are made at intervals along the potentiometer winding) geared to the wavelength drive provides a more sophisticated baseline compensation system. The voltage at each of the taps is controlled by an auxiliary potentiometer and regulated power supply. Each tap will correspond to a wavelength setting of the monochromator and may be spaced at 5–10-nm intervals. The base-

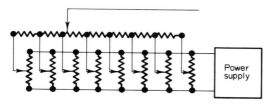

line compensation is set up by adjusting the auxiliary potentiometer for each correction point to give a predetermined reading with no sample or with a scatter reference in the instrument. In this manner a flat baseline can be achieved, provided the curve being compensated is smooth enough to be accommodated by the number and spacing of available correction points.

A motor scan unit for a Bausch and Lomb High Intensity Monochromator with a 38-position multitapped potentiometer baseline compensation system is available commercially from Agricultural Specialty Co., Beltsville, Maryland. This unit was geared to a Bausch and Lomb 500-mm Monochromator in the author's laboratory so that the 38 equally spaced correction points covered either 380 nm or 190 nm. This unit has a number of convenient features which are worth incorporating if such a unit is built: (1) a motor scan mode in which, on pressing a switch, the monochromator automatically scans to the next correction point and stops with a pilot light at each of the 38 auxiliary potentiometers to indicate where the multitapped potentiometer has stopped (this system greatly enchances the ease and speed of setting up a baseline); (2) adjustable limit switches which stop the motor scan at predetermined wavelength limits; and (3) a manual scan mode as well as automatic scans at various predetermined rates. The unit purchased by the author also had switches added at each of the auxiliary potentiometers which, in effect, removed those correction points from the system. A linear current ramp could thus be obtained over a wavelength region by switching out the intermediate correction points and using the auxiliary potentiometers at the end points to adjust the slope.

The ideal compensation system for a single-beam spectrophotometer is obtained by having the photometer on line with a computer. The computerized spectrophotometer developed in the author's laboratory takes readings every 0.12 nm and stores the spectrum in the memory system. At a later time, difference spectra between any of the spectra in the computer can be read out on an oscilloscope or X–Y recorder. Even with this system, however, it is sometimes desirable to use a simple potentiometer geared to the wavelength drive to remove a steep background slope from a spectrum.

Spectral Characteristics. The system response curve of a single-beam spectrophotometer is a combination of the spectral sensitivity of the phototube, the spectral emission of the lamp, the spectral transmission of the monochromator, and any nonlinearities in the photometric system. System response curves recorded as the logarithm of the photocurrent *vs* wavelength and the compensated baselines are presented in Fig. 2 for a Bausch and Lomb 500-mm Monochromator with two different gratings, both with 1200 lines/mm, one blazed at 500 nm and the other at 300 nm, and in Fig. 3 for a Cary Model 14 Monochromator. An EMI 9558C phototube was used with both instruments so that the S20 spectral response of the phototube was a part of the system response. A

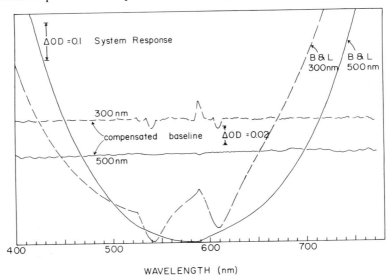

WAVELENGTH (nm)

FIG. 2.   System response (photometer output *vs* wavelength) of spectrophotometer with Bausch and Lomb 500-mm Monochromator with a grating (1200 lines/mm) blazed at 300 nm and at 500 nm. The compensated baseline (2.5-fold greater sensitivity) of the two system responses was obtained with a multitapped potentiometer with correction points every 10 nm.

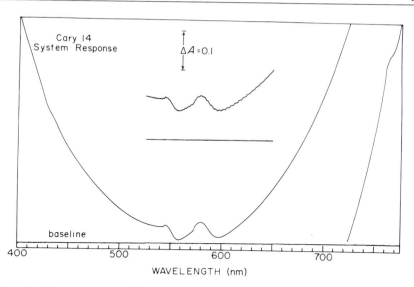

FIG. 3.   System response of spectrophotometer with Cary Model 14 Monochromator. Baseline obtained by storing the system response curve in the computer on two separate scans and plotting the difference of the two curves. Main curve obtained with regular tungsten lamp. Inset curve obtained with 100-W tungsten–iodine lamp.

multitapped potentiometer with correction points every 10 nm was used for baseline compensation with the Bausch and Lomb Monochromator; the computer was used with the Cary Monochromator. The Bausch and Lomb Monochromator with grating blazed at 500 nm gave a smooth system response curve with no points of inflection (Fig. 2). The baseline compensation system with correction points every 10 nm could compensate such a response curve to give a fairly smooth flat baseline. Some scalloping with a period of 10 nm occurs at regions of appreciable curvature in the system response curve because the straight line segments of the compensation system cannot correct for all the curvature. The line structure of the iodine vapor in the 45-W tungsten–iodine lamp (General Electric 6.6A/T2½ Q/Cl-45 W) used with the Bausch and Lomb Monochromator was just discernible in the compensated baseline between 600 and 650 nm. The Bausch and Lomb Monochromator with the grating blazed at 300 nm had "flares" near 540 and 610 nm which could not be adequately compensated with correction points spaced 10 nm apart. The Cary Model 14 Monochromator also had "flares" in the response curve (Fig. 3), but compensation was no problem here since this monochromator was used with a computer which took photometric readings every 0.12 nm. Two system response curves were stored in the computer and subtracted from one another to give the baseline in Fig. 3. The complete system response curve of Fig. 3 was taken with the regular tungsten lamp supplied with the monochromator. A short segment from 525 to 650 nm was also taken with a 100-W tungsten–iodine lamp (General Electric 6.6 A/T2½ Q/Cl-100 W) to demonstrate the iodine absorption lines. The concentration of iodine vapor appears to be greatest in the 100-W lamps. Compensation of these lines is no problem with the computer but would be with the multitapped potentiometer system. With the Cary Monochromator the intensity of light at 550 nm was three- to fourfold greater with a 200-W tungsten–iodine lamp than with the regular tungsten lamp supplied with the monochromator.

Stray Light. Stray light may be defined as any undesirable light measured by the phototube. Common sources of stray light are leaks in the sample or phototube compartments, fluorescence or phosphorescence from the sample, and light of other wavelengths in the monochromatic beam. The effect, regardless of the source, is to limit the maximum density which can be measured and to distort absorption spectra in wavelength regions of high density. A simple pragmatic test to determine the permissible photometric range can be made by measuring the absorbance of a filter which absorbs strongly. The absorption spectra of a Corning 2408 glass filter, which cuts off wavelengths below 620 nm, measured with the 500-mm Bausch and Lomb Monochromator and with the Cary Model 14 Monochromator is shown in Fig. 4. With the Cary Monochromator, measurements were made with the filter both at the entrance optics to the monochromator and in the sample compartment close to the phototube. The system response of the Cary Monochromator with an S11 phototube (EMI 6255 B) is also included in Fig. 4 for comparison to the response with an S20 phototube (EMI 9558C).

No stray light was detected in the Cary Monochromator with the filter at the monochromator entrance; this means that the stray light in this test was less than 1 part in 10^6. The maximum level at a density somewhat above 6 was set by the dark current of the phototube. Reducing the dark current by cooling the phototube might have extended the photometric range to a level where stray light could have been detected. Stray light due to fluorescence and/or phosphorescence by the filter was detected when the filter was placed close to the phototube, and the photometric range was reduced by about 1.5 OD units at 570 nm. Since light scattering samples must be placed in close juxtaposition to the phototube sample, fluorescence can be one of the most troublesome forms of stray light.

With the Bausch and Lomb Monochromator, the maximum absorbance range was limited to 3.3 by the stray light in the monochromatic

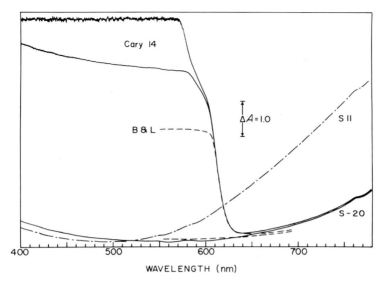

WAVELENGTH (nm)

FIG. 4. Absorption spectrum of Corning 2408 filter measured with the Cary Monochromator (—) and with the Bausch and Lomb Monochromator (---). The filter was
placed at the entrance optics to the Cary Monochromator (upper solid curve) and in the
sample compartment near the phototube (middle solid curve). The system response curves
of the instrument with the Cary Monochromator with an S20 phototube (lower solid curve)
and with an S11 phototube (—–—) are also compared.

pass band. In this test for stray light at wavelengths below 600 nm, the
monochromatic beam included nondispersed energy at longer wavelengths which were transmitted by the filter and measured by the phototube. The test for stray light thus depends upon the spectral response
of the detector.

The photocurrent can be considered to consist of a variable part, i_m,
due to the transmitted measuring beam and a constant part, i_s, due to
stray light or to the dark current of the phototube. The optical density
is log $i_0/(i_m + i_s)$, where i_0 is the photocurrent with no sample. The photometric accuracy is good so long as $i_m \gg i_s$. The measurement comes
into doubt, however, when i_m is less than $10 i_s$ or whenever the density
measurement comes within 1 OD unit of the maximum set either by
stray light or by phototube dark current.

Light Scatter. The close juxtaposition between the sample and the phototube permits measurements to be made on dense, highly scattering
samples. The equations which describe the optics of highly scattering
media have been developed previously.[4] The primary effect of light

[4] W. L. Butler, *J. Opt. Soc. Amer.* **52**, 292 (1962).

scatter in such media is to increase the optical path, thereby increasing the intensity of the absorption bands. The intensification factor β was derived theoretically:

$$\beta = 2 \left(\frac{-4R_\infty^3}{(1-R_\infty^2)^2 \, SX} + \frac{1 + R_\infty^2}{1 - R_\infty^2} \right)$$

where R_∞ is the reflectance of an infinitely thick sample, S is the scatter coefficient of the medium, and X is the thickness of the sample. Good agreement was found between theoretical and experimental values of β in experiments in which small amounts of dye solutions were added to highly scattering media.[4] In practice light scatter may intensify absorption bands as much as 100-fold. With most white scattering media, the scattering power, (SX), is large enough that the second term in the parentheses for β predominates and β may approach $2(1 + R_\infty^2)/(1 - R_\infty^2)$. If the particle size of the scattering powder is much smaller than the wavelength of light, however, the intensification is considerably less than for larger particles of the same material, in part because the scattering coefficient is less.

There is no simple relationship between wavelength or particle size and intensification as might be assumed from single particle scattering theory. In general, the intensification is less at shorter wavelengths because the reflectivity is less. End absorption by slight impurities in a white powder markedly reduce the intensification at shorter wavelengths. With strong absorption bands, the intensification will be less at the maximum of the band than on the sides of the band because the reflectivity is less at wavelengths where the absorption is greater. With weak absorption bands, however, there is remarkably little distortion by the scattering medium.

The use of a light scattering agent as an optical medium is an important technique because it permits spectroscopic measurements to be made on a number of biological materials not otherwise feasible. Addition of light scatter to a dilute solution or suspension may intensify absorption bands to a level where they can be measured. Small pieces of tissue can be ground and mixed into a white powder and measured as a white paste. Heavily pigmented powders, such as lyophilized chloroplasts, which are too dense to be measured even as a thin layer, can be diluted with the dry scattering medium and measured as a dry powder.[5] All these samples can readily be measured at liquid-nitrogen temperature.

The absorption spectrum of a 6-mm-thick sample of a white paste of

[5] W. Menke, C. S. French, and W. L. Butler, Z. *Naturforsch.* **30**, 482 (1965).

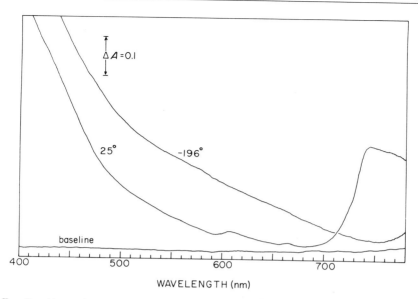

FIG. 5.   Absorption spectrum of 6-mm-thick sample of Al_2O_3 paste (0.5 g of Al_2O_3 in 1 ml of water) at 25° and at −196°. Instrument with Bausch and Lomb 500-mm Monochromator and baseline compensated with multitapped potentiometer.

Al_2O_3 (0.5 g of Al_2O_3 in 1 ml of water) is presented in Fig. 5. The scatter medium gives a steeply rising baseline below 500 nm. Absorption bands of water (overtones of the O–H stretching frequencies) are also apparent because of the strong intensification by the scatter. The water absorption band which has an extinction of 0.01/cm at 750 nm was intensified 40- to 50-fold. Much weaker bands are also apparent near 605 and 665 nm. These absorption bands of water are not present when the sample is frozen at 0° as well as at −196°. The absorption spectrum of 0.5 g of dry Al_2O_3 (not shown) was similar to that of the frozen Al_2O_3 paste.

Applications

The absorption spectrum of a relatively dense suspension of yeast (2×10^9 cells/ml) recorded with the system response compensated by the multitapped potentiometer is presented in Fig. 6. Steeply rising background absorption precludes measurements being made at higher sensitivity. Incorporation of scatter blank such as tissue or filter paper into the baseline compensation would have leveled the spectrum to some degree. For many purposes, however, the absorption spectrum over a limited spectral region can be measured more conveniently with baseline compensation from the simple potentiometer adjusted to give a level

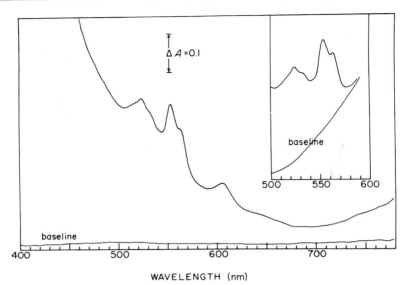

WAVELENGTH (nm)

FIG. 6. Absorption spectrum of yeast suspension (2×10^9 cells/ml) measured with a flat compensated baseline. Inset: Absorption spectrum measured over limited spectral region with the simple potentiometer baseline compensator adjusted to give a horizontal spectrum.

spectrum (inset, Fig. 6). Baseline compensation in this case is very rapid, and the spectrophotometric sensitivity can be increased without increasing the slope of the background absorption.

A suspension of yeast dense enough to give readily measurable absorption spectra, e.g., 2×10^9 cells/ml in Fig. 6 and curve A of Fig. 7, tends to go anaerobic quite rapidly, making measurements on oxidized samples difficult. Diluting the suspension tenfold decreased the rate of oxygen consumption but also decreased the absorption bands to the point where spectral measurements were difficult (curve B, Fig. 7). (The absorption bands were intensified to some extent by the light scatter in the dense suspension.) Adding 0.5 g of Al_2O_3 to 1 ml of the diluted yeast suspension restored the absorption bands by increasing the optical pathlength (curve C, Fig. 7). The sample in this case was a white paste about 6 mm thick. Freezing the sample to $-196°$ intensified the absorption bands further and sharpened the bands so that cytochrome c_1 could be resolved between the cytochrome c and cytochrome b absorption bands (curve D, Fig. 7). For each spectrum recorded in Fig. 7 the current ramp from the linear compensation system was readjusted to give a horizontal spectrum.

A slight extention of the single simple potentiometer compensation system to two simple potentiometers or one center tapped potentiometer permits different current ramps to be used over two spectral regions. The absorption spectrum of beef-heart mitochondria recorded with a linear correction ramp between 495 and 645 nm is presented in Fig. 8, curve A. The absorption bands were on top of a background rising both to longer and shorter wavelengths. By using two correction ramps, however, one from 495 to 585 nm, the other from 585 to 645 nm, the spectrum (curve B) could be adjusted to be approximately horizontal with a discontinuity at 585 nm. Freezing the sample increased the spectral resolution markedly (curve C). The low temperature spectrum shows a clear resolution of cytochrome c, cytochrome c_1, and two forms of cytochrome b.

The magnitude of the absorption bands can be calculated as the difference between the absorbance at the maximum and that at a nearby reference wavelength, provided readings are corrected for the baseline difference between those wavelengths. The magnitude of the cyto-

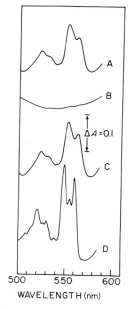

Fig. 7. Absorption spectra of yeast (1-ml sample, 6 mm deep) with simple baseline compensation adjusted to give a horizontal spectrum. Curve A, suspension of 2×10^9 cells/ml; B, suspension diluted tenfold, 2×10^8 cells/ml; C, same sample as curve B but with 0.5 g of Al_2O_3 added to the 1-ml sample; D, same sample as curve C frozen to $-196°$.

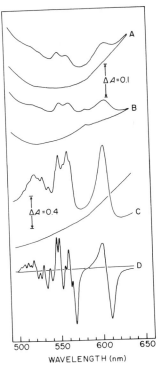

Fig. 8. Absorption spectra of beef-heart mitochondria (1-ml sample, 6 mm deep) and baselines. Curve A, with linear baseline compensation signal between 495 and 645 nm; B, with two regions of linear baseline compensation, 495–585 and 585–645 nm; C, same sample frozen to −196°; D, derivative spectrum of same sample as curve C.

chrome oxidase band at room temperature calculated as the absorbance difference between 607 and 620 nm was 0.037 for both curves A and B. The magnitude of the oxidase band at 77°K calculated as the difference between 604 and 620 nm was 1.04 (or 0.98 between 607 and 617). The light scatter in the frozen sample increased the absorption band of cytochrome oxidase about 27-fold.

The derivative spectrum obtained with the derivative amplifier described in the photometer section emphasizes shoulders on spectra and calls attention to details which might otherwise be overlooked. The derivative spectrum of the mitochondria at 77°K (curve D) shows the main components, such as the two b-type cytochromes, which are adequately resolved in the absorption spectrum. The usefulness of derivative spectroscopy is better exemplified in the region between 500 and 520 nm, where the details might be missed in the absorption spectrum. First derivative spectroscopy was also used to resolve the low-temperature absorption spectrum of spinach chloroplasts[6] into eight absorption bands, most of them due to minor components. The bands persisted through detergent fractionation of the chloroplasts into photosystem 1 and photosystem 2 particles and could be assigned to one or the other photosystem. Spectral resolution of the low-temperature absorption spectrum of chloroplasts by a higher derivative is demonstrated later.

The spectral data in Figs. 7 and 8 were obtained with the simplest form of spectrophotometer. The capabilities of such an instrument for the spectral analysis of biological systems far exceed those of conventional commercial spectrophotometers.

Computerized Spectrophotometer. The output of the photometer was connected to one of the inputs of a 12-bit analog to digital (A–D) converter (Digital Equipment Corp. AF-01) with an input range of 0–5 V. With the sensitivity of the photometer set at 0.5 OD/V, the input range corresponded to a photometric range of 2.5 OD and a ± 1-bit conversion uncertainty corresponded to a photometric noise of ± 0.0006 OD. This noise level was somewhat greater than that inherent in the photometer so that a precision greater than 12 bits (1 part in 2^{12} or 4096) could be used at times if it were available. A 16-bit A–D converter, for instance, would permit a 0 to 10 OD input range with a noise level of 2×10^{-4} OD.

A precision potentiometer (0.025% linearity) geared to the wavelength drive transmitted a signal which was proportional to wavelength to another input of the A–D converter, 0–5 V corresponding to a range of 500 nm. The computer was programmed to start taking readings at a specified wavelength, to take readings at a specified interval, and to take a specified number of readings. Between each wavelength increment the computer added 128 A–D conversions in double precision, divided by 128, and stored the averaged reading sequentially in the memory core of the computer. Both the number of conversions added and the division factor in the averaging subroutine are under program control. Signal averaging can also be achieved by repeatedly scanning a spectrum and adding the spectra together at the same curve locations in the core.

With our present memory capacity (8192 locations), up to seven curves can be stored in the computer. Absolute spectra, difference spectra between any two curves, and the derivative of the absolute and difference spectra can be displayed on an oscilloscope or *X–Y* recorder, punched out on tape, or moved to a new location in the core. The derivative is obtained by taking the difference between readings a specified interval apart. Second derivative spectra can be obtained by placing the first derivative curve at a new location in the core in order to differentiate the curve again. The second derivative curve has the advantage that the

[6] W. A. Cramer and W. L. Butler, *Biochim. Biophys. Acta* **153**, 889 (1968).

minima correspond to the maxima in the original curve. Higher derivative spectra can be obtained by sequentially repeating the derivative process.

Low-temperature absorption spectra of the colorless alga *Prototheca zopfii* reduced by dithionite in the presence of KCN and methanol (curve A) or KCN alone (curve B) are shown in Fig. 9. The direct absorption spectra of the two samples (2.5×10^8 cells in 1 ml of buffer with 0.5 g of Al_2O_3) at 77°K and of the scatter medium (0.5 g of Al_2O_3 in 1 ml of buffer) at 77°K were stored in the computer and read out on the X-Y recorder as the difference between the sample and the blank or as the difference between the two samples (curve A − B). The absorption spectra in the 400–440-nm region were displaced to avoid compressing the photometric scale. The difference spectrum was recorded at twice the sensitivity.[7]

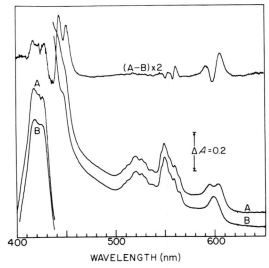

FIG. 9. Absorption spectra of *Prototheca zopfii* (1 ml of suspension of 2.5×10^8 cells/ml with 0.5 g of Al_2O_3) at −196° with baseline compensation by computer. Curve A, cells reduced with dithionite in the presence of 2.5×10^{-4} M KCN and 2.5% methanol; B, cells reduced with dithionite in the presence of 2.5×10^{-4} M KCN; A − B, difference spectrum from computer.

[7]The data in Fig. 9 demonstrate a spectral resolution of cytochromes a and a_3 in the α-band region obtained when cells of *P. zopfii* are reduced in the presence of KCN and methanol. Methanol shifted the α band of cytochrome a from 598 to 603 nm and permitted dithionite to reduce the cyanide–cytochrome a_3 complex to give a 595-nm absorption band.[8] Methanol also shifted the double-peaked Soret band of cytochrome a 4–5 nm to longer wavelength and caused small shifts in the α bands of some of the other cytochromes.

[8]B. Epel and W. L. Butler, *Plant Physiol.* **45**, 723 (1970).

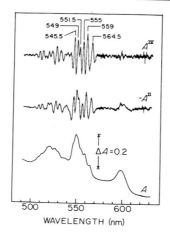

WAVELENGTH (nm)

FIG. 10. Absorption spectrum, second derivative spectrum, and fourth derivative spectrum of the same sample used in Fig. 9, curve B. Sequential differentiations all used the same interval of 1.75 nm.

The absorption spectrum of curve B from Fig. 9 from 490 to 630 nm as well as the second derivative (run as the negative so that maxima would correspond to maxima in the original curve) and the fourth derivative of the spectrum as shown in Fig. 10. These results showing increased resolution in the higher derivative curves stimulated a computer analysis of higher derivative spectra. The computer (PDP 8-I) was programmed to sum together a series of absorption bands specified by wavelength maximum, half-width, magnitude, and band shape (any specified mixture of Gaussian and Lorentzian band shapes) and to sum the derivatives of these bands up to the fourth derivative. A computer-simulated spectrum generated as the sum of the absorption bands shown under the spectrum is given in Fig. 11. A band shape of 50% Gaussian–50% Lorentzian was arbitrarily selected. Absorption bands of 5-nm half-width were placed at 545, 549, 551, 555, 559, and 564 nm in intensity ratios of 0.7:1.0:0.5:0.7:0.7:0.5. The purpose was not to provide a curve fit to the experimental data, but rather to provide a similarly shaped spectrum with an absorption band ($\lambda_{max} = 551$) which was not clearly resolved. The question to be asked is: Can such a band be resolved in the second or fourth derivative of the spectrum, and if so, how does the position of the derivative band correspond to the position of the absorption band? It is shown in Fig. 11 that the 551-nm band was clearly resolved in the fourth derivative curve and that the position agreed closely with the wavelength maximum of the absorption band. A more detailed treatment of spectral analysis by higher derivatives has been

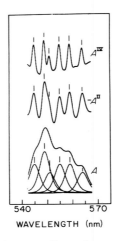

540   570
WAVELENGTH (nm)

Fig. 11.   Computer-generated curves. Curve A, sum of absorption bands specified as 50% Gaussian and 50% Lorentzian band shapes with 5-nm half-width with the following wavelength maxima and relative peak heights: 545 nm, 0.7; 549 nm, 1.0; 551 nm, 0.5; 555 nm, 0.7; 559 nm, 0.7; and 564 nm, 0.5; $-A^{II}$, sum of the negatives of the second derivatives of these bands; A^{IV}, sum of fourth derivatives of these bands. The short vertical lines indicate the wavelength maxima of the component absorption bands.

presented elsewhere.[9] The data in Fig. 10, however, demonstrate that spectral data (even from dense light scattering samples) can be subjected to higher derivative analysis without losing all information in the noise and that information as to the location and numbers of absorption bands can be obtained which is not apparent in the original spectrum.

The absorption spectra of spinach chloroplasts at room temperature and at liquid-nitrogen temperature and the fourth derivatives of these spectra are shown in Fig. 12. The eight bands resolved previously by Cramer and Butler[6] in the first derivative of the low-temperature spectrum of chloroplasts are shown as sharp maxima in the fourth derivative curve.

Several aspects of the measurement of higher derivative spectra are demonstrated in Fig. 13(a) with another sample of chloroplasts at $-196°$. A more dilute suspension was used ($A_{676} = 0.2$), and the sample was scanned four times and added together at the same curve location. Measurements were made every 0.125 nm in the 600–725-nm range. A baseline was also scanned and added four times. Scanning four times should improve the signal-to-noise ratio by a factor of 2. Curve A is the absorption spectrum corrected for the baseline. The fourth derivative of the absorption spectrum with each sequential differentiation taken

[9]W. L. Butler and D. W. Hopkins, *Photochem. Photobiol.* **12,** 439 (1970).

over a 2-nm interval (sixteen 0.125-nm intervals) is shown in curve B. The shorter wavelength interval [2 nm *vs* 4 nm in Fig. 12(b)] improves the resolution and resolves the 677 nm fourth derivative band in Fig. 12(b) into two bands but is less effective in showing broad weak bands at 638 and 661 nm.

The signal-to-noise ratio in the fourth derivative curve can be appreciably improved by using slightly different wavelength intervals for the four differentiating processes. Curve C shows the fourth derivative curve where the first derivative was taken with a spectral interval of 1.625 nm, the second derivative with 1.75 nm, the third with 2.0 nm, and the fourth with 2.5 nm (representing multiples of 13, 14, 16, and 20 times the basic 0.125-nm interval). The analysis of noise in the higher derivative curves has been treated in greater detail elsewhere.[10] However, it can be shown by sequentially numbering the digitized points taken every 0.125 nm along the absorption spectrum that the first point in curve B corresponds to absorption measurement points $65 - 4(49) + 6(33) - 4(17) + 1$ and that the first point in curve C corresponds to points $64 - 51 - 50 - 48 - 44 + 37 + 35 + 34 + 31 + 30 + 28 - 21 - 17 - 15 - 14 + 1$. Subsequent points along each fourth derivative curve are comprised of similar sets of absorption points and are plotted at the middle of the spectral range indicated by the spread of points. Each measured point

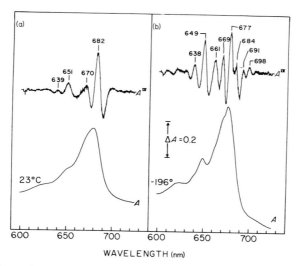

Fig. 12. Absorption spectra and fourth derivative spectra of chloroplasts. Panel a, at room temperature; panel b, at liquid-nitrogen temperature. Sequential differentiations all used the same wavelength interval of 4.0 nm.

[10] W. L. Butler and D. W. Hopkins, *Photochem. Photobiol.* **12**, 451 (1970).

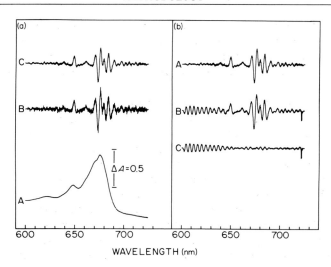

Fig. 13. Panel (a): Curve A, absorption spectrum of chloroplasts at −196°; 1 ml of chloroplasts was measured in the frozen suspension four times and the spectral curves were added together to give spectrum four times the density of the sample. A blank for 1 ml of frozen buffer was measured and summed four times. The absorption spectrum, curve A, is the difference between the accumulated spectra of the sample and the blank; curve B, fourth derivative of curve A obtained by four sequential differentiations using an interval of 2.0 for each; curve C, fourth derivative of curve A using an interval of 1.625 nm for the first derivative, 1.75 nm for the second, 2.00 nm for the third, and 2.50 nm for the fourth.

Panel (b): Curve A, same as curve C, panel (a). Curve B, fourth derivative of chloroplast spectrum without baseline correction. Curve C, fourth derivative of baseline curve. The same sequence of spectral intervals were used for the sequential differentiations in curves A, B, and C [i.e., those used for curve C, panel (a)].

has a certain random noise. It is apparent that the summation of points for curve C gives a random summation of the noise in the 16 measured points, whereas the summation for curve B is not random. Thus it is possible to incorporate a significant signal-to-noise improvement in the higher derivative processes.

In many cases derivative spectra would not require correction for the baseline, provided the spectral region did not include flares from the grating or line structure from the lamp because the baseline slopes are too small to give an appreciable derivative signal over the wavelength interval generally used. The curves in Figs. 12 and 13(a), however, were corrected for the baseline because the iodine lines from the tungsten−iodine lamp gave an appreciable signal in the single-beam spectra. The fourth derivative of the absorption spectrum corrected for the baseline (curve A) and without baseline correction (curve B) is given in Fig. 13(b). Curve C is the fourth derivative of the baseline. The iodine lines are apparent in the 600–650-nm region.

Summary

A single-beam recording spectrophotometer capable of measuring absorption and first-derivative spectra on dense, light-scattering samples at low temperature can be constructed largely from commercially available components. Such an instrument permits spectral analysis of intact biological tissue as well as suspensions of cells and cellular homogenates. A single-beam spectrophotometer with a simple potentiometer baseline compensation system will provide most of the spectral information available. A more sophisticated instrument capable of difference spectra and higher derivative spectra can be obtained by putting the photometer on line with a digital computer via an analog-to-digital converter. Theoretical and practical aspects of these instruments and measurements are discussed, and examples of spectral analysis with these methods are presented.

[2] Flash Kinetic Spectrophotometry[1]

By Bacon Ke

Flash kinetic spectrophotometry is the technique for following rapid reactions induced by a short, intense flash of light. The course of reaction, i.e., the nonequilibrium condition created by the short flash and the return to the equilibrium, is usually followed by measuring the accompanying absorbance change as a function of time. The flash photolysis technique originally developed by Norrish and Porter[1a] records the entire absorption spectrum of the reaction solution at a chosen instant after flash excitation. This is done by passing a second (measuring) flash through the solution to a spectrograph. The time separation between the excitation and measuring flashes is controlled electronically. A series of such measurements at various time intervals provides information on the reaction kinetics.

In photosynthetic reactions the magnitude of the relative absorbance

[1] Contribution No. 447 from the Charles F. Kettering Research Laboratory, Yellow Springs, Ohio 45387. This work was supported in part by a National Science Foundation Grant GB8460.
[1a] R. G. W. Norrish and G. Porter, *Nature (London)* **164**, 658 (1949).

changes is usually so small that this technique becomes impractical. Alternatively, the kinetics of the reaction induced by a short flash can be followed by measuring the change in transmission of a steady measuring beam set at a wavelength appropriate for the compound of interest. The modes of operation of conventional flash photolysis and kinetic spectrophotometry are shown in Fig. 1. Kinetic spectrophotometry has the advantage of being not only more sensitive and rapid in response, but quantitative data on rate constants are more readily obtained.

A typical version of the flash kinetic spectrophotometer suitable for photosynthesis studies is shown in Fig. 2. The excitation flash may come from a xenon flash lamp generated by capacitor discharge, a Q-switched ruby laser, or secondary emissions induced by Q-switched lasers, e.g., stimulated Raman scattering, or organic dye lasers. In principle, the excitation flash should be of short duration compared with the lifetime of the transient species of interest and should have sufficient energy to produce an adequate concentration of transient species. The steady measuring beam should have high stability. The method of detection should be rapid so as to not introduce a lag or distortion in the transient waveform.

This section deals with the elementary principles involved and some general background information, and where possible, there is supplementation with examples of some commonly used components. The presentation to follow may be divided into four general areas (cf. Fig. 2).

1. Excitation flash; light sources; methods of pulse generation; and wavelength production and selection.

2. Measuring light; some common sources; power supply for the sources; and wavelength production.

3. General operation; geometric arrangement of the measuring and excitation light sources; timing; modulation of the measuring light; and the temperature control of the sample.

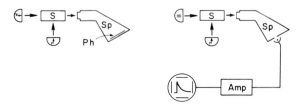

Fig. 1. Schematic presentation of conventional flash photolysis (left) and kinetic spectrophotometry (right). Bent arrow, flash source; double bar, steady light source; S, sample; Sp, spectrograph; Ph, photographic plate; Amp, amplifier.

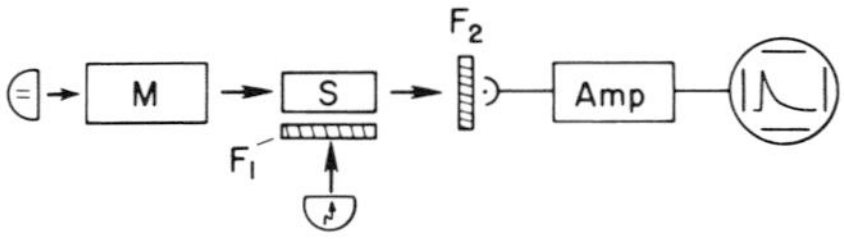

FIG. 2.   A typical version of flash kinetic spectrophotometer for photosynthesis studies. M, monochromator; F_1 and F_2, complementary filters; other symbols as in Fig. 1.

4. Signal detection and readout; detectors and their uses; amplification and manipulation of signals. The last item refers to the signal averaging and sampling techniques.

Excitation Flashes

Xenon Flash Lamps

The conventional method for generating a flash is to discharge a condenser through a gas discharge lamp. The electrical energy stored in the condenser is $\frac{1}{2}(CV^2)$, where V is the voltage in kilovolts and C is the capacitance in microfarads. A representative value of the efficiency of conversion of the electric energy into light energy is about 1%. The energy requirement for photosynthetic reactions depends on the concentration of the active species, its molar extinction coefficient, and the quantum efficiency of the absorbed energy. There is a further restriction on energy requirement due to wavelength utilization imposed by the absorption bands of the molecules being illuminated. Special situations, such as high optical density of the sample, low extinction coefficient, etc., often require a greater energy.

Commercially available flash lamps with durations of 10^{-6}–10^{-4} sec and a flash energy input of the order of 100 J are sufficient for the study of a wide variety of photosynthetic reactions. The half duration of a flash is expressed by the width at 50% of the peak intensity. The improvement in flash duration is usually achieved at the expense of energy output. Porter[2] has compiled a table listing such characteristics as the energy and duration of the flash and the nature of the gas used for flash lamps developed up to 1960. These data have recently been reexamined by Boag.[3] In general, a 10-fold reduction in flash duration requires a 100-fold decrease in energy. Flash duration of the above-stated range would allow reactions with rate constants of 10^6–10^5 or less to be studied.

The discharge of the electrical energy stored in the capacitor through the flash lamp is triggered (switched) by applying a high-voltage trigger

[2]G. Porter, Z. Elektrochem. 64, 59 (1960).
[3]J. W. Boag, Photochem. Photobiol. 8, 565 (1968).

pulse to ionize the gas, thus making it conductive. The storage capacitor can then discharge its energy into the flash lamp, causing a brilliant flash of light.

While the voltage and capacitance together affect the energy as well as the duration of the flash, other factors, such as the inductance and resistance in the discharge path, the physical design of the flash lamp, and the gas composition determine the duration and profile of the flash. The arc resistance in the discharge path limits the rate of capacitor discharge since the resistance in other parts of the circuit is usually small. The arc resistance increases with the length and decreases with the diameter of the arc chamber. Inductance distributed among various circuit components — the condenser, connecting cables, trigger, and lamp — all contribute significantly to the flash duration. To obtain flashes of short duration, lamps of short arc gaps, low-inductance capacitors, high voltage, and low capacitance should be used: further, the capacitors should be mounted close to the lamp. An excellent recent discussion on the various factors affecting the flash characteristics is given by Boag.[3]

Another factor affecting the flash duration is the "afterglow" arising from long-lived excited states in the gas. The delayed emission forms a long tail in the flash profile. Common methods for reducing the tailing in the flash are: addition of a quencher gas such as nitrogen or oxygen into the rare-gas composition; or increasing the quenching surface within the flash lamp by packing the lamp with quartz chips or quartz wool,[3] or by using special lamp geometry by constraining the flash in the annular space between a tube and a concentric rod.[4]

Xenon discharge lamps with a wide range of duration and intensities are available from General Electric, EGG (Boston, Massachusetts), Xenon Corp. (Medford, Massachusetts), PEK (Sunnyvale, California), and many others. We have used a GE type-503 coiled flash lamp with input energy of 100 J and a half duration of 15 μsec, a Suntron-1 linear flash tube (Xenon Corp.) at 45 J and 2-3-μsec duration, and an XE-5 helical flash tube at 80 J and 100 μsec.

"Z-Pinch" Flash Lamp

One recently developed method for producing short-duration flashes with high energy in both the visible and ultraviolet regions is achieved by the so-called "Z-pinch" flash lamp.[5,6] Here the condenser is dis-

[4]C. J. Hochanadel, J. A. Ghormley, J. W. Boyle, and J. R. Riley, *Rev. Sci. Instrum.* 39, 1144 (1968).
[5]E. E. Daby, Doctoral Dissertation, University of Detroit, 1967.
[6]E. G. Niemann and M. Klenert, *Appl. Opt.* 7, 295 (1968).

charged in a low-pressure gas whose ions have been contracted by an electromagnetic pinch toward the center axis. The purpose of the electromagnetic confinement is to keep the ions from contacting the wall of the chamber and thus from rapidly recombining and cooling. At maximum contraction a very high temperature is generated and part of the kinetic energy is emitted in the form of light. One "Z-pinch" device reported by Niemann and Klenert[6] consisting of argon at 0.03 Torr and operated at 3.3 kV and a 4.85-kJ condenser bank produces flashes of 0.2 μsec with total energy of about 40 J/flash measured in the 200–600-nm region. This represents a reduction in flash duration of more than two orders of magnitude for equal light energy output compared with the conventional xenon flash lamp.

Rapid kinetic spectrophotometry using flashes of microsecond durations is suitable for observing triplet-state formation and primary products formed in chemical reactions. However, direct observation of more primary events following singlet-state formation by the absorption of light would require further improvement of the time resolution and further decrease in finite flash duration. As mentioned earlier, improvement in the flash duration of the discharge lamps is usually accompanied by a decrease in the output intensity. At the present moment it is generally agreed that a practical limit in these two parameters has nearly been reached. Although generation of submicrosecond flashes by Z pinch is possible, it has not yet been widely used in kinetic spectrophotometry.

Q-Switched Ruby-Laser Flash

The discovery of the pulsed ruby laser a few years ago was indeed timely since it can readily replace the conventional flash source in rapid kinetic spectrophotometry. Aside from its short duration, the laser radiation is monochromatic. The highly collimated laser beam also allows its use at a distance, thus reducing possible electrical disturbance in the measuring circuit.

Application of 30-nsec pulses at 694 nm from a Q-switched ruby laser to the photochemical activation of cytochrome–chlorophyll interaction in photosynthetic organisms was first reported by Chance and DeVault.[7] Chance and co-workers[8] also first employed the pulse train of 1 msec duration generated from the regular-mode ruby laser in the study of cytochrome kinetics in a *Chlamydomonas* mutant.

The principal component in the ruby laser is a ruby rod (Al_2O_3 doped

[7] B. Chance and D. DeVault, Z. *Elektrochem.* **68**, 722 (1964).
[8] B. Chance, H. Schleyer, and V. Legallais, *in* "Microalgae and Photosynthetic Bacteria" (Japanese Soc. Plant Physiologists, eds.) p. 337. University of Tokyo Press, 1963.

with 0.05% Cr_2O_3) with flat parallel ends. The chromium ions in the ruby are the active component for lasing. When the ruby rod is optically "pumped" by a high-energy flash lamp, the blue and green light absorbed by the ruby raises the chromium ions from the ground levels to higher excited levels. From these excited levels there is a rapid radiationless transition to intermediate, metastable levels. Because of the relatively long lifetime of the metastable levels, with sufficient blue and green light, it is possible to populate more chromium ions in these metastable levels than in the ground levels. Since according to Boltzmann distribution there are normally more chromium ions in the ground than in any excited levels, the above situation is called "population inversion." Population inversion does not lead to laser action unless the ruby rod is placed in a resonant cavity. The resonant cavity consists of two mirrors, one at each end of the rod, and carefully aligned parallel to the rod end faces. One of the mirrors is totally reflecting and the other is partially reflecting. These mirrors may also be coated directly on the polished rod ends. When population is inverted, fluorescence is given off in all directions. Some of this fluorescence light reaches one of the two mirrors forming the cavity and is reflected back through the ruby rod. As it passes through the ruby, it is capable of stimulating other atoms to emit photons at the same frequency and in the same direction as the incident photons. Since the stimulating photon can just as easily be absorbed by an atom in going from the ground level to the higher excited level, it is important that population inversion be maintained, so that absorption is less likely to occur than stimulation. More stimulated emission and amplification occur by repeated reflection between the two mirrors. At threshold the amplification is sufficient to overcome the losses in the Fabry-Perot resonant cavity, and above the threshold an intense burst of red light is produced by the ruby laser (Lengyel[9] is recommended for further details). The beam produced by ruby laser has a wavelength of 6943 Å with a bandwidth of 0.1 Å. The highest efficiency in energy conversion for ruby lasers is about 1%. Typical output of a laser beam consists of a train of pulses or spikes with a total duration of about 5×10^{-4} sec. Commonly available ruby lasers have output energy ranging from 1 to 50 J at power levels of about 10^5 W.

The duration of the laser pulse may be shortened and its power level increased by several orders of magnitude by a technique called "Q spoiling" of the resonant cavity.[10] In the Q-switching process, the quality factor Q of the resonant cavity which is necessary for lasing to

[9]B. A. Lengyel, "Introduction to Laser Physics." Wiley, New York, 1966.
[10]R. W. Hellwarth, *in* "Lasers" (A. K. Levine, ed.), Vol. 1, p. 253. Dekker, New York, 1966.

occur is "spoiled" for a controlled period of time while the ruby is being pumped. This prevents the laser action to occur and the population inversion reaches an extraordinarily high level. By instantly restoring the reflection, thus completing the resonant cavity, the excited ruby releases its stored energy as a single giant pulse with extremely short duration (about 2×10^{-8} sec) and an instantaneous power of 10^7 W or higher.

There are many Q-switching devices. The most commonly used among them are the rotating mirror and the Pockels cells (*cf.* Fig. 3 and 4). The rotating mirror operates by replacing the totally reflecting mirror in the resonant cavity. The resonant cavity is formed only during the short time when the rotating mirror is parallel to the partially reflecting mirror at the other end. By synchronizing the rotation of the mirror, the time when the cavity is completed may be selected to optimize the power of the laser output.

The Pockels cell utilizes an electrooptic rotation of the polarized light in certain crystalline material, such as potassium dihydrogen phosphate. When a Pockels cell is used as a Q switch it is placed inside the resonant cavity between the totally reflecting mirror and one end of the ruby rod. When a high voltage is applied to the Pockels cell, the polarized light output from the ruby (an additional polarizer may be needed in certain lasers) is rotated to spoil the optical cavity. When the voltage is switched off, a process which may be accomplished very rapidly, the resonant cavity is completed by restoring the reflection. Again, as in the rotating mirror device, the time may be optimized to obtain desired power output.

Kerr effect in certain liquids may also be used in a manner similar to that of the Pockels cell. One magnetooptic effect of particular interest to laser switching is the Faraday effect, which is the rotation of the plane of polarization of a light beam as it passes through a material in the presence of a magnetic field.

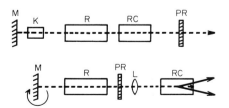

FIG. 3. Raman laser action stimulated by Q-switched ruby laser pulses. Two methods of ruby laser Q switching are shown: by Kerr cell or Pockels cell (top) or by rotating prism (bottom). M, mirror; PR, partial reflector; K, Kerr cell; R, ruby laser rod; RC, Raman cell; L, lens.

Thus far we have restricted our discussions to the ruby laser, which has indeed been the only laser source for inducing photochemical reactions in photosynthesis. It has been shown to be effectively absorbed by chlorophyll *a* in the green alga *Chlamydomonas*.[11] It has been more widely used in examining photochemical reactions in photosynthetic bacteria.[12-14] It is exceedingly effective even though the wavelength of ruby laser falls in a region of low absorption of the bacteriochlorophyll. In fact, even with a relatively low-power ruby laser, it is often necessary to spread the beam and to attenuate the intensity. However, it is obvious that a wider choice of available wavelengths would broaden the possibility of exciting different regions of the absorption spectrum of the different chlorophylls. For instance, it might be useful to have a wavelength for chlorophyll *b* excitation at 630 nm, for bacteriochlorophyll in the near-IR region, or for the exclusive excitation of green plant photosystem 1, say at 720 nm.

Stimulated Raman Emission

Although it is feasible in principle to use individual pulsed gas or solid-state lasers to produce appropriate wavelengths for particular regions of excitation, it is certainly inconvenient and impractical. Several methods for producing variable wavelengths have become available recently. For instance, tuning the wavelength by temperature variation, frequency doubling, stimulated coherent Raman scattering, and finally, the most recently developed and possibly the most useful, organic dye laser. Temperature tuning is still in a stage of development, but it is usually confined to a narrow wavelength range. Frequency doubling in a ruby laser and neodymium glass laser produce light of wavelengths 347 and 530 nm, respectively. Although these wavelengths are not particularly suitable for direct excitation, they may be used in producing coherent Raman scattering and in stimulating the organic dye lasers to be discussed below.

In classical Raman scattering, the frequency of the scattered light differs from the original light by an amount of some internal motion of the molecules of the scattering medium. Usually this internal motion is a vibration, but rotational and electronic shifts can also occur. The lower-frequency Raman lines are called Stokes lines. At higher temperatures where kT becomes comparable to the energy of the internal motion, anti-Stokes Raman frequency, namely, frequency equal to the sum of

[11]W. W. Hildreth, *Plant Physiol.* **43**, 303 (1968).
[12]D. DeVault and B. Chance, *Biophys. J.* **6**, 825 (1966).
[13]W. W. Parson, *Biochim. Biophys. Acta* **153**, 248 (1968).
[14]B. Ke, *Biochim. Biophys. Acta* **172**, 583 (1969).

the incident light and the vibrational energy, may also be observed. The Raman scattered light is always distributed randomly in all directions, and its intensity is very low.

Because of the high power level of the giant pulse ruby laser, the intensity of the Raman scattered light caused by the giant pulse can be greatly enhanced over that of the ordinary Raman scattering. In fact, the Raman emission can even be amplified if the Raman-active medium is placed into a condition of population inversion and stimulated emission in a resonant cavity (Fig. 3). Woodbury and Ng[15] first observed just such a Raman laser action when they placed a cell containing nitrobenzene inside the cavity of a Kerr-cell Q-switched ruby laser (Fig. 3, top). That the exciting radiation from the rotating-prism Q-switched ruby laser is focused on the Raman medium is shown in Fig. 3 (bottom). The efficiency in this arrangement is very high because of the high exciting power. A more detailed discussion on stimulated Raman scattering and various pumping geometries is given by Bhaumik.[16]

In giant laser stimulated Raman scattering, the conversion efficiency can approach 20–30%, and the beam collimation is the same as that of the ruby radiation. A number of liquids, including some aliphatic, aromatic, and heterocyclic compounds, are known to exhibit coherent Raman scattering. More recently, coherent Raman scattering has also been reported for gases and solids. Thus, with the giant ruby laser as the pumping source, many wavelengths, both longer and shorter than 694 nm, may be obtained. For instance, benzene, pyridine, and 1-bromo-naphthalene all give 882-nm emission, and cyclohexane, the 866-nm emission, both of which are suitable for exciting the far-red bacterio-chlorophyll band *in vivo*.

Organic Dye Laser

Another means of obtaining laser radiations covering a wide spectral range is the newest class of lasers—the organic dye lasers. The first true optically pumped dye laser which utilizes the singlet–singlet transitions in an organic dye molecule as the source of coherent light was achieved in 1966 by Sorokin and Lankard.[17] Here, as in the case of ruby laser, the basic requirements of population inversion and stimulated emission also apply. In the case reported by Sorokin and Lankard, a solution of chloroaluminum phthalocyanine in ethanol was excited by a giant pulse ruby laser, yielding a coherent radiation at 755 nm. Shortly thereafter, a number of dyes yielding laser emissions at various wavelengths were

[15]E. J. Woodbury and W. K. Ng, *Proc. IRE* **50,** 2367 (1962).
[16]M. L. Bhaumik, *Amer. J. Phys.* **35,** 330 (1967).
[17]P. P. Sorokin and J. R. Lankard, *IBM J. Res. Develop.* **10,** 162 (1966).

discovered. The bandwidth of dye laser emissions typically measures 5–20 nm.

To obtain nanosecond dye laser pulses, giant pulses from Q-switched ruby (694 nm), the second harmonic of ruby (347 nm), or neodymium (530 nm) have been used. Two commonly used pumping geometries are shown in Fig. 4. In the transverse arrangement (Fig. 4, top), the dye cell is placed in its own cavity with appropriate end reflectors, and the giant laser beam excites at a right angle. In the longitudinal- or end-pumped arrangement [Fig. 4 (bottom)], the dye cell is placed in a secondary cavity, aligned with the primary ruby cavity. As shown in Fig. 4 (top) for the dye 3,3′-diethylthiatricarbocyanine (emission at 816 nm) reported by Sorokin and co-workers,[18] a secondary mirror reflecting 40% at 694 nm and 99% at 816 nm separates the two cavities, and the end mirror reflects 75% at both wavelengths. The longitudinal arrangement achieves a more symmetrical pumping and results in a narrower beam divergence. A conversion efficiency as high as 40–50% has been reported.[18]

Sorokin and co-workers[19] subsequently discovered that dye lasers can also be pumped by fast-rising flash lamps. A fast-rising lamp pulse is necessary in order to prevent alternate paths for relaxation of the excited singlet states, e.g., internal conversion, intersystem crossing, and triplet–triplet absorption, which all tends to raise the lasing threshold. How-

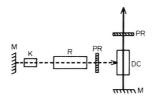

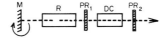

FIG. 4. Two pumping geometries used for organic dye lasers. The transverse (top) and the longitudinal (bottom) arrangements. DC, dye cell; all other symbols as in Fig. 3. Two types of Q switches are shown. However, they are independent of the dye-cell pumping geometry.

[18]P. P. Sorokin, J. R. Lankard, E. C. Hammond, and V. L. Moruzzi, *IBM J. Res. Develop.* **11**, 130 (1966).
[19]P. P. Sorokin, J. R. Lankard, V. L. Moruzzi, and E. C. Hammond, *J. Appl. Phys.* **48**, 4726 (1968).

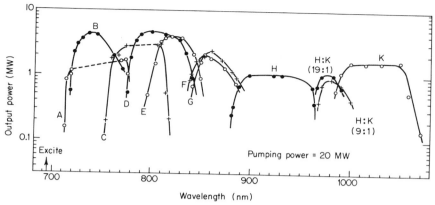

FIG. 5.   Plot of output peak power as a function of wavelength (taken from Miyazoe and Maeda[21]). Dotted-line portion, no emission. Dyes: Curve A: 3,3'-diethyl-10-chloro-2,2'-(4,5,4',5'-dibenzo)thiadicarbocyanine iodide; B: 3,3'-dimethyl-2,2'-oxatricarbocyanine iodide; C: 1,1'-diethyl-11-bromo-2,2'-quinocarbocyanine iodide; D: 1,3,3,1',3',3'-hexa-methyl-2,2'-indotricarbocyanine iodide; E: 3,3'-diethyl-2,2'-thiatricarbocyanine iodide; F: 3,3'-diethyl-2,2'-(5,6,5',6'-tetramethyoxy)thiatricarbocyanine iodide; G: 3,3'-diethyl-2,2'-(4,5,4',5'-dibenzo)thiatricarbocyanine iodide; H: 1,1'-diethyl-4,4'-quinotricarbocya-nine iodide; K: 1,1'-diethyl-4,4'-quinotricarbocyanine iodide. Solvents: glycerin for dye C; methanol for dye D; acetone for all others.

ever, giant pulse laser pumping is required to obtain pulses of high energy, short duration, and for dyes of low quantum yields.

The most distinctive feature of using dye laser pulses for inducing photosynthetic reactions is perhaps the wavelength tunability of the dye-laser emissions. Thus any chosen absorption band of a molecule may be excited to bring about the excitation of a particular quantum state. This would be invaluable in the selective excitation of the different near-infrared bands of bacteriochlorophyll of bacterial chromatophores for possible elucidation of the variable effectiveness of the different bacteriochlorophyll forms in catalyzing the oxidation of cytochromes (cf. Morita[20]).

By selecting a dye with a desired fluorescence spectrum, it is possible to cover the spectral range from the short wavelength end of the visible region to the near infrared. Miyazoe and Maeda[21] used a series of poly-methine dyes to cover completely the spectral range 710–1060 nm (Fig. 5). Thus the wavelengths at the low end may be used for the exclusive excitation of the green plant photosystem I reaction center, whereas

[20]S. Morita, *Biochim. Biophys. Acta* **153**, 241 (1968).
[21]Y. Miyazoe and M. Maeda, *Appl. Phys. Lett.* **12**, 206 (1968).

other wavelengths in the near-infrared region may be used for the excitation of a number of bacteriochlorophyll forms. The peak power of the dye laser as a function of wavelength when pumped in the transverse mode with a Q-switched ruby laser of 20 MW peak power is plotted in Fig. 5.

There are many methods for further wavelength tuning. The simplest is by changing the dye concentration.[18,22] For instance, in the case of 3,3'-diethylthiatricarbocyanine iodide, a center wavelength can occur over a range of 53 nm.[18] For a given dye solution, the lasing wavelength also depends on the type of solvent[18] (range of wavelength shift 20 nm), the cavity Q,[22] path length of the dye cell in the longitudinal pumping mode[23] (shift range 50–60 nm), and temperature.[24] An ingenious method for tuning the dye-laser wavelength was discovered by Soffer and McFarland,[25] who replaced the totally reflecting cavity mirror with a diffraction grating. By rotating the grating, it is possible to vary the output wavelength by 30–40 nm for a given concentration of rhodamine 6G solution. An additional interesting effect resulted from using the diffraction grating is a dramatic bandwidth narrowing of the dye-laser emission. In the case of rhodamine 6G, the bandwidth of the emission was narrowed from 60 to 0.6 Å when the grating was employed in place of the broadband dielectric reflector. Morover, there is no significant loss in the total power output accompanying the bandwidth narrowing.

Laser-Induced Breakdown Sparks

Another laser-induced secondary emission that may be used as a short excitation flash is the gas breakdown spark continuum. Novak and Windsor[26] have recently reported the use of an air breakdown spark induced by a Q-switched ruby laser giant pulse as a spectroscopic flash in conventional flash photolysis where a Q-switched laser pulse is also used as the photolysis flash. Because the induced spark has a duration as short as that of the giant pulse laser flash, the time resolution of flash photolysis can be improved 100-fold. The spark duration varies with the type of gas used and the pressure. For instance, oxygen at 1 atm gives a duration of 30 nsec; xenon at 1 atm, several microseconds. The intensity of the breakdown spark is said to be adequate for single-shot flash spectroscopy. However, no quantitative estimation of the intensity was given.

[22]F. P. Schäfer, W. Schmidt, and J. Volze, *Appl. Phys. Lett.* **9**, 306 (1966).
[23]G. I. Farmer, B. G. Huth, L. M. Taylor, and M. R. Kagan, *Appl. Opt.* **8**, 363 (1969).
[24]G. T. Schappert, K. W. Billman, and D. C. Burnham, *Appl. Phys. Lett.* **13**, 124 (1968).
[25]B. H. Soffer and B. B. McFarland, *Appl. Phys. Lett.* **10**, 266 (1967).
[26]J. R. Novak and M. W. Windsor, *Proc. Roy. Soc. Ser. A* **308**, 95 (1968).

Ramsden and Savic[27] earlier used a 20 nsec, 5–10 MW ruby laser giant pulse to induce a spark in air and observed the spark rising to a maximum within 15 nsec and decaying thereafter in about 150 nsec. They estimated that approximately 0.2 J, representing 60% of the energy of the laser beam is absorbed in the spark.

In summary, it may be anticipated that Q-switched ruby laser and laser-induced secondary emissions will gain broader use in photosynthesis studies. The usefulness of the Q-switched ruby laser pulse has already been demonstrated. Among the laser-induced secondary emissions, the stimulated Raman emission offers a large number of fixed frequencies and the organic dye laser offers a broad, continuous coverage. Both appear promising because of their high efficiency and simplicity.

A dye-laser cavity which can be used in conjunction with a Q-switched ruby laser as the pumping source has recently become available (Xenon Corp., Bedford, Massachusetts). Also, flash-lamp-pumped dye lasers are presently available from half a dozen manufacturers.

Measuring Light

Tungsten Lamp

Since a separate, more comprehensive treatment on light sources can be found elsewhere in this volume,[28] the discussion here will be brief and limited to the more practical aspects. Only a small number of light sources is sufficient to cover the spectral range of importance to photosynthesis. Tungsten lamps are the most versatile and useful artificial sources for the visible and near-IR regions. The spectral characteristics of radiation emitted by tungsten lamps depend upon the thermal radiation properties of the metal with limits imposed by the envelope. The color temperature and luminous efficiency increase with the filament temperature, so does the metal evaporation. The latter results in decreased filament cross section, blackening of the lamp envelope, and ultimately shortening of the lamp life. The rate of evaporation also depends on the filament geometry; it is relatively higher for a thin wire than for a large wire, a coil, or a coiled coil. The newer tungsten–iodine lamps are designed for effectively controlling the metal evaporation. The iodine vapor serves as a regenerative getter and causes the evaporated tungsten to redeposit on the filament. Consequently, the tungsten–iodine lamps with quartz envelopes allow a higher operating temperature. Two examples of commonly used quartz tungsten–iodine lamps are GE type-1958 (30 V, 150 W) and Sylvania type DXM (30 V, 250 W, 3400°K).

[27]S. A. Ramsden and P. Savic, *Nature (London)* **203**, 1217 (1964).
[28]R. W. Treharne, this volume [23].

Deuterium Lamp

Because of the inherently weak ultraviolet emission of thermal radiators, the tungsten lamps are inefficient ultraviolet sources. More suitable for the ultraviolet region are the hydrogen and deuterium lamps. The more recently available deuterium lamp is an excellent high-intensity source of continuous radiation in the 170–350-nm region, with only weak visible emission. Because of the high radiation density and stability, the deuterium lamps are now widely used in ultraviolet spectrophotometry. Commonly available air-cooled deuterium lamps are: DE-50 (Sylvania, 40 V, 0.8 A), D-102 (Quarzlampen, Germany, 90 V, 0.3–0.6 A), and D-100 (Kern, Göttingen, Germany, 60–90 V, 0.3 A). The latter two manufacturers also furnish higher-intensity deuterium lamps which require water cooling: D150W (Quarzlampen, 125 V, 0.8–1.2 A) and WHS-200 (Kern, 100–140 V, 1.0–1.5 A). Liebman and Entine[29] have recently compared the relative radiance of three deuterium lamps: D-102-S, D-150W, and WHS-200.

Xenon and Mercury Lamps

Although the above-described lamps cover practically the entire useful spectral range for photosynthetic reactions, special needs of higher-intensity measuring sources as dictated by the experimental conditions may arise. In this respect, two other light sources may be mentioned. The xenon spectrum, which has an apparent temperature of about 6000°K, is practically continuous in the ultraviolet, visible, and near-IR region, with some particularly strong lines in the 800–1000-nm region. In high-pressure mercury-arc lamps, the main spectral emission consists of the characteristic mercury lines plus relatively weak background radiation. As an illustration, mention may be made of the possible application of these two lamps: the xenon arc lamp can provide good intensity in the blue region and also a strong line at 881 nm for following the reaction-center bacteriochlorophyll P880[13]; mercury lines at 254, 278, 405–408, 436, and 546 nm would be suitable for following plastoquinone, ubiquinone, the blue shift of the Soret band of cytochromes, P700, and the α-band of c-type cytochromes, respectively.

Power Supply

In kinetic spectrophotometry where the measuring beam is used in the single-beam mode the stability of the beam source is of utmost importance. Fortunately highly regulated power supplies are now readily available. For instance, the Harrison Lab Model 6267A (0–36 V, 0–10

---

[29]P. A. Liebman and G. Entine, *Appl. Opt.* **8**, 1502 (1969).

A), which has good constant voltage and constant current regulation, has been found suitable for the tungsten–iodine lamps mentioned above. A Kepco Model HB-250M power supply (0–300 V, 0–1 A) has been satisfactorily used for several deuterium lamps mentioned above as well as a high-pressure mercury arc lamp (ST-75, Quartzlampen).

Most of the arc lamps require a high voltage and high frequency ignition pulse for starting (for instance, 300–500 V dc for the deuterium lamps and 800 V dc for the mercury arc lamps ST-75). A simple starter circuit[30] has been devised for xenon arc lamps (e.g., Hanovia 901C1 or Osram VBO-450) powered by a current-regulated power supply such as Harrison Lab Model 6268A (0–40 V, 0–30 A).

General Operation

Measuring Geometry

For suspensions of chloroplasts, chromatophores, or some algal cells, conventional 1-cm square glass or quartz cuvettes are most suitable. In this case the measuring light can pass through the cuvette in one direction, and the excitation flash may impinge at a right angle to the measuring beam (see Fig. 6A). The optical density of the sample relative to the excitation flash should be such that the sample can be homogeneously irradiated and activated.

Sometimes one may find it convenient to use a more concentrated suspension. In this case, a short-path cuvette may be constructed from two glass or quartz plates with a rubber, Teflon, or even metal gasket forming the spacer. The entire stack may be clamped in a metal frame. Such a cuvette is especially convenient for examining suspensions at liquid-nitrogen temperature. The frame and the cuvette can be conveniently cooled by attaching a fin to it and immersing the fin into liquid nitrogen in a Dewar (Fig. 7). With short-path cuvettes, the excitation flash also comes from a direction at right angle to the measuring beam direction, but slightly off the cuvette center, and, after reflection by a

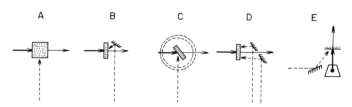

Fig. 6.   Some typical measuring geometries. Thick solid line: incident measuring beam; thin solid line: transmitted measuring beam; dotted line: excitation beam; dots, sample in cuvette or deposited on glass plate. See text for further descriptions.

[30] M. Green, R. H. Breeze, and B. Ke, *Rev. Sci. Instrum.* **39**, 411 (1968).

mirror, falls on the cuvette face at an angle of $30°^{12}$ (Fig. 6B). With a sufficiently thin and wide cuvette, it may simply be oriented 45° with respect to the measuring and excitation beams (Fig. 6C). Another possible arrangement is shown in Fig. 6D, which offers broadside excitation at normal incidence. The mirror has a center hole which allows the passage of the measuring beam.[31] A similar arrangement of excitation with an incident angle of 60° has been used by Porter and Strauss[32] for their microbeam flash-photolysis apparatus.

For algal cells, which settle easily, a reasonably thick algal suspension may be placed on a microscope slide which is mounted horizontally; appropriate filters and the photomultiplier are mounted immediately above (Fig. 6E). For highly turbid samples that scatter light, the proximity of the photomultiplier cathode face to the sample helps collect most of the scattered transmission. Alternatively, a large-diameter lens may be used for the efficient transfer of the scattered transmission to the photomultiplier tube.

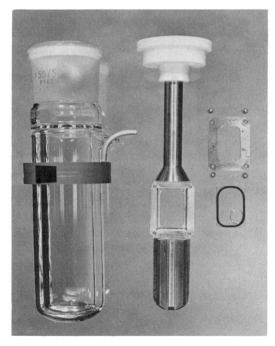

FIG. 7.   A typical cold-finger cuvette and Dewar assembly useful for low-temperature studies.

[31] J. R. Huber, R. P. Widman, and K. Weiss, *Rev. Sci. Instrum.* **40**, 1033 (1969).
[32] G. Porter and G. Strauss, *Proc. Roy. Soc. Ser. A* **295**, 1 (1966).

Modulation of Measuring Light

As discussed in a subsequent section in connection with the improvement of signal-to-noise ratio, one of the most effective ways to accomplish this is to use high measuring light intensity. This may be permissible for certain wavelength regions where an intense measuring light is itself unlikely to produce photochemical reactions. For instance, in the α-band region of cytochromes, or even in the Soret region of bacterial cytochromes, an intense measuring light may be used. However, in those spectral regions where photobleaching of reaction-center chlorophyll is followed, e.g., near 700 nm for P700 or in the near-IR region for the reaction-center bacteriochlorophylls, a slow bleaching would take place when an intense measuring light is used. Also, at liquid-nitrogen temperature where some secondary reactions are irreversible, the measuring light itself may cause a substantial amount of reaction and thus a steady drift of the transmission level. Such situations may be avoided or minimized if the measuring beam can be modulated or gated so that the sample is exposed to it for the minimum period just to cover the time course during which the absorbance change occurs. Two methods may be used to achieve this purpose. One is to use a flash of longer duration as the measuring light source and the other is to use a shutter in conjunction with an intense steady light source.

In his measurement of P870 photooxidation of *Chromatium* chromatophores excited by ruby laser flashes, Parson[13] used a xenon flash lamp as the measuring light source. The xenon lamp provides a strong line at 882 nm which is ideal for following the photobleaching of P870. The incident intensity from the xenon flash lamp was measured to be about 1 nanoeinstein $cm^{-2} \cdot sec^{-1}$, 44 times that available from a tungsten lamp. The sample was exposed to this measuring flash for about 100 μsec before and during the measurement. However, when Q-switched ruby laser flash was used, the photochemical effect caused by the measuring flash was estimated to be only about 0.05% that caused by the excitation laser flash. A choke in series with the flash gives enough of a plateau in the discharge to allow a flat portion to be used for a measuring period of 20 μsec.

A similar device has been used by Yip[33] in a double-beam flash photolysis apparatus, where a 100-J flash lamp with a peak duration of 500 μsec (third width = 6 msec) was used as the measuring source. The flash is flat within 2% over the 500-μsec peak period.

DeVault and Chance[12] boosted the voltage on a 34-W tungsten lamp to twice its rated voltage for a few milliseconds during the measuring period as a means of improving the signal-to-noise ratio.

[33]R. W. Yip, *Rev. Sci. Instrum.* **40**, 1035 (1969).

A rapid shutter constructed of a stepping motor has been in routine use in our laboratory in the measurement of photooxidation of reaction-center chlorophylls. For this purpose we use a high-intensity 250-W quartz tungsten–iodine lamp yielding an intensity of 1 to 2×10^5 ergs $cm^{-2} \cdot sec^{-1}$ at 870 nm (half-bandwidth, 8 nm, $cf.$ Fig. 8). The shutter has been used also in conjunction with Q-switched ruby laser flash excitation. Gated circuit is provided to open the shutter 10 msec before the flash and close it 10 msec after the flash. Shutter-modulated light offers convenience in timing and synchronization, as well as a stable light intensity level during the measurement.

Monochromatic Light Production

Unlike conventional flash photolysis, where white flashes are used for both excitation and measurement, in flash kinetic spectrophotometry, narrow-band monochromatic or broad-band polychromatic flashes are used for excitation and relatively narrow-band monochromatic light beam is used for measurement. In addition, the physical arrangement in flash kinetic spectrophotometry requires that the excitation and measuring wavelengths are sufficiently separated that the photomultiplier which receives the transmitted measuring light is shielded from the excitation light by appropriate filters (cf. Fig. 2).

For excitation flashes, narrow-band or broad-band interference filters plus appropriate cutoff filters are most suitable. Transmission profiles of some interference filters and their uses in exciting the various absorption bands of chloroplasts and bacterial chromatophores are shown in Fig. 8.

To generate monochromatic light for the measuring source, interference filters or monochromators may be used. The grating monochro-

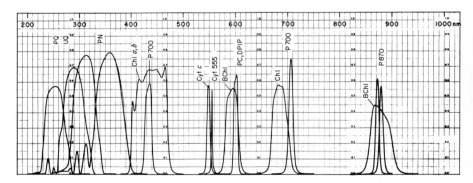

FIG. 8.   Transmission profiles of interference filters and their uses. See text for further descriptions.

mator is convenient for this purpose since a wide range of wavelengths can be readily scanned. To obtain high transmission efficiency over widely different wavelength ranges, several gratings blazed at different wavelengths may have to be used. The continuous filter wedges covering the visible (400–700 nm) (Schott Veril S-200) and near-IR region (700–1200 nm) (Schott Veril SL-200) are also convenient and inexpensive filters for isolating monochromatic measuring light as well as for shielding the photomultipliers. The visible wedge filter has a transmission value of about 30%, the near-IR filter 40%. Both have a half-bandwidth of 12 nm for a slit width of 1 mm.

Although the filter wedge is usually furnished with a wavelength-*vs*-distance calibration curve, so that a wavelength scale can be constructed, it is most convenient when used in conjunction with a monochromator. For instance, a grating monochromator may be used for isolating the measuring light, with a filter wedge as a detector shield. We have also used a near-IR filter wedge for isolating the monochromatic measuring light and a grating monochromator for shielding the photomultiplier (see below). In this case, the wavelength location on the filter wedge can easily be located by synchronizing with the monochromator and checking the maximum transmission.

When measuring light of high intensity is needed, narrow-band interference filters of high transmission may be used. Transmission profiles of several narrow-band filters and their uses in measuring some components of the photosynthetic apparatus are shown in Fig. 8.

In the ultraviolet region, the monochromator is most convenient for isolating the measuring light since interference filters in this region have rather low transmission. The recently available reflection-type interference filters for the ultraviolet region are most suitable for shielding photomultipliers. These filters have rather broad bandwidths and transmission values as high as 90%. Transmission profiles of reflection-type filters covering the range 230–390 nm are also included in Fig. 8.

Suppression of Optical Interference

Besides the leakage of excitation flash by scatter, leakage of chlorophyll fluorescence when measurements are made in the long-wavelength absorption band region of reaction center chlorophylls presents a much more difficult problem. For instance, in the peak wavelength region of P700, P870, etc., the strong fluorescence is also transmitted through the shield filters.

The fluorescence leakage can be substantially reduced, but not completely eliminated, if a monochromator is used to shield the photomultiplier instead of filters. In this way the collimated measuring light transmitted through the sample cuvette will be much more efficiently transferred to the photomultiplier than the uncollimated fluorescence emission. It should be noted that the use of a monochromator for photomultiplier shielding is feasible only for optically clear suspensions of chloroplasts or chromatophores where no scattering occurs.

Signal Detection

Photomultipliers as Detectors

Because of the low-level changes in light transmission associated with physical or chemical changes caused by excitation flashes, detection of such changes is most suitably accomplished by multiplier phototubes. In a multiplier phototube, electrons released from the photocathode by incident photons are accelerated toward the first electrode, called dynode, with considerable energy. The dynode is coated with a substance of low work function so that absorption of a high-energy electron liberates secondary electrons. By proper shaping of the electrodes, these secondary electrons are directed toward the next dynode having a still more positive voltage, which releases a larger number of secondary electrons. This process is repeated at dynodes (typical multiplier phototubes have 6–14 stages) until the initial photoelectron has produced an avalanche of many millions of electrons. The secondary emission from the last dynode is collected at an anode from which the multiplied current can be measured as a voltage across a resistor.

The more practical aspects of the use of these detectors will be discussed with reference to some commonly used photomultiplier tubes for the spectral range useful in photosynthesis. Actually, with two photomultiplier tubes such as EMI type 9558 and DuMont type 6911, the entire range from 200 to 1000 nm can be adequately covered. These end-on type tubes are commonly used because of the diffuse and scattered nature of the transmitted measuring light beam often encountered in photosynthetic samples. Thus, a close coupling with a large photocathode can be more easily achieved. The EMI 9558 tube has a trialkali (Na-KSbCs) cathode with good response in the blue region and extends to about 850 nm. When provided with a quartz window the spectral range is 165–850 nm. The 11-stage dynodes are coated with CsSb secondary emitting surfaces. The DuMont type 6911 photomultiplier is also a 2-in. end-on tube whose AgOCs cathode surface has a maximum spectral response near 800 nm and extends to about 1000 nm or slightly higher. It is much less efficient than the trialkali surface but is the only photoemissive surface which is readily available that is useful in the 800–1000-nm range. The 10-stage dynodes are coated with AgMg surfaces.

Dark-Current Noise, Shot Noise, and Noise Improvement

The sensitivity of the multiplier phototube depends on the efficiency of the cathode material producing photoelectrons. At low intensity, the sensitivity is also limited by the spurious impulses from electrons that are emitted from the cathode spontaneously, mainly by thermal effects. The dark current caused by thermionic emission is a source of "dark noise" and is especially pronounced in the S1-type cathode. Unfortunately, the cathode material with low work function necessary for the far-red response also makes it a good thermionic emitter. Operationally, dark current arising from thermionic emission can be suppressed by cooling the tube.

Care should be taken to distinguish between dark current and noise. Dark current comprises electrons entering dynodes through agencies other than photoemission. Noise is associated only with signal and is present only when signal is present. Noise is a consequence of a statistical fluctuation in the rate of emission of photoelectrons from the cathode and secondary emission from the dynodes, and consequently in the instantaneous current. This current fluctuation, or noise, is a main factor in determining the signal-to-noise ratio (S/N) and thus the sensitivity or resolution of the measuring technique. The cathode noise current due to shot noise, Δi, is given by $\sqrt{2eI\Delta f}$, where I is the cathode current in amperes, e the charge of an electron, 1.6×10^{-19} coulomb, and Δf the bandwidth of the measuring equipment. Then $S/N = \Delta I/\Delta i = \Delta I/\sqrt{2eI\Delta f} = \text{const } (\Delta I/I) \sqrt{I/\Delta f}$.

From this equation it is obvious that a lower S/N is always associated with a larger bandwidth. There are several practical ways of improving the signal-to-noise ratio. Whenever possible, photomultiplier tubes with high cathode efficiency should be chosen, and the spectral range being measured should be matched with the peak spectral response, E, of the cathode surface. It is also clear from the above formula that for a given bandwidth Δf and for a given size transmission-change signal $\Delta I/I$, the S/N can be improved by using a high cathode current consistent with the tube rating. Since the cathode current depends on the incident intensity and the quantum efficiency of the cathode, as much incident light as is available should be used. In this connection, it is well to remember that the measuring light intensity should be maintained at a low level so that it does not itself cause appreciable photochemical action. This is especially true for wavelengths falling in the range of the absorption bands of the reaction center chlorophylls.

Care should be taken that the intensity of the measuring light is well below the specified maximum cathode current rating for the tube.

Above the maximum rating, the response of the cathode current ceases to be a linear function of the light intensity because of the cathode resistance. The light intensity above which the tube may become nonlinear can be estimated by dividing the specified maximum cathode current by the cathode sensitivity. In general, the cathode current must be well below this value in order to satisfy the maximum rating of the anode current.

Another way to improve the S/N ratio is to keep the voltage ratio between the cathode and the first dynode as high as is practical. This is particularly important at low light levels and high bandwidth. It is quite common to make this potential difference twice or more the interdynode voltage. By maintaining the cathode-to-dynode-1 voltage high, the effect of magnetic field is also diminished. In order to obtain an adequate voltage, regardless of changes in the dynode chain current, some forms of voltage stabilizer such as Zener diodes may be used.[34] It is also possible to improve the S/N ratio by operating with voltage between dynodes 1 and 2 and dynodes 2 and 3 at a higher value than for succeeding interdynode voltages.

Amplifiers and Readout Devices

Observation of transient phenomena such as flash-induced rapid absorbance changes can best be made on oscilloscopes which use an electron beam as a pointer that is deflected by an electrostatic field. The multiplied current from the photomultiplier tube is converted into voltage across a resistor and is passed to a dc amplifier which has a bandwidth compatible with the flash duration and sufficient gain, e.g., the Tektronix type-D differential preamplifier. In using the differential amplifier, an external voltage input from a battery source through a potentiometer may be applied to one channel to offset the dc signal from the photomultiplier. In an earlier arrangement,[35] a split measuring beam was used and the output from a second auxilliary photomultiplier was used as the balancing source. Balancing with two photomultipliers has the advantage that changes due to fluctuations in the power supply or light source will cancel out. However, with the availability of well-regulated power supplies, single-beam operation is simpler and quite adequate. For a given required bandwidth, the input capacitance must be kept low for a given anode resistance to yield a low input RC time constant. The bandwidth can readily be increased by decreasing the anode

[34]"EMI Photomultiplier Tubes," issued by Gencom Division, Whittaker Corp., Plainview, L.I., N.Y.
[35]B. Ke, R. W. Treharne, and C. McKibben, *Rev. Sci. Instrum.* **35**, 296 (1964).

resistance of the photomultiplier. Of course, such a resistance decrease also changes the amplification.

The Tektronix type 1A5 differential amplifier has also been used. It is a wide-band (dc to 50 MHz), high-gain (1 mV/cm) and high common-mode-rejection-ratio unit.

Signal Averaging and Sampling Technique

The magnitude of absorbance changes occurring during *in vivo* photosynthesis caused by flash excitation is usually small. For instance, with a sample of green plant chloroplasts with an optical density of unity at the red peak, complete photoconversion of, say P700, would expect to yield a maximum absorbance change of $\sim 2.5 \times 10^{-3}$, assuming the extinction of P700 is nearly the same as that of the bulk chlorophyll, and the absorbance change associated with the cytochrome at the α band would only be $\sim 5 \times 10^{-4}$. At an adequate bandwidth for good time resolution, the expected S/N ratio would be so low as to make observation by single flash excitation marginal without resorting to other means such as high-intensity measuring light to improve the S/N ratio. To circumvent the problems of either using a high-intensity measuring light or introducing waveform distortion by electrical filtering, a commercially available Computer of Average Transient (CAT) has been used since 1963 to extract signals from a high amplitude of background noise.[35] The CAT computer is a portable, on-line digital computer which applies the averaging technique to improve the S/N ratio by repeated additions of data samples. The Model-400 CAT computer, for instance, when triggered by a synchronizing signal, samples the measured signal at 400 intervals, converts the amplitude of the signal at each interval into a digital number of counts. These counts are then stored in a particular address in a magnetic core memory, after being added to the number previously stored in this address. Each addition cycle is initiated by a synchronizing signal so that coherent portions of the data sample, which are phase-locked to the synchronizing signal, sum algebraically with each repetition cycle. The random noise components do not add in reproducible phase, and therefore tend to diminish as the number of repetitions increases. The improvement is proportional to the square root of the number of signal repetitions. After the data are accumulated, they may be read out in either digital or analog form. For analog presentation of the memory contents, the digital data are converted through current summing networks, and analog voltages are derived, one voltage representing the arithmatic information and one voltage representing address information, which in turn are used to deflect the beam in the cathode ray tube or to drive an *X–Y* recorder. The data can also be taken in digital form, which is particularly valuable if they are to be processed further. An example of improving the S/N ratio of the 515-nm absorbance change in *Chlorella* is shown in Fig. 9.

The Model 400 CAT-computer has a dwell time of 56 μsec/data point. The later Model 1000 has a dwell time of 31.25 μsec/address. Sweep expanders utilizing a sampling technique may be used in conjunction with the CAT computers to extend the upper frequency response and reduce the dwell time per address point of the above two computers to 2 and 1 μsec, respectively. The sweep expander operates on a sampling technique. At a selected dwell time per address, the CAT computer records 4–8 amplitude samples during each sweep of the transient waveform. These sampling points are stored $100/N$ or $128/N$ addresses apart in the computer memory, where N is the effective dwell time per address point. At each successive sweep, the sampling times are delayed by one sample interval, with respect to the CAT trigger. Thus $100/N$ or $128/N$ triggers must occur before a single complete set of 400 or 1024 points can be sampled, digitized, and data accumulated in each CAT memory address. The degree of S/N improvement is directly proportional to an integral number of complete digitizing sequences applied. Thus, noise reduction by this sampling technique would involve very long measurements.

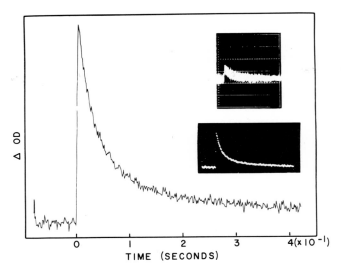

FIG. 9. Transient absorption change at 515 nm in *Chlorella* initiated by 20-μsec red flashes. Upper inset: oscilloscope trace; lower inset: CAT computer display; main curve: transcription of the computer display on an *X–Y* recorder.

Another more recently available signal averager unit is that introduced by Fabri-tek Instruments. These units also handle measurement problems by using an analog-to-digital converter, a memory, arithmetic circuits, and a timing unit to sequence the operations. One of the features of this averager (e.g., Model 1062) that is useful for medium-speed measurements is the provision for signal preconditioning to reduce noise level at the digitizer input. It uses an adjustable low-pass filter to allow rejection of high-frequency noise components. Another plug-in unit which is especially useful for rapid kinetic spectrophotometry is the rapid digitizer (Model 952). This plug-in unit itself also has 1024 addresses and can digitize measurements at 1 μsec/address point. The rapid digitizer is essentially a fast memory buffer and performs a write-only operation. By using this separate memory for recording the digital values measured for each signal, the limitation caused by the long time required for memory and addition operations is largely avoided since the digitizing process itself can be extremely fast. Thus, unlike the sweep expander unit for the CAT computer described above, which uses a sampling technique, with the Model 952 unit, the entire 1024 address points are digitized during one single sweep of the signal when initiated by a trigger pulse. The quantity recorded is the average voltage during a data-point interval. After a signal is measured, the buffer digitizer automatically reads out at 50 μsec/data point and transfers the recorded data to the main averager unit (e.g., Model 1062). This process clears the memory for receiving information from the next signal waveform. The averager receives the information and adds it to the previously accumulated data. An example of extracting the ubiquinone photoreduction signal by the Fabri-tek Model 1062 signal averager and the Model 952 rapid digitizer is shown in Fig. 10.

For submicrosecond resolution, it is necessary at the present time to resort to the sampling technique. As described earlier (*cf.* the sweep expander used in conjunction with the CAT computer), during each signal course only one address or sample point is taken. To improve the S/N ratio, a large number of samples are taken and the sample values are averaged by a simple RC integrating circuit and then recorded. After a phase shift, the next sample point is repeated in the same fashion. The samplings are repeated until the entire time course of the signal is covered. Application of the sampling technique to photosynthesis studies has recently been described by Wolff *et al.*[36] for the determination of the risetime of the 515-nm absorbance increase. They have used a

[36]C. Wolff, H. E. Buchwald, H. Rüppel, K. Witt, and H. T. Witt, *Z. Naturforsch.* **24B,** 1038 (1969).

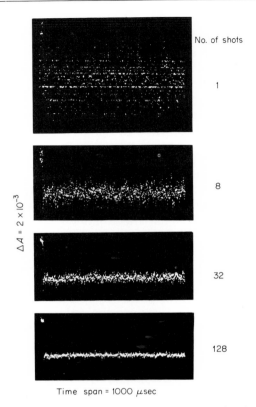

No. of shots

1

8

32

128

$\Delta A = 2 \times 10^{-3}$

Time span = 1000 μsec

FIG. 10. Signal-to-noise ratio improvement of a 275-nm absorbance-change signal in a photochemically active particle isolated from *Chromatium*. A Fabri-tek Model 1062 signal averager and a Model 952 rapid digitizer were used in these experiments.

Q-switched ruby laser as the excitation source at a repetition rate of 5 pulses/sec. The Hewlett-Packard Model 185A sampling oscilliscope has a gate width of 0.4 nsec and can sample at 5 pulses/sec without decrease in overall sensitivity or signal deterioration during sampling because of a long holding time of the sampling diode gate. In the 515-nm absorbance change experiments, each sampling sequence requires 3000 giant pulse ruby laser flashes. By the sampling technique, an upper limit of 20 nsec has been estimated for the risetime of the 515-nm absorbance increase. They have previously proposed that the 515-nm change represents a light-induced electrical field change across the thylakoid membrane. Thus, the field formation takes place within 20 nsec by a light-induced transfer of electrons across the thylakoid membrane.

Additional Methods for Elimination of Optical Interferences

One method discussed earlier utilizes a monochromator for shielding the photomultiplier in order to suppress or eliminate interferences from fluorescence emission from the sample under study. Buchwald and Rüppel[37] recently reported an electric method for suppressing the interfering fluorescence signal by chopping the measuring beam at a high frequency (14 MHz) by means of an electrooptic modulator. The ac voltage change from the photomultiplier resulting from the flash-induced absorbance change was fed through a high-frequency narrow bandpass (2 MHz) filter, and then detected by a phase-sensitive detector and finally put into a signal averager.

When using a signal averager such as Fabri-tek Model 1062, certain interferences may be eliminated by a simple subtraction in the computer. This may best be illustrated with an example, as shown in Fig. 11. The Q-switched ruby-laser-induced photooxidation of the reaction center P870 from mutant *spheroides* was originally recorded over the entire span of 1000 μsec. As can be seen, the photooxidation had a risetime faster than 1 μsec and a rather slow decay. At the point of the laser flash, fluorescence leakage causes an additional downward spike superimposed on top of the signal. After the completion of the signal recording, the second half of the memory was erased, and the experiment was repeated but without the measuring light. As shown in the third quarter, an amount of fluorescence equivalent to that encountered in the first experiment was recorded. Again, the memory in the fourth

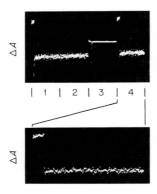

FIG. 11.   An illustration of one method for elimination of optical interference. See text for further explanation.

[37] H. E. Buchwald and H. Rüppel, *Nature (London)* **220**, 57 (1968).

quarter was erased, and the information in the first and third quarters were transferred into it in the add and subtract mode, respectively, and the resultant signal free from fluorescence leakage is presented in the fourth quarter. An expanded form of this same signal is also presented in the bottom half of the figure. It is obvious that this method is applicable only when the intensity of the excitation flashes is constant, and the leakage is not unreasonably large that overloading or clipping of the signal occurs.

Double-Flash Excitation

Two excitation flashes applied in succession and separated by a short time interval can be useful for probing the reaction mechanism and kinetics of a given species as well as its reaction partners. An example is given in Fig. 12 in which two excitation flashes separated by 100 μsec were applied to *Chromatium* chromatophores. At the measuring wavelength of 422 nm, the light-induced absorbance changes represent the oxidation of cytochrome-422. The risetime of cytochrome-422 oxidation is 2–3 μsec[13,14] and the reaction is known to be coupled to the bacteriochlorophyll reaction center P870.[13] In the 100-μsec dark period, P870 should be regenerated, but the oxidized cytochrome is not. In order for the second reaction to take place, there must be more than one equivalent of cytochrome to P870, which is actually the case in the *Chromatium* chromatophores. According to the current concept, the primary photoact is the formation of the oxidized P870 and the concomitant formation of a primary reductant by accepting the electron from P870. Under these circumstances, P870, although fully reduced by cytochrome-422, cannot be expected to undergo a second photooxidation unless the primary reductant had enough time to lose its electron and to be ready

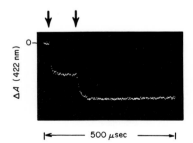

FIG. 12. Double-flash excitation for cytochrome-422 oxidation in *Chromatium* chromatophores. First flash from Q-switched ruby laser; second flash from a xenon flash lamp. Flash separation 100 μsec. Time interval between flash pairs 5 sec. Sixteen flash pairs averaged.

to participate in the primary photoact again. By means of double-flash excitation, Parson[38] found that it takes 60–80 μsec before a second laser flash is able to oxidize half of the P870 oxidized by the first flash. This time has been interpreted as being the turnover time of the primary acceptor molecule.

An absorbance change with maximum decreases at 435 and 682 nm and with decay time of 2×10^{-4} sec has recently been reported for the photosystem-II particles isolated from spinach chloroplasts by digitonin.[39] The absorbance has been assigned to the reaction center chlorophyll of photosystem II (Chl a_{II}). More recently, the nature of the Chl a_{II} absorbance change has been reinvestigated by means of double-flash excitation.[40] The flashes were separated by 2.5×10^{-4} sec, which is slightly longer than the half-decay time of the 690-nm signal. At the application of the second flash, it was found that an almost full-scale absorbance decrease could be observed. It has been shown by measurement of oxygen evolution and of the absorbance change associated with a semiquinone that electron transfer from water across the reaction center chlorophyll to the semiquinone requires a half-time of 6×10^{-4} sec. The inference from the double-flash experiments was that the Chl a_{II} absorbance change was not due to a redox change with the Chl a_{II} itself participating in the electron transfer, but rather that Chl a_{II} serves as a sensitizer for the electron-transfer reaction.

Several approaches are available for producing double giant pulses from the ruby laser. One successful method which provides a time separation of 100 nsec–400 μsec is to Q switch the resonator by different amounts so that the first pulse does not completely cancel the population inversion. Wetzels and Alfs[41] have designed a Pockels-cell driving circuit for this application.

[38]W. W. Parson, *Biochim. Biophys. Acta* **189,** 384 (1969).
[39]G. Döring, J. L. Bailey, W. Kreutz, and H. T. Witt, *Naturwissenschaften* **55,** 220 (1968).
[40]G. Döring, G. Renger, J. Vater, and H. T. Witt, *Z. Naturforsch. B* **24,** 1139 (1969).
[41]W. Wetzels and A. Alfs, *Rev. Sci. Instrum.* **40,** 1642 (1969).

[3] Hydrogen Ion Buffers

By N. E. GOOD and S. IZAWA

General Considerations

In the study of biological reactions, it is important to maintain a constant, but often exceedingly low, concentration of hydrogen ions. For instance, when the pH of a reaction medium should be maintained at

8.0 ± 0.1 the permissible range of hydrogen ion concentrations lies between 1.25×10^{-8} and 8.0×10^{-9} M, a range of only 4×10^{-9} M. Yet all too frequently the reaction being studied produces or consumes many thousands of times this amount of hydrogen ion. Even in systems which in theory neither produce nor consume hydrogen ions, unintentional gain or loss of atmospheric CO_2 or minor side reactions can cause disastrously wide fluctuations in pH. Moreover, we are constantly discovering more biological processes associated with the movement of ions across membranes. Obviously local pH changes occur whenever the movement of cations across a membrane does not precisely equal the movement of anions. For these reasons it is always desirable and often necessary to employ some means of stabilizing the pH of media.

This stabilization of the pH is usually accomplished by utilizing one of the following freely reversible reactions:

$$H^+ + A^- \ \rightleftharpoons\ HA; \qquad pH = pK_a + \log \frac{[A^-]}{[HA]} \tag{1}$$

$$H^+ + N\equiv\ \rightleftharpoons H^+N\equiv; \qquad pH = pK_a + \log \frac{[N\equiv]}{[H^+N\equiv]} \tag{2}$$

where A^- is a readily protonated anion and $N\equiv$ is a readily protonated primary, secondary, or tertiary amine. Such substances are known as hydrogen ion buffers since they tend to prevent pH changes. In their presence, the addition or removal of hydrogen ion simply shifts concentrations to restore equilibrium, increasing or decreasing by a small amount the relatively large reservoir of protonated buffer. It should always be remembered, however, that the buffering capacity diminishes rapidly as either the protonated or nonprotonated form of the buffer becomes depleted; the log terms in Eqs. (1) and (2) above then change a great deal with the addition or removal of small amounts of hydrogen ion. An arbitrary but useful rule is the following: for reasonable buffering capacity the ratio of the protonated to the nonprotonated form should not exceed 10 or be less than 0.1. However, the nearer the pH to the pK_a of the buffer, the lower the total buffer concentration for a given amount of pH stabilization—and a low buffer concentration is usually safer than a high buffer concentration.

Since the total buffering capacity of commonly used buffers is always the same (on a molar basis) and buffers covering a wide range of pK_a's are now available, the choice of a buffer for any particular system usually depends on factors that have nothing to do with buffering, i.e., on the presence or absence of undesirable or (more rarely) desirable side effects. A few of these side effects can be predicted. For instance, phosphate buffer at pH 7.5 contributes approximately seven times as many ions to the medium as the zwitterionic buffer tricine at the same pH and therefore has a more marked effect on the ionic strength of the medium. Some buffers do not form complexes with metal ions whereas others, such as phosphate, tend to precipitate polyvalent metals and still other buffers form soluble complexes with metal ions, lowering and stabilizing the concentrations of free Mg^{2+}, Ca^{2+}, Mn^{2+}, Fe^{3+}, and Cu^{2+}, etc. (These latter hydrogen ion buffers can also be used as metal ion buffers.) Some buffers, notably certain carboxylic acids, phosphate, and pyrophosphate, can serve as metabolic substrates, yielding or consuming energy or displacing equilibria. Borate is notorious for forming complexes with a number of metabolites. Nevertheless many side effects cannot be predicted, and buffers may uncouple or inhibit or otherwise modify reactions by mechanisms not yet understood. Clearly a buffer which is adequate or even superior in one system may not be very good in another system. For these reasons, the selection of an appropriate buffer will always depend on extensive testing.

Until recently the evaluation of buffers was difficult because so few were available for comparison, especially covering the range pH 6–9. The great popularity of tris(hydroxymethyl)aminomethane (Tris) was a measure of the shortcomings of alternatives since Tris itself is far from ideal in most situations. Accordingly, we undertook to design and prepare a number of new buffers[1] most of which are now commercially available. We designed the buffers with the following properties in mind:

1. In order to remain in the aqueous medium and not dissolve in the biological phase of the reaction media, the buffers should have maximum solubility in water and minimum solubility in other solvents. Moreover, to keep the concentration of potentially harmful buffer as low as possible in cellular organelles, the buffer employed should pass through membranes with difficulty, if at all.

2. Ions contributed by the buffer to the medium should be as few as possible. If specific ions or specific ionic strengths are required for reactions, appropriate salts can be added.

3. The pK_a's of the buffers should be influenced as little as possible by buffer concentration, temperature, and the ionic constitution of the medium.

4. If the buffers form complexes with metals, these complexes should be soluble and the binding constants should be known.

5. The buffers should resist enzymatic and nonenzymatic alterations, and they should not resemble enzyme substrates enough to

[1] N. E. Good, G. D. Winget, W. Winter, T. N. Connolly, S. Izawa, and R. M. M. Singh, *Biochemistry* **5**, 467 (1966).

act as analog inhibitors. Neither should they react with any metabolite or with any component of the medium.

6. They should not absorb light at wavelengths longer than about 230 nm lest their presence interfere with spectrophotometric assays.

7. They should be easy to prepare and easy to purify. Ease of purification is particularly important, since biological systems can be extremely sensitive to minute traces of some kinds of impurities.

The proved usefulness of a high proportion of buffers possessing these properties suggests that such properties may be regarded as general criteria of excellence. However, each buffer application poses special problems and stresses different criteria. For instance, control of ionic strength may be critical or a matter of little concern. There are many other considerations relevant to specific applications. Bicine (see below) is a poor buffer for Hill reaction studies, but only because it is oxidized slowly by ferricyanide. Tricine is so rapidly photooxidized by flavins that room light can cause reduction of flavin enzymes or rapid oxygen uptake.[2] HEPES, which is in many ways an ideal buffer, interferes with the Folin protein assay. Thus it is obvious that buffers must be tailored to the needs of the system to be studied. Moreover, in exploring new phenomena one must never be content with a single buffer or even with the comparison of two buffers; if two do not give the same results, one cannot be sure which is introducing the grosser artifacts. This point is illustrated in Figs. 1 and 2. Tris buffer displaces the pH curve of electron transport and phosphorylation in chloroplasts by almost a pH unit when compared with at least nine other buffers (which all give the same pH optimum) and in the electron transport of chloroplasts uncoupled by the removal of coupling factor, phosphate lowers the pH optimum by a similar amount. Obviously, the buffer used in reaction mixtures is one of the important variables which must be thoroughly investigated as a matter of routine. It follows that several buffers should be available for every pH range.

Available Buffers

In this discussion, little mention is made of the multitude of carboxylic acids, most of which are useful only below pH 6. These are of limited applicability in studies of photosynthesis or in studies of the reactions of isolated plant organelles. Protonatable anions which do provide buffering ranges between pH 6 and pH 10 are bicarbonate, carbonate, phosphate, pyrophosphate, various phosphonates, arsenate, substituted phenols, various barbiturates, borate, and borate–mannitol

[2] R. K. Yamazaki and N. E. Tolbert, *Biochim. Biophys. Acta*, 197, 90 (1970).

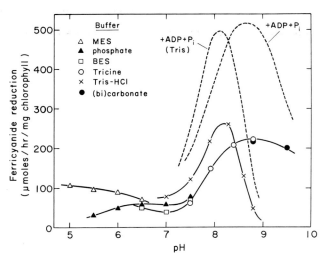

FIG. 1.   Rates of nonphosphorylating electron transport as influenced by the pH of the medium and the buffer employed. Spinach chloroplasts were suspended in a medium containing potassium ferricyanide (0.8 mM) and the indicated buffer (50 mM) at the indicated pH; the chloroplasts were illuminated with a saturating intensity of red light at 19°. Broken lines show the rates of electron transport to be expected during phosphorylation, i.e., after addition of ADP (0.5 mM), Na_2HPO_4 (15 mM), and $MgCl_2$ (1.0 mM) to the same system. From N. E. Good, S. Izawa, and G. Hind, *Curr. Top. Bioenerg.* 1, 75 (1966).

complexes. Of these only bicarbonate–carbonate and the several phosphorus-containing ions can be considered widely useful. The remainder are inhibitory, unstable, or otherwise unsatisfactory for most systems. Cations providing a reservoir of protons are amines—primary, secondary, or tertiary, but not quaternary. Fortunately, aliphatic amines with pK_a's ranging from below 2 to above 11 are numerous and readily obtained. Moreover the amino groups can be combined in a single molecule with anionic groups to form zwitterions—and zwitterions possess many of the desirable properties itemized above.

We present below a list of buffers that are of general interest to biologists, particularly to those involved in photosynthesis research. They are arranged in order of increasing pK_a, and together they cover, with a good deal of overlapping, every buffering range likely to be used in the study of photosynthetic reactions. We have not attempted to include all buffers since we have no direct knowledge of the desirability of many, new or old. Rather we have drawn heavily on our own experiences and on our own introductions.

The pK_a represents the midpoint of the buffering range. In general, the protonated (ionized) forms of the amine buffers are less likely

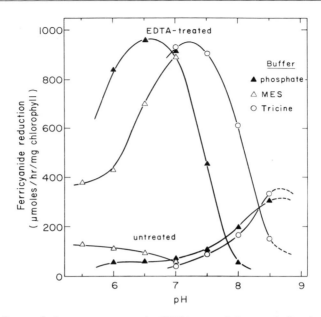

FIG. 2. Rates of electron transport in EDTA-treated (uncoupled) spinach chloro-plasts as influenced by the pH of the medium and the buffer employed. The chloroplasts were preincubated for 30 minutes at 0° in 0.01 M tricine buffer (pH 7.3) containing 1.0 mM EDTA. Reaction mixture as in Fig. 1. Temperature 15°. From N. E. Good, S. Izawa, and G. Hind, *Curr. Top. Bioenerg.* 1, 75 (1966).

to be inhibitory than the nonprotonated forms. Therefore with Tris and the zwitterionic buffers it is usually better to work at a pH below the pK_a than above it. The converse is true of anionic buffers, especially the carboxylic acids; with these it is usually better to work at a pH above the pK_a where the greater part of the buffer is ionized.

The pK_a of a buffer, and therefore the pH of the buffered solu-tion, changes with temperature. If one adjusts a buffer at one tempera-ture and uses it at another temperature, the extent of this change must be known. For most of the buffers described, we have given approximate values to use in computing the change. However, it is probably best to adjust the pH at the desired temperature.

All the buffers described are highly soluble in water, and most are insoluble in alcohol. None of those recommended absorbs in the ultraviolet to the extent of interfering with standard spectrophoto-metric assays.

In each instance we have shown the equilibrium reaction which accounts for the buffering. The required pH must be obtained by choos-ing a buffer with appropriate pK_a (near the desired pH) and then adjusting the molar ratios of the protonated and nonprotonated forms. These adjustments may be accomplished by mixing the computed amounts [see Eqs. (1) and (2)] of the free buffer and one of its salts, or by titrating a solution of the free buffer with base or acid while the pH is being monitored on a meter. The numerous zwitterionic buffers: MES, ACES, BES, MOPS, TES, HEPES, HEPPS, tricine, bicine glycyl-glycine, TAPS, and glycine are isolated, purified, and sold in their protonated, zwitterionic forms. Buffer solutions are normally prepared by titrating a solution of the zwitterions with, e.g., NaOH, KOH, or LiOH, or if metal ions are to be avoided, with Tris or with other amines.

pK_a 4.25   Succinate

$$H^+ + {}^-OOC-CH_2CH_2-COOH \rightleftharpoons HOOC-CH_2CH_2-COOH$$

Comments: Buffering range much too low for most photosynthesis research. Mentioned here because it has been useful in "acid-bath" phosphorylation studies with chloroplast lamellae.[3]

pK_a 5.6   Succinate

$$H^+ + {}^-OOC-CH_2CH_2-COO^- \rightleftharpoons HOOC-CH_2CH_2COO^-$$

Comments: See above.

pK_a 5.8   Pyrophosphate

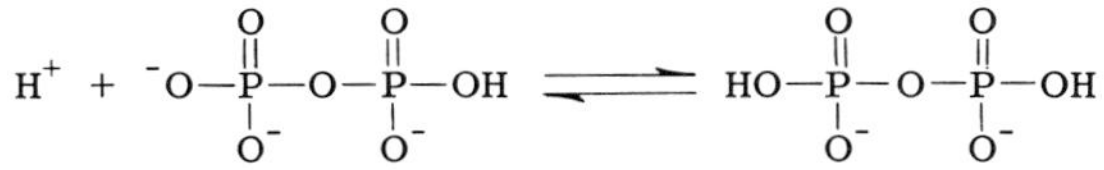

$\Delta pK_a/degree = -0.01$

Binds or precipitates many polyvalent cations.
Comments: Can be useful but its use is complicated by its tendency to form complexes and by the fact that it is directly involved in many biochemical reactions. Thus there is always the possibility that it may alter equilibria or serve as an energy source. Has been reported to stimulate CO_2 fixation in isolated chloroplasts.[4]

pK_a 6.0   Maleate

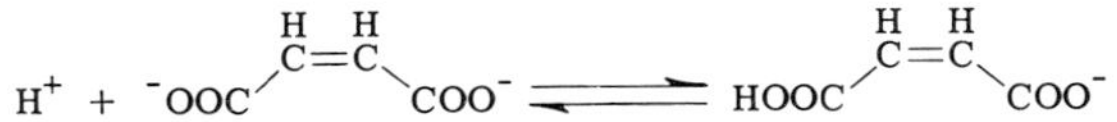

Comments: A buffer which has little merit. Has been used with Tris in an ill-advised attempt to provide wide-range buffering—a mistake because the two buffering ranges do not overlap significantly and both buffers are sometimes inhibitory. MES is a better buffer for pH levels near 6.

[3] A. T. Jagendorf and E. Uribe, *Proc. Nat. Acad. Sci. U. S.* **55**, 170 (1966).
[4] R. G. Jensen and J. A. Bassham, *Proc. Nat. Acad. Sci. U. S.* **56**, 1095 (1966).

pK_a 6.15   MES; 2-(N-morpholino)ethanesulfonic acid

$$H^+ + O\!\!-\!\!\bigcirc\!\!-\!\!N-CH_2CH_2SO_3^- \rightleftharpoons O\!\!-\!\!\bigcirc\!\!-\!\!\overset{+}{HN}-CH_2CH_2SO_3^-$$

ΔpK_a/degree $= -0.011$

Very little tendency to bind metal ions.

Comments: MES has been used successfully in the isolation of photosynthetically active, intact chloroplasts.[4] At pH 6.15 (0.05 M) it does not inhibit the autotrophic growth of *Chlorella*. It does not interfere with the Folin protein assay. Undergoes some decomposition when autoclaved in the presence of glucose.

pK_a 6.3   Bicarbonate

$$H^+ + HCO_3^- \rightleftharpoons H_2CO_3 \rightleftharpoons CO_2 + H_2O$$

Comments: This is an excellent buffer for some reactions, but unfortunately it requires a closed system to prevent the escape of CO_2. Moreover the optimal concentrations of CO_2 for photosynthesis are too low to afford much buffering and the removal of CO_2 by photosynthesis leads to a rise in pH. With the decreasing use of manometers for the study of gas exchanges the need for a reliably closed system has become a nuisance. Not recommended unless, for other reasons, a high level of CO_2 and a closed system are mandatory.

pK_a 6.5   Orthophosphite

$$H^+ + {}^-O-\overset{\overset{O}{\|}}{\underset{\underset{H}{|}}{P}}-O^- \rightleftharpoons HO-\overset{\overset{O}{\|}}{\underset{\underset{H}{|}}{P}}-O^-$$

ΔpK_a/degree $= 0.003$[5]

Forms complexes with, and may precipitate Mg^{2+} and Ca^{2+}, but not as readily as does orthophosphate.

Comments: This buffer probably deserves wider testing, although claims that it is biologically inert[5] have not been wholly substantiated; it inhibits oxygen uptake and nitrogen fixation in *Azotobacter*.[6]

pK_a 6.8   Orthophosphate

$$H^+ + {}^-O-\overset{\overset{O}{\|}}{\underset{\underset{OH}{|}}{P}}-O^- \rightleftharpoons HO-\overset{\overset{O}{\|}}{\underset{\underset{OH}{|}}{P}}-O^-$$

ΔpK_a/degree $= -0.005$

Comments: Unless care is exercised phosphate may precipitate Mg^{2+}, Ca^{2+}, Fe^{3+} and other polyvalent cations. Moreover, its usefulness as a buffer

[5] H. E. Robertson and P. D. Boyer, *Arch. Biochem. Biophys.* **62**, 396 (1956).
[6] W. A. Bulen and D. S. Frear, *Arch. Biochem. Biophys.* **66**, 502 (1957).

is limited by the fact that it is an important metabolite; for this reason one often wishes to study phosphate concentration as one of the experimental variables. It is nevertheless an extremely useful buffer. Not infrequently the phosphate required for other reasons provides, incidentally, all the buffering required.

pK_a 6.9   ACES; N-(2-acetamido)-2-aminoethanesulfonic acid

$$H^+ + H_2N-\overset{\overset{O}{\|}}{C}-CH_2NH-CH_2CH_2SO_3^- \rightleftharpoons H_2N-\overset{\overset{O}{\|}}{C}-CH_2NH_2^+-CH_2CH_2SO_3^-$$

$\Delta pK_a/\text{degree} = -0.02$

Binds Cu^{2+} but not Mg^{2+}, Ca^{2+}, or Mn^{2+} significantly.

Comments: Satisfactory buffer for uncoupled electron transport in chloroplast lamellae but not otherwise tested.

pK_a 7.05   Imidazole

Comments: This buffer has little to recommend it except its buffering range, and many better buffers covering the same range are now available. It has been used in the past when phosphate was not acceptable. It is much too reactive and unstable to be a satisfactory buffer. It uncouples electron transport from phosphorylation in chloroplast lamellae preparations.

pK_a 7.15   BES; N,N-bis(2-hydroxyethyl)-2-aminoethanesulfonic acid

$H^+ + (HOCH_2)_2{=}N-CH_2CH_2SO_3^- \rightleftharpoons (HOCH_2)_2{=}NH{\pm}CH_2CH_2SO_3^-$

$\Delta pK_a/\text{degree} = -0.016$

Binds Cu^{2+} somewhat but not Mg^{2+}, Ca^{2+}, or Mn^{2+}.

Comments: Satisfactory for electron transport studies in chloroplast lamellae preparations but otherwise not tested.

pK_a 7.15   MOPS; 3-(N-morpholino)propanesulfonic acid

$\Delta pK_a/\text{degree} = -0.013$

Negligible metal binding.

Comments: This buffer is satisfactory for studies of electron transport and phosphorylation in chloroplast lamellae preparations if one wishes to use a pH near neutrality. It does not inhibit the growth of *Chlorella* cells in concentrations up to 0.2 M. It does not interfere with the Folin or biuret protein assays. MOPS is very easily prepared from inexpensive materials and may soon replace other buffers for most applications in this pH range. It is not completely stable when autoclaved in the presence of glucose.

pK$_a$ 7.5   TES; N-tris(hydroxymethyl)methylaminoethanesulfonic acid

$$H^+ + (HOCH_2)_3{\equiv}C{-}NH{-}CH_2CH_2SO_3^- \rightleftharpoons (HOCH_2)_3{\equiv}C{-}NH_2^+CH_2CH_2SO_3^-$$

$\Delta pK_a/degree = -0.02$

Binds Cu^{2+} slightly but not Mg^{2+}, Ca^{2+}, or Mn^{2+}.

Comments: This is an excellent buffer for many biological reactions. It has been used to advantage in isolating chloroplast lamellae, in studies of CO_2 fixation by intact chloroplasts,[7] in studies of oxidative phosphorylation in plant mitochondria, and in protein synthesis studies, but it may not be quite as good for photophosphorylation studies as the next three buffers. It does not inhibit the autotrophic growth of *Chlorella*, and it does not interfere with the Folin protein assay.

pK$_a$ 7.55   HEPES; N-2-hydroxyethylpiperazine-N'-ethanesulfonic acid

$$H^+ + HOCH_2CH_2N\overbrace{}NCH_2CH_2SO_3^- \rightleftharpoons HOCH_2CH_2\overset{+}{N}H\overbrace{}NCH_2CH_2SO_3^-$$

$\Delta pK_a/degree = -0.014$

Does not bind Mg^{2+}, Ca^{2+}, Mn^{2+}, or Cu^{2+}.

Comments: One of the best general purpose buffers available for biological research. It has been used to advantage in tissue culture, in oxidative phosphorylation studies with plant mitochondria, in protein synthesis studies with cell-free bacterial systems, in photophosphorylation studies, and in studies of CO_2 fixation by isolated chloroplasts.[4] It does not inhibit the autotrophic growth of *Chlorella*. Unfortunately the Folin protein assay cannot be used in the presence of HEPES. The biuret protein assay is unaffected. (HEPES, by the rules of English phonetics, should be pronounced in the same ways as "heaps.")

pK$_a$ 8.1   HEPPS; N-2-hydroxyethylpiperazine-N'-propanesulfonic acid

$$H^+ + HOCH_2CH_2N\overbrace{}NCH_2CH_2CH_2SO_3^- \rightleftharpoons HOCH_2CH_2N\overbrace{}\overset{+}{H}NCH_2CH_2CH_2SO_3^-$$

$\Delta pK_a/degree = -0.015$

Does not bind Mg^{2+}, Ca^{2+}, or Cu^{2+}.

Comments: Similar to HEPES in many properties but less tested. Probably will not prove to be as widely useful. Its higher buffering range makes it very good for photophosphorylation studies, especially when the metal binding characteristics of tricine must be avoided. Interferes with the Folin, but not with the biuret, protein assay.

[7] D. A. Walker, *Phytochemistry* **6**, 495 (1967).

pK_a 8.15   Tricine; N-tris(hydroxymethyl)methylglycine

$$H^+ + (HOCH_2)_3\equiv C\!-\!NHCH_2COO^- \rightleftharpoons (HOCH_2)_3\equiv C\!-\!NH_2^+CH_2COO^-$$

pK_a/degree $= -0.021$

Binds Mg^{2+} weakly, Ca^{2+} and Mn^{2+} moderately, and Cu^{2+} very strongly. May protect sulfhydryls from the action of mercury compounds.

Comments: Becoming the standard buffer for work with chloroplast lamellae. Probably could replace Tris to advantage in many systems, but it is not as widely useful as HEPES. Does not inhibit the growth of *Chlorella* cells but rather stimulates, perhaps by acting as a heavy metal buffer. Does not interfere with the Folin protein assay if additional copper is added. Very subject to the photooxidation by excited flavins and flavoproteins.[2]

pK_a 8.2   Pyrophosphate

ΔpK_a/degree $= -0.006$

Comments: See above under pK_a 5.8

pK_a 8.3   Tris; tris(hydroxymethyl)aminomethane

$$H^+ + (HOCH_2)_3\equiv C\!-\!NH_2 \rightleftharpoons (HOCH_2)_3\equiv C\!-\!NH_3^+$$

ΔpK_a/degree $= -0.031$

Does not bind Mg^{2+}, Ca^{2+}, or Mn^{2+}.

Comments: In spite of its faults this has been the biochemist's standard buffer for 20 years. Because it is readily available in highly purified form and is relatively inexpensive it continues to be the buffer of choice for many uncritical applications. As a primary amine it has undesirable reaction characteristics and its appreciable solubility in organic solvents permits it to penetrate membranes and accumulate in the biological phase of reaction systems. Its use should be avoided in photosynthesis research. It is a weak uncoupler of electron transport. It distorts the pH curve of photophosphorylation toward lower values (see Fig. 1) and, at high concentrations or at high pH, it inhibits electron transport in chloroplasts by destroying the oxygen-producing system.

pK_a 8.35   Bicine; N,N-bis(2-hydroxyethyl)glycine

$$H^+ + (HOCH_2CH_2)_2\!=\!N\!-\!CH_2COO^- \rightleftharpoons (HOCH_2CH_2)_2\!=\!NH^+CH_2COO^-$$

ΔpK_a/degree $= -0.018$

Binds Mg^{2+} weakly, Ca^{2+} and Mn^{2+} moderately, and Cu^{2+} very strongly.

Comments: More easily oxidized by ferricyanide than tricine, otherwise satisfactory for the Hill reaction. No known advantages over tricine.

pK_a 8.4   Glycylglycine

$$H^+ + H_2N-CH_2\overset{O}{\overset{\|}{C}}-NHCH_2COO^- \rightleftharpoons H_3\overset{+}{N}-CH_2\overset{O}{\overset{\|}{C}}-NHCH_2COO^-$$

$\Delta pK_a/degree = -0.028$

Binds Mn^{2+} slightly and Cu^{2+} moderately strongly. Binding of Mg^{2+} and Mg^{2+} and Ca^{2+} negligible.

Comments: An excellent buffer for its range, but relatively difficult to make and therefore will probably continue to be expensive. Satisfactory for studies of electron transport and phosphorylation in chloroplast lamellae preparations, but for this purpose it offers no advantage over the more easily prepared tricine and TAPS.

pK_a 8.55   TAPS; N-tris(hydroxymethyl)methyl-3-aminopropanesulfonic acid

$$H^+ + (HOCH_2)_3{\equiv}C-NH-CH_2CH_2CH_2SO_3^- \rightleftharpoons (HOCH_2)_3{\equiv}C-NH_2^+CH_2CH_2SO_3^-$$

$\Delta pK_a/degree = -0.027$

Does not bind Mg^{2+}, Ca^{2+}, Mn^{2+}, or Cu^{2+}.

Comments: Satisfactory for studies of electron transport and phosphorylation in chloroplast lamellae preparations. Otherwise not widely tested. Potentially inexpensive.

pK_a 9.9   Glycine

$$H^+ + H_2NCH_2COO^- \rightleftharpoons H_3^+NCH_2COO^-$$

Comments: A good buffer at this very high range. Satisfactory for studies of electron transport and phosphorylation in chloroplasts, but one rarely employs such high pH levels in photosynthesis studies.

pK_a 10.25 Carbonate

$$H^+ + CO_3^{2-} \rightleftharpoons HCO_3^-$$

Comments: Can be used for studies of photophosphorylation and electron transport at the highest pH's where amine buffers such as Tris are exceedingly inhibitory. Carbonate-bicarbonate hydrogen ion buffers have the additional advantage of serving also as CO_2 buffers, producing constant low levels of CO_2 which may be computed.[8] Many algal cells carry on photosynthesis quite happily when directly suspended in such media. Alternatively the CO_2-regulating carbonate-bicarbonate mixture may be placed in the center well of a vessel, preferably with some carbonic anhydrase or arsenite to catalyze the hydration-dehydration of CO_2.

[8] O. Warburg and G. Krippahl, Z. Naturforsch. B 15, 364 (1960).

Synthesis and Purification of Buffers

The methods of preparation described below were designed to give good yields by simple procedures accessible to all. However primary emphasis has been placed on methods that consistently yield a very pure product. Since buffers are typically used in massive amounts, the importance of purity cannot be overemphasized.

Preparation of Tricine

A solution of the Tris salt of chloroacetic acid is prepared by dissolving 378 g (4 moles) of chloroacetic acid in 300 ml of H_2O and adding 484 g (4 moles) of Tris during the process of solution. Heat of neutralization raises the temperature to about 45°. Another 484 g of Tris is suspended in 350 ml of H_2O in a 4-liter beaker. Approximately one-third of the Tris chloroacetate solution is added. The suspension is heated on a hot plate with continuous mechanical stirring. As the temperature increases, the Tris dissolves. At 80° the reaction becomes rapid and the temperature begins to rise quickly. The hot plate is turned off. Residual heat in the hot plate and the heat of reaction raise the temperature to 107°, where the mixture begins to boil. After about 5 minutes of boiling another third of the Tris chloroacetate is added, whereupon the temperature drops to 95° then quickly rises to 108°, where boiling recommences. After 5 minutes of boiling the remainder of the Tris chloroacetate is added and washed in with 25 ml of H_2O. The hot plate is turned on again, and the reaction mixture is boiled gently for 15 minutes; during this time the temperature rises from 109° to 111°. The reaction mixture is removed from the hot plate and allowed to cool to about 85°. Total reaction volume is then about 1500 ml. After 2600 ml of 95% alcohol has been added, the solution is seeded with a few crystals of tricine (if necessary) and cooled slowly, with continuous stirring until it becomes too thick. The reaction mixture is then left overnight at 2°–4°. The precipitate is filtered off on a large Büchner funnel and pulled almost dry with a plastic sheet over the surface of the crystals to maintain a vacuum and squeeze out the solution. The precipitate is then washed with 1 liter of 95% alcohol and again pulled almost dry. The precipitate is redissolved in 750 ml of H_2O, and the solution is heated to 70°. Thirty-five grams of acid-washed charcoal (Norit) is added together with 25 g of Celite filteraid, and the suspension is again filtered. (Caution: To prevent any charcoal getting through, it is well to prepare the filter with a thin coating of filteraid.) The charcoal on the filter is washed with 250 ml of hot H_2O. The combined filtrate and washings are heated to 70° while 1500 ml of 95% alcohol is being added. The solution is then slowly cooled with continuous mechanical stirring to about 4°. The precipitate is filtered off, washed with 500 ml of 95% alcohol and dissolved in 650 ml of H_2O. The solution is again heated to 70° while 1500 ml of 95% alcohol is added. It is then cooled slowly with continuous stirring to about 10°. The precipitate is filtered off as before on a Büchner funnel, washed with two lots of 250 ml of 95% alcohol, and dried on the filter overnight with continuously applied vacuum. (Caution: The plastic seal over the top of the funnel must be tight or dust will be carried into the precipitate in the air stream.) The final precipitate is coarsely granular and free flowing. Yield 365–385 g (51–55%) on the basis of the chloroacetic acid.

Preparation of MOPS

Two moles (174 g) of morpholine is dissolved in 400 ml of 50% alcohol. The solution is cooled to about 20° and 122 g (1 mole) of propane sultone is added at a rate which does not allow the temperature to rise over 40°. (Caution: Overcooling the reaction mixture may prevent the reaction from starting until too much propane sultone has been added and there is a danger of the reaction becoming uncontrolled.) After all the propane sultone has been added, the temperature is slowly raised on a hot plate to 70°. Glacial acetic acid (66 ml = 1.1 mole) is added and enough absolute alcohol to make the total volume 2.1 liters. The solution is then cooled slowly with continuous mechanical stirring (seeding with crystals of MOPS if necessary) to 0°. The precipitate is filtered on a Büchner funnel, and washed with 300 ml of 95% alcohol at room temperature. It is then redissolved in 600 ml of 50% alcohol at 65°. Twenty-five grams of acid-washed charcoal (Norit) and 25 g of Celite filteraid are added, and the resulting suspension is filtered on a Büchner funnel prepared with a layer of filteraid. The charcoal on the filter is washed with a little hot 50% alcohol. To the combined filtrate and washings is added 1000 ml of 95% alcohol at 65°. The solution is then cooled with continuous mechanical stirring to 0°. The precipitate is collected as before on a Büchner funnel, washed with 200 ml of 95% alcohol, and redissolved in 200 ml of H_2O plus 1500 ml of 95% alcohol at 70°. The solution is cooled to 0° with stirring, the precipitate is collected on a Büchner funnel, washed with 250 ml of 95% alcohol and dried on the filter (see preparation of tricine). The free-flowing crystals weigh 125–144 g (60–69% on basis of the propane sultone).

Preparation of MES

To a 4-liter beaker are added 1 liter of H_2O, 333 g (2.0 moles) of sodium chloroethanesulfonate and 400 g (4.6 moles) of morpholine.

The resulting solution is stirred continuously with a mechanical stirrer and heated to a gentle boil for 3 hours. If during this time the boiling point rises above 112°, a little water should be added. Final volume should be about 1 liter. The reaction mixture is cooled to 85°, and 2200 ml of 95% alcohol is added. After warming to about 70° a small amount of sodium chloride is filtered off. To the filtrate is added 300 ml (5 moles) of glacial acetic acid. The resulting suspension of MES is heated to 70°, and water is added very cautiously until the precipitate just dissolves. The solution is then cooled slowly with continuous mechanical stirring to about 0°. The precipitate is filtered off on a Büchner funnel, sucked almost dry with a plastic membrane sealing the surface of the precipitate, washed with 300 ml of 95% alcohol, and sucked almost dry again. It is then suspended in 1 liter of 95% alcohol at 45°, and water is added again very cautiously until the precipitate just dissolves. The solution is heated to 70°, 35 g of acid-washed charcoal (Norit) and 25 g of Celite filteraid are added. The charcoal suspension is filtered hot through a Büchner funnel prepared with a thin layer of filteraid. The charcoal on the filter is washed with 100 ml of hot 50% ethanol. To the combined filtrate and washings is added 1 liter of boiling 95% ethanol. The mixture is cooled with continuous stirring to 0°. The precipitate is removed by filtration on a Büchner funnel, washed with 250 ml of 95% alcohol, and resuspended in 1500 ml of 95% alcohol. The suspension is heated with stirring to 70°, and water is added cautiously until the MES just dissolves. The solution is then cooled slowly with continuous mechanical stirring to 0°. The final precipitate is collected on a Büchner funnel, washed with 250 ml of 95% alcohol, and sucked dry with a plastic membrane over the top of the precipitate. Yield 300 g (85% on the basis of the sodium chloroethanesulfonate).

Preparation of HEPES

To a 4-liter beaker are added 520 g (4 moles) of N-(2-hydroxyethyl)piperazine, 240 ml of H_2O, and 333 g (2 moles) of sodium chloroethanesulfonate. The slurry is heated with stirring on a water bath. At about 90° the reaction becomes vigorous and the temperature rises to 110°–115°. (Caution: it is well to heat the waterbath slowly lest the reaction become too vigorous.) Within about 15 minutes the temperature begins to fall and the reaction is held at 105° for 30 minutes by placing the beaker on a hotplate. During this time continuous vigorous stirring of the viscous mixture is required to prevent local overheating. After cooling the reaction mixture slightly, 75 ml of triethylamine and 3 liters of hot absolute alcohol are added. Fifteen grams of acid-washed charcoal (Norit) and 15 g of Celite filteraid are stirred in, and the resulting suspension is filtered on a Büchner funnel. The mixture of charcoal, filteraid, and sodium chloride on the filter is washed with 200 ml of hot absolute alcohol. To the combined filtrate and washings are added 230 ml of glacial acetic acid. (A few drops of the resulting mixture when greatly diluted with water should give a pH of 5.0 ± 0.2). The acidified filtrate is allowed to stand at $0°–4°$ overnight and the precipitate is filtered off on a Büchner funnel. The precipitate is then washed with 500 ml of 95% alcohol (room temperature) and resuspended in 1.5 liters of 95% alcohol to which has been added 300 ml of H_2O. The resulting solution is heated to $75°$. Twenty-five grams of Norit and 25 g of Celite are added and the hot suspension is filtered through a Büchner funnel prepared with filteraid. The filter is washed with 250 ml of boiling 95% alcohol. To the combined filtrate and washings are added 1.5 liters of hot absolute alcohol. The solution is cooled slowly with continuous stirring to $5°$ then left overnight at $0°–4°$. The precipitate is filtered off on a Büchner funnel, washed with 300 ml of 95% alcohol, and resuspended in 2.5 liters of 95% alcohol. The suspension is heated to $70°$, and water is added very slowly and cautiously until the precipitate just dissolves. The solution is then cooled slowly with continuous stirring to $5°$ and left overnight at $0°–4°$. The final precipitate is sucked dry on a Büchner funnel with a plastic seal over the surface. Yield 280–375 g (60–80% on the basis of the sodium chloroethanesulfonate).

[4] Ion Transport
(H^+, K^+, Mg^{2+} Exchange Phenomena)

By RICHARD A. DILLEY

Ion transport, particularly of H^+ ions, has been shown to be intimately involved with electron transport and the energy conversion system of chloroplasts.[1–3] Most of this work derived its impetus from the Mitchell chemiosmotic hypothesis, which assumes that chemical potentials of ion gradients could be the driving force for phosphorylation.[4]

[1] A. T. Jagendorf and E. Uribe, *Brookhaven Symp. Biol.* **19**, 215 (1966).
[2] M. Schwartz, *Nature (London)* **219**, 915 (1968).
[3] R. A. Dilley, *in* "Progress in Photosynthesis Research" (H. Metzner, ed.), Vol. III, p. 1354. H. Laupp, Tübingen, 1969.
[4] P. Mitchell, *Biol. Rev. Cambridge Phil. Soc.* **41**, 445 (1966).

The basic observations have been that light-induced electron flow results in H^+ ion uptake into the grana membrane of chloroplasts,[1] usually in exchange for K^+ and Mg^{2+} ions.[5] The cation exchange for protons has not been studied in great detail, but it is known that the efflux processes are variable in circumstances where the proton uptake is consistently observed to be fairly constant in rate and extent.[5,6] The purpose of this article is to outline the experimental techniques used in measuring these ion transport processes.

Measurement of pH Changes in Chloroplast Suspensions

Preparation of Chloroplasts. Chloroplast preparation may be carried out by a variety of methods, all giving basically the same type of proton uptake activity.[2,5,6] The routine preparation consists of grinding about 50 g of deveined spinach leaves for 30 seconds in a Waring Blendor with 100 ml of 0.4 M cold sucrose (or sorbitol) containing 0.01 M KCl, 0.02 M tricine-NaOH, pH 7.8, and 3 mM Na ascorbate. The brei is strained through at least eight layers of cheesecloth, and the filtrate is centrifuged for 1.5 minutes at 200 g. The pellet is discarded and the supernatant solution is centrifuged for 10 minutes at 1000 g. The resulting pellet is resuspended in about 20 ml of the grinding medium and centrifuged again at 1000 g for 5 minutes. The pellet is resuspended in a medium consisting of 0.4 M sorbitol, 0.01 M KCl, 3 mM Na ascorbate, and 9 mg of bovine serum albumin per milliliter. Chlorophyll is determined in the standard way.[5]

Apparatus. The recording of pH changes can be carried out using one of a variety of commercially available pH meters connected to an appropriate recorder. All that is required is that the pH meter-recorder pair provide full-scale deflection for pH changes of 0.05 to 0.5 pH unit. Depending on the buffering capacity of the reaction mixture, the light-induced pH change can be as large as 0.4 pH unit, i.e., from an initial pH of 6.0 to a pH of 6.3 or 6.4 in the light. Because temperature changes result in about a 0.02 pH unit decrease per degree of increase, the reaction should be carried out in a jacketed and thermostatted vessel.

It is important to stir the reaction mixture so that ion diffusion is not unduly complicated by unstirred layer effects. Stirring also aids uniform illumination. Magnetic stirrers have been found not to introduce noise into pH recording.

Illumination can be provided by a variety of light sources, such as a high intensity microscope light. Near saturating white light intensity

[5]R. A. Dilley and L. P. Vernon, *Arch. Biochem. Biophys.* 111, 365 (1965).
[6]A. R. Crofts, D. W. Deamer, and L. Packer, *Biochim. Biophys. Acta* 131, 97 (1967).

for a reaction mixture containing 20 μg of chlorophyll per milliliter is about 2×10^5 ergs cm^{-2} sec^{-1} with the infrared radiation removed by a Corning 1-69 filter. The infrared filter (or 5 cm thickness of water containing a trace of copper sulfate) serves to alleviate most of the heating effects of the light beam.

Reaction Mixture. A standard reaction mixture contains the following components in a total volume of 10 ml: 100 mM KCl, 5 mM MgCl$_2$, 20 μg chlorophyll per milliliter as chloroplasts, and 15 μM pyocyanine or phenazine methosulfate as the electron transport cofactor. [Methyl-viologen (0.4 mM) can serve in place of pyocyanine, as can trimethyl-hydroquinone (0.13 mM).] The pH should be adjusted to around pH 6 with standard acid or base. The light-induced pH changes can be observed at pH values from about 5.5 to 8.5.

An initial pH of around 6 is a convenient pH. A typical trace for the light-induced pH changes of chloroplast reaction mixtures of the type described above is shown in Fig. 1. The rate and the extent of the pH changes may be estimated from the pH deflection due to the addition of standard acid or base. Recently Polya and Jagendorf[7] showed that chloroplast suspensions have a greater buffering power in the light than in the dark, although in my experience this is observed only with reaction mixtures of very low buffering capacity. For a variety of reasons, it is advisable to have a buffer concentration of about 0.3–1 mM in the reaction mixture. Interestingly, the response time for detecting changes in pH is markedly increased by the presence of such buffer concentrations. The reason, as first pointed out by Mitchell, has to do with the effect of buffering compounds on the rate of propagation of

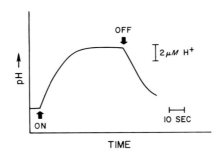

Fig. 1.   Light-induced pH changes observed with chloroplast suspensions. The reaction mixture consists of 100 mM KCl, 5 mM MgCl$_2$, 20 μg of chlorophyll per milliliter as chloroplasts, and 15 μM pyocyanine. The initial pH was about 6, and other conditions were as mentioned in the text.

[7]G. M. Polya and A. T. Jagendorf, *Biochem. Biophys. Res. Commun.* 36, 696 (1969).

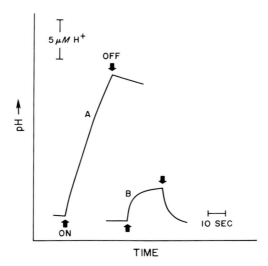

Fig. 2. pH changes observed at pH 8 in the presence and the absence of an ATP-forming system. The reaction mixture was similar to that given in Fig. 1 with the following additions: Curve A, 0.8 mM ADP, and 0.8 mM K$_2$HPO$_4$; initial pH 8.0. Curve B, the reaction mixture was similar to that for curve A except that K$_2$HAsO$_4$ replaced K$_2$HPO$_4$. In the case of curve B a typical light-induced proton uptake reaction is obtained, leading to a steady state level of proton accumulation. In curve A, steady-state pH changes are not obtained owing to the continued pH increase due to esterification of ADP with phosphate. The ongoing pH changes obtained in the case of curve A may be used as a measure of the rate of ATP formation. [M. Nishimura, T. Ito, and B. Chance, *Biochim. Biophys. Acta* **131**, 97 (1967).

ΔpH from the bulk solution to the vicinity of the glass electrode.[8] The initial rate of pH change may be 3–5 times faster with 1 mM tricine present in a standard reaction mix (at an initial pH near 8) compared to the rate in the absence of buffer. The KCl used as a suspending medium does not substitute for the buffer in this effect.

In addition to measuring the reversible light-induced hydrogen ion uptake of chloroplasts, this technique is also readily used to measure the rate of photophosphorylation in chloroplasts.[9] For this assay, in addition to the above components of the reaction mixture, are added the following: 1 mM ADP and 3 mM potassium or sodium phosphate; the pH of this system is adjusted to near pH 8. If the chloroplasts are capable of carrying out photophosphorylation, the type of pH trace obtained is that shown in Fig. 2A. Again, the rate can be determined by

[8]P. Mitchell and J. Moyle, *Biochem. J.* **104**, 588 (1967).
[9]M. Nishimura, T. Ito, and B. Chance, *Biochim. Biophys. Acta* **59**, 177 (1964).

a comparison to the pH change noted upon addition of standard acid or base. Under these conditions, it is difficult to evaluate how much of the initial pH change observed upon illumination is due to the hydrogen ion pump and how much reflects ongoing photophosphorylation. To check this point, one can substitute arsenate in place of phosphate and obtain a trace of the type shown in Fig. 2, curve B. In this case, since there is no net (phosphate) esterification, the only pH changes observed are those due to reversible hydrogen ion uptake activity.

Measurement of Potassium, Sodium, Magnesium, and Calcium Transport

Atomic Absorption Spectrophotometry

As stated above, hydrogen ion uptake activity in chloroplasts is usually accompanied by the extrusion of either potassium or magnesium. (In some cases, sodium and calcium may also be exchanged outward for hydrogen ions taken up.) Atomic absorption spectrophotometry is the most convenient method for measuring the concentrations of these cations, and typical measurements are described in detail by Dilley and Vernon.[5] The technique is very straightforward and will not be discussed here. Rather, various methods to separate chloroplasts from the suspending medium are discussed, thereby allowing assays to be performed on the supernatant solution.

Millipore Filter Technique. This technique consists simply of passing the chloroplast suspension over a Millipore filter and thus separating the chloroplasts from the suspension. For this, various types of Millipore filters and filter holders are available. One convenient type, called the Swinny Adapter, readily fits on the end of standard syringes, and the suspension is filtered through the Millipore filter in less than 2 seconds. Larger filters can be used with a Büchner funnel to reduce the pressure below the filter and draw the suspension onto the filter. It is convenient to use the glass prefilters over the filter proper so as to reduce the probability that the chloroplasts will clog the filter holes. Millipore filters with a pore size of 0.45 μ are very effective in separating chloroplasts from a suspension. Aliquots of the filtrate may then be treated according to the directions for atomic absorption spectrophotometry. Comparison of the data to that of standard curves of various salt solutions will allow one to determine the concentration of the given ion in the filtrate.

Centrifugation Technique. If rapid separation of the chloroplasts from the suspending medium is not necessary or if Millipore filters are not available, centrifugation may be used to separate chloroplasts from the suspension. The resulting supernatant fluid can then be handled as outlined above for the Millipore filter filtrate.

The ion content in the chloroplast fraction may be assayed following digestion of the organic matter with nitric acid, after which one suitably dilutes the digestate with distilled water.

Ion Specific Electrodes for K^+ and Na^+

In some circumstances concentrations of potassium and sodium ion can readily be measured with electrodes in a manner similar to the measurement of H^+ ion concentration.[10] Electrodes made from special glass show relative selectivity, in some cases with high potassium sensitivity and lower sodium and other monovalent cation sensitivity; other glasses are available with high sodium sensitivity and relatively lower potassium and other monovalent cation sensitivity. Most of these electrodes are also quite sensitive to hydrogen ion concentration changes.

The apparatus used in measuring ion concentration with the cation electrodes is similar to that used for measuring hydrogen ion concentration. The reference electrode can be the same reference electrode as that used with the pH measurement, and the cation electrode may be plugged into the pH meter in place of the hydrogen ion glass electrode. The potential generated by the cation electrode is a log function of cation concentration; hence much greater sensitivity is obtained if the concentration of the cation to be measured is fairly low in the suspending medium. In the case of chloroplast potassium efflux, it is difficult to detect the potassium efflux if the ambient concentration is greater than about 5×10^{-4} M. For this reason, chloroplasts should be prepared in the absence of either sodium or potassium ions and also resuspended in the absence of these ions. A typical reaction mixture, which we have used with good success for measuring light-induced potassium efflux, consists of the following: 200 mM sucrose, 20 mM Tris·HCl, and an electron transport cofactor of the type mentioned above in the discussion on hydrogen ion measurements. A typical experimental result for light-induced potassium efflux from chloroplasts is shown in Fig. 3. The lack of potassium reentry into the chloroplasts after the light was turned off is attributed to a competition between the Tris cation, which is at a much higher concentration, and the K^+ ion, which is extruded into a low K^+ medium.

Many types of ion selective electrodes are now becoming available

---

[10]C. Moore and B. Pressman, *Biochem. Biophys. Res. Commun.* **15**, 562 (1964).

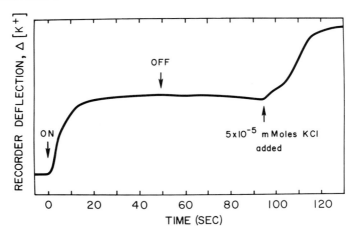

Fɪɢ. 3.   Light-induced K^+ efflux in exchange for H^+ influx with spinach chloroplasts. K^+ concentration changes were measured with a cation sensitive electrode as described in the text. The reaction mixture contained 200 mM sucrose, 20 mM Tris·HCl, pH 6, and 0.13 mM trimethylhydroquinone. The final volume was 10 ml, and other conditions were as described in the text.

commercially, including magnesium, calcium, chloride, and many others. Information on these various electrodes is readily available from the manufacturers.

[5] Techniques for Studying Light-Induced Paramagnetism in Photosynthetic Materials by Means of Electron Paramagnetic Resonance Spectroscopy

By ELLEN C. WEAVER and HARRY E. WEAVER

Introduction and Scope

This article is concerned with methods relevant to performing electron paramagnetic resonance (EPR) experiments on light-induced resonances in photosynthetic materials. To this end, we consider the experimental aspects of working with aqueous suspensions of necessarily complex systems: algae, chloroplasts, subchloroplast fractions, bacteria, chromatophores, and "reaction-center" preparations. The emphasis is on experiments at ambient temperature. Techniques for quantitative and qualitative determination of the information content of the light-induced EPR signals is outlined. It appears virtually certain that one of these, the single light-induced signal in photosynthetic bacteria, and the corresponding signal I in plants, is a property of the reaction center, and as such is evidence of the first chemical act following absorption of a photon.[1,2]

Space limitations make it impossible to deal with the basic phenomenon of paramagnetic resonance. The biologically oriented reader will find Bersohn and Baird[3] an excellent introduction. Some other useful references are listed.[4-8] We shall also omit mention of paramagnetic transition elements which are often present as minor constituents in photosynthetic systems.

There are two main advantages of the EPR spectrometer as a detector of light-induced effects in any photobiological process. First, the energy required to induce allowed transitions between quantized magnetic levels of unpaired electrons responsible for a signal is sufficiently low (approximately 1 cal/mole vs 100 Kcal/mole for transitions in ultraviolet spectroscopy) that the organism is totally unaffected by the scanning microwave radiation, which commonly has a wavelength of the order of 3 cm. Second, substances such as bulk chlorophyll, which do not become paramagnetic when illuminated, are transparent in the EPR spectrometer. The problem faced by the optical spectroscopist of detecting the small optical density changes resulting from oxidation-reduction reactions of one absorbing molecule in the presence of several hundred absorbing but inert ones does not confront the EPR spectroscopist.

EPR Spectroscopy

The basic observation is the fact that photosynthetic materials become paramagnetic when illuminated with visible light. By definition, this tells us that absorption of a photon has resulted in the unpairing of previously paired electrons. However, there can be considerably more

[1] E. C. Weaver, *Annu. Rev. Plant Physiol.* **19**, 283 (1968).

[2] E. C. Weaver and H. E. Weaver, *Science* **165**, 906 (1969).

J. R. Bolton, D. W. Reed, and R. K. Clayton, *Photochem. Photobiol.* **9**, 209 (1969).

[3] M. Bersohn and J. C. Baird, "An Introduction to Electron Paramagnetic Resonance." Benjamin, New York, 1966.

[4] M. S. Blois, Jr., and E. C. Weaver, "Photophysiology" (A. C. Giese, ed.), Vol. 1, p. 35. Academic Press, New York, 1964.

[5] G. E. Pake, "Paramagnetic Resonance." Benjamin, New York, 1962.

[6] C. P. Slichter, "Principles of Magnetic Resonance." Harper, New York, 1963.

[7] G. Schoffa, "Elektronenspinresonanz in der Biologie." Braun, Karlsruhe, 1964.

[8] M. S. Blois, Jr., H. W. Brown, R. M. Lemmon, R. O. Lindblom, and M. Weissbluth (eds.), "Free Radicals in Biological Systems." Academic Press, New York, 1961.

information gained from the position, or "g value," of the resonance, its shape, its intensity, temperature dependence, the kinetics of its formation and decay, and its response to exciting light of varying wavelength, duration, and intensity. Computer techniques for signal-to-noise ratio (S/N) enhancement enable the researcher to extract information otherwise lost in random noise. Instructions and admonitions follow.

Basic EPR Procedure. An aqueous suspension, when mounted in a microwave cavity of rectangular geometry, must be contained in a flat quartz sample holder (*ca.* 0.03 cm deep) in order to minimize the absorption of the microwave electric field by the polar solvent and the concomitant decrease of S/N ratio, while providing as large a sample volume as feasible. About 0.4 ml of sample can be used to fill the cuvette.

The EPR cavity must provide an entrance for light. This can be done by an opening at one end of the cavity, which consists of a grid of parallel conductors whose spacing is close enough so as not to degrade the quality factor, Q, of the cavity. Such an arrangement can be constructed with about 50% transmission of the incident parallel illumination. A mesh of fine gold wire can be substituted, and although it admits considerably more light, the cavity Q is considerably lower than with the grid. The sensitivity required to detect unpaired spins in aqueous biological material is so great that a high cavity Q is necessary; factors which degrade Q seldom can be tolerated. Ideally, the area of the cuvette which can be illuminated should correspond closely to the full sensitive volume of the sample.

A second type of opening to a resonant cavity has "chimneys" on both the front and back faces of the cavity; the advantage of this arrangement is that the sample can be illuminated from either or both sides, or that EPR and optical measurements can be made simultaneously. However, here again the quality factor, Q, of such geometries is likely to be lower than that of the single-ended gridded cavity, especially for aqueous samples. Figure 1 illustrates these cavities.

Since useful EPR signals stem from the absorption of energy from the microwave *magnetic (H)* field component within the cavity and the significant losses from the interaction of the solvents and sample holder (cuvette) with the microwave *electrical (E)* field component, special care must be taken in positioning the sample in the cavity to correspond to the region of minimum *E*-field and thus the region of maximum *H*-field to ensure maximum S/N for a given sample. Instructions are usually provided by the manufacturer of commercial equipment on the proper positioning of the sample holder for various possible cavity geometries within the general operating instruction manual for the spectrometer. However, contrary to the statements of those researchers, who are accustomed to dealing with solutions in nonpolar solvents or with solid samples, useful signals *can* be obtained from very wet, even dilute, samples.

It is wise for the neophyte to locate the resonance in a sample with a strong signal such as DPPH (2,2-diphenyl-1-picrylhydrazyl), either as a solid or in benzene solution, in pitch, or in an aqueous solution of manganous (Mn^{2+}) ions. The light-induced free radical will be found at about the same magnetic field strength as any of these, since all free radicals have a *g* value close to 2 (see next section). With the sample in place, the magnetic field is scanned in the region of the expected resonance. The light illuminating the sample is turned on, and the scan is repeated. For this preliminary work, a modulation amplitude of 8–10 gauss may be used. A power setting of 50–100 mW of microwave power, and a response time of 0.1–1 second with a scanning speed of 20–40 gauss per minute should provide nearly optimal instrumental settings to enable the experimenter to determine whether light-induced resonance exists.

Finer adjustments to experimental procedure will follow. Assuming one has indeed found a light-induced resonance, the following parameters are of importance.

The Spectroscopic Splitting Constant, or g Value. The *g* value corresponds in a qualitative way to the wavelength of absorption peaks in optical spectroscopy. It is defined by the ratio:

$$g = \frac{h\nu}{\beta H_0} \tag{1}$$

where h = Planck's constant, $6.62517 \pm 0.00023 \times 10^{-27}$ erg/sec; β = Bohr magneton, 0.92731×10^{-20} erg/gauss, ν = klystron frequency at resonance, Hz; H_0 = dc magnetic field, gauss.

Blois, Brown, and Maling[9] have discussed the problems and techniques of precise *g*-value determinations for molecules of biological interest. For the purposes of most experimenters, a careful comparison of the unknown resonance with a suitable known one can give an estimate of the *g* value good to the third decimal place.

The most satisfactory standards currently available for the type of experiment discussed here are of the compounds incorporating nitroxide radicals. Peroxylamine disulfonate (Fremy salt) is commercially available[10] and has a precisely determined *g* value of 2.0054.[11] Other nitroxides, such as the 2,2,6,6-tetramethyl piperidine nitroxides (known

[9]M. S. Blois, Jr., H. W. Brown, and J. E. Maling, *in* "Free Radicals in Biological Systems" (M. S. Blois, Jr., H. W. Brown, R. M. Lemmon, R. O. Lindblom, and M. Weissbluth, eds.), p. 117. Academic Press, New York, 1961.
[10]Aldrich Chemical Co., 2369 N. 29th St., Milwaukee, Wisconsin.
[11]J. J. Windle and A. K. Wiersema, *J. Chem. Phys.* **39,** 1139 (1963).

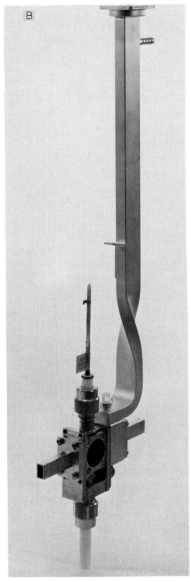

FIG. 1. Different types of electron paramagnetic resonance cavities. (A) The multi-purpose, rectangular (TE_{102}) cavity showing the grid for admission of light. The sample cuvette is held vertically as shown. (B) An optical transmission cavity showing the two "chimneys" for passage of light (Courtesy Varian Associates). (C) A dual cavity illustrating the position of the unknown sample in the front part and the known sample in the rear.

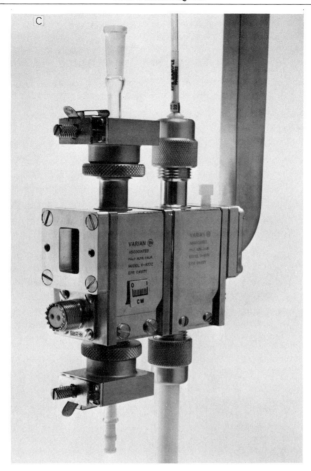

FIG. 1C

as "spin labels"[12]) have an identical g value and are far more stable. These compounds are water soluble and thus in solution have about the same effects on the intrinsic performance of the EPR spectrometer as biological samples. Another commonly used standard is DPPH with a g value of 2.0036 in benzene. The distance between the peaks at high and low magnetic fields in Fremy salt is 26.0 gauss (see Fig. 2). By measuring the distance in gauss between the point where the central peak crosses the baseline and where the spectrum of the unknown crosses the baseline, the unknown g value may then be computed according to Eq.

[12]Available from Syva Assoc., 3181 Porter Drive, Palo Alto, California 94304.

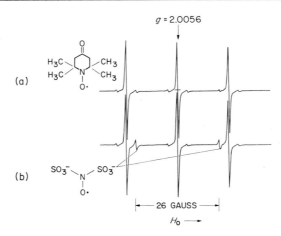

Fig. 2. (a) Spectrum of one of the stable nitroxide radicals (2,2,6,6-tetramethyl piperi-dine nitroxide), 10^{-4} M in water. Instrumental settings were: modulation amplitude, 0.2 gauss; power, 1 mW; magnetic field (H_0) sweep rate, 25 gauss/minute; response time, 0.1 second. (b) Another sweep of the same sample and with the same settings as (a) but with the addition of a capillary of Fremy salt (peroxylamine disulfonate) into the cavity. The two small peaks just proximal to the outer ones are due to Fremy salt, and the central ones are coincident. The peaks of the Fremy salt are 26.0 gauss apart. The point where either spectrum crosses the baseline defines $g = 2.0056$.

(1). An X–Y recorder whose X axis is directly coupled to the scanned dc field (H_0) is useful for this purpose. Care should be taken to position each sample in the cavity similarly, because the cuvette affects not only the cavity Q, but also its resonant frequency. If there is an expendable aliquot of the unknown, a spin label can be added directly to it and the relative peak positions determined on the mixture.

If a double cavity is available, the determination can be made with more accuracy than with a single one in which the samples must be measured successively. A double cavity consists of two rectangular single cavities joined together (see Fig. 1C). Both known and unknown samples may be placed in a double cavity simultaneously; since they are literally in the same cavity, sources of error in g-value measurements are auto-matically compensated for. Factors which could contribute to such an error would be cavity Q, microwave phase, and incident power. The same is true of intensity measurements (see the section on intensity determination, below). A given g value is not necessarily unique to a single molecule and by itself does not provide proof of a radical's identity.

Line Shape Analysis: Qualitative and Quantitative Aspects. In order to take advantage of the improved instrumental performance by means of baseline stabilization through detection of an ac modulated microwave carrier, the magnetic dc field applied to the sample (3.4 kG for a microwave frequency of about 9.5 GHz) has superimposed on it an ac field of variable amplitude and a fixed frequency of 100 kHz. The frequency need not be precisely chosen and usually is a compromise of several competing instrumental factors; the application of this standard experimental technique will have a marked effect on the form of the absorption (or dispersion) line shape presented to the recording device. The exact nature of the effect and the assumptions underlying the mathematical analysis have an important role in the quantitative determinations from which spin concentrations are deduced.

The light-induced resonances in photosynthetic material are very nearly gaussian in form (see Fig. 3).[13] One of the components of the total signal in plants (signal II) shows some partially resolved hyperfine structure. An analysis of the form of a given EPR absorption line is not discussed in detail here, except for a brief comment; a gaussian line shape indicates that the unpaired electron is influenced by a statistically large number of weak (relative to the electronic Zeeman interaction at 3.4 kG) magnetic interactions, stemming usually from neighboring protons unless a specific isotope substitution, e.g.,^2H, has been made. Resolvable, or partially resolvable, hyperfine structure indicates that a relatively small number of protons is influencing the electrons.

As a consequence of the phase-sensitive detection of the 100-kHz modulation signal which is normally present during the scan of the mag-

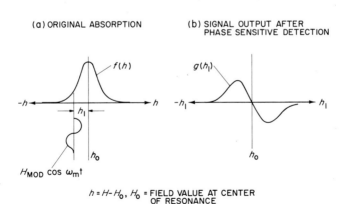

(a) ORIGINAL ABSORPTION   (b) SIGNAL OUTPUT AFTER PHASE SENSITIVE DETECTION

$h = H - H_0$, H_0 = FIELD VALUE AT CENTER OF RESONANCE

FIG. 3.   (A) A symmetrical "bell-shaped" curve, $f(h)$, with its absorption maximum centered at H_0. (B) The first Fourier component, $g(h_1)$, of the function $f(h)$, after phase-sensitive detection.

[13]E. C. Weaver and H. E. Weaver, *Photochem. Photobiol.* **2**, 325 (1963).

netic field through the resonance, the resonance absorption will display a dc voltage on the recording device with both positive and negative excursions, the form of which will resemble a derivative of the original absorption signal (see Fig. 3). However, this voltage vs field presentation on a recorder will be an exact derivative only in the special case where the modulation amplitude is small compared to the natural width of the line. The relationship between the signal amplitude and the peak-to-peak width of the signal can be calculated in a special case in closed analytical form;[14] the results for a Lorentzian line are displayed here by way of illustration in Fig. 4. The effect of the modulation amplitude on the natural linewidth is initially negligible until a maximum in the dependence of signal height on modulation amplitude is reached, at which time the apparent linewidth has increased significantly. Thus, the choice of modulation amplitude is somewhat analogous to the choice of slit width in optical spectroscopy. For modulation amplitudes that are large compared with natural linewidth, the separation between the resonance maxima and minima is proportional to the modulation amplitude. For this instrumental reason and for other experimental reasons, a method of calculation of intensity should be used which does not depend on the detailed line shape. This is especially important when comparing simple lines with those having resolved or partially resolved structure, or those that may be asymmetrical.

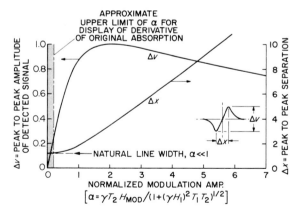

FIG. 4.   Output signal height, ΔV, and peak separation, ΔX, vs the modulation amplitude, H_{mod}, presented as a dimensionless parameter, α. The effect on these results due to the rf field amplitude, H_1, is taken into account. While the above analysis is quantitatively valid only for a Lorentz line shape (based on Bloch's phenomenological equations for magnetic resonance), the general aspects give a qualitative feeling for modulation effects on the more complex line shapes encountered in photoinduced free radicals in aqueous solutions.

[14]H. E. Weaver, Ph.D. Thesis, Stanford University, Physics Department, 1952.

Intensity Determination The intensity of an EPR signal is proportional to the area under the absorption curve. The simplest, but least reliable, method of line intensity comparison is that of estimating peak amplitudes. This is valid only for the case of identical line shapes and widths. In practice, no two free radical species are identical, so that comparing peak amplitudes is admissible only when, for instance, one is comparing light-induced signal I under different conditions of illumination. Microwave power levels well below saturation must also be employed. A slightly better method, which at least compensates for differences in width (but not in shape), is the method of calculating the height × (width)^2 for the derivative line. This method, although making use of additional curve parameters, still has the shortcoming of not utilizing much of the information content of the total curve. Again, the estimate is significant only if the signal behavior is linear with respect to both the modulation amplitude and rf levels. A third, and by far the most reliable, numerical method of intensity measurement on the typical signal is that of the first moment calculation.[15] The method allows determination of the intensity of the original absorption line with only one numerical integration step and yields a value independent of the detailed line shape. Furthermore, it takes into account the value of the modulation amplitude [*cf.* Eq. (4) below], and thus it does not matter if one of them is overmodulated.[16] Again, the only restriction is that the signals to be compared must be linear in amplitude with the rf field amplitude. Referring to Fig. 3, the first moment is defined as follows:

$$M_1 = \int_{-\infty}^{+\infty} h_1 g(h_1) dh_1 \tag{2}$$

where the intensity is defined as

$$M_0 = \int_{-\infty}^{+\infty} f(h) dh \tag{3}$$

then the following relationship holds

$$M_1 = -H_{mod} M_0 \tag{4}$$

where H_{mod} is the modulation amplitude.

The units of horizontal axis of Fig. 3 are related to the swept magnetic field in the experiment. However, in an actual experiment, the units selected will probably be the arbitrary intervals of the linear graph paper on which the signal is recorded. The fineness of the interval dh_1 is likely to be a function of the patience of the operator. The power of this

[15] E. R. Andrew, *Phys. Rev.* **91**, 425 (1953).
[16] K. Halbach, *Phys. Rev.* **119**, 1230 (1960).

numerical method lies in the fact that standards and unknowns of very different line shape and structure may safely be compared; furthermore, it lends itself readily to adaptation for computer computation.

In order to obtain a measure of the absolute spin concentration, some suitable paramagnetic standard must be selected. It should be obtainable in pure crystalline form or as a standardized, absolutely stable solution, or be readily titratable; it should be soluble in water in order to present the same lossy characteristics to the microwave field as the biological samples treated here. Solutions of a manganous salt can be used, but the peaks are far broader than the light-induced resonances in photosynthetic material. The most satisfactory standards are the tetremethyl piperizine nitroxides[12] which are stable for periods of months in a refrigerator.

A few words of precaution: standard and unknown should be measured successively in the same cuvette, since the sensitive volume may vary from cuvette to cuvette; the gain setting should be calibrated on the spectrometer, or the concentration of the standard be adjusted so that the same gain settings may be used for both; a scanning rate should be used which is slow enough to ensure that the limitations of the recorder writing speed do not distort the line shape; both known and unknown should be positioned several times, since minute changes in orientation can make rather large changes in signal amplitude, and the maximum amplitudes should be used for the first moment calculations. As is the case for g-value determinations (see section on spectroscopic splitting constant, above), a double cavity is an accessory which aids in accurate determinations of signal intensity. When a light-induced signal is the unknown, all possible spin sites should be illuminated. If light intensity is limiting, then the suspension should be diluted until it can be light saturated.

We have mentioned the dependence of signal characteristic on microwave power. The output of the klystron can be attenuated, and the signal amplitude is a function of the amount of attenuation. The amplitude dependence of signal I in plants on power at three temperatures is illustrated in Fig. 5. Since klystrons and attenuators differ, it is again wise to calibrate. This can be done directly with a suitable meter, for example, a Hewlett-Packard 431B power meter. This is attached to the waveguide in place of the cavity, and direct readings are taken for different attenuator settings and at various klystron frequencies. Note that saturation occurs at a far lower power when low temperatures are employed.

Kinetic Analysis. The time course of the formation and disappearance of a light-induced signal may be determined as follows: with the light

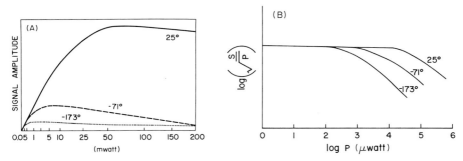

FIG. 5. (A) Saturation with microwave power of light-induced electron paramagnetic resonance signal I in a chloroplast preparation. Signal amplitude in arbitrary units is plotted against the square root of microwave power (in milliwatts) incident on the cavity. The values of milliwatts given in the figure refer to the values actually measured, i.e., before formation of the square root. The values for the signal amplitude at the three temperatures were set equal at 0.2 milliwatt (outside of the saturation range), and the other values were normalized accordingly. In case saturation is not observed, a straight line should result in this kind of a plot. After Fig. 1 of H. Beinert and B. Kok, *Biochim. Biophys. Acta* **88,** 278 (1964). (B) These are the same data as those plotted in Fig. 5(A), but treated in the manner described in Fig. 1, termed a "Vänngård plot," in H. Beinert and W. H. Orme-Johnson [*in* "Magnetic Resonance in Biological Systems" (A. Ehrenberg, B. G. Malmström, and T. Vänngård, eds.), pp. 221–247. Pergamon, Oxford, 1967]. The ordinate represents log (signal height/ $\sqrt{P}$); and on the abscissa, log P is plotted (P in microwatts).

on, the point of maximum deflection of the signal from the baseline is located and the magnetic field fixed at that point. Using a time-based recorder, the light may be switched on and off, flashed, or shuttered. A time constant considerably shorter than the expected rise times should be employed. The optimum modulation amplitude may be used to improve the S/N ratio; a slight adjustment of the dc field may be necessary to compensate for small shifts in the signal maximum or minimum due to the effect of the modulation. This broadening due to high modulation does not affect the time course of signal formation.

The order of the reaction can be determined by fitting points on the curve to standard physicochemical formulas.[17]

Signal-to-Noise Enhancement. If one is dealing with a signal which is near or in the noise level and if it can be maintained in a steady state, then a very slow scanning rate coupled with a long time constant can sometimes make it visible. However, if the reaction is fast, such as the signal which results from an impulse of light but which is reversible and can be repeated except for statistically random variations, a com-

[17]W. J. Moore, "Physical Chemistry" 3rd ed. Prentice-Hall, Englewood Cliffs, New Jersey, 1962.

puter of average transients (CAT) can be useful. Theoretically, one can get a S/N ratio improvement of $n^{1/2}$, where n equals the number of scans, provided the EPR parameters are not changed. If the scanning rate is faster or the response time shorter with the CAT than without it, then the improvement will be less than $n^{1/2}$. Depending on the type of information sought, the CAT can be triggered by a field sweep or by the same circuit which times the light pulse.

Variable Temperature Experiments. In many experiments involving the complex systems studied by EPR and in the realm of this chapter, temperature-dependent effects are an essential part of the investigation. In order to provide a controlled temperature enclosure for the sample while keeping the microwave cavity at room temperature, a quartz Dewar insert is supplied as an accessory with the cavity. Associated with the Dewar is a temperature sensor and regulating unit which supply a controlled flow of gas to the Dewar at a prescribed temperature. By this means a continuously variable temperature, regulated to within $\pm 1°$ over a range of $-190°$ to $+200°$, can be provided for the spectrometer. Equipment required to go beyond these temperature limits is special in nature and will not be treated here.

While commercially available systems are relatively simple in principle, a few words of caution regarding the use of a variable temperature accessory may be useful to the reader new to EPR.

1. Effect of the Dewar insert on the cavity quality factor and resonant frequency. Since efficient thermal isolation is required, the Dewar enclosing the sample is usually a single two-walled evacuated Dewar which contains a "window" along the portion within the cavity that is free of silvering. All parts not exposed to the microwave rf fields are silvered to reduce thermal losses via radiation. To achieve adequate isolation, the dimensions of the Dewar are such as to place the quartz walls in significant electric fields of the cavity. This will have some adverse effect on the Q of the cavity but, depending on the quality of the quartz and wall thickness used, is usually not serious. The dielectric constant of the quartz, however, will cause a substantial downward shift in the resonant frequency, often as much as 10% of the intrinsic cavity frequency. Therefore, readjustment of the mechanical tuning of the reflex klystron as well as the reflector voltage (and usually the sample cavity iris tuning screw) will be required. A proportional reduction in the magnetic field will be required to again observe resonance at $g = 2$.

2. Effect on the rf field distribution. An attendant, often unappreciated, effect on the spatial distribution of the rf fields within the cavity will have a significant influence on the performance of the EPR spectrometer. Therefore, samples which exhibit a given room temperature

saturation behavior with incident microwave power in the absence of a Dewar will have different characteristics when measured in the presence of a Dewar at room temperature. This stems from the fact that the magnitude of the rf magnetic field is no longer related simply to the cavity Q and incident power. In the case of a sample such as Mn^{+2} ion (e.g., 10^{-6} M) in water, there will appear to be somewhat greater S/N for the same rf power incident on the cavity. Therefore, a quantitative comparison of the saturation behavior of an unknown material as a function of temperature should always be carried out with the Dewar in place.

3. Quartz Dewars and UV irradiation. Extremely pure quartz is used in the construction of Dewars to minimize detectable background signals stemming from paramagnetic impurities in the quartz. However, extended irradiation of samples through the quartz Dewar walls by UV radiation will often result in a gradual growth of permanent background signal in the $g = 2$ region. This "color center" when it becomes objectionable can usually be thermally annealed by heating the Dewar in an oven. It is recommended that this annealing be done under the manufacturer's directions.

Light Sources and Calibration. The problem of illuminating an EPR cuvette is not intrinsically different from that of illuminating any other type of reaction vessel in photosynthetic studies. One wishes light of known wavelength and intensity distributed evenly over the sensitive area. It is often useful to be able to superimpose or alternate light of two wavelengths. This can be done with conventional lens and filter systems. However, a particularly useful light source for EPR work[18] uses bundles of fiber optics which are designed to fit onto the cavity window and are secured with a collar. The fibers are then branched into two arms, at the end of which are two light sources, each focused onto the end of the fiber optics, which are then intermingled in a random way so that the two lights can be completely superimposed at the cavity window. There is provision for insertion of neutral density and color filters into the optical paths. The solenoid-activated shutters can be switched by hand or with a programmed timer. A trigger pulse is provided to actuate the CAT sweep (see the section on signal-to-noise enhancement, above). Direct calibration of the lights is possible with a commercial radiometer, for which a special probe designed to fit inside the EPR cavity can be obtained (Yellow Springs Instrument Co., Yellow Springs, Ohio).

A pulsed ruby laser (wavelength 694 nm) is a convenient source of short, intense flashes of light. The wavelength is suitable for green plants. If the experimenter wishes to use only the laser as illumination, he must

---

[18] A. Margozzi, M. Henderson, and E. Weaver, *Photochem. Photobiol.* 9, 549 (1969).

exclude the considerable amount of light from the high-intensity lamp used to excite the laser. This can be done by taping an interference filter with a narrow transmission band at 694 nm directly on the EPR cavity. Attenuation with a minimum of scattering can be accomplished by using reflections off an optical wedge, or with neutral filters specially designed to withstand the laser beam. A flat bottle containing copper sulfate is a useful filter, and its attenuation can be measured in a spectrophotometer.

Calibration of the number of photons actually hitting the sample in the EPR cuvette can be done with a sensitive calorimeter available from companies dealing with laser accessories. The elements actually interposed between the light source and the sample may be directly measured or calculated from other information. The number of photons per flash is a function of the temperature of the laser as well as of other factors. With the present pace of laser technology, this source of light should be an increasingly useful tool in photosynthesis research.

Gas lasers which emit a continuous beam of monochromatic light (for example, He·Ne with 633 nm) also have the advantage over white light sources of being collimated and of needing no heat filters. High-intensity incandescent and arc lamps have a high heat content, which must be attenuated by $CuSO_4$ solutions, running water, or some other heat filter.

Miscellaneous. It is sometimes desirable to know the optical density of a sample being used in EPR studies. It is possible to make an adapter which allows the experimenter to use the EPR cuvette in a spectrophotometer and thus obtain a direct spectrum of the material.

The cavity must be kept strictly clean. Do not allow smoking in the room with the spectrometer. Occasionally the cavity should be checked by running a scan with no sample at all. If there are deviations from baseline, especially in the $g = 2$ region, the manufacturer should be consulted for cleaning instructions.

[6] ATP Formation

By David Geller

Numerous methods are available for the measurement of ATP formation. In systems involving the phosphorylation of ADP at the expense of inorganic phosphate (P_i), the simplest procedure is to measure P_i

concentration by a colorimetric method.[1-7] Although most convenient, this offers only an indirect estimate of ATP formed. Circumstances may require the measurement of small changes in P_i at relatively high P_i concentrations.

A second measure of ATP formation is that of the pH change associated with phosphorylation.[8] Although especially useful for rapid kinetic measurements, this procedure suffers from a lack of specificity. The biological systems of interest show pH changes associated with ion transport which are not directly related to ATP formation. At low levels of ATP formation, the pH change induced by ion transport may indeed exceed that caused by phosphorylation.

Therefore it is preferable to measure ATP directly. For example, this has been accomplished by measurement of the incorporation of radioactive P_i into ATP. Labeled ATP is isolated by adsorption on charcoal[9] or by extraction of the remaining P_i by isobutanol–benzene in the presence of excess molybdic acid.[10] These radiochemical procedures suffer the disadvantage that exchange reactions, which do not represent net phosphorylation, may introduce large errors. Furthermore, the isolation procedure may fail to differentiate between ATP (or glucose-6-P) and other phosphate esters (such as pyrophosphate); this is especially true of the solvent extraction method.

Direct chemical estimation of ATP is provided by the specific sensitive procedures of luminescence[11,12] and fluorometry.[13] Although more sensitive to ATP than direct fluorometry, the luminescence of the fire fly luciferin–luciferase system is limited to the measurement of ATP. The fluorometric procedures in some instances may be more useful since they utilize a variety of enzymatic methods involving phosphate esters

[1]C. H. Fiske and Y. SubbaRow, *J. Biol. Chem.* **66**, 375 (1925).
[2]O. H. Lowry and J. A. Lopez, *J. Biol. Chem.* **162**, 421 (1946).
[3]J. Josse, *J. Biol. Chem.* **241**, 1938 (1966).
[4]P. S. Chen, Jr., T. Y. Toribara, and H. Warner, *Anal. Chem.* **28**, 1756 (1956).
[5]H. H. Taussky and E. Schorr, *J. Biol. Chem.* **202**, 675 (1953).
[6]O. H. Lowry, Vol. IV [17].
[7]B. N. Ames, Vol. VIII [10].
[8]B. Chance and M. Nishimura, Vol. X [99].
[9]M. D. Kamen, Vol. VI [38].
[10]S. O. Nielsen and A. L. Lehninger, *J. Biol. Chem.* **215**, 555 (1955).
[11]B. L. Strehler, *in* "Methods of Enzymatic Analysis" (H. U. Bergmeyer, ed.), p. 559. Academic Press, New York, 1965.
[12]F. Welsch and L. Smith, *Biochemistry* **8**, 3403 (1969).
[13]O. H. Lowry, J. V. Passonneau, F. X. Hasselberger, and D. W. Schulz, *J. Biol. Chem.* **239**, 18 (1964).

derived from ATP. For example, in many situations maximal rates of phosphorylation (ATP formation) require the addition of an accessory trap for ATP, such as hexokinase and glucose. Measurement of ATP formation is provided by fluorometric estimation of glucose-6-P concentration.

Radiochemical Method

Principle. Glucose (or mannose)-6-P is synthesized by incubating the phosphorylating preparation with labeled P_i, ADP, hexokinase, and glucose (or mannose). The sugar phosphate formed is isolated by paper electrophoresis and counted.

Reagents

Citrate buffer: 40 mM sodium citrate, pH 3.6. A mixture of 20 ml of 2 M citric acid and 33.3 ml of 1 N NaOH is diluted to 1 liter with water.

Aniline phthalate reagent. A mixture of 930 mg of aniline and 1.6 g of phthalic acid is dissolved in 100 ml of n-butanol (saturated with water at 25°).[14]

Scintillation reagent. In 1 liter of toluene are dissolved 4 g of 2,5-diphenyloxazole and 0.1 g of 1,4-bis-2-(5-phenyloxazolyl)benzene (Packard Instrument Co., Inc., La Grange, Illinois).

Procedure. In the course of an experiment, labeled glucose (or mannose)-6-P is formed from labeled P_i,[9] ADP, hexokinase, and glucose (or mannose). The radioactive mixture is acidified with trichloroacetic acid and centrifuged. To each acid supernatant, 2% in trichloroacetic acid, is added sufficient carrier P_i and glucose(or mannose)-6-P to assure a concentration of 20–50 mM of each. A 10-μl aliquot is electrophoresed at 5° on Whatman 3 MM paper in citrate buffer for 30 minutes at 80 V/cm. Then the paper is dried for several minutes in an oven at 100°. Radioactive spots may be visualized by exposure of the paper to Kodak No-Screen Medical X-ray Film. Routinely, however, the sugar phosphate spots are located with aniline phthalate reagent. The paper is sprayed with aniline phthalate and placed in the 100° oven for several minutes. The brown sugar phosphate spots are cut out, and the paper squares are placed in vials of scintillation reagent and counted in a liquid scintillation spectrometer. The radioactivity of the sugar phosphate (corrected by comparison with an appropriate blank) is compared to that of the

---

[14]R. J. Block, E. L. Durrum, and G. Zweig, "A Manual of Paper Chromatography and Paper Electrophoresis," p. 133. Academic Press, New York, 1955.

total P_i, determined by counting a spot of a diluted aliquot of unfractionated acid supernatant.[15]

Electrophoresis of the three adenine nucleotides and P_i in the citrate system separates all four components. All migrate toward the anode. AMP, ADP, and ATP migrate at 0.3, 0.6, and 0.8 (respectively) of the distance traversed by P_i. The latter moves 15–16 cm toward the anode (80 V/cm, 30 minutes). Glucose(and mannose)-6-P migrate slightly ahead of (but are not separated from) ADP. The presence of label in the sugar phosphate (but not ADP) may be assured by checking the coincidence of position and shape of radioactive spot (on X-ray film) with the spot produced by treatment of the electrophoretogram with aniline phthalate.

Fluorometric Method

Principle. Glucose-6-P, derived from ATP by means of hexokinase and glucose, is measured by estimation of TPNH generated in the presence of glucose-6-P dehydrogenase and TPN. TPNH is measured by its fluorescence. If PMS[16] or TMPD are present, the fluorometric determination is preceded by extraction of these dyes with bromobenzene under strongly alkaline conditions.

Reagents and Equipment

Farrand Ratio Fluorometer or equivalent, equipped with Corning filters 5860 (primary filter) and 4303 and 3387 (combined secondary filter).[13,17] The lamp used for excitation is a mercury arc.

Bromobenzene. Redistilled, saturated with water at 0°.

Buffer for fluorometry. Tris·HCl, 20 mM, containing 5 mM MgCl$_2$ and 50 μM TPN. This is adjusted to pH 7.8 (25°) by the addition of the sample to be analyzed.

Glucose-6-P dehydrogenase. The suspension of the enzyme (Boehringer, 5 mg/ml) is diluted 30-fold in 0.02% bovine serum albumin − 0.1 M Tris · HCl, pH 7.5 (25°). The enzyme solution, stored at 0°, is stable for at least 24 hours.

Procedure. In an experiment glucose-6-P is formed from ATP generated by the test system in the presence of hexokinase and glucose. The

[15]D. M. Geller, *in* "Bacterial Photosynthesis" (H. Gest, A. San Pietro, and L. P. Vernon, eds.), p. 161. Antioch Press, Yellow Springs, Ohio, 1963.

[16]The abbreviations used are: PMS, N-methylphenazonium methosulfate; TMPD, $N,N,N',$-N'-tetramethyl-p-phenylenediamine.

[17]O. H. Lowry, N. R. Roberts, and J. I. Kapphahn, *J. Biol. Chem.* **224,** 1047 (1957).

incubation is terminated by the addition of perchloric acid (final concentration, 0.3 M). One volume of each acid extract (the supernatant solution following centrifugation of the acidified samples) is mixed vigorously with 1 volume of 1 N KOH and 8 volumes of cold bromobenzene at 0°. After centrifugation of this mixture at 0°, the aqueous phase is promptly analyzed for glucose-6-P.[18] A 10-μl aliquot (10-1000 μM in glucose-6-P) is added to 1 ml of buffer in a selected 10 × 75 mm Pyrex glass tube. The increment in fluorescence is recorded, reading 5 minutes after the addition of 5 μl of glucose-6-P dehydrogenase solution. The reading is repeated 5 minutes later to assure completion of the reaction. The concentration range of glucose-6-P in this assay is from 0.1 μM (this produces a doubling of fluorescence) to a maximum of 10 μM.

The alkaline treatment, followed by extraction with bromobenzene, removes such dyes as TMPD and PMS, with a minimum recovery of 95% of glucose-6-P. It should be noted, however, that glucose-6-P is labile under strongly alkaline conditions. The temperature of the alkaline medium should be low (0°) and the time of exposure to alkali kept to a minimum (several minutes). In the absence of the dyes, the bromobenzene treatment is omitted.

Several methods may be used to enhance the sensitivity of the fluorometric assay. The intensity of fluorescence is increased about 10-fold by the addition of strong alkali in the presence of peroxide.[17] Furthermore, the sensitivity of the fluorometric method may be enhanced by as much as several orders of magnitude by the procedure of enzymatic cycling.[19] These results are comparable to that of the luciferin–luciferase assay.

[18]D. M. Geller, *J. Biol. Chem.* **244**, 971 (1969).
[19]O. H. Lowry and J. V. Passonneau, Vol. VI [111].

[7] Light-Triggered and Light-Dependent ATPase Activities in Chloroplasts

By C. CARMELI and M. AVRON

Chloroplasts prepared from leaves of plants such as lettuce or spinach catalyze the hydrolysis of ATP to ADP and P_i at very low rates. However, in the presence of sulfhydryl reagents, such as cysteine, lipoic acid, or dithiothreitol, light triggers an ATPase activity in chloroplasts.[1-3] This light-triggered and magnesium-dependent ATPase activity is maintained in the dark after light-triggering. Light also induces a calcium-activated, light-dependent ATPase activity which ceases as soon as the light is turned off.[4] Both the light-triggered and the light-dependent ATPase activities are believed to represent a reversal of some of the reactions involved in the latter stages of photophosphorylation.[5] ATPase activity is assayed either by a colorimetric determination of the P_i released from ATP or by the more sensitive determination of $[^{32}P]P_i$ released from $[^{32}P]ATP$.

Reagents
    Tricine-NaOH, 0.3 M, pH 8
    KCl, 0.4 M
    MgCl₂, 0.05 M
    CaCl₂, 0.01 M
    ATP, 0.05 M, pH 8
    $[^{32}P]ATP$, 2 mM, pH 8, containing 5×10^6 cpm/ml
    NaK phosphate, 0.04 M, pH 8
    DTT (dithiothreitol), 0.1 M, pH 8
    PMS (phenazine methosulfate), 0.3 mM
    TCA (trichloroacetic acid), 30%
    Pyruvate kinase (free of NH_4^+ ion since NH_4^+ inhibits the reaction[1]; dialysis against 2 mM KCl plus 5 mM Tris, pH 7.5, can be utilized)
    PEP (phosphoenolpyruvate), 0.04 M, pH 8
    Chloroplasts, containing about 0.5 mg of chlorophyll per milliliter prepared as described in Vol. 23 [18].

Light-Triggered ATPase Activity

Materials. Reaction can be carried out in simple test tubes placed in a thermostatted water bath with glass walls to permit illumination of the sample. We have utilized specially designed glass vessels fastened to a bar which is attached by its axis to a stand (see Fig. 1). The stand was placed in a temperature-controlled water bath with glass walls. Illumination is provided by 300-W flood lamps. This arrangement provides

[1]B. Petrack, A. Craston, F. Sheppy, and F. Farron, *J. Biol. Chem.* **240**, 906 (1965).
[2]G. Hoch, and I. Martin, *Biochem. Biophys. Res. Commun.* **12**, 223 (1963).
[3]R. H. Marchant and L. Packer, *Biochim. Biophys. Acta* **75**, 458 (1963).
[4]A. Bennun and M. Avron, *Biochim. Biophys. Acta* **79**, 646 (1964).
[5]A. Bennun and M. Avron, *Biochim. Biophys. Acta* **109**, 117 (1965).

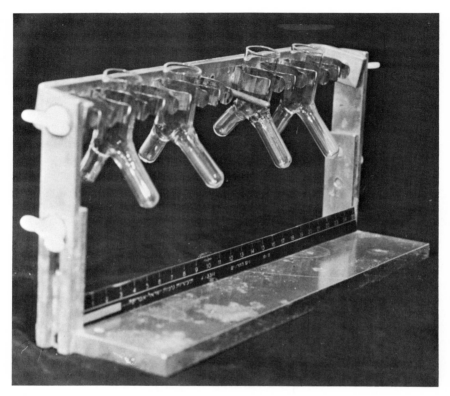

Fig. 1.   Glass vessels designed for use in reaction for light-triggered ATPase activity.

for the easy addition of the reagents placed in the side arms into the main compartments of ten vessels simultaneously, by rotating the bar.

Reaction Mixture. The reaction mixture contains 0.1 ml of each of the following stock reagents giving the indicated final concentrations: tricine, 30 mM; KCl, 40 mM; MgCl$_2$, 5 mM; DTT, 10 mM; ATP, 5 mM; PMS, 30 μM; and chloroplasts containing about 50 μg chlorophyll in a total volume of 1.0 ml.

Procedure. All components of the reaction mixture except ATP are placed in small test tubes (or the main compartments of the special vessels) except ATP (which is added to the side arms in the special tubes). After 5 minutes of equilibration in the water bath at 25° in the dark, the light is turned on for 5 minutes at an intensity of about 100,000 lux. The light is turned off, and the ATP is immediately added and mixed with the reaction components (the sidearm content is mixed with the contents of the main compartment in the special tubes). Rapid adding and mixing is important to avoid errors due to the rapid decay of the light-triggered activity in the dark in the absence of ATP.[1-3] The reaction is allowed to proceed 10–15 minutes in the dark and is stopped by the addition of 0.1 ml of TCA. The tubes are immediately placed in ice (the contents of the special tubes are transferred to small centrifuge tubes) and centrifuged at 500 g for 15 minutes at 0–4°. Aliquots from the supernatant solutions are taken for P_i determination according to the method of Ames.[6]

ATPase activity in continuous illumination can also be measured if, after light-triggering, light intensity is lowered and sufficient pyruvate kinase and PEP are added to compete with photophosphorylation for the rephosphorylation of ADP released by the ATPase activity. The experimental procedure is the same as described above except for the addition of 0.1 ml of PEP and 1 unit of pyruvate kinase. Light intensity is lowered to 10,000 lux after light-triggering.

Specific activity is defined as micromoles of P_i released from ATP per milligram of chlorophyll per hour.

Properties

Light-triggered ATPase activity hydrolyzes ATP, GTP, or ITP. The highest rates were obtained with ATP. The apparent K_m for ATP in 9×10^{-5} M. ADP is a competitive inhibitor with an apparent $K_i = 4 \times 10^{-5}$ M.[5] Specific activity is 50–200. Uncouplers of photophosphorylation, such as 50 μM atebrin, 0.5 mM gramicidin, and 3 mM NH_4Cl, inhibited the light-triggering stage. The same uncouplers stimulate initially and inhibit later the light-triggered activity if present during the dark reaction stage. The ATPase activity is stimulated only by these uncouplers if the reaction stage takes place in the light.[7] The reaction stage of light-triggered ATPase activity is inhibited by DIO-9 (3 $\mu g/ml$) and phlorizin, 2 mM.[7]

Light-Dependent ATPase

Reaction Mixture. Into small test tubes the following amounts of the stock reagents are added to give the indicated final concentrations: 0.1 ml tricine, 30 mM; 0.05 ml KCl, 20 mM; 0.1 ml $CaCl_2$, 1 mM; 0.1 ml K,Na phosphate, 4 mM; 0.1 ml [^{32}P] ATP, 0.2 mM containing 5×10^5 cpm; 0.1 ml PMS, 0.03 mM; 0.1 ml chloroplasts containing about 50 μg of chlorophyll per milliliter in a total volume of 1.0 ml.

Procedure. The tubes containing the reaction mixture are allowed to equilibrate in a temperature-controlled water bath at 15°c in the dark.

[6]B. N. Ames, Vol. VIII [10].
[7]C. Carmeli, *Biochim. Biophys. Acta* **189**, 256 (1969).

Illumination at 100,000 lux for 20 minutes is provided by the same arrangement as described for light-triggered ATPase. The reaction is terminated by turning the light off followed by the addition of 0.1 ml TCA. The chloroplasts are removed by centrifugation at 500 g for 10 minutes at 0–4°c. Aliquots from the supernatant solutions are taken for [^{32}P] P_i determination. [^{32}P] P_i is separated from [^{32}P] ATP by the isobutanol–benzene extraction method,[8] and P_i is determined from the radioactivity in the organic phase. Specific activity is defined as micromoles of P_i released per milligram of chlorophyll per hour.

Properties

Light-dependent ATPase activity has substrate specificity for ATP, GTP, and ITP with the highest rate obtained with ATP (apparent K_m = 4 × 10^{-5} M). ADP is a competitive inhibitor with an apparent K_i = 3 × 10^{-5} M.[5] The specific activity is in the order of 10–30. With white light and PMS as cofactor, saturation is approached at 200,000 lux.[9] The activity is inhibited by Mg^{2+} ions and by uncouplers of photophosphorylation such as 5 × 10^{-6} M atebrin, 1.6 × 10^{-5} M CCP, and 7 × 10^{-6} M gramicidin S.[7]

[8]M. Avron, *Anal. Biochem.* **2**, 535 (1961).
[9]M. Avron, *J. Biol. Chem.* **237**, 2011 (1962).

[8] Light-Influenced ATPase Activity: Bacterial

By T. Horio, Y. Horiuti, N. Yamamoto, and K. Nishikawa

Chromatophores prepared from *Rhodospirillum rubrum* are able to synthesize ATP from ADP and P_i coupled with photosynthetic electron flow in a cyclic fashion.[1] Besides photosynthetic ATP formation, chromatophores can catalyze the oxidation of NADH by molecular oxygen coupled with ATP formation.[2] In addition, chromatophores have activities for ATP–P_i exchange and ATPase either in the light or in dark-

[1]T. Horio and M. D. Kamen, *Biochemistry* **1**, 144 (1962).
[2]J. Yamashita, S. Yoshimura, Y. Matuo, and T. Horio, *Biochim. Biophys. Acta* **143**, 154 (1967).

ness, which are at least in part brought about because of the reversibility of the ATP-forming reaction.[3,4]

ATPase activity by chromatophores is firmly associated with particle structure and influenced by illumination, redox dyes, and extraction of the quinones from chromatophores. Conceivably, the ATPase activity is catalyzed at least in part by an enzyme system containing redox components associated with chromatophores.

Preparations and Assay Method

The culture of *R. rubrum* cells and preparation of chromatophores from them are performed in the same manner as described elsewhere in this series.[5] ATPase activity is assayed according to the methods described elsewhere,[5] except for the following addition. In some cases, the reactions are carried out under continuous illumination incident on the reaction test tubes of approximately 1000 ft-c from a bank of tungsten lamps. Reactions in the dark are carried out in test tubes covered with aluminum foil.

ATPase Activity Influenced in the Light[6]

The rate of ATPase reaction by chromatophores is slower in the light than in the dark, with most preparations of chromatophores, and without addition of redox dyes. On the contrary, ATPase activity is significantly stimulated by 4×10^{-4} M phenazine methosulfate (PMS) in the light. When a Lineweaver-Burk plot is made of the reciprocals of the ATP concentrations and the rates of the ATPase reaction, the light appears to be a competitive inhibitor against ATP in the absence of PMS and to be a noncompetitive stimulant against ATP in the presence of PMS.

The inhibition of ATPase activity in the light is significantly neutralized by adding antimycin A, whereas the inhibitor hardly influences the activity in the dark. The neutralization is maximal at 0.1 μg/ml of antimycin A, where the extent of the light inhibition is diminished by approximately 70%. It has been shown that antimycin A at 0.1 μg/ml completely inhibits photosynthetic ATP formation, unless PMS is present.[7] The inhibitor, 2-heptyl-4-hydroxyquinoline-*N*-oxide (0.67 μg/ml),

[3]T. Horio, K. Nishikawa, M. Katsumata, and J. Yamashita, *Biochim. Biophys. Acta* 94, 371 (1965).
[4]T. Horio, K. Nishikawa, Y. Horiuti, and T. Kakuno, *in* "Comparative Biochemistry and Biophysics of Photosynthesis" (K. Shibata *et al.*, eds.), p. 408. Tokyo University Press, Tokyo, 1968.
[5]See Vol. 23 [63].
[6]Y. Horiuti, K. Nishikawa, and T. Horio, *J. Biochem.* 64, 577 (1968).
[7]T. Horio and J. Yamashita, *Biochim. Biophys. Acta* 88, 237 (1964).

shows the same effects as antimycin A, both in the dark and in the light.

Either in the presence or the absence of PMS, ATPase activity decreases when chromatophores are illuminated with monochromatic light at 880 nm, which is the absorption peak of bacteriochlorophyll associated with the chromatophores from wild-type cells. It has been shown that PMS is reduced in white light but not in 880-nm light,[8] an observation indicating that the stimulation of ATPase activity by white light in the presence of PMS is caused by the photochemical reduction of the dye. This is confirmed by the observation that ATPase activity in the presence of PMS in the dark increases and then is decreased by the addition of increasing concentration of ascorbate, capable of reducing PMS, as shown in Fig. 1.

E_h-Dependent ATPase Activity

When 2,6-dichlorophenolindophenol (DCPI) in the oxidized form ($E_{m,7} = +0.217$ V) is added to the reaction mixture for the ATPase activity assay, it is reduced by chromatophores at a slow rate, and the E_h

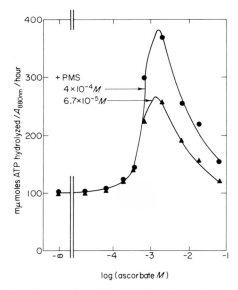

log (ascorbate M)

FIG. 1.   Effect of phenazine methosulfate (PMS) and ascorbate on ATPase activity in the dark. Experimental conditions are as described in the text, except that 4.0×10^{-4} M or 6.7×10^{-5} M PMS and various concentrations of ascorbate are added as indicated. Circles, + PMS (4.0×10^{-4} M); triangles, + PMS (6.7×10^{-5} M).

[8] D. M. Geller, Ph.D. Dissertation, Harvard University, Cambridge, Massachusetts, 1958.

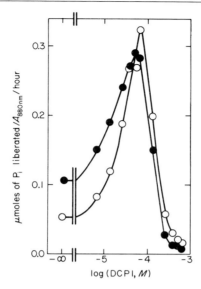

FIG. 2. Effect of 2,6-dichlorophenolindophenol (DCPI) on ATPase activities in the light and in the dark. Experimental conditions are as described in the text, except that DCPI is added as indicated. Open symbols, in the light; filled symbols, in the dark.

value of the reaction mixture becomes more positive with increasing concentrations of the dye. As shown in Fig. 2, ATPase activities in the light and in the dark are accelerated with increasing concentrations of DCPI up to 6.7×10^{-5} M, and retarded at higher concentrations. The effect of DCPI concentrations are fairly similar for the ATPase activities in the light and in the dark, except for the reproducible difference as follows: At concentrations of DCPI lower than 6.7×10^{-5} M, ATPase activity in the light is lower than in the dark, and vice versa at higher concentrations. Antimycin A (0.33 μg/ml) does not influence ATPase activity in the dark in the presence of DCPI at any concentration tested. Further, when ascorbate is added in the presence of 6.7×10^{-4} M DCPI, where ATPase activity is significantly lowered, the activity increases and then decreases with increasing concentrations of ascorbate; the maximum rate is obtained at 5.0×10^{-4} M. The ATPase activity is not altered by repeated washing of chromatophores with various concentrations of DCPI and/or ascorbate, an indication that the effect of redox dyes is reversible. The E_h values of the reaction mixtures containing 6.7×10^{-4} M DCPI and various concentrations of ascorbate are measured with a pH and millivolt meter (Model PHM4) with a platinum electrode (type P 101) available from Radiometer Co., Copenhagen. The assay system is calibrated relative to 10^{-2} M ferri- plus ferrocyanide in the assay buffer mixture. For this redox couple, the value $E_{m,7} = +0.423$ was used. The measured E_h value of a reaction mixture containing 6.7×10^{-4} M DCPI plus 5.0×10^{-4} M ascorbate is approximately $+0.15$ V, where the rate of the ATPase reaction in the dark is maximal. Half-maximal ATPase reaction in darkness is obtained at approximately either $+0.1$ V or $+0.2$ V. Although the E_h value required for the maximum rate of ATPase reaction is independent of the concentration of the redox buffer, the maximum rate attained in the presence of 6.7×10^{-4} M DCPI and an appropriate concentration (5.0×10^{-4} M) of ascorbate is approximately one-fifth the maximum rate attained in the presence of 6.7×10^{-5} M DCPI and an appropriate concentration of ascorbate. Probably two adjacent different oxidation-reduction components in the electron transport system are functional components of one of the coupling sites which lead to ATP formation, and ATPase activity at the site appears when one of the two components is in the oxidized form and the other is in the reduced form.

Ubiquinone-10-Dependent ATPase Activity[9]

Method for Extraction of Quinones from Chromatophores and Their Reconstitution by Addition of Quinones[10]

Chromatophores from the blue-green mutant of *R. rubrum* (G-9) are washed with an excess of water and then lyophilized. Lyophilized chromatophores are suspended in a volume of isooctane such that they would show $A_{873\,nm}/ml = 5$ if suspended in the same volume of water, and then shaken moderately at 4° for 80 minutes, followed by centrifugation. The resulting precipitate is dried at 4° under vacuum (extracted chromatophores). For reconstitution, the extracted chromatophores are suspended in isooctane containing an appropriate amount of pure quinone, and then dried at 4° under vacuum. They are then suspended in a volume of 0.1 M glycylglycine–NaOH buffer containing 10% sucrose (pH 8.0) such that the resulting suspension would show $A_{873\,nm}$ of approximately 50. Essentially the same results are obtainable with chromatophores from wild-type cells; the extraction of quinones is significantly more difficult with chromatophores from wild-type cells than with chromatophores from the mutant cells.

Quinones present in chromatophores are determined as follows. An aqueous chromatophore suspension ($A_{873\,nm}/ml = 50$) is extracted three times with 10 volumes of a mixture of acetone and methanol (8:2, v/v).

[9]N. Yamamoto, H. Hatakeyama, K. Nishikawa, and T. Horio, *J. Biochem.* **67**, 587 (1970).
[10]S. Okayama, N. Yamamoto, K. Nishikawa, and T. Horio, *J. Biol. Chem.* **243**, 2995, (1968).

The three extracts are combined in a separatory funnel, and an equal volume of petroleum ether plus an equal volume of water saturated with NaCl are added thereto. The petroleum ether should be redistilled before use; the fraction boiling between 40° and 60° is collected. After gentle swirling, the petroleum ether phase is collected and then washed with an excess of water to remove the acetone and methanol. The resulting petroleum ether phase is shaken three times in succession with an equal volume of 95% methanol; the bacteriochlorophyll and phospholipids present are almost completely extracted into the methanol phase. The petroleum ether phase is evaporated in vacuum with a rotary evaporator. The dried material is dissolved in a small volume of isooctane, spotted on a silica gel thin-layer plate (TLC-plates Silica Gel, E. Merck AG., Darmstadt) and developed with a mixture of chloroform and benzene (1:1, v/v) at 20° for 2 hours. During chromatography, four zones are formed: a yellow zone with an R_f value of 0.5 (ubiquinone-10), a purple zone with an R_f value of 0.4 (rhodoquinone), a pink zone with an R_f value of 0.2 (bacteriopheophytin), and a pale yellow-green zone with an R_f value of 0 (bacteriochlorophyll). The quinones are eluted separately with spectroscopically pure ethanol. The absorption spectra of the eluates are measured before and after addition of $NaBH_4$. Concentrations of quinones are calculated from the absorption changes for which the molar difference extinction coefficients ("oxidized" *minus* "reduced") of ubiquinone-10 and rhodoquinone are 12.25×10^3 at 275 nm[11] and 7.2×10^3 at 283 nm[12] respectively. Recovery by this procedure is approximately 80% when pure quinones are used as the starting materials. Further identification of the quinones is achieved by means of cochromatography with authentic quinones on a silicone-impregnated paper.[13]

Properties of Extracted and Reconstituted Chromatophores

The chromatophore activities for photosynthetic ATP formation and for ATPase are fairly stable against lyophilization; they are 30–50% of those of untreated chromatophores. Chromatophores contain approximately 3.5–4.0 nmoles of ubiquinone-10 and 0.7 nmole of rhodoquinone per $A_{873 \text{ nm}}$ unit. Other quinone derivatives are not detectable in comparable amounts. Almost all the quinones associated with chromatophores are extracted by shaking in isooctane at 4° for a period longer than 45 minutes. The activity for photosynthetic ATP formation, whether measured in the presence of ascorbate or in the presence of PMS, is com-

[11]See Vol. X [68].
[12]W. Parson and H. Rudney *J. Biol. Chem.* **240**, 1855 (1965).
[13]See Vol. VI [36].

pletely depressed when all the quinones are extracted, and is restored to the original level when ubiquinone-10 is added back at the same level as that originally present. When lyophilized chromatophores are freed of the quinones, the activities in the dark for ATPase reaction in the presence and absence of either DNP or DCPI are significantly, but not completely, reduced in rate, in contrast to the activity for ATP formation, as shown in Fig. 3. The activities thus reduced are restored to the original level when ubiquinone-10 is provided. The ATPase activity is therefore composed of two types, one dependent and the other independent of the quinone. With lyophilized and reconstituted chromatophores, the activity is approximately 80% inhibited by oligomycin (3.3 μg/ml). The oligomycin-insensitive activity is not affected by extraction of the quinones. Most of the oligomycin-sensitive activity is dependent on the quinone. When ubiquinone-10 is added to quinone-free chromatophores, ATPase activities in the presence and in the absence of DNP

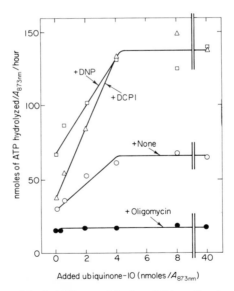

FIG. 3. Restoration of dark ATPase activity by addition of various amounts of ubiquinone-10 to extracted chromatophores. Experimental conditions are as described in the text, except for the following: Chromatophores which have been extracted with isooctane for 60 minutes are reconstituted by addition of various amounts of ubiquinone-10 as indicated. In some cases, the reactions are carried out in the presence of 6.7×10^{-5} M DCPI or 2.0×10^{-3} M DNP or 3.3 μg/ml oligomycin. Open circles, + none (no addition); squares, + DNP; triangles, + DCPI; closed circles, + oligomycin. The rates of ATPase reaction by nontreated chromatophores are 0.064, 0.185, and 0.163 μmole of ATP hydrolyzed/$A_{873\ nm}$/ hour for "+ None," "+ DNP," and "+DCPI," respectively.

or DCPI increase with increasing amounts of added ubiquinone-10 and reach a maximum at approximately 4 nmoles/$A_{873\ nm}$, the same amount as originally present. With quinone-free chromatophores, the quinone-independent activity is markedly stimulated by DNP, but hardly by DCPI, indicating that DNP stimulates both quinone-dependent and independent activities, whereas DCPI stimulates the former, but hardly the latter.

Of various quinones (4 nmoles/$A_{873\ nm}$) added to the extracted chromatophores, ubiquinone-9, -7, -6, hexahydroubiquinone-4, rhodoquinone, plastoquinone, α-tocopherolquinone, and phylloquinone are as effective as ubiquinone-10 for the restoration of ATPase activity. Menadione, 2,3-dimethoxy-5,6-dimethylbenzoquinone, 4-amino-1,2-naphthoquinone, 2-amino-1,4-napthoquinone, 2-acetoamino-1,4-naphthoquinone, and S-(2-methyl-1,4-naphthoquinonyl-3)-β-mercaptopropionic acid are not effective.

[9] Two-Stage Phosphorylation Techniques: Light-to-Dark and Acid-to-Base Procedures

By ANDRE T. JAGENDORF

Light-to-Dark Technique[1,2]

The rationale of the experiment is to separate in time the immediate effects of light, especially electron transport, from steps in the phosphorylation reaction itself. The procedure described below can be used to demonstrate the point of action of given reaction mixture components, and of activators or inhibitors of phosphorylation.

Procedure. Chloroplasts may be isolated from spinach leaves ground in 0.4 M sucrose, 10 mM NaCl, and 50 mM Tris, pH 8.0; or in any of the other usual buffered osmotic media. The chloroplasts (with or without an optional wash in the same medium) are resuspended to a concentration of 0.05 mg of chlorophyll per milliliter or less in 10 mM NaCl, kept for 20 minutes, collected by centrifugation at 10,000 g, washed once, and finally resuspended to a final concentration of 0.50 mg of chlorophyll per milliliter in 10 mM NaCl. All operations are at 0°C.

[1] Y. K. Shen and G. M. Shen, *Sci. Sinica* 11, 1097 (1962).
[2] G. Hind and A. T. Jagendorf, *Proc. Nat. Acad. Sci. U.S.* **49**, 715 (1963).

Two reaction mixtures are prepared, and put into 12×75 mm test tubes. The first one, for the *Light Stage*, contains in a total volume of 0.4 ml:

Dimethylglutaric acid, pH 6.0, 3 μmoles
NaCl, 18 μmoles
Pyocyanine, 18 nmoles

(These result in concentrations of 3.3 mM, 20 mM, and 20 μM, respectively, after addition of the chloroplasts.)

The *Dark Stage* reaction mixture contains, in a total volume of 0.9 ml, the following components:

ADP, pH 8.3, 2 μmoles
Sodium phosphate, pH 8.3, 2 μmoles
Radioactive phosphate, 0.25 μCi
MgCl$_2$, 5 μmoles
Na tricinate, pH 8.3, 50 μmoles

The tube containing the *Dark Stage* reaction mix is wrapped in aluminum foil and covered with an aluminum culture tube metal slip cap 15 mm in diameter (obtainable from Will Corporation) in which has been drilled a $^3/_{64}$-inch hole. In this way the contents are sufficiently dark so that even if the experiment is performed in room light or in strong actinic light, no phosphorylation will occur unless the chloroplasts have been preilluminated. Both sets of tubes are kept in an ice bath, and all operations are carried out in a cold room.

Two 1.0-ml syringes are required — one for the chloroplasts and reaction mixture, and one for the 20% TCA used to denature the chloroplasts at the end. Two special 4-inch long 20-gauge cannulas without a sharpened point (obtainable from Becton-Dickinson Corporation) are used. One of these should be kept in the stock suspension of chloroplasts; the other is used for transferring chloroplasts and reaction mixture.

In dim (preferably green) light, a syringe is attached to the cannula standing in the chloroplast suspension, and 0.5 ml of chloroplasts is drawn into the syringe. The syringe is then removed from the first cannula and attached to the second, standing in the test tube containing the *Light Stage* reaction mixture. The chloroplasts are injected into the test tube, then the mixed chloroplasts and reaction mixture components are drawn back up into the syringe. The syringe with cannula attached is now transferred so that the cannula is inserted into the hole in the cap over the *Dark Stage* reaction mixture tube. The syringe is illuminated with strong actinic light (for instance, from a slide projector, the beam being filtered through a tank of water or of 1% ferrous ammonium sulfate in acid to remove infrared, and through any other desired color filter) for 15–30 seconds. With the light still on, the syringe is depressed, injecting the chloroplasts into the dark reaction mixture. The light may be (but does not have to be) turned off. The syringe is removed, the cannula being left in place. A second syringe, previously loaded with 0.2 ml of 20% TCA is now attached to the end of the cannula, and its contents are injected after the chloroplasts have been in the dark for 10 or more seconds.

The best control consists in performing all operations in complete or near darkness (dim green safelight only). Results from these controls have not usually been appreciably higher than in operations in which the chloroplasts are added to a tube containing a combination of both reaction mixtures and TCA.

The denatured reaction mixture is centrifuged, and a 0.1-ml aliquot of the supernatant solution is placed directly on a planchet for measurement of the total counts. Another aliquot of 1.0 ml is assayed for radioactivity in organic phosphate by the method of Avron.[3] The sample is added to an 18×150 mm test tube containing 1.2 ml of acetone. After 10 minutes, 1.5 ml of water previously saturated with isobutanol:benzene is added. Then, 7 ml of a 1:1 mixture of isobutanol:benzene is added, and the contents are mixed by a Vortex mixer. After phase separation, 0.8 ml of 5% ammonium molybdate in 4 N H_2SO_4 is added, and the water layer is gently mixed. After they have stood for 5 minutes, the tubes are shaken vigorously on a Vortex mixer for 20–30 seconds. After phase separation the upper layer is aspirated off into a trap. Potassium phosphate (0.02 ml, 20 mM) is added, and a second extraction with isobutanol:benzene is performed (the second extraction is not necessary unless very low incorporation must be measured). Of the remaining aqueous layer, 1.0 ml is placed on a planchet and taken to dryness, and its radioactivity is measured. The amount of phosphate esterified in micromoles is calculated by multiplying the total phosphate in the reaction mixture by the proportion of counts transferred to the aqueous phase. Taking dilution steps into account, the formula used here is:

$$\frac{\text{sample cpm} \times 3.3 \times 2.0}{\text{"total" cpm} \times 20} \times 2.0 \ \mu\text{moles phosphate}$$

Observed yields of ATP vary with the season. In the system described above, yields of ATP between 15 and 100 nmoles of ATP per milligram of chlorophyll may be expected (i.e., 4–25 nmoles per reaction tube).

Comments on Procedural Details. The yield has not varied very much because of the nature of the isolation medium, insofar as this has been

[3] M. Avron, *Biochim. Biophys. Acta* **40**, 257 (1960).

examined. Satisfactory results may be seen with chloroplasts washed once in sucrose-tricine-NaCl. The 10 mM NaCl broken chloroplasts were used originally to minimize possible contributions from endogenous substrates, etc.

Severalfold higher yields are obtained by adding to the light stage 20 mM pyridine, or other organic bases uncharged at pH 6, to act as internal proton reservoirs.[3a]

The inorganic phosphate level used here is below the optimum for phosphorylation. It represents a compromise between the concentration needed for good phosphorylation yields and the declining proportion of phosphate, and therefore of radioactive counts esterified, as the total level rises. The use of higher specific activity of inorganic phosphate is not an aid in solving this problem because of background levels of counts which remain in the aqueous phase in spite of repeated extractions. These come from radioactivity in compounds other than inorganic phosphate — most commonly from pyrophosphate, but sometimes from other elements as well. Commercially available phosphate will be found to contain up to 0.5%, or even 1.0%, of the counts in such interfering materials. Preliminary purification of the radioactive phosphate is always needed in order to detect and measure accurately low levels of incorporation (on the order of 0.2% in the case of low-yield experiments). The following procedure has been found to be adequate to reduce background contamination to 0.03% routinely:

To 10 mCi of radioactive phosphate, add 2.0 μmoles of carrier inorganic phosphate. Make to 1.0 N in HCl and boil for 1 hour. This solution is then passed through a 0.3 × 5 cm column of activated charcoal, and the effluent is brought to pH 4.0 with NaOH.

If reducing substances (ascorbate, —SH compounds, etc.) are added during the reaction, they will reduce the phosphomolybdate complex to the blue form which cannot be completely extracted into organic solvents. If reducing compounds are present, they can be removed by adding 0.1 ml of bromine water to the denatured reaction mixture. Extra bromine is then removed by bubbling air or nitrogen through the solution until the bromine coloration disappears entirely.

Separation from inorganic phosphate may be accomplished also by adsorption of ATP on activated charcoal.[4] The charcoal is washed by centrifugation, then plated directly onto a planchet. Although the procedure permits counting directly all the ATP esterified in the reaction

---

[3a]N. Nelson, H. Nelson, Y. Naim, and J. Neumann, *Arch. Biochem. Biophys.* 145, 263 (1971).
[4]M. D. Kamen, Vol. VI, p. 313.

tube, in practice we have found it to be slightly more time-consuming than the isobutanol extraction.

The ATP which is formed may be measured enzymatically, as by fire-fly luciferase.[5] Although this is feasible with some chloroplast preparations, others seem to retain too much adenylic kinase (F. Welsch, personal communication), so that the added ADP is converted into ATP in control tubes to an extent that interferes with the small amount formed experimentally. In this case it is helpful to decrease the level of ADP in the *Dark Stage* reaction mixture, and to decrease the time the chloroplasts are in contact with it to about 5 seconds before denaturation. Phosphorylation occurs only during the first 5 or 6 seconds in the *Dark Stage*.

The incorporation of label from inorganic phosphate into ATP cannot be accomplished through adenylic kinase, but it can occur if both adenylic kinase, forming ATP, and a P_i-ATP exchange reaction are active. Such exchange activity can be aroused by illuminating chloroplasts together with —SH compounds, especially dithiothreitol; however, the rates are usually very low, and they are essentially nonexistent in the absence of —SH compounds.

Actinic illumination should be bright in order to drive maximal rates of electron transport in the *Light Stage*. There is no need to keep chloroplasts and reaction components dark prior to turning on the actinic light, unless a careful time course for illumination is being evaluated. Ordinary room light (up to about 15 or 20 ft-c) is probably insufficient to cause any measurable phosphorylation. Dyes other than pyocyanine (i.e., ferricyanide, ferredoxin plus NADP, FMN, etc.) may be used,[6] but lower yield will be expected generally and the time in the light should be extended. The pH optimum for phosphorylation in the *Dark Stage* is similar to that for the usual photophosphorylation. In the *Light Stage* it is about pH 6.0, probably because of the kinetic characteristics of a competing light-independent back reaction (see below). No extensive evaluation has yet been made of buffers other than dimethylglutaric; however, yields are satisfactory with no buffer (pH adjusted by HCl) or with low concentrations of maleate, and probably other buffers would work also.

The difference in pH between the 2 stages should be remembered when the effect of an added reagent is evaluated. When ineffective if added only in the dark stage, it might possibly be because its penetration or action requires a more acid pH. This possibility should be

---

[5] W. D. McElroy, Vol. VI, p. 445.
[6] G. Hind and A. T. Jagendorf, *Z. Naturforsch. B* **18**, 684 (1963).

checked by observing the pH sensitivity of the usual, single-stage photo-phosphorylation reaction to the particular inhibitor.

Yields of ATP are slightly better at 0–5° than when all components are at room temperature. The temperature during illumination has usually been controlled by performing the experiment in a cold room and keeping both reaction mixtures either in an ice bath or at cold-room temperature before use. However, it is possible to perform the experiment at the ordinary laboratory bench if a water-jacketed cell is constructed, containing open-ended glass tubes into which the syringes can be fitted fairly tightly. This cell (with at least one side glass or Lucite) can be used to maintain a low temperature during the rather brief illumination period.

Variations in the Technique. (i) DARK-DECAY KINETICS. After illumination, the chloroplasts may be injected into an empty, dark test tube. This tube should be fitted with a cap having two holes drilled in it. One hole is for the cannula through which the chloroplasts are injected. The second hole should have another cannula, and a syringe attached in which is contained the 0.9 ml of *Dark Stage* reaction mixture. The second syringe is depressed a given (variable) number of seconds after the chloroplasts are injected. With this procedure it is possible to introduce with reasonable accuracy a dark interval of 2 seconds or more between illumination of the chloroplasts and start of the phosphorylation reaction. This permits determination of the kinetics of decay of the intermediate state or compound(s) which support the phosphorylation process. If different buffers and/or sodium hydroxide are present in the dark tube to start with, the dark-decay kinetics can be measured at different pH values.

(ii) DILUTION TECHNIQUE. In the procedure described above, an inhibitor which affects light steps only (including electron transport) will fail to have any action if added in the *Dark Stage*. However, an inhibitor of the phosphorylation steps will decrease the yield even when added in the *Light Stage,* because it is still present in the subsequent *Dark Stage*. This pattern will also be shared by an inhibitor which affects both the light and the dark steps in phosphorylation. To overcome this ambiguity, Gromet-Elhanan and Avron introduced[7] the technique of diluting the *Light Stage* reaction mixture 20-fold at the time the ADP and phosphate are added. Inhibitors present in the *Light Stage* should thus be reduced to noninhibitory concentrations, at which point they would not be acting on the *Dark Stage* steps. Validity of the results requires that the inhibitor be one that does not bind irreversibly to chloroplast com-

[7] Z. Gromet-Elhanan and M. Avron, *Plant Physiol.* **40,** 1053 (1965).

ponents and also does not cause irreversible damage when present in the illuminated reaction mixture. These points may be examined in control experiments, where preillumination in the *Light Stage* reaction mixture is followed by ordinary photophosphorylation in a similarly diluted reaction.

In the procedure of Gromet-Elhanan and Avron[7] the *Light Stage* contained 0.1 mg of chlorophyll in a total volume of 0.5 ml. This was injected into a *Dark Stage* with total volume equal to 9.5 ml, containing (in micromoles) Tris·HCl, pH 7.8, 120; ADP, 2.5; Na,K phosphate pH 7.8, 7.5; and approximately 3×10^7 cpm in radioactive phosphate. The dark tube was shaken in a Vortex test tube mixer both before and after the injection to ensure thorough mixing of the light and dark systems. After 30 seconds in the dark, the chloroplasts were denatured by the addition of 1 ml of 30% perchloric acid. After centrifugation, 2.4 ml was taken for radioactive ATP assay.

Significance. Results obtained by this procedure were originally interpreted as giving indirect evidence for the existence of a high-energy compound or intermediate formed in the light, capable of supporting the phosphorylation of ADP in subsequent darkness. The hypothetical intermediate was designated "X_E," being both energetic and unknown. Subsequent interpretations have veered toward the possibility that the energy of light and electron transport drives proton uptake and other ion movements in the light[8,9] and that the stored intermediate consists of ionic electrochemical activity gradients. This interpretation is based on concepts developed by Mitchell[10] and on the observation of proton movements under conditions similar to those used in the *Light Stage* of this experiment. A third possibility is that electron transport continues for several seconds after the light is turned off[11]; however, spectrophotometric observations have not indicated this to be the case, and the failure of electron transfer inhibitors to act when added in the *Dark Stage* only argues against it as well.

Acid-to-Base Technique

Chloroplasts were found[12] to phosphorylate ADP without illumination if they were placed at pH 4 for a short time, then transferred to pH 8.5. This result has been taken as evidence confirming phosphorylation of

[8]J. Neumann and A. T. Jagendorf, *Arch. Biochem. Biophys.* **107,** 109 (1964).
[9]R. A. Dilley and L. P. Vernon, *Arch. Biochem. Biophys.* 111, 365 (1965).
[10]P. Mitchell, *Biochem. Soc. Symp.* 22, 142 (1962).
[11]W. S. Lynn, *Biochemistry* 7, 3811 (1968).
[12]A. T. Jagendorf and E. Uribe, *Proc. Nat. Acad. Sci U.S.* **55,** 170 (1966).

ADP driven by a pH gradient. The technique has sometimes been called ATP formation due to a "pH jump."

Procedure. Chloroplasts are swollen, washed, and stored in 10 mM NaCl as described above. Two reaction mixtures are prepared and put into 12 × 75 mm test tubes stored in an ice bath. The first one, for the *Acid Stage,* contains in 0.4 ml:

>   Succinic acid, pH 4.0, 12.5 μmoles
>   DCMU, 27 nmoles

(The concentrations are 14 mM and 30 μM, respectively, after addition of chloroplasts.)

The second tube, for the *Base Stage,* contains in 0.9 ml:

>   Tricine, pH 8.3, 100 μmoles
>   ADP, pH 8.3, 0.2 μmoles
>   Inorganic phosphate, pH 8.3, 2.0 μmoles
>   MgCl$_2$, 5.0 μmoles
>   NaOH (to neutralize the succinic acid), 19.6 μmoles

The whole mixture contains components brought to pH 8.3 before addition of the NaOH.

Two syringes and two cannulas are required. Chloroplasts (0.5 ml) are taken up through a cannula into the first syringe, which is withdrawn and reattached to the second cannula standing in the *Acid Stage* tube. The chloroplasts are injected into the reaction mixture and allowed to stay in the tube for 10–15 seconds. At this point, the mixture is drawn up into the syringe, and the syringe and cannula are transferred to the *Base Stage* tube so that the mixture can be injected into the phosphorylation mixture exactly 20 seconds after chloroplasts were first added to the acid. The syringe is then removed, but the cannula is left in place; a second syringe containing 0.2 ml of 20% TCA is fitted onto the end of it, and the reaction is terminated by injection of the acid 15 seconds later.

Measurement of ATP synthesis is performed in the same way as for light-to-dark phosphorylation. Yields of ATP have been observed between 30 and 240 nmoles of ATP per milligram of chlorophyll, depending on the season and possibly other unidentified factors.

The control for this experiment is one in which chloroplasts are added to a test tube containing a mixture of both the *Acid Stage* and *Base Stage* reaction mixtures, so that no pH transition occurs. They should be left in this mixture for 35 seconds before TCA is added.

Comments on Details of Procedure. In this experiment the chloroplasts should definitely be swollen in hypotonic solutions before use, since osmotic shrinkage as by 0.03 M sucrose or above has been found to be inhibitory.[13] The succinic acid appears to act as an internal buffer or reservoir of hydrogen ions,[13,14] and the swollen thylakoid discs may be needed to provide sufficient room to store a large amount of this acid.

The "substrate level" (10 mM or higher) concentration of succinic acid is required for a good yield of ATP. A fairly large number of organic acids may be used in place of succinic acid. Highly effective organic acids include itaconic and glutaric acids, and nonphysiological acids which might be used include o-phthalic, p-phthalic, and barbituric acids. The main requirements seem to be an acid that can penetrate into the inside of the thylakoid disk fairly rapidly at pH 4; this involves having at least one pK (usually out of two, since mostly dicarboxylic acids are effective) near pH 4.2–4.4.[15] Nonpenetrability of the completely charged dianion at pH 8 may be another requirement.

A major complicating factor in these experiments is the drop in ATP yield if chloroplasts are held too long at pH 4.0 (which is unphysiological). The rate at which this drop occurs varies markedly with different organic acids present in the acid stage, with the concentration of one particular organic acid (i.e., the drop is more rapid with 10 mM succinic acid than with 3 mM), and with the particular batch of chloroplasts. Recent experiments show that the chloroplast phosphorylation enzyme ("CF-1") tends to be denatured at pH 4. Both the acid denaturation of CF-1, and the loss in ability to make ATP after an acid–base transition, are remarkably potentiated by the presence of 0.1 millinormal or higher levels of polyanions, including ferricyanide, ferrocyanide, polygalacturonic acid, pectic acid, heparin, and polyvinyl sulfate.

The existence of acid denaturation sets the lower pH limit for effective ATP synthesis. It is usually possible to have the acid mixture at pH 3.8 for 15 or 20 seconds, but below this the yields drop off or the chloroplasts coagulate and so cannot be easily syringed, or both.

The optimum pH for the *Base Stage* varies a little, depending on how acid the chloroplasts have been previously. With lesser degrees of acidity (i.e., pH 5 instead of pH 4) in the *Acid Stage*, the *Base Stage* has to be more alkaline for the threshold of activity to be seen, and the optimum is also shifted as much as 0.5 pH unit higher.[12]

The experiment may be done in complete darkness, with chloroplasts stored for up to 3 hours in complete darkness ahead of time. However, since no redox dye "cofactor" is present, and with DCMU added, light

[13]E. Uribe and A. T. Jagendorf, *Arch. Biochem. Biophys.* **128**, 351 (1968).
[14]E. Uribe and A. T. Jagendorf, *Plant Physiol.* **42**, 697 (1967).
[15]E. Uribe and A. T. Jagendorf, *Plant Physiol.* **42**, 706 (1967).

has absolutely no effect on the amount of ATP formed either in the controls or in the experimental tubes. This has been confirmed for both room light and strong actinic light. It is possible, therefore, to perform the experiment under ordinary room light.

The considerations of concentrations of phosphate, purity of radioactive phosphate, and methods of ATP measurement are the same as for the light-to-dark transition noted above.

Other Activations. The same acid-to-base transition, if performed with chloroplasts preincubated for a short time in 80 mM dithiothreitol, substitutes for light in activation of the Mg-dependent hydrolysis of ATP,[16] and also in the arousal of a P_i–ATP exchange reaction.[17] The activation phenomenon is likely to involve a change in conformation of the terminal coupling enzyme or "CF-1," brought about by some interaction between the reduced form of the enzyme and a stressed condition of the membrane to which the enzyme is attached. Light, and the acid–base transition, probably lead to the same stressed condition of the thylakoid membrane, perhaps consisting of a transmembrane gradient in hydrogen ion electrochemical activity.

The acid–base transition was observed to cause chloroplasts to emit light,[18] and it was originally thought that the energy storage of the pH jump might be somehow translated into chlorophyll activation to the excited state level, followed by fluorescence. However, it was soon found that chloroplasts must be recently preilluminated for this to occur,[19] and thus the acid-base transition is only a trigger for appropriate recombination and translation into fluorescence of other forms of stored energy. Furthermore, ionic transitions of other sorts (low to high salt, neutral pH to pH 2.5) are equally or more effective in triggering light emission, but not effective in causing ATP formation.[20] Thus the high energy state caused by the acid-to-base transition is not necessary for triggering light emission.

Significance. ATP synthesis in this system has been taken to be due to an artifically imposed gradient in hydrogen ion concentration from inside to outside the thylakoid disks. Several correlations between ATP synthesis and amount of internal acid support this concept.[13,14] The need for a large difference in pH between the two stages, rather than simply sufficient acidity in the first stage and a definite pH optimum in the

[16]J. Kaplan, E. Uribe, and A. T. Jagendorf, *Arch. Biochem. Biophys.* **120**, 365 (1967).
[17]R. Bachofen and K. Specht-Jurgensen, *Z. Naturforsch. B* **22**, 1051 (1967).
[18]B. C. Mayne and R. K. Clayton, *Proc. Nat. Acad. Sci. U.S.* **55**, 494 (1966).
[19]B. C. Mayne, *Photochem. Photobiol.* **8**, 107 (1968).
[20]C. D. Miles and A. T. Jagendorf, *Arch. Biochem. Biophys.* **129**, 711 (1969).

second stage, was shown most clearly in the case of ATPase activation.[21] Again, for ATPase activation and inferentially for ATP formation, the *Base Stage* was found to be effective with a fairly large range of cations, hence reentry of the cation activating a cation pump does not appear to be part of the mechanism.[22] The possibility has been raised[11] that the acid–base transition, by successively inverting and restoring the usual redox positions of internal carriers, arouses exergonic electron transport, which then makes ATP in the usual way. The stoichiometry (maximum of 1 ATP per 4–5 chlorophyll molecules) and the failure of electron transport inhibitors (DCMU, antimycin A, HOQNO) to inhibit acid–base ATP formation argue against this interpretation.

[21] J. Kaplan and A. T. Jagendorf, *J. Biol. Chem.* **243**, 972 (1968).
[22] Y. Nishizaki and A. T. Jagendorf, *Arch. Biochem. Biophys.* **133**, 255 (1969).

[10] Oxygen Electrode

By David C. Fork

The application of polarographic techniques to the study of O_2 exchange in photosynthetic organisms has made possible a number of investigations that were either impossible or too difficult to undertake by conventional manometric methods. Thus many investigators have applied this technique successfully to measurements of action spectra, enhancement and chromatic-transient effects, Hill reactions, early transients of O_2 evolution using whole cells, algal thalli, or chloroplasts. Simultaneous comparative measurements of oxygen evolution and of fluorescence[1,2] or of absorbance changes[3] have also been made.

The O_2 electrode has many advantages over the manometric technique, particularly in the great sensitivity of the method and the ease and rapidity with which the measurements can be made. In addition, the optical setup can be made much more easily since only a very small sample need be illuminated. In certain instances, a sample can be left on the electrode for extended periods of time and used for subsequent comparative measurements. The polarographic technique can be used to follow rapid transients of O_2 exchange that would be completely unnoticed with the manometric technique. Finally, electrodes are relatively easy devices to make.

[1] T. T. Bannister and G. Rice, *Biochim. Biophys. Acta* **162**, 555 (1968).
[2] C. Bonaventura and J. Myers, *Biochim. Biophys. Acta* **189**, 366 (1969).
[3] T. Marsho, Ph.D. Thesis, Univ. of North Carolina, 1968.

This article will treat the use of the oxygen electrode in photosynthesis research and will not cover its extensive application to studies of respiration, although in some instances respiration can be used to follow photosynthesis indirectly. For example, light produces an inhibition of respiration in the nonsulfur purple bacterium *Rhodospirillum rubrum* that can be measured with an electrode and that has been found to be sensitized by light absorbed by bacteriochlorophyll.[4]

Principle

An electrical measurement of oxygen is possible since it is reduced at the negative platinum cathode, probably by reaction 1[5,6]

$$O_2 + 2H_2O + 4e^- = 4OH^- \tag{1}$$

With a silver anode and KCl as the supporting electrolyte, the anode reaction is

$$4Ag + 4Cl^- = 4AgCl + 4e^- \tag{2}$$

The resulting current flow is proportional to the concentration of oxygen in the sample.

For these reactions to proceed, the cathode is usually held at about -0.5 V with reference to the anode. Conventional circuits suitable for this purpose have been described.[7-10]

The potential to be applied to the electrode is best determined by measuring a polarogram for the particular conditions employed. Ideally, this polarogram of current *vs.* electromotive force has a flat plateau, where variations of voltage applied to the electrode produce no variation of current flowing. In practice, however, these polarograms usually do not have flat regions.[8] It is necessary, therefore, to make sure that the voltage applied to the electrodes does not vary. As pointed out by Myers and Graham,[8] this voltage will vary if a large resistance exists in the circuit, since changes in current produced by oxygen evolution by the sample would then give a change of the potential at the platinum electrode. To avoid this, the resistance of the reference elec-

[4]D. C. Fork and J. C. Goedheer, *Biochim. Biophys. Acta* 79, 249 (1964).
[5]D. T. Sawyer and L. V. Interrante, *J. Electroanal. Chem.* 2, 310 (1961).
[6]J. J. Lingane, *J. Electroanal. Chem.* 2, 296 (1961).
[7]F. T. Haxo and L. R. Blinks, *J. Gen. Physiol.* 33, 389 (1950).
[8]J. Myers and J. R. Graham, *Plant Physiol.* 38, 1 (1963).
[9]J. M. Pickett and J. Myers, *Plant Physiol.* 41, 90 (1966).
[10]Y. de Kouchkovsky, Ph.D. Thesis, Univ. of Paris, 1963.

trode, as well as the resistance of the measuring circuit, must be low. For this purpose, a resistance of 100 ohms in series with the electrodes has been used; the voltage was measured across this with a sensitive amplifier of suitably low impedance: e.g., Beckman Model 14, Keithley Model 150.[8,11] If larger resistors are used in the measuring circuit, the voltage at the cathode can be kept at a constant value by means of a potentiostat arrangement.[2,9]

It is important when using an electrode to maintain a constant temperature or to provide temperature compensation, described below as done by Carritt and Kanwisher.[12] The increase of electrode current with increasing temperature[12] results from the effect of temperature on the diffusion and solubility of oxygen as well as on the electrode reactions themselves. de Kouchkovsky[10] found the log of the ratio of current flow per mole of dissolved oxygen to increase linearly with the reciprocal of the absolute temperature. He obtained a Q_{10} around 2.8 from these data. Carritt and Kanwisher[12] provided automatic compensation of their electrode by taking advantage of the negative temperature coefficient of thermistors. By embedding an appropriate thermistor in the electrode, they found it possible to obtain satisfactory compensation over a limited temperature range. Commercial electrodes employing this principle are available that are compensated in the range from 15° to about 45°. Even when these compensated electrodes are used, it is important to remember that the dissolved oxygen in the sample still varies with temperature. Probably the simplest solution to problems caused by these temperature effects is to provide, if possible, close thermoregulation of the experimental setup.

Depending upon how the electrode is used, measurements of rates or of changes in concentration of oxygen can be obtained.

The Oxygen Electrode

The Bare Electrode

Blinks and Skow[13] measured oxygen evolution from the marine alga *Ulva* by pressing a piece of the thallus to a bare platinum electrode with an agar disk. With this arrangement they could detect oxygen evolution from actinic light flashes as short as 20 msec. For their action spectra measurements, Haxo and Blinks[7] modified the Blinks and Skow method by holding a marine alga tight against the electrode with a

[11]C. S. French, D. C. Fork, and J. S. Brown, *Carnegie Inst. Washington Yearb.* **60**, 362 (1961).
[12]D. E. Carritt and J. W. Kanwisher, *Anal. Chem.* **31**, 5, (1959).
[13]L. R. Blinks and R. K. Skow, *Proc. Nat. Acad. Sci. U. S.* **24**, 413 (1938).

strip of permeable cellophane and immersing this assembly in a relatively large reservoir of seawater. The reservoir was connected to a calomel anode by means of a seawater and KCl agar bridge. The signal produced upon illumination of the cells with this setup is proportional to the *rate* of oxygen evolution because the current is due to the steady-state diffusion of oxygen in the reservoir through the sample to the surface of the cathode. Very early measurements obtained immediately after illumination are not true rates, however, but a mixture of both rate and concentration measurements.

While the use of an electrode immersed in a large reservoir of medium was an improvement over the original method used by Blinks and Skow, this method frequently gives difficulty with drifts and random fluctuations of the dark current. A flowing system, introduced by Myers and Graham[8] gave a significant improvement in stability. With the flowing system, as with the Haxo and Blinks system, a rate measurement is obtained because a solution of constant oxygen tension is passed at a constant rate over the sample held on the electrode by means of a dialysis membrane. Thus again, a signal produced as a result of oxygen exchange in the sample is superimposed on the constant signal produced by the steady-state diffusion of oxygen from the flowing stream. The constant flow required can be produced by simply allowing an aerated solution in a reservoir to flow past the electrode, after which it is discarded or returned to the reservoir. Pumps driven by synchronous motors may also provide this flow.[14,15]

Certain marine algae are well suited for use with an electrode flush with a flat surface since the fronds are flat and relatively large and are one cell thick (*Porphyra*) or two cells thick (*Ulva*). By splitting the tubular thallus of *Enteromorpha* lengthwise, it is possible to obtain a piece of green alga one cell thick.

Single-celled algae and chloroplasts are less convenient to use because a cell paste must be spread as thinly and uniformly as possible over the electrode surface and then covered with a damp piece of cellophane or dialysis tubing, at the same time trying to avoid gross disturbance or loss of the sample altogether—a process that takes considerable skill. Since it is virtually impossible to obtain reproducible samples by this procedure, one is restricted to comparative measurements of the effects of varying treatments on one sample, such as the

---

[14]D. C. Fork, *Carnegie Inst. Washington Yearb.* **61**, 343 (1962).
[15]R. W. Hart, *Carnegie Inst. Washington Yearb.* **61**, 365 (1962).

effect of different wavelengths of light, inhibitors, changes of the gas phase.

In order to avoid this difficulty, a horizontally positioned bare electrode can be recessed an appropriate amount (usually a few tenths of a millimeter) so that a known number of cells can be applied and left to settle out on the cathode surface. In this way, fairly reproducible samples can be applied. Microscopic examination of a sample deposited on a recessed electrode by sedimentation revealed that a layer one cell thick could be obtained.[2]

Calibration of the flush-type electrode is not practical. However, Bannister and Rice[1] calibrated an electrode recessed 0.2 mm by assuming that almost all (about 98%) of the photosynthetically evolved O_2 is reduced at the surface of the cathode. This assumption is probably valid since a single-cell layer of *Chlorella*, for example, is only about 5μ thick when the cells have settled out on the platinum surface. Bannister and Rice's calibration based on this assumption lead to rate measurements for O_2 production by *Chlorella* in saturating light that were similar to those obtained by other methods.

If almost complete reduction of the photosynthetically produced O_2 occurs at the platinum surface, then the rate of O_2 production, r, in microliters of O_2 per microliter of cells per hour can be determined from the equation:

$$r = 0.582IT/VP \qquad (3)$$

where I is the net photocurrent in microamperes, T is the absolute temperature, V is the volume (in microliters) occupied by the cells in the electrode chamber, and P is the pressure in millimeters of mercury.

Myers and Graham[8] suspected silver poisoning of *Chlorella* from the large-area silver electrodes used to overcome a drifting cathode potential produced by diffusion of KCl out of the agar bridge connected to the calomel reference electrode. To overcome this difficulty they used large-area platinum anodes (1.5 meter of tightly coiled platinum wire) in contact with a calomel electrode in the effluent of a flow-through system. Silver poisoning was apparently not a problem in the experiments of Bannister and Vrooman[16] since their large area Ag/AgCl reference electrode was kept from direct contact with the experimental solution by means of an agar plug 5 mm thick and 20 mm in diameter which was intended to trap the silver ions.

[16]T. T. Bannister and M. J. Vrooman, *Plant, Physiol.* **39**, 622 (1964).

An important modification in the use of the bare electrode has been developed by Joliot[17-19] and is discussed in this volume (see [11]).

The main advantage of using a bare electrode instead of a membrane-covered electrode is that rapid response times are obtained because the diffusion path between the site of oxygen production and its reduction at the cathode is kept to a minimum. A disadvantage of the bare electrode (but not the Joliot type) is that it frequently reacts with substances added to the circulating solution. For example, measurement of oxygen evolution with ferricyanide as the Hill oxidant cannot be made with the conventional bare electrode since ferricyanide reacts with the electrode. Also, cyanide cannot be added since it reacts with the silver anode forming a silver–cyanide complex. A bare electrode is frequently "poisoned" by certain dissolved organic substances. The sensitivity of a bare electrode used in seawater will usually decrease with time, apparently because metals and other ions plate out on the surface of the platinum. A white deposit is commonly seen on the platinum after it has been used in seawater for some time. In addition, since both oxygen consumption and OH^- production occur as a result of reactions (1) and (2), it is possible that unfavorable physiological conditions are produced for cells resting directly on the surface of a bare electrode.

Since bare electrodes become "poisoned" by substances present in fluids, such as blood, Clark and co-workers[20] made a very important contribution to oxygen electrode technique by covering the platinum with a protective membrane permeable to oxygen but almost impermeable to other substances in solution. In a later modification, Clark[21] made an electrode in which both the cathode and the silver/silver chloride reference electrode were covered by a polyethylene membrane 1 mil (25.6μ) thick. This electrode provides the basic design for the membrane-covered electrodes currently in use.

The Membrane-Covered Electrode

The membrane-covered electrode (Figs. 1 and 2) can be used as a rate-measuring device in the same manner as described above for the bare electrode. Since this electrode has both the platinum and reference electrodes as well as the electrolyte solution confined under a thin film of Teflon or polyethylene permeable only to gases but not to ions, the silver poisoning mentioned above is not a problem. Further, many substances can be added to the surrounding or circulating solution

[17]P. Joliot, M. Hofnung, and R. Chabaud, *J. Chim. Phys.* **63**, 1423 (1966).
[18]P. Joliot, *Brookhaven Symp. Biol.* **19**, 418 (1967).
[19]P. Joliot and A. Joliot, *Biochim. Biophys. Acta* **153**, 625 (1968).
[20]L. C. Clark, Jr., R. Wolf, D. Granger, and Z. Taylor, *J. Appl. Physiol.* **6**, 189 (1953).
[21]L. C. Clark, Jr., *Trans. Amer. Soc. Artif. Intern. Organs* **2**, 41 (1956).

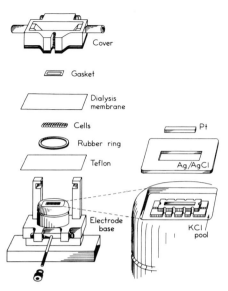

FIG. 1. A Teflon-covered platinum electrode assembly for measuring the rate of O_2 production in a flowing system shown in Fig. 2. After D. C. Fork, *Plant Physiol.* **38,** 323 (1963).

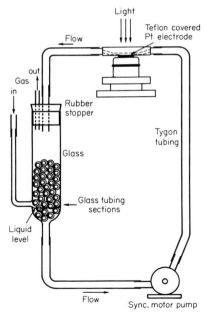

FIG. 2. The flowing system of J. Myers and J. R. Graham [*Plant Physiol.* **38,** 1 (1963)] provides a constant flow of aerated medium over a sample held on the platinum electrode and permits rate measurements with the oxygen electrode. Circulation is produced by a centrifugal pump [R. W. Hart, *Carnegie Inst. Washington Yearb.* **61,** 365 (1962)] driven by a synchronous motor. After D. C. Fork, *Plant Physiol.* **38,** 323 (1963).

without the difficulty of a reaction occurring at the surface of the electrode.

Other advantages also result from isolating the sample both physically and electrically from the electrode surface. The possible deleterious effect of a high pH discussed above is avoided. Also, the membrane and electrolyte films serve to remove the sample from immediate contact with the electrode surface, probably avoiding a highly anerobic condition in the layer of sample nearest the cathode.

Removal of the sample from immediate contact with the electrode surface must slow down the response time of the membrane-covered electrode in comparison to the bare electrode, but no comparative measurements have been made on the speed of response of the two types of electrodes. However, rapid transients of O_2 evolution have been detected from illuminated chloroplasts with the membrane-covered electrode.[22] A Teflon membrane 6.4 μ (0.25 mil) thick was used with this electrode. Teflon film[23] of this thickness is relatively easy to handle without tearing or stretching. Thinner films are available, but they are difficult to apply over the electrode surface without deforming them. It is advisable to test thin membranes for holes with an ohmmeter using agar-KCl contacts.

The flush type of membrane-covered electrode suffers the same disadvantage discussed above for the bare electrode since successive reproducible samples cannot be applied. Pickett[24] developed a recessed membrane-covered electrode in order that reproducible samples could be obtained. A sample compartment 0.1 mm thick was produced by inserting a black polyethylene spacer between the dialysis membrane and the Teflon membrane (Figs. 3 and 4). A flow rate of 5 ml of aerated medium per minute over the dialysis membrane forming the top of the cell chamber was provided by a gas-operated lift pump with gravity flow from a constant height. Pickett used a Ag/AgCl reference electrode in a buffered electrolyte solution containing 0.5 M KCl and 0.5 M $KHCO_3$ to prevent loss of CO_2 from the sample chamber by diffusion through the Teflon and neutralization by the OH^- produced in the electrode reaction.

Weiss and Sauer[25] have used this type of electrode with *Chlorella* incubated in darkness for varying periods to measure O_2 evolution from 20 nsec flashes from a Q-switched ruby laser.

[22]D. C. Fork, *Plant Physiol.* **38**, 323 (1963).
[23]Dilectrix Corp., Farmingdale, New York.
[24]J. M. Pickett, *Carnegie Inst. Washington Yearb.* **65**, 487 (1967).
[25]C. Weiss, Jr., and K. Sauer, *Biophys. J.* (Society Abstracts) **9**, A-29 (1969).

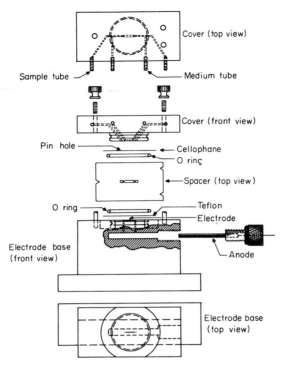

FIG. 3. The recessed membrane-covered electrode developed by Pickett to permit measurements on successive reproducible samples. From J. M. Pickett, *Carnegie Inst. Washington Yearb.* **65,** 487 (1967); reproduced by permission of the author.

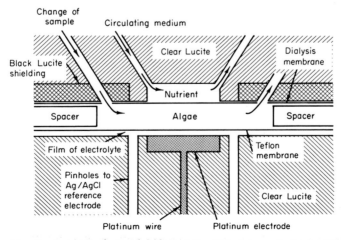

FIG. 4. Details of the electrode chamber of the Pickett-type electrode. (Provided by courtesy of Dr. Charles Weiss, Jr.)

A comparative-type calibration of the recessed membrane-covered electrode was made by Heber and French.[26] The current produced by oxygen evolved from chloroplasts in a ferricyanide Hill reaction was measured. The oxygen production was calculated for various incident light intensities by assuming near quantitative reduction at the electrode surface described above. These values were then compared with rates of ferricyanide reduction measured spectrophotometrically on another aliquot diluted so as to give the same optical density. The values agreed surprisingly well (within 8%), considering the many kinds of error that can enter into comparative measurements of this type.

By immersing a small Clark-type electrode in a constantly stirred solution contained in a closed system such as in a spectrophotometer cuvette, measurements of the concentration of dissolved oxygen are obtained. Commercially available electrodes of this type are available.[27] Some have temperature compensation employing the method of Carritt and Kanwisher.[12] A description of a simply constructed electrode suitable for lecture or student laboratory use was given by Foster.[28]

For a concentration-measuring electrode, it is necessary to have very constant and effective stirring. Frequently a synchronous motor operating a small magnetic stirrer can be used for this purpose. It is sometimes more convenient to attach, by a flexible shaft, a small stirring bar directly to the shaft of the synchronous motor.

Calibrations for electrodes used to measure concentration of oxygen can readily be made, since the current produced by the electrode is a linear function of oxygen concentration.[29,30] Since air contains 21% oxygen, at 1 atmosphere the partial pressure due to the oxygen alone is 21% of 760 mm, or 160 mm (Hg). A liquid 100% saturated with air also has a partial pressure of oxygen of 160 mm. Calibration can be achieved by bubbling the experimental solution with air and, if desired for a linearity check, with dilutions of air obtained from a gas mixing device until 100% saturation is achieved. Reference to a table or a determination by an independent method (such as the Winkler technique) gives the quantity of dissolved oxygen in a 100% saturated sample at a particular temperature. The electrode current for zero oxygen concentration can conveniently be determined by adding a concentrated yeast suspension.[30]

[26]U. Heber and C. S. French, *Planta* **79**, 99 (1968).
[27]Yellow Springs Instrument Co., Inc., Yellow Springs, Ohio, 45387; Beckman Instruments, Inc., 2500 Harbor Blvd., Fullerton, California 92634.
[28]J. M. Foster, *Bioscience* **19**, 541 (1969).
[29]P. W. Davies and F. Brink, Jr., *Rev. Sci. Instr.* **13**, 524 (1942).
[30]B. Chance and G. R. Williams, *J. Biol. Chem.* **217**, 383 (1955).

[11] Modulated Light Source Use with the Oxygen Electrode

By Pierre Joliot

Photosynthetic production of oxygen is usually detected by amperometric titration. This problem is discussed by D. Fork in this volume [10].

The modulated electrode represents a special adaptation of the stationary electrode.[1,2] The main limitation of the stationary rate electrode is that it is difficult to interpret quantitatively the amperometric response during fast transitory phases of photosynthetic activity. The transfer function which establishes a relation between the amperometric current and the rate of O_2 evolution is complex, the proportionality between these two factors being satisfied only during the stationary phase of photosynthetic activity.

The use of the same type of electrode with modulated illumination solves a great part of these difficulties, the response of the modulated electrode being strictly proportional to the rate of O_2 evolution over a very large range of time. In this method, photosynthetic material is illuminated by light modulated in intensity at a frequency that can be adjusted between 5 and 500 cps. As the dark reactions between photoreaction II and O_2 formation are very fast, the rate of O_2 formation is modulated at the same frequency as the light. If the photosynthetic material is sufficiently close to the platinum electrode, the amperometric current presents a modulated component. This modulated current is amplified and the dc component is rejected by an amplifier tuned to the frequency of the light. At any time, the amplitude of the modulated current is proportional to the average rate of O_2 production, the time response of the apparatus being limited to the period of modulation of the light. The main technical characteristics of the modulated electrode are described first; the biological meaning of the modulated response follows.

Description of the Apparatus

Simplified Scheme. The apparatus consists of at least two compartments. The bottom of the lower compartment is formed by a bare platinum electrode, and the top by a stretched dialysis membrane. Algae or chloroplasts are introduced inside this compartment, which can be 0.1–0.3 mm

[1] L. R. Blinks and R. K. Skow, *Proc. Nat. Acad. Sci. U. S.* **24**, 413 (1938).
[2] F. T. Haxo and L. R. Blinks. *J. Gen. Physiol.* **33**, 389 (1950).

thick. In a few minutes, algae or chloroplasts settle on the surface of the platinum electrode. Above the dialysis membrane is a second compartment about 5 mm thick, the vertical walls of which are silver. The top of this chamber is closed by a glass window. The silver wall of the chamber forms the reference electrode after electrolytic deposit of silver chloride.

The upper compartment is connected to a reservoir of buffer. O_2 and solute are continuously renewed by gravity flow inside the top chamber and, by diffusion through the membrane, inside the lower chamber. The geometry of the apparatus does not appear to be very critical, and only a few important characteristics must be satisfied. First, the photosynthetic material must be able to settle directly onto the surface of the platinum electrode. In fact, the sensitivity of the apparatus increases rapidly when the distance between the electrode and the source of modulated O_2 decreases. Second, the electrochemical resistance of the circuit must be very low ($<20\ \Omega$). Any type of high resistance link between the two electrodes, an agar bridge for instance, reduces considerably the sensitivity, especially at high frequencies.

Detailed Scheme of the Modulated Electrode. As an example, the detailed diagram of a modulated electrode now in use in our laboratory is presented in Fig. 1. This electrode can be used for a variety of purposes including detection of O_2 emission, reduction of Hill oxidants, or photooxidation of hydroxylamine.

This apparatus consists of three compartments separated by two dialysis membranes. The middle compartment, which is in an intermediate position between the two compartments previously described, is 0.5 mm thick. This third compartment is used when a solute able to interact with the silver electrode is introduced in the medium. Buffer + solute flows only in the middle compartment. By diffusion through the membrane the solute can penetrate into the lower chamber, which contains the photosynthetic material. Pure buffer flows through the upper compartment and prevents solute from reaching the silver electrode. The third compartment is also used in presence of a colored solute. The light absorption is considerably less than if the solute is introduced into the upper compartment, which is about 10 times thicker than the middle one.

The apparatus itself is formed by two main stainless steel parts (A and B), which are thermostatted by water circulation. Stainless steel appears to be sufficiently electrochemically inert and allows very efficient temperature regulation of the electrode. A piece of solid silver is included inside the upper part (A). The upper compartment is formed by a window (10 mm × 20 mm) inside the silver block. The dimensions of this window are larger than dimensions of the platinum electrode in order

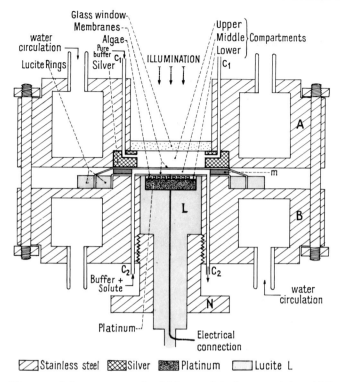

FIG. 1.   Diagram of the apparatus: the thickness of the lower and the middle compartments have been increased. For explanation, see text.

to prevent direct illumination of the silver by the modulated beam which is focused on the platinum electrode. So it is possible to avoid photoelectric or thermal effects on the silver which could produce a perturbation of the modulated current.

The top of the upper compartment is formed by a glass window. Pure buffer can flow through the upper compartment by two channels (C_1). Two dialysis membranes are stretched on the top of the lower part (B) by two concentric Lucite rings. These two membranes are separated by a stainless steel spacer (m) 0.5 mm thick. Inside this spacer is an oblong cavity, the dimensions of which exceed slightly those of the platinum electrode. Through holes inside the lower membrane, this cavity communicates with two channels (C_2) inside the lower stainless steel part (B), and the content of the middle compartment is continuously renewed by gravity flow.

The platinum electrode (3 mm × 5 mm) is formed by a solid piece of platinum included inside a piece of Lucite (L). Tops of the platinum electrode and of the Lucite support are optically polished. Upon this flat surface is fixed a disk of black electrical tape with a cutout of the same dimensions as the platinum electrode, to obtain a 0.1 mm-deep cavity retaining the algae or chloroplasts. The Lucite support of the platinum electrode fits into a cylindrical hole inside the lower stainless steel part. After a drop of chloroplasts or algal suspension has been placed on the platinum surface, the electrode support is pushed upward and secured against the lower membrane with a nut (N). During insertion, excess suspension is drained out by two tiny grooves inside the Lucite support of the electrode. Depending upon the size of the algal or chloroplast particles, complete settling, i.e., a steady-state sensitivity, is reached in 5–20 minutes.

To illuminate the sample, two beams are available. One beam is modulated by a rotating disk driven by a small variable speed dc motor (Brion Leroux Birotax, revolutions per minute proportional to voltage). The modulation frequency can be varied continuously between 1 and 200 cps by varying the voltage. The other beam provides a continuous background which can be superimposed on the modulated beam.

Electrical Circuit (Fig. 2)

The platinum electrode can be polarized either positively or negatively (+1 to −1 V) with respect to the reference electrode. In most of the cases, polarization voltage can be furnished by a battery connected to a low-resistance potentiometer. Since the potential of the reference electrode can vary, especially with high dc amperometric currents or for positive polarization, a reference calomel electrode can be used to monitor the potential between platinum and solution. This calomel electrode is immersed in the buffer reservoir feeding the middle compartment and connected to the high impedance input of a servoamplifier. The potential of the silver electrode is regulated to maintain the input voltage at a selected value. The time constant of the amplifier is about 3 seconds to eliminate noise of higher frequencies.

Signal Detection

A specific requirement of the circuit used to detect the modulated current is the low value of the load impedance of the amplifier (10–100 Ω). The platinum electrode presents a large capacitance (about 10 μF) and it is necessary to keep the time constant, RC, of the circuit small enough compared to the period of modulation of the amperometric current. To increase the sensitivity a transformer with a turn ratio of 10:100

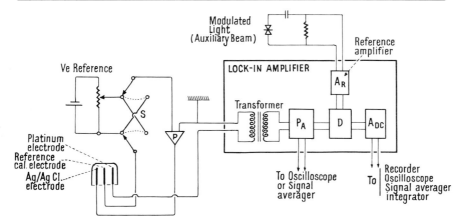

FIG. 2.   Electrical circuit. S, Polarity switch; P_A, preamplifier; D, detector; A_{DC}, dc amplifier; P, polarization amplifier.

is generally necessary to match the impedance between the amperometric circuit and the input of the amplifier. Depending upon the type of experiment, two types of amplification can be used.

1. For experiments that do not require phase measurements or short-time responses, the modulated current can be amplified and the dc component rejected by a simple filter amplifier tuned to the frequency of the modulated current. Time response is a decreasing function of the frequency and increasing function of the selectivity of the amplifier; time response is generally 10–20 times longer than the period of modulation of the current.

2. Demodulation and amplification can be carried by a lock-in amplifier. The detector must be synchronized with light modulation and the lock-in signal is obtained from a photodiode which is illuminated by a light beam chopped, as is the detecting beam, by the same rotating disk. In normal conditions, phase angle between the reference signal and the modulated signal is adjusted for maximum response. Output of the lock-in amplifier is connected either to a pen recorder, to an oscilloscope for viewing fast phenomena, or to a signal averager to increase the signal-to-noise (S/N) ratio of repeatable phenomena. Time response and S/N ratio can be adjusted by varying the integration time after demodulation.

Characteristics of the Modulated Response

For a negative polarization of the electrode, the amplitude of the modulated current is proportional to the amplitude of the modulated rate of O_2 production. The linearity between current and rate has been proved experimentally[3] and justified theoretically.[4] The modulated current is specific for photosynthetic activity; dc current due to the reduction of O_2 diffusing from the upper compartment or current due to other substances which could react at the electrode are not detected. So, this bare electrode can be used in presence of various electroactive solutes. For the same reasons, respiration which is not modulated by the light is not detected by the modulated electrode.

Sensitivity and Time Response. Sensitivity of the modulated electrode is exceptionally high. In the case of *Chlorella* cells, it is possible to detect the effect of red light modulated at 5 cps, with an incident energy of 0.1 erg/cm^2·sec. Such illumination produces oxygen at 10^{-17} moles/sec on the totality of the surface of the electrode. Sensitivity decreases rapidly with the frequency of modulation of the light. For instance, in the case of a layer of *Chlorella* 30 μ thick deposited on the electrode, sensitivity drops by a factor of 8 between 10 and 100 cps.

As the sensitivity is limited by the amplitude of the noise, an increase in sensitivity is obtained by increasing the time constant introduced after detection by the lock-in amplifier. The time response of the apparatus is generally limited by this time constant. Nevertheless when fast responses are required, best results are obtained without any integrating time constant. In this case, the time response is approximately equal to the period of modulation of the light. The limit which can be reached is about 0.05 second for frequency of modulation of 200 cps.

Polarogram. The polarogram in modulated light can be defined by the variation of the amplitude of the modulated current as a function of the polarization voltage. In the case of O_2, the polarogram presents an inflexion at -0.6 V and a plateau at -1.1 V. The first wave corresponds to the reduction of O_2 in H_2O_2, and the second wave to the reduction of O_2 in H_2O. The very high dc current due to the reduction of hydrogen ions, which appears for voltage higher than 0.7 V, does not perturb these measurements.

Experiments are generally performed at -0.6 V. At this voltage the concentration of O_2 at the level of the electrode is not equal to zero, and algae in direct contact with the electrode are kept in aerobic condition.

Other substances that react with the electrode can be produced by photosynthetic activity. The concentration of these substances can be modulated and yield a modulated current. Polarograms, which are spe-

[3] P. Joliot. *C. R. Acad. Sci.* **260,** 5920 (1965).
[4] P. Joliot, M. Hofnung, and R. Chabaud, *J. Chim. Phys.* **10,** 1423 (1966).

cific for the substance detected, can be used to identify the substance as the origin of the modulated current.[5]

Diffusion of Oxygen from Modulated Sources to the Electrode

This problem has been studied in detail both from theoretical and experimental points of view.[4] Herein are presented only some conclusions that are useful for correct use of the modulated electrode.

If we consider an infinitely thin layer of modulated source at the distance, p, from the platinum electrode, the amplitude, A, of the modulated current is:

$$A = v \cdot k \cdot e^{-p \sqrt{2\pi\nu} / \sqrt{2D}} \tag{1}$$

in which v is the modulated flow of O_2 produced in the layer, ν the frequency of modulation, and D the diffusion constant of O_2. It can be concluded from this equation that modulated sources placed too far from the electrode do not contribute to the modulated response. For each value of the frequency, one can define the thickness of an active layer in which modulated sources participate in the modulated response. This thickness varies between 30 and 6 μ for frequencies varying from 5 to 40 cps. So, for frequencies higher than 40 cps, a monolayer of algae, such as *Chlorella*, is sufficient to give a maximum response. A thicker layer increases the light absorption without any gain in sensitivity. Another consequence of Eq. (1) is the importance of the geometrical characteristics of the biological material. For instance, the modulated response obtained with algae surrounded by a mucilaginous shell, as with *Porphyridium*, decreases very fast as a function of increasing frequency. The shell represents a "dead" layer which separates the chloroplasts from the electrode. Attenuation by this dead layer becomes very important for a frequency higher than 50 cps.

Biological Significance of the Modulated Response

The rate of a photoreaction is generally proportional both to the intensity of absorbed light and to the concentration of the photosubstrate. The rate v_{II} of photoreaction II, which is at the origin of O_2 formation, is:

$$v_{II} = k \cdot i \cdot [E] \tag{2}$$

in which [E] represents the concentration of the photosubstrate. In fact, v_{II} is not exactly proportional to $[E]^6$, but Eq. (2) represents a good enough approximation for the following discussion.

[5]P. Joliot and A. Joliot, *Biochim. Biophys. Acta* **153**, 625 (1968).
[6]A. Joliot and P. Joliot, *C. R. Acad. Sci.* **258**, 4622 (1964).

If the intensity, i, is modulated with an amplitude, I, and if the concentration, [E], of the photosubstrate is not appreciably modulated by the light, the rate of formation of the primary photoproduct is modulated with an amplitude:

$$V_{II} = k \cdot I \cdot [E] \tag{3}$$

If [E] is modulated, the response depends upon the product of two sinusoidal functions of different phases, and the interpretation of the modulated response becomes extremely complex. The modulation of E can be neglected if the amount of E destroyed during one period of modulation of the light remains always small in respect to the total concentration of E ($< 2\%$). These conditions are always satisfied in weak light illumination. In strong light, it may be necessary to increase the frequency of modulation in order to decrease the amount of substrate E destroyed during a period. Absence of modulation of the concentration E can be checked easily by measuring the relative value of the phase angle between the modulated light and the modulated amperometric current. If E is not modulated, the phase shift must be independent of the light intensity. A light-dependent shift of phase angle reveals generally an appreciable modulation of the concentration of the photosubstrate.

The value of v_{II} is not always equal to the rate of O_2 formation because O_2 is not the primary product of photoreaction II. Between this photoreaction and O_2 formation, there occurs a chain of dark reactions. Modulation is transmitted through this chain, and O_2 production is modulated at the same frequency as the primary photoproduct. Depending upon the constant rates of the intermediary reactions and of the frequency of modulation, the amplitude of the modulated rate of O_2 production is attenuated and the phase is shifted. The constant rate of the limiting reaction can be calculated from measurements of attenuation and of phase shift of the signal as a function of frequency.[4] The application of this method to the problem of O_2 formation shows that the limiting reaction is very fast ($t_{1/2} = 1$ msec) and, for frequencies between 1 and 100 cps, the attenuation due to the limiting step is negligible. The filtering effect on modulation of a slow dark-limiting step can be demonstrated in the case of the reaction of photoabsorption for O_2. This reaction described by different authors[7,8] is directly sensitized by system I photoreaction. Theoretically, photoabsorption must be at the

[7]C. S. French and D. C. Fork, *Carnegie Inst. Washington Yearb.* **60**, 351 (1961).
[8]P. Joliot, *Biochim. Biophys. Acta* **102**, 116 (1965).

origin of a modulated amperometric current interfering with the modulated current due to the O_2 production. Contrary to photoemission of O_2, photoabsorption involves a slow dark reaction ($t_{1/2} = 1$ sec).[8] For frequencies higher than 10 cps, this dark reaction eliminates completely the modulation of O_2 concentration due to photoabsorption.

Experimental Results

In negative polarization of the platinum electrode, modulated electrode allows a specific measurement of the rate of photoreaction II. As pointed out before, perturbing effects of respiration and photoabsorption are eliminated. This electrode is especially well adapted to the measurement of transitory kinetics in weak light illumination.

The O_2 gush at the onset of illumination is illustrated in Fig. 3. The biological material was a mutant of *Chlamydomonas*[9] in which photoreaction I is completely blocked. Oxygen evolution presents two phases: a first phase (0–0.5 sec) during which the rate of O_2 production increases and a second phase during which the rate of O_2 production decreases. The first phase corresponds to the activation reaction, the second phase corresponds to the O_2 gush which is induced by the emptying of the pool of oxidant A.[8] With this mutant, the pool of A is not reoxidized by photoreaction I and a good isolation of the O_2 gush is obtained even in weak light.

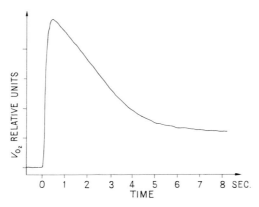

FIG. 3.   Oxygen gush at the onset of illumination. *Chlamydomonas reinhardi* mutant H 54 F [see P. Bennoun and R. P. Levine, *Plant Physiol.* **42**, 1284 (1967)]. Frequency of modulation, 25 cps; temperature, 20°.

[9]P. Bennoun and R. P. Levine, *Plant Physiol.* **42**, 1284 (1967).

The activation phase measured in weak light with *Chlorella pyrenoidosa* is shown in Fig. 4. The small lag observed at the beginning of the activation period is significant.

Chromatic Effects. Equation (3) shows that for a modulated beam of fixed intensity, the amplitude of the modulated current depends only upon the concentration of the photosubstrate E. If a continuous beam is superimposed on the modulated beam, this last one can be considered as a detecting beam which measures the variation of concentration of E.

In Fig. 5 is shown an application of this principle to the study of Emerson stimulation.[10] At the beginning of the recordings, *Chlorella* cells are illuminated with 480 nm light modulated at 10 cps. (This light sensitizes preferentially photosystem II.) At zero time, a continuous beam of 695 nm light is superimposed on the modulated beam. This continuous beam does not give a direct response, and, in fact, no instantaneous variation of the signal can be observed. During the first few seconds, the modulated current increases. This stimulation is due to the oxidation of the electron acceptor of photosystem II induced by the 695 nm beam. Using a classical O_2 electrode, it is difficult to separate the two effects occurring simultaneously: increase of the light factor ki [see Eq. (2)] due to the fraction of 695 nm light absorbed by photosystem II pigments, and increase of the concentration factor E.

Action Spectrum. If we refer to Eq. (2), the action spectrum of photoreaction II can be defined by the variation of the factor ki which is pro-

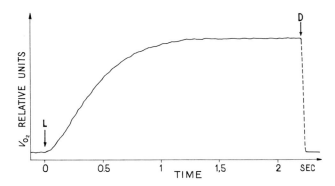

FIG. 4.   Oxygen evolution at the onset of weak illumination (activation phase). *Chlorella pyrenoidosa*. Frequency of modulation, 25 cps; temperature, 20°; *L*, light on; *D*, light off.

[10]P. Joliot, *Brookhaven Symp. Biol.* **19,** 418 (1966).

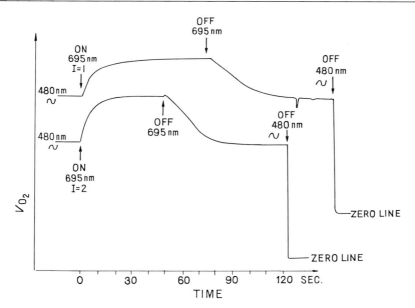

Fɪɢ. 5.   Emerson enhancement. *Chlorella pyrenoidosa*. *I*, intensity of the continuous beam (695 nm); temperature, 3°C. From P. Joliot, *Brookhaven Symp. Biol.* **19**, 418 (1966).

portional to the number of photons absorbed by photosystem II pigments as a function of wavelength. Determination of ki requires that the concentration factor [E] remain constant and the wavelength of illumination be varied. To keep [E] constant, algae and chloroplasts are illuminated by a continuous beam of rather strong intensity (about 2000–4000 ergs sec/cm²). The concentration of E reaches rapidly a steady-state value. Wavelength of this continuous beam is not important. Action spectrum measurements are performed with a weak detecting beam, the intensity of which is about 1–2% of the intensity of the continuous beam. This weak detecting beam does not modify the steady-state concentration of E. The action spectrum of photosystem II (Fig. 6, curve *2*) is obtained by calculating for each wavelength the ratio of the amplitude of the modulated current to the number of incident photons.[11]

Detection of Other Photoproducts of Photosynthetic Activity. The modulated electrode is able to detect any photoreaction which produces or consumes a substance reacting at the platinum electrode.[5] An important

[11]P. Joliot, A. Joliot, and B. Kok, *Biochim. Biophys, Acta* **153**, 635 (1968).

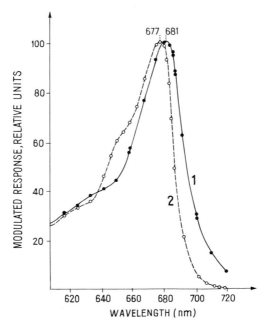

FIG. 6. Action spectra of photosystem I and photosystem II, Spinach chloroplasts. Temperature, 20°; frequency of modulation, 90 cps. Curve 1: Action spectrum of photosystem I; polarization of the platinum electrode, +0.6 V; electron acceptor, methylviologen, 5×10^{-5} M; continuous background, 650 nm. Curve 2: Action spectrum of photosystem II; polarization voltage, −0.6 V; electron acceptor, NADP, 10^{-3} M, ferredoxin (catalytic amount); continuous background, 720 nm. From P. Joliot, A. Joliot, and B. Kok, *Biochim. Biophys. Acta* **153,** 635 (1968).

application is the measurement of the rate of reduction of different Hill oxidants (ferrycyanide, dichlorophenolindophenol, methylviologen, quinone, etc.). In the case of a low potential oxidant, such as methylviologen, the modulated response gives a specific measurement of the rate of photoreaction I. An action spectrum of photosystem I obtained with spinach chloroplasts in the presence of methylviologen[11] is given in Fig. 6, curve 1.

Another example of an application of the modulated electrode is the study of the photooxidation of hydroxylamine performed by Bennoun and Joliot.[12] Hydroxylamine is able to replace water as electron donor of photoreaction II. The product of oxidation, which is not yet identified, reacts with a negatively polarized platinum electrode.

[12]P. Bennoun and A. Joliot, *Biochim. Biophys. Acta* **189,** 85 (1969).

[12] Measuring the Delayed Fluorescence and Chemiluminescence of Chlorophyll *in Vivo*

By RODERICK K. CLAYTON

This article describes simple ways to measure delayed light emission and chemically induced luminescence of chlorophyll (Chl) in green plants and algae, using relatively inexpensive equipment (total cost less than 2000 dollars in 1970). The basic needs are: (1) a detector (photomultiplier) that is sensitive enough in the red, from about 650 to 750 nm, operated in conjunction with an amplifier and a strip chart recorder or an oscilloscope; (2) a means of first subjecting the plant material to a program of illumination and/or chemical change, and then quickly presenting it to the detector. The provision for illumination is essential, even when chemiluminescence is to be measured, because the chemically induced luminescences of Chl *in vivo* generally require that the system first be activated by light.[1]

In the ensuing description, some specific commercial components are mentioned. These are given for the convenience of the reader, but are only representative of a great variety of comparable and competitively priced components.

A very simple and useful arrangement is shown in Fig. 1. It is only necessary to purchase a photomultiplier microphotometer from American Instrument Co., Silver Springs, Maryland, plus a Bausch and Lomb VOM-5 recorder from any major scientific supplier, and a small spotlight, such as is used as a microscope illuminator. The microphotometer should be chosen with a type 1P22 photomultiplier, for sensitivity in the red. The photometer head has a filter housing in which a sample cell (cuvette or small test tube) can be supported to receive material from an external syringe. A red filter, such as a Corning No. CS2-58, can be placed between the sample cell and the photomultiplier in order to reduce the response to "false" light.

In operation, the plant material, a suspension of chloroplasts or algae, is held in the syringe and a reagent can be placed, if desired, in the sample cell. After a suitable illumination program the material is injected into the sample cell and its luminescence is recorded. A second syringe could be fitted so as to deliver a second chemical perturbation after the plant material has been injected into the sample cell.

---

[1]B. C. Mayne, *Photochem. Photobiol.* **8,** 107 (1968).

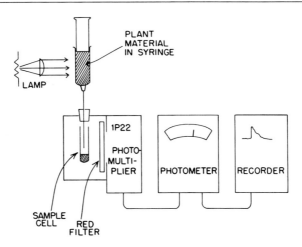

FIG. 1.   A simple arrangement for measuring the luminescence of chlorophyll in chloroplasts and algae. The photomultiplier microphotometer is from American Instrument Co. (see the text).

The foregoing instrument is limited because of its marginal sensitivity in the red and its time constant of about 0.1 second. With a little effort one can make an apparatus designed more specifically for the purpose at hand, at a cost of less than 1200 dollars including shop charges but excluding the recorder or oscilloscope.

As shown in Fig. 2, the optical arrangement can be extended to allow illumination of the material in the sample cell as well as in the syringe. This permits measurements with leaf sections and other structures too large to pass through a syringe. It also allows measurement of the prompt fluorescence, using a combination of blue exciting light and a red filter in the photomultiplier window to intercept scattered exciting light. A suitable combination of filters is Corning Nos. 4-96 (blue) and 2-64 (red).

The best choice of photomultiplier is one having the S20 spectral response, which affords high sensitivity in the red together with a reasonably low dark current. While an improvement can be had by cooling such a photomultiplier to dry ice temperature, this is not necessary in most applications. Suitable photomultipliers are: E. M. I. Type 9558B or 9698B (Whittaker Corporation, Gencom Division, 80 Express St., Plainview, New York), and RCA Type 7265 or 7326 (Radio Corporation of America, Electron Tube Division, Harrison, New Jersey).

A satisfactory way to apply high voltage to the photomultiplier is shown in Fig. 3 A. There are many regulated dc high voltage supplies on the market, designed specifically for use with photomultipliers. One such is the Model 245 made by Keithley Instruments, Inc., 28775 Aurora Road, Cleveland, Ohio.

The photocurrent can be delivered to a commercial "electrometer amplifier" or "picoammeter," costing in the neighborhood of 600 to 1000 dollars — for example, the Model 602 Electrometer made by Keithley Instruments. However, a simple and completely suitable electrometer amplifier can be made at much less cost, using a high quality operational amplifier such as the type PP25C from Philbrick Researches, Inc. (Allied Drive at Route 128, Dedham, Massachusetts). This little feedback amplifying module is powered by about 5 mA at ±15 V dc, obtained from batteries or from Philbrick's Model PR-30 power supply. The amplifying circuit is shown in Fig. 3 B. With a resistance of 10^9 ohm in the feedback loop as shown, an input of 10^{-9} A is transformed into an output of 1 V. A "precision" 10^9 ohm resistor can be obtained from Keithley Instruments or from Victoreen Instrument Co. (5806 Hough Ave., Cleveland, Ohio).

The output of the amplifier can be fed into any suitable recorder, such as the Bausch and Lomb VOM-5 mentioned earlier, or into an oscilloscope. The circuit of Fig. 3 B includes a variable resistance-capacitance smoothing network by which faster components of noise can be eliminated (at the cost, of course, of a slower response). The values of R and

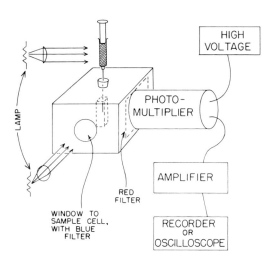

Fig. 2.   A more elaborate arrangement for measuring chlorophyll luminescence and fluorescence *in vivo*. Components as described in the text and in Fig. 3.

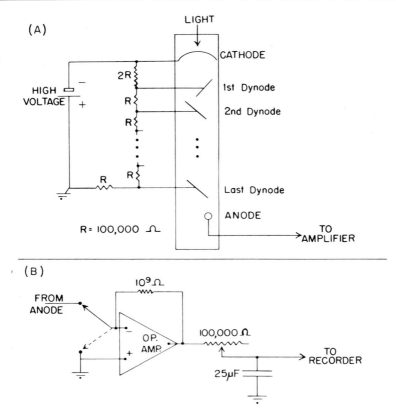

Fɪɢ. 3.   (A) A generally satisfactory way to connect a photomultiplier. The cathode receives negative high voltage up to about −1000 V. (B) An electrometer circuit using an operational amplifier (op. amp.) (see the text) to convert a small photocurrent to a voltage signal compatible with a recorder or oscilloscope. An adjustable smoothing filter is included.

C in this network were chosen to be compatible with a recorder or oscilloscope of high input impedance (about 1 megohm or more).

The foregoing instruments have been chosen for their simplicity. The investigator who wishes to construct a Becquerel phosphoroscope,[2] or a stop-flow apparatus fitted for optical work,[3] or a pumped circulating or "continuous squirt-mixing" arrangement will enjoy certain advantages, but the simple method described here is useful for a great variety of experiments.

[2] R. K. Clayton, *Biophys. J.* **9**, 60 (1969).
[3] Q. H. Gibson and L. Milnes, *Biochem. J.* **91**, 161 (1964).

[13] Quantum Yield Determinations of Photosynthetic Reactions

By Martin Schwartz

By definition, photosynthetic reactions involve endothermic photochemical reactions brought about by the absorption of light quanta. It follows that the determination of the quantum yield or quantum efficiency of photosynthesis can be important in elucidating the overall photochemical mechanism of this biological process.

A brief description of the photochemical principles involved in quantum yield determinations is presented below. The description of these principles is followed by a discussion of various techniques and calculations used in quantum yield determinations of photosynthetic reactions.

Photochemical Principles

Radiant energy is absorbed in discrete units called quanta (or photons). Only absorbed light quanta induce photochemical events (Grotthuss' law). In the primary photochemical process, a single absorbed photon excites a single photoreactive molecule (Einstein's law of photochemical equivalence). In conventional terms, 1 mole of light quanta (N photons) excites 1 mole (N molecules) of photochemically reacting substance, where the Avogadro number N equals 6.02×10^{23}. One mole of light quanta (6.02×10^{23} photons) is called an Einstein.

The *quantum yield*, ϕ, or quantum efficiency of a photochemical reaction is defined by the ratio of the number of molecules reacting chemically to the number of light quanta absorbed

$$\phi = \text{quantum yield of quantum efficiency}$$

$$\phi = \frac{\text{number of molecules reacting chemically}}{\text{number of light quanta absorbed}} \tag{1}$$

$$\phi = \frac{\text{number of moles of molecules reacting chemically}}{\text{number of Einsteins absorbed}} \tag{2}$$

The *quantum requirement* is defined as the reciprocal of the quantum yield,

$$1/\phi = \text{quantum requirement}$$

$$1/\phi = \frac{\text{number of light quanta absorbed}}{\text{number of molecules reacting chemically}} \tag{3}$$

The energy of a quantum, ϵ, is proportional to the frequency of radiation, ν, expressed in $\sec^{-1}$.

$$\epsilon = h \times \nu \tag{4}$$

Planck's constant h has the value 6.626×10^{-27} erg $\times$ sec.

The energy content of a photochemical equivalent (1 Einstein) is defined by the equation:

$$E = Nh\nu = Nhc/\lambda \tag{5}$$

where c is the velocity of light (2.998×10^{10} cm $\times \sec^{-1}$), λ is the wavelength in cm, and N is the Avogadro number (6.02×10^{23}). Thus,

$$E = \frac{6.02 \times 10^{23} \times 6.626 \times 10^{-27} \times 2.998 \times 10^{10}}{\lambda(\text{cm})} \text{ ergs/Einstein}$$

or

$$E = \frac{1.197 \times 10^{8}}{\lambda(\text{cm})} \text{ ergs/Einstein} \tag{6}$$

If λ is expressed in nanometers (nm), then E is

$$E = \frac{1.197 \times 10^{15}}{\lambda(\text{nm})} \text{ ergs/Einstein} \tag{7}$$

where (1 nm $= 10^{-7}$ cm). The corresponding E value in gram-calories (cal) per Einstein is

$$E = \frac{2.86 \times 10^{7}}{\lambda(\text{nm})} \text{ cal/Einstein} \tag{8}$$

where 1 erg $= 2.389 \times 10^{-8}$ cal.

Technical Considerations

The essential technical requirements for quantum yield determinations are: (a) a source of monochromatic light, (b) a method for measuring the incident quantum intensity of the monochromatic light source, (c) a method for measuring either directly or indirectly the fraction of the incident quantum intensity absorbed by the reacting system, and (d) a method for assaying the resulting photochemical process under light-limiting conditions.

Monochromatic Light

In most instances accurate quantum yield determinations are performed with relatively narrow spectral bands (monochromatic) of actinic light. Prism and diffraction grating monochromators are the instruments of preference for obtaining narrow spectral bands of light.

However, transmission characteristics of moderately priced monochromators do not yield sufficient light intensities for most practical quantum yield measurements of photosynthetic systems. Presently, narrow single pass-band interference filters or combinations of absorption filters are employed for most quantum yield studies of photosynthesis. A number of companies provide a large selection of interference filters for the ultraviolet and visible region.[1] One caution should be emphasized in employing interference filters. It is necessary to place the filter in the optical system so that it intercepts an essentially parallel light beam. Otherwise the wavelength transmission characteristics of the interference filter will be shifted.

Measurement of Incident and Absorbed Quantum Intensity

Both physical and chemical methods are used for determining the incident actinic intensity. Physical methods of preference employ either bolometers or thermopiles.[2] Chemical methods involve the use of chemical actinometers.

The Lummer-Kurlbaum bolometer[3,4] with a large receiving surface is employed in the writer's laboratory. It has the advantage that the electrical response characteristics are constant over the entire receiving surface ($\sim$ 10 cm^2), allowing one to measure the entire *cross-sectional* intensity of an actinic beam. Use of a small receiving surface bolometer (or thermopile) introduces difficulties because of the inhomogeneity of the *cross-sectional* intensity of most actinic light beams, necessitating an integration of intensities over the *cross-sectional* area of the actinic light beam.

An example illustrating the use of a bolometer in determining the absolute light intensity in Einstein units follows.

Example

The bolometer, previously calibrated against a standard lamp, reads 4.56×10^3 ergs $\times$ min^{-1} per microvolt deflection. The wavelength of the actinic light source, isolated with a 650 nm interference filter, blocked

[1]Excellent interference filters for the visible and ultraviolet region can be purchased from Thin Film Products, Division of Infrared Industries, P.O. Box 557, Waltham, Massachusetts 02154, or Baird-Atomic, Inc., 33 University Road, Cambridge, Massachusetts.

[2]F. Weigert, "Optische Methoden der Chemie," Chapter X. Akad. Verlagsges., Leipzig, 1927.

[3]O. Warburg, "New Methods of Cell Physiology." Wiley (Interscience), New York, 1962.

[4]W. Bladergroen, "Problems in Photosynthesis." Thomas, Springfield, Illinois, 1960.

on the short wavelength side to X-rays and on the long wavelength side to infinity, was found to yield a bolometer reading of 2000 μV. Recalling Eq. (7), the energy content in ergs per Einstein of a mole quanta (1 Einstein) at 650 nm is

$$\frac{1.197 \times 10^{15}}{650} = 1.84 \times 10^{12} \text{ ergs/Einstein}$$

Thus, the incident actinic intensity (I_0) is

$$I_0 = \frac{\text{bolometer deflection} \times \text{calibration factor}}{\text{energy content for specific } \lambda \text{ in ergs/Einstein}}$$

$$I_0 = \frac{2000 \times 4.56 \times 10^3}{1.84 \times 10^{12}}$$

$$I_0 = 4.96 \ \mu\text{Einsteins/min}$$

where $10^6 \ \mu$Einsteins = 1 Einstein.

The determination of the incident actinic intensity with a chemical actinometer relies on the constancy within certain spectral regions of the quantum yield of the photochemical reactions. The potassium ferrioxalate actinometer developed by Hatchard and Parker[5] is reliable and convenient for determining light intensity at wavelengths shorter than 440 nm. Our experience with this actinometer confirms their results, indicating an essentially constant quantum yield for this system of unity ($\phi = 1 \pm 0.1$) for wavelengths between 405 nm and 440 nm.

Essentially, the ferrioxalate actinometer is assayed by monitoring the photochemical reduction of the Fe^{3+} complex ion to the Fe^{2+} complex ion. The concentration of Fe^{2+} ions is measured spectrophotometrically by the o-phenanthroline reaction.

An example illustrating the use of the chemical actinometer in determining the absolute light intensity in Einstein units follows.

Example

In a typical intensity determination employing the ferrioxalate actinometer, the o-phenanthroline assay for Fe^{2+} ions is standardized first. The solutions required for standardizing and assaying the system are:

(A) Actinometer solution: 7.4 g $K_3Fe(C_2O_4)_3 \cdot 3H_2O$ in 100 ml of 0.1 N H_2SO_4 (dark bottle)
(B) 0.1 N H_2SO_4
(C) 2.5 mM $FeSO_4$ in 0.1 N H_2SO_4
(D) Development agent: 0.1% o-phenanthroline (1 mg/ml) dis-

[5] C. G. Hatchard and C. A. Parker, *Proc. Roy. Soc. Ser. A* **235**, 518 (1956).

solved in 1.8 N Na acetate and 1.0 N H_2SO_4 (store in dark bottles)

PROCEDURE

Step 1. Fill 4 cuvettes (1 cm optical path) to a final volume of 2.5 ml ($V_F = 2.5$) as described in Table I below.

Step 2. Wait 10 minutes in darkness, read at 510 nm, and calculate the millimolar extinction coefficient ($\epsilon_{510}^{Fe^{2+}}$) for use in the subsequent actinometer determination of light intensity.

TABLE I
FILLING OF CUVETTES FOR INTENSITY DETERMINATION

Cuvette No.	(A) (ml)	(B) (ml)	(C) (ml)	(D) (ml)		OD_{510}	ΔOD_{510}	$\epsilon_{510}^{Fe^{2+}} \left(\dfrac{cm^2}{\mu mole} \right)$
1	2.0	—	0.10	0.4		1.26	1.23	12.3
2	2.0	0.05	0.05	0.4	$\xrightarrow[\text{dark}]{10 \text{ min}}$	0.64	0.61	12.2
3	2.0	0.08	0.02	0.4		0.28	0.25	12.5
4	2.0	0.10	—	0.4		0.03	—	—
							Av.[a]	12.3

[a]Where

$$\epsilon_{510}^{Fe^{2+}} \frac{\Delta OD_{510}}{c \times d}$$

and $d = 1$ cm and $c = \mu$moles Fe^{2+}/ml.

As shown, the standardization of the actinometer solution for determining the concentration of the Fe^{2+} ion indicates that at 510 nm, 1 μmole Fe^{2+} ions/ml yields an optical density reading (ΔOD_{510}) of 12.3.

The procedure for determining the incident actinic intensity (I_0) with this actinometer is based on the photochemical reduction of the complex Fe^{3+} ion (solution A) to a complex Fe^{2+} ion, assuming a constant quantum yield of unity ($\phi = 1$). The concentration of the complex Fe^{2+} ion which is formed photochemically is measured against the control system as described above.

ACTINOMETER PROCEDURE

Step 1. Expose 2.0 ml actinometer solution (A) to the monochromatic light source (λ range between 400 and 440 nm).

Step 2. After light is turned off, add 0.1 ml (B) and 0.4 ml (D).

Step 3. Wait 10 minutes in darkness and read at 510 nm against an unexposed sample (cuvette 4 above).

The o-phenanthroline determination of two exposed actinometer solutions yielded ΔOD_{510} values of 0.58 and 1.21 when irradiated for 1 and 2 minutes, respectively. Thus, the incident actinic intensity (I_0) is

$$I_0 = \frac{\Delta OD_{510} \times V_F}{\epsilon_{510}^{Fe^{2+}} \times \phi \times \text{time}}$$

(1 minute exposure)

$$I_0 = \frac{0.58 \times 2.5}{12.3 \times 1 \times 1} = 0.118 \frac{\mu\text{moles Fe}^{2+}}{\text{minute}} \implies 0.118 \frac{\mu\text{Einsteins}}{\text{minute}}$$

(2 minutes exposure)

$$I_0 = \frac{1.21 \times 2.5}{12.3 \times 1 \times 2} = 0.123 \frac{\mu\text{moles Fe}^{2+}}{\text{minute}} \implies 0.123 \frac{\mu\text{Einsteins}}{\text{minute}}$$

It is to be emphasized that equating the rate of Fe^{2+} production (μmoles Fe^{2+}/minute) and the incident intensity (μEinsteins/minute) is based on the assumption that the actinometer maintains a quantum yield of unity,

$$\phi = \frac{\mu\text{moles Fe}^{2+}/\text{min}}{\mu\text{Einsteins/min}} = 1$$

Determination of Light Absorption

In optically dense suspensions of biological materials, such as concentrated algal or chloroplast suspensions, the incident actinic intensity can be equated with the absorbed actinic intensity (I_{Abs}). In quantum yield determinations employing such systems, it is sufficient to determine the incident intensity (I_0) with a bolometer, thermopile, or chemical actinometer and to compute ϕ values directly for the reaction under investigation. However, for most quantum yield determinations with biological systems there is an advantage in employing optically thin suspensions in order to expose the entire sample to approximately equivalent intensities. Under these conditions only a fraction of the incident intensity (I_0) is absorbed. A determination of the fraction of the incident intensity absorbed by the biological system can be carried out with an integrating sphere.[6] In Table II, percentage absorption measurements obtained with an integrating sphere for spinach chloroplast suspensions are presented for a number of wavelength regions between 640 nm and 730 nm.[7]

[6] B. Kok, *Enzymologia* 13, 1 (1948).
[7] M. Schwartz, *Biochim. Biophys. Acta* 131, 548 (1967).

TABLE II
PERCENT ABSORPTION MEASUREMENTS FOR SPINACH CHLOROPLASTS[a]

Wavelength[b] (nm)	5 μg Chl$_t$[c] per ml	10 μg Chl$_t$[c] per ml	20 μg Chl$_t$[c] per ml	40 μg Chl$_t$[c] per ml
730	0.49	0.98	1.96	3.90
720	1.23	2.48	4.90	9.53
710	3.06	6.0	11.65	22.0
700	6.77	13.1	24.5	43.0
690	18.4	33.4	55.6	80.3
680	34.5	57.1	81.6	96.6
670	22.2	39.5	63.3	86.6
650	18.95	34.3	56.8	81.3
640	14.65	26.6	46.2	71.0

[a]Integrating sphere, 1.0 cm light path.
[b]Half-band, 5 nm.
[c]Chl$_t$ = total chlorophyll ($a + b$).

The determination of the fractional absorption (α) at a specified wavelength and the determination of the incident intensity (I_0) at this wavelength allow one to calculate the fraction of the incident light intensity absorbed by the biological system

$$\alpha I_0 = I_{Abs} \qquad (9)$$

Assaying Photochemical Processes under Quantum Yield Conditions

The simple form of the Einstein law of photochemical equivalence as stated earlier implies that in photosensitive systems one molecule decomposes for each quantum absorbed. This form of the law is rarely realized experimentally, although evidence indicates that it applies to the primary process in photochemical reactions. That is, the rate of the primary photochemical process is the rate at which the photochemically reactive molecules absorb radiation. Quantum yield studies of complex photochemical systems such as photosynthesis (e.g., O_2 production, CO_2 assimilation, cytochrome oxidation, and reduction processes) require knowledge of experimentally variable parameters such as (1) the effect of concentration in establishing the order of reaction, (2) the effect of temperature in ascertaining whether thermal reactions are important, and (3) the influence of light intensity on the rate of reaction. If the quantum yield changes with light intensity, it follows that either light is not the limiting factor or that more than one photochemically reacting molecule is taking part in the sequence of reactions resulting in the turnover of the quantity that is being monitored experimentally.

Fortunately, most photosynthetic processes can be assayed under conditions where the rate of the reaction under investigation is directly proportional to light intensity and independent of temperature, thereby lending themselves to quantum yield studies.

A typical quantum yield determination of Hill reaction activity with ferricyanide as the oxidant is shown in the example below.

Example

Spinach chloroplast suspensions containing 5 μg of chlorophyll per milliliter are exposed to an actinic light source ($\lambda = 640$ nm). The reaction cuvettes have an optical path of 1 cm. Potassium ferricyanide is employed as the Hill oxidant. The rate of the reaction is followed by determining the production of ferrocyanide produced photochemically concomitant with oxygen production according to the stoichiometry:

$$4 \ Fe(CN)_6^{3-} + 2H_2O \xrightarrow[\text{chloroplasts}]{\text{light}} 4 \ Fe(CN)_6^{4-} + 4H^+ + O_2$$

The results are tabulated in Table III.

TABLE III
QUANTUM YIELD DETERMINATION OF HILL REACTION ACTIVITY WITH FERRICYANIDE

I_0 (nEinstein/min)	I_{Abs}[a] (nEinstein/min)	$V_{Fe^{2+}}$[b] (nmoles/min)	ϕ $V_{Fe^{2+}}/I_{Abs}$
27	3.96	1.32	0.33
55	8.07	2.64	0.33
109	16.0	5.38	0.34
218	31.9	10.2	0.32
437	64.0	15.2	0.24 ⎫ [c]
874	128	22.7	0.18 ⎭

[a] $I_{Abs} = I_0 \times \alpha_{640}$ (see Tabel II).
[b] $V_{Fe^{2+}}$= photochemical rate of ferrocyanide production.
[c] Decreasing ϕ values at these intensities indicate that light intensity is no longer the rate-limiting factor for this reaction.

[14] Measurement of Hill Reactions and Photoreduction

By A. TREBST

I. Introduction

In 1937 R. Hill[1] discovered that a leaf homogenate would evolve oxygen when illuminated in the presence of a ferric salt (ferric oxalate). In 1944 Warburg[2] showed that also p-benzoquinone would be reduced

[1] R. Hill, *Nature (London)* **139**, 881 (1937).
[2] O. Warburg and W. Lüttgens, *Naturwissenschaften* **38**, 301 (1944)

with oxygen evolution by a crude chloroplast preparation on illumination. Some seven years later, three groups demonstrated photosynthetic NADP reduction by chloroplasts.[3-5] After the successful preparation of intact chloroplasts[5a] in 1954,[6] it could be shown that the oxygen evolution system resides in the green particulate fraction obtained from intact chloroplasts by osmotic shock and that the CO_2-fixation system (Calvin cycle) can be separated from the light-dependent reactions of photosynthesis.[7] The particulate fraction, i.e. the thylakoid system of whole chloroplasts, is able to catalyze two types of photosynthetic reactions: photophosphorylation and photosynthetic reduction of an electron acceptor concomitant with oxygen evolution (Hill reaction). A Hill reaction is defined as the photoreduction of an electron acceptor at the expense of water, which is oxidized to oxygen.

$$H_2O + acceptor_{oxid} \xrightarrow[\text{chloroplasts}]{hv} \frac{1}{2} O_2 + acceptor_{reduced}$$

The final Hill acceptor in photosynthesis *in vivo* is CO_2. In whole chloroplasts it is accepted now that the physiological Hill reaction is the photoreduction of NADP. Numerous artificial compounds instead of NADP were tried and found to serve as an electron acceptor in the Hill reaction in whole chloroplasts, among them ferric salts and *p*-benzoquinones, whose photosynthetic reduction led to the discovery and definition of the reaction.

It is clear now that several redox reactions and redox carriers besides the actual light-accepting pigments are localized in the thylakoid system of the chloroplasts, and participate in photosynthetic NADP reduction at the expense of water photooxidation. The term electron transport chain is used to express the view that there is an electron flow from the redox potential level of water to that of NADP via a number of redox catalysts. We tend to believe at present that there are two light reactions involved in photosynthetic electron transport, and that there is an electron transport chain between water and photosystem II, one link-

[3]W. Vishniac and S. Ochoa, *Nature (London)* **167**, 768 (1951).

[4]L. J. Tolmach, *Nature (London)* **167**, 946 (1951).

[5]D. I. Arnon, *Nature (London)* **167**, 1008 (1951).

[5a]Intact chloroplasts are defined as chloroplasts that have an intact outer membrane and are therefore able to fix CO_2. Whole chloroplasts are osmotically shocked chloroplasts with an intact thylakoid system but a disrupted chloroplast membrane, and therefore are unable to fix CO_2. In broken chloroplasts the thylakoid system has been fragmented by various treatments, e.g., by detergents or sonication.

[6]D. I. Arnon, *Science* **122**, 9 (1955).

[7]A. Trebst, H. Y. Tsujimoto, and D. I. Arnon, *Nature (London)* **182**, 351 (1958).

ing the two photosystems, and one between photosystem I and NADP. The total reaction:

$$H_2O + NADP^+ \xrightarrow{h\nu} \frac{1}{2} O_2 + NADPH + H^+$$

may be written in an abbreviated form to indicate electron flow from water to NADP via two photosystems.

Several compounds have been identified as carriers in the electron transport system in spinach chloroplasts: ferredoxin, ferredoxin-NADP reductase, cytochrome c reducing substance (CRS), ferredoxin reducing substance (FRS), plastocyanin, cytochrome f, cytochrome b_3, cytochrome b_6, and plastoquinone. Other unidentified carriers, e.g., a quencher of photosystem II, a donor for photosystem II, and another carrier M between the two light reactions, have been implicated by spectroscopic methods or by results with mutants. Only the location of the carriers after photosystem I (ferredoxin, ferredoxin-NADP reductase, CRS, FRS) are reasonably well clarified, whereas the exact role and location of the carriers between the two light reactions are still uncertain. No information is available on the nature of the carriers before photosystem II. Because of these uncertainties no carriers are identified in the electron flow schemes used herein.

The electron transport system from water to an acceptor for photosystem I is coupled to ATP formation in a stoichiometric way. Electron flow is controlled by the phosphorylating system. The basal electron flow rate in the absence of the phosphorylation cofactors (ADP, P_i, Mg) is increased usually by 3- to 5-fold, by addition of these compounds or of an uncoupler. The measured stoichiometry of this coupled electron flow, or noncyclic photophosphorylation, is 1 ATP per atom of oxygen evolved or 2 electrons transferred. However Izawa and Good[8] have questioned whether the ATP ratio in the coupled electron flow system is really only 1 ATP:2 electrons transferred. By subtracting the uncoupled basal rate from the coupled rate, and relating the measured ATP values to the difference, a calculated ATP ratio of 2 ATP:2 electrons transferred is obtained. This calculated ratio has been observed

[8]S. Izawa and N. E. Good, *Biochim. Biophys. Acta* **162**, 380 (1968).

experimentally with various Hill reactions as well as in a photoreduction system where an artificial electron donor is oxidized.[9]

The physiological Hill reaction in whole chloroplasts, i.e., photosynthetic NADP reduction, has the largest cofactor requirement and utilizes the longest electron transport chain. Not necessarily is the complete electron transport chain participating when an unphysiological acceptor is reduced. Even two light reactions are not necessarily required, if the acceptor used has a positive redox potential. In theory, an electron transport chain could reduce an acceptor at any of several points along the chain, the point of reduction being determined by the redox potential and the accessibility of the acceptor.

It seems convenient to divide the different Hill reactions according to reductions of electron acceptors by photosystem I or by photosystem II (reaction type I or II vs III). Acceptors reduced by photosystem I may be further subdivided as dependent or independent of ferredoxin (reaction type I and II). Since all Hill reactions in whole chloroplasts are coupled to ATP formation and seem to be reduced after the quencher (Q) of photosystem II, the following scheme may be written in which A shall be an unidentified carrier in the chain interconnecting the two photosystems.

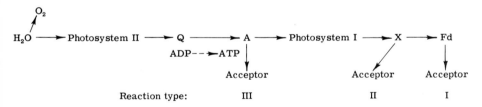

As before, a Hill reaction is defined as the reduction of an acceptor accompanied by oxygen evolution. Water is therefore the electron donor in a Hill reaction, and various electron acceptors may be reduced at different points along the interconnecting electron transport chain. Similarly, unphysiological electron donors may replace water and thereby donate electrons into the electron transport chain at various points. Under the latter conditions, an electron acceptor is reduced in the light at the expense of the artificial electron donor system. It is convenient again to divide these reactions with artificial electron donors, termed photoreductions, according to the participation of photosystem II or of photosystem I. For the latter case, they may be subdivided further into those coupled to ATP formation and those that are not.

[9]H. Böhme and A. Trebst, *Biochim. Biophys. Acta* **180,** 137 (1969).

$$O_2$$

$$H_2O \longrightarrow Photosystem\ II \longrightarrow Q \longrightarrow A \longrightarrow Photosystem\ I$$

ADP--►ATP

Acceptor

Donor   Donor   Donor

A summary of the various types of photosynthetic reactions in chloroplasts is given in Table I, and further discussion is provided later.

II. General Conditions

An introduction into the principal methods for measuring the Hill reaction was provided earlier in this series by Vishniac.[10] The manometric techniques for measuring oxygen evolution or uptake are summarized by Umbreit, Burris, and Stauffer in the volume entitled "Manometric Techniques" (Burgess Publ. Co., Minneapolis, Minnesota, 1959). The oxygen electrode and measurement of ATP formation are described in this volume (see [6] and [10]).

Hill reactions are usually measured in cell-free preparations of chloroplasts from higher plants or of a particulate fraction from blue-green or green algae. A Hill reaction in intact algal cells is possible only with lipophilic compounds, e.g., quinones, as acceptors, which will penetrate the outer cell wall.

Usually spinach chloroplasts are used, but some other plants also yield active chloroplast preparations. In general the chloroplasts are simply suspended in a buffer, and the acceptor is added. In whole chloroplasts, the rate of electron flow is controlled by the phosphorylating system; therefore either Mg, ADP, and P_i or an uncoupler, such as NH_4Cl, should be added for maximal activity. Low salt concentrations harm the chloroplasts. Addition of serum albumin protects the chloroplasts and therefore stimulates the rates.[11] The reaction may be run under aerobic conditions in some cases (ferricyanide, NADP) when the reduced form of the acceptor is not autoxidizable. If the acceptor is autoxidizable by oxygen, the reaction should be run in nitrogen or argon. In some cases, as will be discussed later, the reoxidation of the reduced acceptor by oxygen is beneficial and is used in the measurement; therefore, the reaction is run in air purposely.

Previously Tris·HCl buffer was generally used, but since the report of Good on the beneficial effects of other buffers, HEPES/NAOH buffer

[10]W. Vishniac, Vol. IV, p. 342.
[11]M. Friedländer and J. Neumann, *Plant Physiol.* 43, 1249 (1968).

or tricine buffer is employed.[12] The pH of the reaction mixture depends on the kind of Hill reaction tested. Due to the control of the rate of electron flow by the phosphorylating system in Hill reactions coupled to ATP formation, the pH optimum of the phosphorylating system determines the pH optimum of electron transport. In the presence of ADP + P_i + Mg, there is a rather sharp pH optimum around 8 in Tris, and near 8.3 in HEPES buffer. In the presence of an uncoupler like NH₄Cl there is a broad optimum for the Hill reaction between pH 7.2 and 8.3, and good electron transport activity is obtained even at lower pH. In sonicated chloroplasts, the pH optimum of the ferricyanide Hill reaction is even below pH 7. The pH optimum for electron donor systems for photosystem I is less sharp, and the reaction may be measured between pH 7.2 and 8.3. Also there is less strict dependence on conditions — e.g., on whether or not the phosphorylation cofactors or an uncoupler are present. Electron donor systems for photosystem II have the same properties as Hill reactions if the acceptor is reduced beyond photosystem I. However, if the acceptor is reduced by photosystem II, the pH optimum seems to be around 7.

Depending on the light intensity and the sensitivity of the assay, whole chloroplasts equivalent to about 5–10 μg of chlorophyll are required to measure a Hill reaction with a spectrophotometer or an oxygen electrode; whereas, 50–100 μg of chlorophyll are needed for manometric measurements because of the lesser sensitivity of this method. White light may be used. Monochromatic light is desirable, if a donor system for photosystem II is to be defined.

The cofactor requirement for the various Hill reactions discussed here are those for spinach chloroplasts. There is little indication, though, that cell-free preparations from algae or chloroplasts from other plants show a different requirement. However, in some algae certain of the endogenous redox carriers are more soluble than those of spinach. For example, the cytochrome f (cytochrome-552) of *Euglena* is solubilized during preparation of the chloroplasts.[13] Therefore the photoreduction of NADP by *Euglena* chloroplasts shows an absolute cytochrome f dependence.

The midpoint redox potential of the primary acceptor for photosystem I has been measured by Black,[14] by Zweig and Avron,[14a] and by

[12]N. E. Good, G. D. Winget, W. Winter, T. N. Connolly, S. Izawa, and R. M. Singh, *Biochemistry* **5**, 467 (1966).
[13]S. Katoh and A. San Pietro, *Arch. Biochem. Biophys.* **118**, 488 (1967).
[14]C. C. Black, *Biochim. Biophys. Acta* **120**, 332 (1966).
[14a]G. Zweig and M. Avron, *Biochem. Biophys. Res. Commun.* **19**, 393 (1965).

SIX DIFFERENT PRINCIPAL TYPES OF HILL REACTIONS AND PHOTOREDUCTIONS
CATALYZED BY CHLOROPLASTS UPON ILLUMINATION

Described in section	Participating portion of electron transport chain of chloroplasts (PS = photosystem)	Electron donor	Electron acceptor	Measured as	ATP formation
Reaction Type I					
III, A, 1, a	$H_2O \xrightarrow{O_2} PS\ II \longrightarrow PS\ I \rightarrow X \rightarrow Fd \rightarrow acceptor$	H_2O	NADP	Oxygen evolution or NADPH formation	Coupled stoichiometrically, controlled by phosphorylation system
III, A, 1, b		H_2O	Cytochrome c	Cytochrome c reduction	
Reaction Type II					
III, A, 2, a	$H_2O \xrightarrow{O_2} PS\ II \longrightarrow PS\ I \rightarrow X \rightarrow acceptor$	H_2O	Quinone	Oxygen evolution or oxygen uptake	Coupled stoichiometrically, controlled by phosphorylation system
III, A, 2, b		H_2O	Methylviologen	Oxygen evolution or oxygen uptake	
III, A, 2, d		H_2O	Methylred	Disappearance of methyl red	
Reaction Type III					
III, B, 1	$H_2O \xrightarrow{O_2} PS\ II \rightarrow acceptor$	H_2O	K ferricyanide	Oxygen evolution, Avron method	See discussion in text

Reaction Type IV					
IV, A, 2 and 3 and 4	Donor⟶PS I→X→Fd→acceptor	TMPD DAD or DCPIP	NADP NADP	NADPH formation NADPH formation	Not coupled, no control Coupled, but no control
IV, A, 3 and 4	Donor⟶PS I→X⟶acceptor	DAD or DCPIP	Quinone or viologen	Oxygen uptake	Coupled, but no control
Reaction Type V					
IV, B, 1	Donor→PS II⟶PS I→X→Fd→acceptor	Ascorbate or phenylene diamine	NADP	NADPH formation	Coupled, control not fully investigated
IV, B, 1	Donor→PS II⟶PS I→X→acceptor	Ascorbate	Quinone or methylviologen	Oxygen uptake	Coupled stoichiometrically, controlled by phosphorylation system
Reaction Type VI					
IV, B, 2	Donor→PS II→acceptor	Semi-carbazide or benzidine	DCPIP	DCPIPH$_2$ formation	Not investigated
		Diphenylcarbazide	K ferricyanide	Avron method	Not investigated

Kok *et al.*[15] to be around −550 to −650 mV. The midpoint of the acceptor of Photosystem II according to recent results of Cramer and Butler[16] is around −60 mV.

III. Hill Reactions with Water as Electron Donor

A. Reduction of an Acceptor by Photosystem I

$$H_2O + A \longrightarrow \frac{1}{2} O_2 + AH_2$$

$$H_2O \xrightarrow{\quad} \text{photosystem II} \xrightarrow{\quad} \text{photosystem I} \longrightarrow A$$

(with O_2 evolving from H_2O)

All Hill reactions may be followed by the oxygen evolution which accompanies the reduction of the acceptor, the oxygen being measured manometrically or with an oxygen electrode. In some cases described below, the specific change in spectral properties of the acceptor upon reduction may be measured.

1. Hill Reaction Depending on Ferredoxin: Reaction Type I

a. Reduction of NADP. The reaction is measured at the wavelength of maximum absorbance of reduced NADP, i.e., 340 nm. One may observe directly the absorbance at 340 nm, using a spectrophotometer with an attached illumination source, or measure the absorbance change after the illumination period. For the latter, the sample is quickly centrifuged at high speed (10,000 g), and an aliquot of the supernatant solution is used to measure the absorbance at 340 nm before or after the addition of 3 μg of PMS. PMS quickly destroys any NADPH and thereby permits an easy means to assess the background absorbance in each sample. Dilute HCl may also be used in the determination of NADPH when the protein content of the sample is low. Addition of 10^{-3} M benzylviologen and a small amount of the ferredoxin–NADP reductase will destroy specifically NADPH but not NADH.[17]

If the pH of the illuminated chloroplast preparation is below 7, NADPH is slowly decomposed. Therefore after illumination the sample is quickly brought up to a higher pH by the addition of KOH or, better, of tertiary sodium phosphate. After centrifugation, the sample is adjusted to pH 8.5, and its absorption (due to NADPH) at 340 nm is read.

[15]B. Kok, H. J. Rurainski, and O. V. H. Owens, *Biochim. Biophys. Acta* **109**, 347 (1965).
[16]W. A. Cramer and W. L. Butler, *Biochim. Biophys. Acta* **172**, 503 (1969).
[17]G. Ben Hayyim, Z. Gromet-Elhanan, and M. Avron, *Anal. Biochem.* **28**, 6 (1969).

The NADPH formed in NADP photoreduction can be oxidized by glutathione and glutathione reductase, which is specific for NADPH, and the Hill reaction is then followed by measuring the appearance of SH groups (see footnote 10).

The reduction of NADP by whole chloroplasts is dependent on ferredoxin and ferredoxin–NADP reductase; 5 μM ferredoxin is a saturating concentration. Whole chloroplasts, prepared in the usual manner in sucrose or NaCl, usually contain saturating amounts of the ferredoxin–NADP reductase. Fragmented (broken) chloroplasts are, however, depleted of the reductase. Only purified reductase should be added since a crude preparation may actually inhibit, particularly if the reaction is run under air, because of contaminating enzymes. Ferredoxin from higher plants may be replaced by phytoflavin, flavodoxin, or ferredoxin from bacteria.

b. Cytochrome c Reduction. The reduction of horse heart cytochrome c is measured spectrally at a wavelength of 550 nm. Because of the high absorbance of the oxidized cytochrome, less than 0.2 μmole (per 3 ml) is added to the chloroplast suspension. Cytochrome c reduction by whole chloroplasts requires the addition of ferredoxin; 5 μM ferredoxin is saturating. Ferredoxin–NADP reductase is not required.

Recently, Fujita and Myers isolated another compound which stimulates cytochrome c reduction by *Anabaena* particles.[18] This cytochrome c reducing substance (CRS) has also been found in spinach, and a role for it as a component of the primary acceptor for photosystem I has been proposed.[19] With spinach chloroplasts this spinach CRS substitutes for ferredoxin in the reduction of cytochrome c.

In broken chloroplasts (fragmented by detergents), reduced cytochrome c becomes an electron donor for photosystem I via plastocyanin. Because of possible cycling, therefore, special precautions are necessary to measure cytochrome c photoreduction in such preparations.

c. Reduction of Ferredoxin. Catalytic amounts of ferredoxin are required for photosynthetic NADP and cytochrome c reduction. When substrate amounts of spinach ferredoxin are added, it serves as the terminal acceptor in a Hill reaction. Its reduction is a one-electron process and may be followed spectrophotometrically at 420 nm. Cyclic electron flow with ferredoxin as cofactor or autoxidation of the reduced ferredoxin by oxygen may interfere.

---

[18]Y. Fujita and J. Myers, *Arch. Biochem. Biophys.* 119, 8 (1967).
[19]G. Regitz, R. Berzborn, and A. Trebst, *Planta* 91, 8 (1970).

2. Hill Reactions Independent of Ferredoxin: Reaction Type II

a. *Reduction of Quinones.* All benzo-, naphtho-, and anthraquinones are reduced by chloroplasts in the light. However, the hydroquinones formed may be electron donors for the electron transport chain of chloroplasts (see Section IV, B). Furthermore low redox potential hydroquinones are easily reoxidizable by oxygen. Therefore the type of measurement used for a Hill reaction with quinone depends on the redox potential of the quinone used. Only with high potential quinones (redox potentials more positive than +100 mV, at pH 7) can oxygen production be used as an assay method. The decrease in absorbance of the quinone around 400 nm (the maximum depends on the structure of the quinone) may be measured with a sensitive spectrophotometer, preferably a double-beam instrument. The reaction must then be performed under strictly anaerobic conditions. The easiest method for measuring the reduction of low potential quinones (below zero volt) is to run the reaction in air and to follow oxygen uptake when the hydroquinone formed is reoxidized by oxygen. This is described below.

b. *Reduction of Dipyridylium Salts or Viologens.* These compounds, e.g., benzylviologen, diquat, methylviologen or paraquat, triquat, have very low redox potentials (more negative than −400 mV) and are used in part commercially as herbicides.[14,15] Their biochemical usefulness comes from the fact that methylviologen, with a redox potential similar to ferredoxin replaces it in certain enzymatic reactions. In photosynthesis, dipyridylium salts have been used to measure the redox potential of the primary acceptor of photosystem I.[14-15] Because of their low redox potentials, photosynthetic reduction of dipyridylium salts seem to assure that the reduction occurs by photosystem I.

The radical formed on reduction of methylviologen may be measured at 396 nm[20] or at 604 nm under strictly anaerobic conditions. Since the radical is readily reoxidized by oxygen, the reduction of dipyridylium salts may be more easily measured as oxygen uptake (see below).

c. *Oxygen as Acceptor.* When the reduced form of an acceptor used in a Hill reaction is autoxidizable, oxygen is taken up if the reaction is run in air. Light-dependent oxygen uptake by chloroplasts is often called the Mehler reaction. Mehler had observed already in 1951 the stimulation by quinoid compounds of light-dependent oxygen uptake by chloroplasts, and subsequently it was followed by Walker and Hill.[21] The stoichiometry of the reaction and its analogy to the Hill reaction was

[20]G. Zweig and M. Avron, *Biochem. Biophys. Res. Commun.* **19**, 397 (1965).
[21]D. A. Walker and R. Hill, *Plant Physiol.* **34**, 240 (1959).

settled in 1961. The reaction sequence in a Mehler reaction is a light-dependent Hill reaction followed by a dark autoxidation of the reduced acceptor.

$$H_2O + A \xrightarrow{hv} AH_2 + \frac{1}{2} O_2$$

$$AH_2 + O_2 \longrightarrow A + H_2O_2$$

Sum: $$H_2O + \frac{1}{2} O_2 \xrightarrow{hv} H_2O_2$$

A need not be a two-electron acceptor; it can be a quinone or a dipyridylium salt, for example (see more examples in the paper of Hill and Walker[21]) reduced in the light to the radical level only.

The mediating acceptor should have a redox potential below zero in order to ensure reduction by photosystem I rather than photosystem II and to ensure a fast and not rate-limiting autoxidation of the reduced form. The stoichiometry of oxygen uptake is 0.5 mole of oxygen per two electrons transferred. In order to obtain the correct stoichiometry, either the endogenous catalase activity of chloroplasts has to be inhibited by the addition of 10^{-3} KCN or of NaN_3 or the H_2O_2 has to be used to form acetaldehyde by adding excess catalase and ethanol. If the endogenous catalase is not inhibited, oxygen evolution and oxygen uptake may cancel each other. This is often called "pseudocyclic electron transport." Since the reduction of the acceptor is coupled to ATP formation, the complete reaction sequence with an electron flow from water as electron donor to oxygen as electron acceptor is coupled to a stoichiometric ATP formation (= pseudocyclic photophosphorylation).

Oxygen uptake in well prepared spinach chloroplasts without the addition of any acceptor is low. The small oxygen uptake is probably due to the reoxidation of the reduced primary acceptor of photosystem I. Oxygen uptake, however, as already stressed, is highly stimulated (4- to 20-fold, depending on the structural state of the chloroplast) by the addition of a low potential quinone, a dipyridylium salt, an autoxidizable dye and also by the addition of excess solubilized primary acceptor.[19] The stimulation is a measurement of the reactivity of the compound provided. As stated previously, the oxygen uptake is a convenient and composite measure of the Hill reaction together with the acceptor activity of the compound provided. The cofactor is added in only catalytic amounts since it is continuously re-formed; e.g., 10^{-4} M anthraquinone or methylviologen support maximal oxygen uptake.

In cell-free preparations of blue-green algae the primary acceptor is either more accessible to oxygen or present in higher amounts than in spinach chloroplasts. Therefore oxygen uptake by illuminated preparations of blue-green algae is stimulated less by the addition of a quinone, i.e., the basal rate is already high.

The measurement of oxygen uptake, rather than oxygen evolution, catalyzed by a low potential compound, is actually easier to do with the oxygen electrode (and manometrically), and is now the common method to evaluate Hill activity of chloroplast preparations.

An additional measure of the reaction is the amount of H_2O_2 formed. The H_2O_2 accumulated when the endogenous catalase is inhibited is reacted with titanylsulfate, and the absorbance at 397 nm of the peroxide complex is measured.

d. *Reduction of Methyl Red.* Azocompounds seem to be reduced to the amines by whole chloroplasts.

The reduction of methyl red described by Ash *et al.*[22] is measured by the disappearance of absorbance at 440 nm. A concentration of 20 mM methyl red was used in the reaction mixture. The disadvantage of this otherwise very convenient method is that the amine formed in the case of methyl red is an electron donor, as discussed later. Cycling therefore should be expected. Also methyl red can be reduced by chlorophyll solutions in a Krasnovsky reaction, and furthermore there is a substantial dark reduction.[22]

e. *Reduction of Tetrazolium Salts.* Ash *et al.*[22] described the photosynthetic reduction of tetrazolium blue by chloroplasts. The reaction was followed at 580 nm. A concentration of 10 mM tetrazolium blue in the reaction mixture was used.

f. *Reduction of Dithiodinitrobenzoic Acid (DTNB).* The reduction of DTNB can be followed by measuring the appearance of SH groups. It appears, though, from the cofactor requirements that illuminated chlorophyll alone is able to reduce DTNB so that its use is limited.[23]

g. *Reduction of K Ferricyanide.* As will be discussed subsequently, K ferricyanide may accept electrons from the electron transport chain at various sites depending on the structural state of the chloroplasts. Although it has been proved that K ferricyanide is an acceptor for photosystem II in sonicated and detergent-treated chloroplasts, it is the opinion of the author that K ferricyanide accepts electrons from photosystem I in whole chloroplasts, unless electron flow through photosystem I

[22]O. K. Ash, W. S. Zaugg, and L. P. Vernon, *Acta Chem. Scand.* 15, 1629 (1961).
[23]L. P. Vernon, personal communication, 1969.

is prevented. The measurement of this Hill reaction is described in detail below.

B. Photochemical Reduction of Electron Acceptors for Photosystem II at the Expense of Water as Electron Donor: Reaction Type III

The sum of this reaction is the same as the Hill reaction described above.

$$H_2O + A \xrightarrow{h\nu} \frac{1}{2} O_2 + AH_2$$

However only one light reaction participates. The electron flow is

$$H_2O \xrightarrow{} \overset{\displaystyle O_2}{\nearrow} \text{photosystem II} \longrightarrow A$$

The overall stoichiometry of the reaction therefore is the same as for Hill reactions in which the acceptor is reduced by photosystem I. Unfortunately, neither the properties of the electron transport system as presently known nor the inhibitors presently at hand permit an easy distinction between a reduction by photosystem II or by photosystem I.* It is possible to evaluate this question by measuring the possible enhancement effect by two wavelengths of light (see [24]). However, at present there is little consensus that even NADP reduction shows an enhancement effect. In the Hill reaction with ferricyanide, some authors report an enhancement effect in whole chloroplasts, but others do not. However, there is no doubt that in fragmented chloroplasts, particularly sonicated chloroplasts, a photoreduction of certain acceptors is possible by photosystem II.[24] This may be concluded from the plastocyanin requirement for photoreductions by photosystem I; in contrast, reductions by photosystem II are plastocyanin independent. Also mutants lacking photosystem I are still able to reduce ferricyanide, indicating that only photosystem II is needed in this Hill reaction.

1. Reduction of K Ferricyanide

The Hill reaction with ferricyanide as electron acceptor seems to be most popular. However, as stressed above, in whole chloroplasts the complete electron transport system is not necessarily involved, and the

---

*Note added in proof: Trebst et al.[42] and Böhme et al.[43] recently described a new inhibitor, dibromothymoquinone, a plastoquinone antagonist which inhibits only photoreductions by photosystem I, not those by photosystem II.

[24]S. Katoh and A. San Pietro, in "The Biochemistry of Copper" (J. Peisach, P. Aisen, and W. E. Blumberg, eds.), p. 407. Academic Press, New York, 1966.
[42]A. Trebst, E. Harth, and W. Draber, Z. Naturforsch. 25b, 1157 (1970).
[43]H. Böhme, S. Reimer, and A. Trebst, Z. Naturforsch. 26b, 341 (1971).

exact point in the interconnecting electron transport chain where ferri-cyanide accepts electrons is not yet established. Besides measuring the oxygen evolved in the reaction, the disappearance of ferricyanide absorbance at 420 nm may be used. Since the molar absorption coefficient is only about 1000, this method is rather insensitive. A very sensitive method has been developed by Avron and Shavit[25] based on a suggestion by Krogmann and Jagendorf. The ferrocyanide formed in the Hill reaction is allowed to react with a ferric salt to form the ferrous salt which complexes with 4,7-diphenyl-1,10-phenanthroline sulfonate. The complex is measured at 535 nm and has a molar absorbance coefficient of 20,500.

2. Reduction of Indophenol

The reduction of indophenol dyes like dichlorophenolindophenol (DCPIP) or trichlorophenolindophenol (TCPIP) may be followed by measuring the disappearance of the blue dye at 600 nm. This is one of the easiest and most sensitive methods to measure a Hill reaction. Reduced DCPIP is an electron donor for photosystem I, as described in reaction type IV.

3. Reduction of Quinones

As already discussed, quinones with a redox potential more positive than 100 mV are possibly reduced by photosystem II. Kok has measured the redox potential of photosystem II in this way.[26] The disappearance of absorption may be followed at about 400 nm. Since the autoxidation of the hydroquinones formed is not sufficiently fast, the Mehler-type reaction (discussed earlier) recommended for lower potential quinones accepting electrons from photosystem I is not applicable.

IV. Photoreduction of Electron Acceptors at the Expense of Artificial Electron donors

$$DH_2 + A \xrightarrow{h\nu} D + AH_2$$

A. Electron Donors for Photosystem I (and Acceptors for Photosystem I): Reaction Type IV

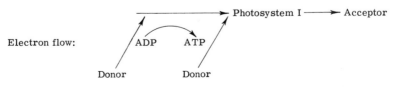

[25]M. Avron and N. Shavit, *Anal. Biochem.* **6,** 549 (1963).
[26]B. Kok and E. A. Datko, *Plant Phsyciol.* **40,** 1171 (1965).

A principal property of such a system in which only photosystem I is involved is that it is DCMU insensitive and proceeds in light of wavelengths greater than 700 nm. Electron flow from the donor to the acceptor may or may not be coupled to ATP formation, indicating at least two possible points of entry of electrons into the chain before photosystem I (see, however, discussion under DAD). The first artificial electron donor system was described by Vernon and Zaugg,[27] who showed that the addition of DCPIP + ascorbate would overcome the inhibition of photosynthetic NADP reduction by DCMU.

All photoreductions of reaction type IV are measured as the reduction of the acceptor. As acceptors for photosystem I, most of the compounds described in Section III (photoreduction at the expense of water) may be used, i.e., NADP, low potential quinones, viologens, oxygen, and methyl red. Cytochrome c has a redox potential too close to that of the electron donors and therefore cannot be used as acceptor. As a matter of fact reduced cytochrome c may be used as electron donor (see below). Light-induced oxygen uptake via a low-potential quinone (e.g., anthraquinone sulfonate, which has a redox potential of -250 mV, or methylviologen) at the expense of reduced DCPIP or phenylenediamines is the easiest check on photochemical activity of photosystem I. To assure, however, that the natural photosystem I is indeed participating, the measurement of NADP reduction via ferredoxin and ferredoxin-NADP reductase with an artificial donor system is recommended rather than oxygen uptake when highly fragmented chloroplasts are used.

The donors for photosystem I are described below. Usually only catalytic amounts of the donors (10^{-4} M) are added to the reaction mixture and kept reduced by excess ascorbate (10^{-3} to 10^{-2} M).

1. p-Phenylenediamine as Electron Donor

At a concentration of 10^{-4} M and above, the unsubstituted and N-substituted p-phenylenediamines are electron donors for photosystem I.[28] The reaction is insensitive to DCMU. At lower concentrations (10^{-5} M), phenylenediamines are electron donors for photosystem II, and the reaction is then DCMU sensitive (see Section IV, B). Electron flow from phenylenediamine via photosystem I to an acceptor is not coupled to ATP formation at concentrations below 10^{-2} M.

2. TMPD as Electron Donor

N,N,N',N'-Tetramethyl-p-phenylenediamine (TMPD) is the best electron donor among the N-substituted phenylenediamines.[28] Its properties are the same as those described under p-phenylenediamine.

[27]L. P. Vernon and W. S. Zaugg, *J. Biol. Chem.* **235,** 2728 (1960).
[28]A. Trebst, *Z. Naturforsch. B* **19,** 418 (1964).

3. DAD as Electron Donor

Diaminodurene, or C_2,C_3,C_5,C_6-tetramethyl-p-phenylenediamine (DAD) is, so far, the best electron donor for photosystem I in whole chloroplasts.[29] It reverses the DCMU inhibition of NADP reduction to more than 95% at a concentration of 10^{-4} M. Electron flow is coupled to ATP formation which would suggest a point of entry of electrons into the interconnecting electron transport chain before the coupling site. Since, however, electron flow is only slightly stimulated by ADP + P_i and by an uncoupler,[30] it is possible that the ATP is formed in a superimposed cyclic electron flow. The redox potential of DAD is 220 mV.

4. DCPIP, TCPIP as Electron Donor

Reduced DCPIP or TCPIP were the first artificial electron donors used[27] at a concentration of 10^{-4} M. Electron flow is coupled to ATP formation.[31] A stimulation of electron flow by uncouplers and by phosphorylation cofactors, though small, has been reported.[32] However, the stoichiometry of electron flow to ATP formation does not reach one, and the same question of whether or not ATP is formed in a superimposed cyclic flow (as discussed above under DAD) applies.[33]

5. Cytochrome c as Electron Donor

Reduced cytochrome c is an electron donor for photosystem I only in fragmented chloroplasts, but not in whole chloroplasts.

6. Phenazine Methosulfate (PMS) as Electron Donor

PMS is a cofactor for cyclic electron flow and therefore an electron donor for photosystem I, as all cofactors of cyclic photophosphorylation must be. PMS probably interacts directly with photosystem I.[41] Since PMS catalyzes cyclic electron flow even in the presence of excess ascorbate and since it chemically oxidizes NADPH, its use for measuring photosystem I activity in a noncyclic electron flow system is limited.

7. Ascorbate as Electron Donor

As mentioned already, the electron donors for photosystem I described so far are used only in catalytic amounts and are kept reduced

[29] A. Trebst and E. Pistorius, *Z. Naturforsch. B* **20**, 143 (1965).
[30] A. Trebst and E. Pistorius, *Biochim. Biophys. Acta* **131**, 580 (1967).
[31] M. Losada, F. R. Whatley, and D. I. Arnon, *Nature (London)* **190**, 606 (1961).
[32] D. L. Keister, *J. Biol. Chem.* **240**, 2673 (1965).
[33] Z. Gromet- Elhanan, *Arch. Biochem. Biophys.* **123**, 447 (1968).
[41] B. Rumberg and H. T. Witt, *Z. Naturforsch.* **19b**, 693 (1964).

by substrate amounts of ascorbate. This implies that ascorbate itself is not an electron donor for photosystem I. This is true for whole chloroplasts. In fragmented chloroplasts, notably in sonicated and detergent-fragmented chloroplasts, ascorbate does become an electron donor for photosystem I.[34] This is indicated by the observation that in such chloroplasts electron flow is only slightly stimulated by the further addition of DCPIP or DAD to ascorbate. Ascorbate, however, is an electron donor for photosystem II as described in Section IV, B,1,a.

Plastocyanine Dependence of Electron Donors for Photosystem I

p-Phenylenediamines, DCPIP, TMPD, DAD, ascorbate, and cytochrome *c* donor systems have been shown to be plastocyanin (or cytochrome *f*) dependent in detergent or sonicated chloroplasts.[34] DCPIP at concentrations above 10^{-3} *M* bypasses plastocyanin to a certain extent; i.e., the reaction proceeds in the absence of plastocyanin. The plastocyanin requirement for the donor systems in fragmented chloroplasts should not be taken as evidence that the point of entry of electron donors into the electron transport chain in whole chloroplasts is at the plastocyanin functional site.

B. Electron Donors for Photosystem II

1. Using an Electron Acceptor of Photosystem I: Reaction Type V

Donor → photosystem II → photosystem I → acceptor

In contrast to electron donor systems for photosystem I (reaction type IV), electron donor systems for photosystem II are DCMU sensitive and require light of wavelengths below 700 nm. They show[35] a red drop phenomenon and increased fluorescence of photosystem II. All the electron acceptors described in Section III, A may be used for photosystem I, i.e., NADP via ferredoxin, low potential quinones, viologens, and oxygen. The reaction is followed most conveniently by measuring NADPH formation or oxygen uptake.

Numerous compounds have now been described as electron donors for photosystem II,[36] indicating that photosystem II, which *in vivo* oxidizes water to oxygen, can nonspecifically oxidize many nonphysiological donors. The site at which the nonphysiological electron donors described below interact with the electron chain before photosystem II is not yet established. It could be that photosystem II itself oxidizes the

---

[34]E. Elstner, E. Pistorius, P. Böger, and A. Trebst, *Planta* 79, 146 (1968).
[35]T. Yamashita and W. L. Butler, *Plant Physiol.* 43, 1978 (1968).
[36]T. Yamashita and W. L. Butler, *Plant Physiol.* 44, 435 (1969).

added donor or that a precursor of oxygen evolution is the oxidizing component.

a. Ascorbate as Electron Donor. Ascorbate (10^{-3} to $10^{-2} M$) in the absence of a dye is a convenient electron donor for photosystem II.[9] The DCMU-sensitive ascorbate photooxidation, mediated by quinone (oxygen via a quinone as electron acceptor for photosystem I) gave the first indication for electron donor systems for photosystem II.[37]

b. o-Hydroquinones and p-Hydroquinones as Electron Donors. o-Hydroquinones as well as p-hydroquinones in concentrations between 10^{-3} and $10^{-2} M$ are oxidized by photosystem II. p-Hydroquinone has been shown by Yamashita and Butler[35] to restore fluorescence for photosystem II and NADP reduction in Tris-treated chloroplasts. The photooxidation of o-hydroquinones, like DOPA, by oxygen has been shown by Trebst *et al.*[37] to require photosystems II and I.

c. p-Phenylenediamines. p-Phenylenediamines are electron donors for photosystem II at low ($10^{-5} M$) concentrations.[35,36] The reaction is then inhibited by DCMU. As discussed in Section IV, A, phenylenediamines are also electron donors for photosystem I in a DCMU-insensitive reaction.

d. Cysteine as Electron Donor. Cysteine has been shown to be an electron donor for photosystem II by Katoh and San Pietro.[38] However, even at high concentrations ($10^{-2} M$), cysteine is not a very good donor.

e. Benzidine as Electron Donor. Benzidine has been shown by Yamashita and Butler[36] to be a good electron donor for photosystem II. Its use is particularly valuable in the measurement of photosystem II in fragmented chloroplasts, as discussed in Section IV, B, 2.

f. Phenylhydrazines as Electron Donors. Substituted phenylhydrazines and semicarbazides have recently been introduced by Yamashita and Butler[36] together with a large number of additional compounds, and shown to be oxidized by photosystem II. Their particular usefulness is discussed in the next section.

Some of the electron donor systems for photosystem II have been shown to be coupled to ATP formation.[9,39] Ascorbate photooxidation (i.e., ascorbate as donor and oxygen as acceptor via a quinone) is stimulated by the phosphorylating system and by an uncoupler as is the normal Hill reaction.[9]

[37]A. Trebst, H. Eck, and S. Wagner, *in* "Photosynthetic Mechanisms in Green Plants" *Nat. Acad. Sci.—Nat. Res. Counc. Publ.* 1145, (1963).
[38]S. Katoh and A. San Pietro, *Arch. Biochem. Biophys.* 122, 144 (1967).
[39]T. Yamashita and W. L. Butler, *Plant Physiol.* 43, 1978 (1968).

2. Using Electron Acceptor of Photosystem II: Reaction Type VI

Donor $\longrightarrow$ photosystem II $\longrightarrow$ acceptor

This is the simplest reaction sequence for measuring the activity of photosystem II. It is comparable to the measurement of the activity of photosystem I described as reaction type IV.

When photosystem I participates in photosynthetic electron flow system there is a large redox potential difference between the electron donor (whether an electron donor for photosystem I, reaction type IV, or for photosystem II, reaction type V) and the electron acceptor. Therefore the dark rate is negligible. If, however, only photosystem II participates, i.e., reaction type VI, the redox potential of donor and acceptor are in a comparable range. Therefore a number of combinations of electron acceptors (Section III, B) and electron donors (Section IV, B, 1) are not possible because of fast chemical dark reactions. For example, ferricyanide and positive potential quinones are immediately reduced by ascorbate or cysteine in the dark. For an easy check on the activity of photosystem II, therefore, the donor and the acceptor must not react with each other chemically.

Yamashita and Butler measured the activity of photosystem II by the reduction of ferricyanide (electron acceptor) with benzidine or semicarbazide as electron donors.[36] Under the conditions employed, ferricyanide did not oxidize these compounds in the dark. Vernon and Shaw[40] recommended DCPIP as electron acceptor and 1,5-diphenylcarbazide as electron donor for a more satisfactory measurement of the activity of photosystem II; DCPIP reduction is measured at 590 nm.

[40]L. P. Vernon and E. R. Shaw, *Biochem. Biophys. Res. Commun.* **36**, 878 (1969).

[15] Simultaneous Measurement of Photosynthesis and Respiration Using Isotopes

By G. E. HOCH

Concomitant measurement of photosynthesis (oxygen evolution, carbon dioxide fixation) and respiration (oxygen consumption, carbon dioxide evolution) may be accomplished utilizing isotopic oxygen and carbon dioxide. The technique was first used by Weigel *et al.*,[1] who em-

[1]J. W. Weigel, P. M. Warrington, and M. Calvin, *J. Amer. Chem. Soc.* **73**, 5058 (1951).

ployed ^{14}C-labeled CO_2 to assay respiratory carbon dioxide evolution during photosynthesis of algal cells. Because satisfactory radioactive isotopes of oxygen are not available, mass spectrometric analysis using ^{18}O (or ^{17}O) is required for the oxygen determinations. This technique was introduced by Brown and co-workers[2] and has the advantage of being applicable to both oxygen and carbon dioxide (using stable ^{13}C).

Theory

The technique relies upon different isotopic distributions between the sources of uptake and evolution. For example, the oxygen evolved from cells illuminated in normal water will have an isotope concentration equal to that of the water providing no isotopic discrimination occurs (discrimination generally appears as a faster rate of reaction of the lighter isotope). Since normal water is $\sim 99.8\%$ ^{16}O and 0.2% ^{18}O (^{17}O will be neglected here), the oxygen evolved will have this distribution of isotopes. Since oxygen is a diatomic gas, three species will be evolved $^{16}O = {}^{16}O$ (mass 32), $^{16}O = {}^{18}O$ (mass 34), and $^{18}O = {}^{18}O$ (mass 36) and they will appear in a ratio governed by statistical laws

$$32:34:36 = (0.998)^2:2(0.998 \times 0.002):(0.002)^2$$

Mass 36 may be neglected as insignificant, and the total rate of oxygen production (^{T}P) will be

$$^{T}P = {}^{32}P + {}^{34}P$$

since

$$^{34}P = 0.004 \ ^{32}P$$

$$^{T}P = 1.004 \ ^{32}P$$

in the absence of respiration.

The above equation will be valid if isotope effects do not occur and if the pool from which oxygen is evolved does not change isotopic content. Although isotope effects do occur,[3] their magnitude is too small to be of significance in these experiments. If oxygen is evolved from water or from some compound in isotopic equilibrium with water, the pool should remain unchanged during the course of the experiment. This is because the high concentration of liquid water ($55 M$) outside the cell, and presumably not drastically different inside the cell, can be considered an "infinite reservoir." The assumption of unchanged pools is much less enticing in the case of carbon dioxide, for which no comparable reservoir seems to be available.

[2]A. H. Brown, A. O. Nier, and R. W. Van Norman, *Plant Physiol.* **27**, 320 (1952).
[3]P. H. Abelson and T. C. Hoering, *Proc. Nat. Acad. Sci. U.S.* **47**, 623 (1961).

The rates of uptake (U) of the isotopic species (again assuming no isotope effect) will be in proportion to their concentrations. Hence

$$^{32}U:^{34}U:^{36}U = [^{32}O_2] : [^{34}O_2] : [^{36}O_2]$$

which leads to (ignoring mass 36)

$$^{T}U = {}^{34}U\left(1 + \frac{[^{32}O_2]}{[^{34}O_2]}\right)$$

and

$$^{32}U = {}^{34}U \times \frac{[^{32}O_2]}{[^{34}O_2]}$$

Finally

$$^{T}P = ({}^{32}net - {}^{32}U)\ 1.004$$

$$= \left({}^{32}net - {}^{34}U\ \frac{[^{32}O_2]}{[^{34}O_2]}\right)\ 1.004$$

Similar calculations may be made for carbon dioxide (with due consideration to the carbonate–bicarbonate equilibrium).

Methods

Brown and co-workers[4] have employed a conventional Warburg flask with the mass spectrometer leak built into the plug (Fig. 1). The plug was ground into the neck of the Warburg flask and connected to the mass spectrometer analyzer tube by a length of copper tubing containing a coil. This allowed the flask to move and permitted normal respirometer type agitation. The atmosphere was thus sampled above the cell suspension. This method has the advantages and disadvantages of conventional Warburg manometry, about which much has been written. Primarily, errors may arise because of the gas–liquid interface, which presents a diffusion barrier. As a result, appreciable concentration gradients may develop and cause the isotopic ratio at the cells to be different from that measured in the atmosphere. This has been the subject of intensive investigation by Good and Brown.[5]

Direct sampling from the liquid phase is possible if a semipermeable membrane is used. The permeability of Teflon, cellophane, rubber, etc., is much greater for oxygen and carbon dioxide than for water. The semipermeable properties of these membranes has been used by Hoch and Kok[6] to measure isotopic composition of oxygen dissolved in cell

[4]J. A. Johnston, and A. H. Brown, *Plant Physiol.* 29, 177 (1954).
[5]N. E. Good and A. H. Brown, *Biochim. Biophys. Acta* 50, 544 (1961).
[6]G. E. Hoch and B. Kok, *Arch. Biochem. Biophys.* 101, 160 (1963).

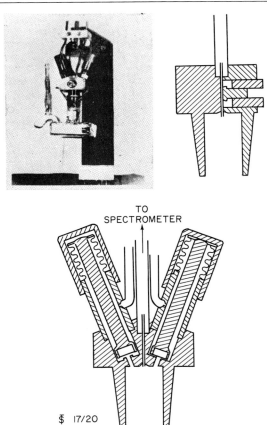

TO
SPECTROMETER

$ 17/20

FIG. 1.   Mass spectrometer leak assembly for use with manometer vessels [J. A. Johnston and A. H. Brown, *Plant Physiol.* **29**, 177 (1954)]. The lower diagram shows solenoid-operated valves for addition of gas mixture and male plug connection. The variable leak is shown in upper right.

suspensions, thus avoiding a gas–liquid interface and the attendant problems. The sample cell and inlet system is shown in Fig. 2. It is similar to Clark-type oxygen electrodes with the mass spectrometer replacing the polarograph system. The time required for the instrument to respond to a concentration change depends on the membrane thickness. Great sensitivity can be obtained in this arrangement because of the absence of a gas phase. On the other hand, the method is not suitable for long-term observations because of the dilution of the dissolved isotopic oxygen with oxygen evolved during photosynthesis. The limiting factor in this arrangement is stirring. If fast response and sensitive measurements are desired, the stirring must be very fast and constant.

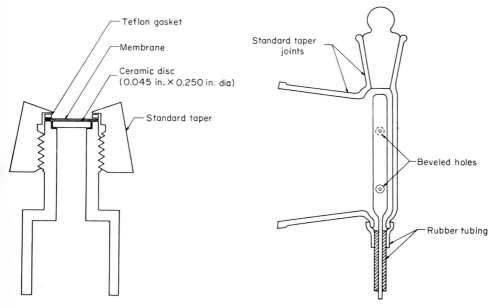

FIG. 2.   Liquid sampling system from G. E. Hoch and B. Kok [*Arch. Biochem. Biophys.* **101**, 160 (1963)]. The left-hand unit attaches to the mass spectrometer ion source. Mechanical support for the membrane is provided by the porous ceramic (alundum) disk. The vessel on the right is stirred by the paddle, which is connected externally to a vibrator.

The isotopic technique has been used with leaves by Ozbon *et al.*[7] The apparatus consisted of a closed system containing about 1 liter of gas, which was circulated over a leaf (Fig. 3) with continuous monitoring of

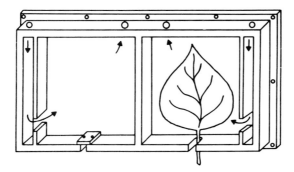

FIG. 3.   Leaf chambers used by R. J. Volk and W. A. Jackson [*Crop. Sci.* **4**, 45 (1964)] for their recirculating system.

[7] J. L. Ozbon, R. J. Volk, and W. A. Jackson, *Plant Physiol.* **39**, 523 (1964).

the atmosphere's content.[8] This convenient technique suffers from the rather unknown diffusional resistance of the leaf.

Instruments

All mass spectrometers depend upon ionization of a gaseous sample and separation of ionized species on the basis of their mass-to-charge ratio. Ionization is usually effected by an electron beam of about 70 V energy emitted thermionically. This suffices to cause predominantly single ionizations, and hence the mass-to-charge ratio is simply the mass. In magnetic deflection instruments, the ions are accelerated and collimated by a series of slotted plates held at a negative potential. Unfortunately, the ionization and collimation process is very inefficient. The ions then enter a magnetic field, where the various masses are dispersed. By adjustment of either the magnetic field or accelerating potential, the mass of interest may be made to fall on an exit slit. Detection may be made either by an ion multiplier or by measuring the current generated when the ions impinge on a plate connected to ground. The detector current is a linear function of the species concentration over many orders of magnitude. Other instruments use ion resonance methods to disperse the masses instead of a magnetic field. By a proper combination of dc and radiofrequency voltages, an ion may be made to traverse a spiral path along an analyzer tube having the right geometry and impinge upon a detector, other ions of different mass (not in resonance) being grounded on the tube walls.

In all cases the measured currents are low, and stability of all elements is crucial for best results. Little has been done to employ the noise-reducing procedures used in optical spectroscopy or to improve efficiency of the instruments. Hoch and Kok[6] reported a signal-to-noise ratio of 500 when measuring oxygen from air dissolved in water. Doubtless this could be improved by at least an order of magnitude.

Difficulties with the instruments arise from the necessity of maintaining a low pressure within the analyzer tube (so that the mean free path of the ions is long enough to minimize collisions) and the usual ones associated with dc measurements.

[8] R. J. Volk and W. A. Jackson, *Crop Sci.* 4, 45 (1964).

[16] Measurement of Size Distribution with the Coulter Counter

By KAZUO SHIBATA

The instrument, the Coulter counter,[1] measures the number of particles above a preset particle volume which are present in a suspension. One can, therefore, determine the particle-size distribution from the counts at different preset threshold volumes as well as the total particle number per unit volume of the suspension. With this instrument, the errors in counting the number of particles can be greatly reduced, and the size distribution can be measured much more accurately and quickly than by light or electron microscopy. The instrument, which was originally designed for routine counting of red blood cells,[1-4] have recently been applied to other cells or subcellular particles as well as to particles produced in industry. In this section the technique for use of the instrument and its applications to isolated chloroplasts and photosynthetic microorganisms are described. Illustrations are made mostly on the technique of using Model A, a simpler model for manual operation. The more advanced instrument, Model B, with a recorder or pulse-height analyzer can be used with minor modification of the technique.

Principle and Operation

The instrument is diagrammatically illustrated in Fig. 1. One electrode (E_2) is placed in a beaker (B, 50–250 ml) containing a sample suspension (F); into the beaker is placed a vertical glass tube (D), sealed from the suspension except at the point where an orifice or aperture (A) is located. The second electrode (E_1) is contained in this tube. The instrument is based upon the principle that each particle or cell in the suspension, when passing through the aperture, causes a momentary increased impedance (or pulse) to the steady flow of electric current through the aperture between the two electrodes. The aperture limits the volume of the medium (electrolyte) in which the change of impedance takes place.

To operate the instrument, stopcock C_1, which is connected to a bottle under a constant reduced pressure, is first opened so that the level of

[1]W. H. Coulter, *Proc. Nat. Electron. Conf.* **12**, 1034 (1956).
[2]G. Brecher, M. Shneiderman, and G. J. Williams, *Amer. J. Clin. Pathol.* **26**, 1439 (1956).
[3]C. F. T. Mattern, F. S. Brackett, and B. J. Olson, *J. Appl. Physiol.* **10**, 56 (1957).
[4]J. L. Grant, M. C. Britton, and T. E. Kurtz, *Amer. J. Clin. Pathol.* **33**, 138 (1960).

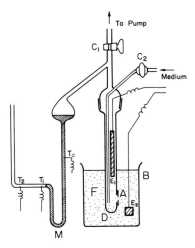

FIG. 1. Simplified drawing of the Coulter counter, showing aperture (A), sample suspension (F) in a beaker (B), stopcocks (C_1 and C_2), aperture tube (D), electrodes (E_1 and E_2) for pulses, electrodes (T_0, T_1, and T_2) for switching, and manometer (M).

mercury in the right arm of manometer M is raised and balanced at a higher level. When the stopcock is closed, the mercury falls back, drawing the particle suspension into the tube through the aperture. When particles pass through the aperture, they cause electrical pulses, and the number of pulses during the passage of mercury from switch T_1 to T_2 (0.5 ml in volume between these electrodes for switching) is counted automatically. Stopcock C_2, which is for flushing or filling up the tube with the medium, is kept closed during this procedure.

The wall of the aperture plate cemented on tube D is thicker than the aperture diameter, so that the aperture is, in fact, a cylindrical tunnel. Consequently, the current between the two electrodes depends entirely on the resistance of the fluid in the tunnel. In the absence of particles, this resistance is constant, but, when a particle enters the tunnel, the current between the electrodes drops. The magnitude of this drop is proportional to the volume of the particle, provided that the particle diameter is less than one-third the diameter of the aperture; the measurement of particle volumes below this size is independent of their shape. It is fairly simple to convert the current pulses into voltage pulses and to count the pulses above any given voltage. The voltage of particle size over which counts are made can be limited by means of a threshold control composed of electrical filters. An oscilloscope is provided so that a rough visual check can be made of the particle size distribution, and the threshold level is displayed on the oscilloscope as a brightness modulation; the pulses above the threshold level are brighter than those below it.

A microscope is mounted on the side of the sample beaker to examine the condition of the aperture. Caution has to be exercised to avoid the use of a medium or a buffer containing a large number of dust particles, which would block the aperture. Chemicals or salts for the preparation of buffers or electrolytes often contain such foreign dust particles. Centrifugation, distillation, or filtration of the medium is helpful in removing such large particles as well as small particles which would raise the background count. A large particle blocking the aperture can be removed in the following manner. The inside of tube D is first flushed with the medium by opening both stopcocks, C_1 and C_2, and sudden stoppage of the medium flow by turning off stopcock C_1 builds a shock of positive pressure inside the tube, which will push out the blocking particle. Another way is to open stopcock C_2 quickly while stopcock C_1 is closed, this will release the negative pressure suddenly. Bubbling of air in the tube through the aperture by lowering the container of suspension is helpful to clean the aperture after removal of the blocking particle.

The main variables to be adjusted are the particle concentration, the conductivity of medium (or diluent), the aperture diameter and the degree of amplification of pulses controlled by a sensitivity switch. The sample suspension is diluted appropriately with a medium having a conductivity to provide an optimum electric current or resistivity through the aperture. The best signal voltage is obtained at higher resistivities, so that a relatively weak electrolyte is desirable. However, excessive resistivity raises the noise level. Resistivities of 100–2000 ohm-cm are usually chosen; 1% saline at 20° is about 50 ohm-cm. The use of buffers is recommended because the pH value of the medium greatly affects the resistivity. For example, 0.04 M phosphate buffer at pH 7.0 is suitable as a medium for chloroplasts, and 0.9% NaCl solution is usable for counting blood cells.

The amplification can be varied stepwise from S (sensitivity) = 1 to S = 8 (or 10), and the degree of amplification changes by a factor of approximately 2 on each step ($\Delta S = 1$) of switching. Therefore, the count at $S = 2$ and V_t (the threshold value) = 20, for example, is approximately equal to the count at $S = 3$ and $V_t = 40$. Replaceable tubes with an aperture of 30 μ to 2000 μ in diameter are available, and the diameter determines the counting period as well as the particle size to be measured. Particles in the range between 0.5 μ and 1000 μ in diameter can be measured by appropriate choice of these variables.

Most of the distribution measurements are made on an arbitrary relative scale (expressed by V_r) of volume or in the scale (expressed by V_t) on the threshold dial under certain conditions of the above variables. These relative scales of volume can be converted to an absolute scale (V) of volume by use of data on particles with a known diameter measured under the same conditions. Plastic spheres and pollens, which are listed in Table I, are available for this purpose. A suspension of pollen grains is prepared by putting the grains in a medium with a drop of a detergent solution, followed by shaking at intervals for 1 hour and standing overnight. The addition of detergent is helpful to minimize the formation of doublets or higher aggregates of grains.

TABLE I

PLASTIC SPHERES AND POLLENS FOR STANDARDIZATION

Material	Approximate size (μ)	Aperture to be calibrated (μ)
Plastic spheres (Dow Chemical Co.)	3.49	30 and 50
Puffball spores	4.8	30 and 50
Ragweed pollen	19–20	100–280
Pecan pollen	45–50	280–560
Sweet vernal grass pollen	45–50	280–560
Corn pollen	85–90	400–1000
Crab eggs	400	2000

Measurements and Corrections

Total Cell Number

The total number, N_s, of cells in a suspension can be counted at an appropriate setting of sensitivity, S, and threshold volume, V_t. The value of V_t suitable for this purpose is indicated by vertical arrows in Fig. 2 where the upper curve shows the count, N (the count after the correction described below) or N' (the count before the correction) vs V_t and the lower curve shows -ΔN (the decrease of N for a constant increment of V_t) versus V_t which is proportional to $-dN/dV_t$. The background count, N_b, of small dust or contaminating cells can be thus eliminated.

The direct count, N', of cell number, however, includes a systematic error due to the coincidence counting loss, which is caused by the phenomenon that more than one particle pass through the aperture at the same time when particle concentration is high. The count, N', must

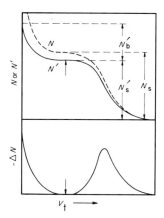

FIG. 2.   The N' (or N) vs V_t curve related to the $-\Delta N$ vs V_t curve.

therefore be corrected for this effect. The correction is made according to the following equation;

$$N = N' + \alpha(N')^2 \tag{1}$$

where α is a function of the aperture diameter and the manometer volume and must be determined for each instrument with the aperture and the manometer to be used.

Determination of α can be made in the following manner. Prepare a concentrated cell suspension, and dilute it to obtain a series of suspensions of known relative concentrations, N_r;

$$N = \beta N_r \tag{2}$$

where β is the proportionality constant for the series of suspensions. From these equations, we obtain

$$\beta N_r = N' + \alpha(N')^2 \tag{3}$$

Dividing this equation by $\beta N'$

$$N_r/N' = (1/\beta) + (\alpha/\beta)N'. \tag{4}$$

Therefore, plotting (N_r/N') against N' gives a straight line, from which the value of α can be determined. Tables showing the values of N versus N' for commonly used apertures are very useful. A high count of N' for a concentrated cell suspension provides a high statistical accuracy but requires a greater correction, so that the best count is a compromise between these opposite conditions. With a 100-μ aperture for the 0.5 ml manometer, for example, the counts between 20×10^3 and 80×10^3 are preferable.

Size Distribution

A volume-distribution curve can be obtained by simple subtraction of successive counts, N_i and N_{i+1}, for a constant shift of V_t, and by plotting the drop, $N_{i+1} - N_i = -\Delta N$, against V_t, which is proportional to the particle volume, V. In this case, again, each count has to be corrected for the coincidence count loss before subtraction. The $-\Delta N$ vs V_t curve can be converted to the $-\Delta N$ vs V curve in an absolute unit of volume if we know the conversion factor of V/V_t, which can be estimated from data for particles with known volumes.

In the measurement of a distribution curve in terms of volume, an appropriate value of sensitivity, S, has to be chosen. Curves A, B, and C in Fig. 3 illustrate the data that would be obtained for the same sample at different sensitivities, $S = 2$, 3, and 4, for the same shift of V_t. Since the degree of amplification of the pulse height is approximately doubled by raising the sensitivity by one step, the volume-distribution curve is flattened by a factor of ½ by the increase of sensitivity by one step. When the sensitivity is lowered, on the other hand, the curve is sharpened and shifted toward smaller values of V_t. Therefore, a small difference in setting the value of V_t introduces a greater error in the reading of $-\Delta N$ at the lower sensitivity, while the fractional error in the value of $-\Delta N$ is greater at the higher sensitivity. An appropriate setting of sensitivity in the case of Fig. 3 may therefore be $S = 3$.

The data shown in Fig. 4 are the volume-distribution curves obtained by Bassham and Kanazawa[5] during the synchronous culture[6] of Chlorella

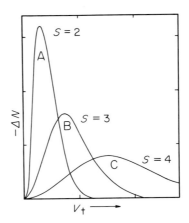

FIG. 3.   Measurements of the $-\Delta N$ vs V_t curve at different sensitivities.

[5]J. A. Bassham and T. Kanazawa, private communication, 1969.
[6]H. Tamiya, Y. Morimura, M. Yokota, and R. Kunieda, Plant Cell Physiol. 2, 383 (1961).

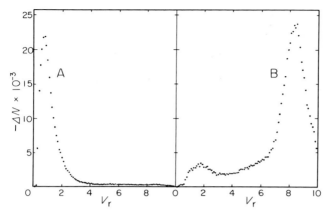

FIG. 4. The $-\Delta N$ vs V_t curves determined by digital recording of $-\Delta N$ and V_t values with a computer connected to the Coulter counter. The points on the left side of figure indicate the distribution obtained for a suspension of daughter cells of *Chlorella*, and the points on the right side indicate the data obtained for a mixture of mother and daughter *Chlorella* cells.

cells. These curves were measured with the Coulter counter of the Model B type connected to a computer for digital recording of $-\Delta N$ and V_t values. The points on the left-hand side of this figure indicate the volume distribution of daughter *Chlorella* cells (the dark cells), and the points on the right side indicates the presence of large mother cells (the light cells) and small daughter cells. Fluctuation of these points indicate the order of error in the measurements.

The measurements of distribution in terms of volume are advantageous when dealing with the measurements in a narrow size range at a single setting of sensitivity or when dealing with rapid measurements with a recorder or a computer. Another way to see the distribution is the measurements in terms of a logarithmic scale, $\log V$, $\log V_r$, or $\log V_t$; the measurements of $-\Delta N$ for a constant increment of $\log V$, $\log V_r$, or $\log V_t$. This method has the following advantages over the measurement in terms of volume. (a) The error in the value of $-\Delta N$ is nearly constant over a wide range of volume to be measured at different sensitivities. This is because roughly the same division of V_t may be used at different sensitivities. (b) Distribution peaks are more symmetrical on the logarithmic scale than on the volume scale. (c) When cells grow uniformly, the distribution peak on the logarithmic scale shifts without any change of its shape and height, while the peak on the volume scale is flattened and deformed. These advantages are seen in the data[7] shown

[7] K. Shibata, Y. Morimura, and H. Tamiya, *Plant Cell Physiol.* **5**, 315 (1964).

in Figs. 5 and 6, which indicate the distribution changes during the synchronous culture of *Chlorella* cells as measured on the logarithmic scale. The values of t in these figures indicate the time in hours during the cultivation. The broading of the peak from $t = 0$ to $t = 7$ and the sharpening after $t = 7$ in Fig. 5 indicate that the growth rate is greater in the middle range of log V, and the data in Fig. 6 indicate that the mother cells shrink uniformly in the dark before division.

A simple way to measure the curves on a logarithmic scale is to set up a table of V_t values for a constant shift of log V_t. An example is shown in Table II which illustrates the measurements over the sensitivity range from $S = 6$ to $S = 2$. In the table, the relative volume, V_r, at V_t (the threshold value) = 10.0 and $S = 6$ is assumed to be $V_r = 10$ (or log $V_r = 1.0$), and the degree of amplification at $S = 2$ or 4 relative to the degree at $S = 6$ is expressed by $1/f$. The value of f at $S = 4$ was estimated from the value of V_t at $S = 4$ giving the same count of N' as that determined at $V_t = 100.0$ and $S = 6$, by taking overlapping counts of a test sample at these different sensitivities. The values of V_t at $S = 4$ for a constant shift of log V_t between log $V_r = 2.0$ and 2.5 in the table were thus determined and, by the same procedure, the values of V_t at $S = 2$ for the range between log $V_r = 2.5$ and 3.2 were determined. Such tables showing the relationship between V_t and log V_r over different sensitivities are very useful for a number of measurements. The figures in the fourth column show the counts (N') obtained for a suspension of *Chlorella* cells, and the 5th and 6th columns include the values (N) corrected for the coincidence count loss and $-\Delta N$ values calculated from these counts, respectively.

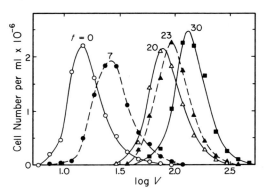

Fig. 5.   Changes of cell-size distribution as measured in terms of log V (in μ^3) during 30 hours of synchronous culture of *Chlorella* cells in the light. The culture time in hours is shown by t.

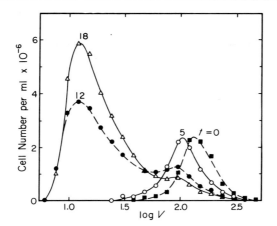

FIG. 6.   Changes of cell-size distribution as measured in terms of log V (in μ^3) during 18 hours of the dark period following the synchronous culture of *Chlorella* cells in the light shown by the data in Fig. 5; the curve of $t = 0$ in this figure shows the same data as shown by the curve of $t = 30$ in Fig. 5.

The result obtained for spinach chloroplasts treated with dodecyl-benzene sulfonate (DBS) is shown in Fig. 7, which indicates the disintegration of chloroplasts into grana by the action of this detergent.[8] Another example[9,10] of volume change is found when chloroplasts kept

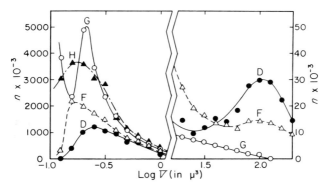

FIG. 7.   Disintegration of spinach chloroplasts into grana by the action of DBS as measured with the Coulter counter; curves D to G indicate disappearance of the peak at log V (in μ^3) = 2.0 of chloroplasts swollen by the effect of DBS and appearance of a peak of grana between log $V = -1.0$ and -0.5.

[8]M. Itoh, S. Izawa, and K. Shibata, *Biochim. Biophys. Acta* **69,** 130 (1963).
[9]M. Itoh, W. Izawa, and K. Shibata, *Biochim. Biophys. Acta* **66,** 319 (1963).
[10]S. Izawa, M. Itoh, and K. Shibata, *Biochim. Biophys. Acta* **75,** 349 (1963).

TABLE II

The $-\Delta N$ versus log V_t Curve Determined for a *Chlorella* Cell Suspension with a 100 μ Aperture and a 0.5 ml Manometer

	log V_r	log(V_r/f)	$V_r/f = V_t$	$N' \times 10^{-3}$	$N \times 10^{-3}$	$-\Delta N \times 10^{-3}$
$S = 6$	1.0	1.0	10.0	70.5	102.3	0
(log $f = 0$)	1.1	1.1	12.6	70.5	102.3	−0.4
	1.2	1.2	15.9	70.5	102.7	−0.2
	1.3	1.3	20.0	70.8	102.9	0.9
	1.4	1.4	25.1	70.4	102.0	−0.7
	1.5	1.5	31.6	70.7	102.7	2.0
	1.6	1.6	39.8	69.8	100.7	4.1
	1.7	1.7	50.1	67.8	96.6	8.4
	1.8	1.8	63.1	63.5	88.2	8.3
	1.9	1.9	79.4	59.1	79.9	9.9
	2.0	2.0	100.0	53.5	70.0	
$S = 4$	2.0	1.444	27.8	53.2	69.4	14.9
(log f	2.1	1.544	35.0	43.9	54.5	17.9
$= 0.556$)	2.2	1.644	44.1	31.5	36.6	14.4
	2.3	1.744	55.5	20.2	22.2	9.2
	2.4	1.844	69.8	12.3	13.0	5.0
	2.5	1.944	87.9	7.7	8.0	
$S = 2$	2.5	1.322	21.0	7.5	7.8	3.7
(log f	2.6	1.422	26.5	4.0	4.1	1.7
$= 1.178$)	2.7	1.522	33.3	2.4	2.4	0.9
	2.8	1.622	41.9	1.5	1.5	0.7
	2.9	1.722	52.7	0.8	0.8	0.4
	3.0	1.822	66.4	0.4	0.4	0.2
	3.1	1.922	83.6	0.2	0.2	0.1
	3.2	2.022	105.3	0.1	0.1	

in darkness are illuminated. Curve A in Fig. 8 is the volume distribution on the logarithmic scale of chloroplasts kept in the dark, and curve B is the distribution in the light. A change as much as 50% in volume was found on the illumination. The count over the distribution maximum on curve A, which is expressed by the area, N_0^*, in the figure, decreases to N_t^* on illumination. The rather rapid change of volume by light could be followed sensitively by measuring the change of N^* at the single setting of V_t at the maximum.

The Coulter counter has been used for the measurements of other species of cells such as bacteria,[11] white and red blood cells[1-4,12] and a

[11] K. G. Lark and C. Lark, *Biochim. Biophys. Acta* **43**, 520 (1960).
[12] W. J. Richar and E. S. Breakell, *Amer. J. Clin. Pathol.* **31**, 384 (1959).

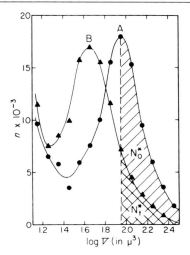

Fig. 8.   The distribution curves measured on the logarithmic scale for the chloroplasts incubated in the dark (curve A) and in the light (curve B).

number of photosynthetic marine organisms.[13] The applications to marine organisms have been reviewed comprehensively by Sheldon and Parsons.[13]

[13] R. W. Sheldon and T. R. Parsons, "A Practical Manual on the use of the Coulter Counter in Marine Research." Coulter Electronics Sales Co., Canada, 1967.

[17] Methods for the Measurement of Chloroplast Volume and Structure *in Vitro* and *in Vivo*

By LESTER PACKER and SATORU MURAKAMI

The measurement of chloroplast volume (size) and other alterations of structure that affect shape (configuration) and internal structure (conformation) can be assessed by a variety of methods that are applicable to one or more of these facets of chloroplast structure. The main techniques that thus far have been applied for assessing facets of chloroplast structure are: gravimetric (packed volume, fresh weight), microscopic (light and electron microscope), and photometric (absorbance or transmission, light scattering) methods and the electronic particle (Coulter) counter.

After available methods, the next most important factor in determining chloroplast volume both *in vitro* and *in vivo* is the establishment of a defined metabolic state. Three aspects seem to be especially important in this regard: (1) the type of electron flow (cyclic or noncyclic; this depends largely upon the presence of artificial or natural cofactors and the light intensity); (2) the ionic environment in which the chloroplasts are suspended; and (3) the presence of conditions for photosynthetic activity, such as ATP formation and carbon dioxide fixation. The extent to which such activities are present in chloroplasts depends upon the condition in which they are isolated or examined *in vitro*. (The presence or absence of the outer chloroplast envelope membrane will affect measurement of the volume of the chloroplast as a whole and also the extent to which activity in the assimilation of carbon dioxide is retained.) Therefore, the usefulness of the available methods and the effects of metabolic states as they are presently known for chloroplasts *in vitro* and *in vivo* are discussed. Several reviews have recently appeared, in which the metabolic aspects have been considered in greater detail.[1-4]

Clearly it is possible to apply a greater variety of techniques to the analysis of the various facets of chloroplast volume and structure for isolated chloroplasts than for chloroplasts *in vivo*. Since reliable methods for assessing these properties of *in vivo* chloroplasts are needed, and often are indirect (photometric), it becomes especially important to correlate the interrelationships between methods and the types of information that these methods can provide *in vitro*.

Chloroplasts *in Vitro*

Illumination of chloroplasts can lead either to an increase, or swelling,[5] or to a decrease, or shrinkage,[6-8] in volume, depending upon the ionic environment in which chloroplasts are suspended. Deamer *et al.*[9] established that the uptake of hydrogen ions by chloroplasts illuminated in the presence of strongly dissociated ions, such as sodium chloride, lead to swelling because of the large increase in internal pH which oc-

[1]L. Packer and P. A. Siegenthaler, *Int. Rev. Cytol.* **20,** 97 (1966).
[2]L. Packer and D. W. Deamer, *in* "Photophysiology," (A. C. Giese, ed.), Vol. 3, p. 91, Academic Press, New York, 1968.
[3]L. Packer and A. R. Crofts, *Curr. Top. Bioenerg.* **2,** 23.
[4]L. Packer, S. Murakami, and C. W. Mehard, *Ann. Rev. Plant Physiol.* **21,** 271 (1970).
[5]L. Packer, P. A. Siegenthaler, and P. S. Nobel, *J. Cell Biol.* **26,** 593 (1965).
[6]L. Packer, *Biochem. Biophys. Res. Commun.* **9,** 355 (1962).
[7]L. Packer, *Biochim. Biophys. Acta* **75,** 12 (1963).
[8]M. Itoh, S. Izawa, and K. Shibata, *Biochim. Biophys. Acta* **66,** 319 (1963).
[9]D. W. Deamer, A. R. Crofts, and L. Packer, *Biochim. Biophys. Acta* **131,** 81 (1967).

curs in these relatively lightly buffered chloroplasts. Similar results occur when chloroplasts are suspended in the presence of strongly dissociated divalent ions.[10-12] An example of the effect of illumination or its actions on chloroplast volume, under such conditions, is shown in Fig. 1 for isolated spinach chloroplasts (largely with the outer membranes absent). Here a comparison is made between the rate of swelling by three methods: the Coulter counter, absorbance (or transmission) and the measurement of packed volume performed in chlorocrit (capillary) tubes. Clearly, all three methods agree in that they show chloroplast swelling upon illumination and that a slow swelling also occurs in darkness during the relatively long time course of these experiments. The presence of a cyclic electron transport cofactor, phenazine methosulfate (PMS), enhances the rate and extent of light-induced swelling.

The conditions necessary to perform such experiments are simple (see legend to Fig. 1). In the Coulter counter (Model B type), the concentra-

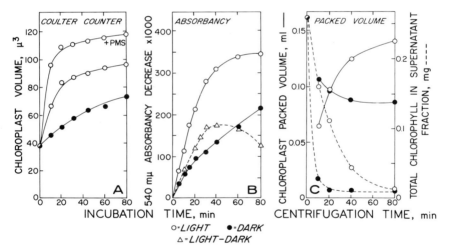

FIG. 1. Time course of the light-induced swelling of spinach chloroplasts *in vitro*. Chloroplast volume changes were determined by the Coulter counter (A), absorbance (B), and packed volume (or chlorocrit) (C) techniques. Reaction mixtures are (A) 50 mM Tris·HCl (pH 8) and 175 mM NaCl with or without 20 μM phenazine methosulfate (PMS), (B) 20 mM Tris·HCl (pH 8), 350 mM NaCl, and 20 μM PMS, and (C) 20 mM Tris·HCl (pH 8) and 35 mM NaCl. From L. Packer, P. A. Siegenthaler, and P. S. Nobel, *Biochem. Biophys. Res. Commun.* **18**, 474 (1965) and *J. Cell Biol.* **26**, 593 (1965).

[10]P. S. Nobel and L. Packer, *Plant Physiol.* **40**, 633 (1965).
[11]P. S. Nobel, S. Murakami, and A. Takamiya, *Plant Cell Physiol.* **7**, 263 (1966).
[12]P. S. Nobel and S. Murakami, *J. Cell Biol.* **32**, 209 (1967).

tion of chloroplasts employed must necessarily be low.[5] Considerably more chlorophyll is required to carry out the absorbance measurements, and an initial absorbance between 0.5 and 1 is adequate. Even higher concentrations of chloroplasts are required for carrying out the packed volume determinations. Each method offers advantages and disadvantages; e.g., in the case of the Coulter counter, it is not absolutely certain what volume is being measured. Although results correlate well with the volume of the inner membrane system, it is not clear how aggregation affects the measurements. Absorbance does not measure volume directly and must be correlated with volume by the use of a direct technique. Packed volume does constitute a direct method, but it suffers from the limitations imposed by the high concentrations of chloroplasts that are required. This makes it more difficult to set conditions for establishing a particular metabolic state. Nevertheless, the salient feature of the comparison of methods shown in Fig. 1 is the clear demonstration that all three methods are adequate to show chloroplast swelling under these experimental conditions.

A further comparison of methods for demonstrating light-dependent swelling of chloroplasts is shown in Table I, where it can be seen that the packed volume, absorbance, and Coulter counter techniques also correlate with measurements of fresh weight. Although determination of

TABLE I

COMPARISON OF METHODS FOR MEASURING LIGHT-DEPENDENT
HIGH-AMPLITUDE CHLOROPLAST SWELLING[a,b]

Time (min)	Packed volume		Fresh weight		Absorbancy (× 1000)		Coulter counter	
	Light (ml)	Dark (ml)	Light (mg)	Dark (mg)	Light $E_{540\ nm}$	Dark $E_{540\ nm}$	Light (μ^3)	Dark (μ^3)
0	0.36	0.34	47.0	47.0	385	395	58	58
30	0.72	0.46	75.0	53.0	240	348	118	71

[a]From L. Packer, P. A. Siegenthaler, and P. S. Nobel, *J. Cell Biol.* **26**, 593 (1965).
[b]Chloroplasts in 100 mM Tris·HCl, pH 7.2, containing 350 mM NaCl were isolated by differential centrifugation between 200 g (5 minutes) and 600 g (15 minutes). The reaction mixture contained Tris·HCl (20 mM, pH 8.0), NaCl (350 mM), PMS (20 μM), and chloroplasts equivalent to 200 μg of chlorophyll per milliliter, except for absorbance measurements; for these samples were diluted 20-fold (just before reading) from an initial chlorophyll concentration (200 μg/ml). For Coulter counter experiments, chloroplasts were isolated in Tris·HCl (50 mM, pH 8.0) and NaCl (175 mM) using successive centrifugations of 200 g (1 minute) and 200 g (10 minutes), and the reaction mixture contained Tris·HCl (50 mM, pH 8.0), NaCl (175 mM), PMS (20 μM), and chloroplasts (15 μg chlorophyll/ml). Just before measurement, the chlorophyll concentration was diluted to 5 mμg/ml, which corresponded to approximately 7500 chloroplasts per milliliter.

TABLE II
INFLUENCE OF IONIC COMPOSITION OF THE MEDIUM
ON CHLOROPLAST VOLUME[a]

Type of medium			Scattering, 546 nm (%)		Absorbance, 546 nm		Volume (μ^3)	
Cation	Anion	Example[b]	Dark	Light	Dark	Light	Dark	Light
Dissociated	Dissociated	NaCl	100	160	0.8	0.4	45	80
Dissociated	Associated	NaAc	100	260	0.8	1.0	45	15
Associated	Dissociated	NH_4Cl	100	80	0.8	0.4	45	120
Associated	Associated	NH_4Ac	60	60	0.3	0.3	120	120

[a]From L. Packer and D. W. Deamer, in "Photophysiology" (A. C. Giese, ed.), Vol. 3, p. 91. Academic Press, New York, 1968.
[b]Data are from spinach chloroplasts suspended in 0.1 M solutions.

fresh weight is somewhat more cumbersome and not dynamic, nevertheless it provides a direct correlation of the extent to which volume changes occur in isolated chloroplast suspensions and is least subject to criticisms of interpretation.

It is now established that the ionic environment greatly influences the effect of illumination upon chloroplast volume.[9,13,14] A main consideration in understanding the interrelation of chloroplasts with the ionic environment depends upon whether the ions in the medium interact with hydrogen ions. In Table II the influence of the ionic composition of the medium on chloroplast volume is summarized; three different methods are used: light scattering, absorbance (or transmission), and volume changes determined by the electronic particle counter. Chloroplasts suspended in the presence of weak acid anions which associate with hydrogen ions contract or shrink upon illumination because the accumulation of hydrogen ions leads to the expulsion or efflux from chloroplasts of uncharged, undissociated acids that carry out osmotically active material, thus leading to osmotic collapse. The opposite situation occurs when chloroplasts are suspended in the presence of weak base cations, such as ammonium chloride, where, upon illumination, the hydrogen ions taken up become associated with the uncharged ammonia, leading to the accumulation of the charged species inside. Since this species cannot penetrate the membrane, the result is an increase of osmolarity inside and chloroplast swelling. When both the anion and the cation can exist in appreciable concentration in their undissociated form, as with the ammonium acetate, then chloroplast swelling readily occurs

[13]A. R. Crofts, D. W. Deamer, and L. Packer, Biochim. Biophys Acta, 131, 97 (1967).
[14]D. W. Deamer and L. Packer, Arch. Biochem. Biophys. 119, 83 (1967).

in the dark. All methods shown agree with one another except in one instance—the use of light scattering. Light-scattering measurements show that increases in scattering occur both when the chloroplasts swell upon illumination or when they contract, as in the case when they are suspended in either sodium chloride or sodium acetate solutions.

Thus, while absorbance (and/or transmission) and light-scattering measurements are valuable as a means of assessing dynamic changes in chloroplast volume, it is clear that only absorbance or transmission change is directly correlated with volume change. Therefore both light scattering and transmission changes should be monitored simultaneously to provide more precise information on the configurational and conformational changes of chloroplasts by photometric methods. More detailed examinations of the correlation between light scattering and transmission changes in illuminated chloroplasts are shown in Figs. 2

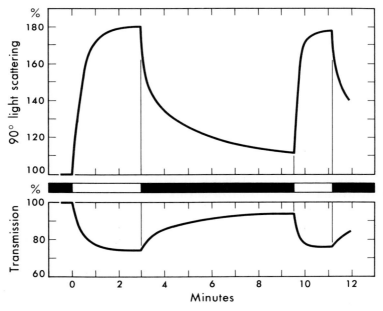

FIG. 2. Kinetics of light-induced, large-scale light scattering and transmission changes which accompany photoshrinkage in isolated spinach chloroplasts suspended in a sodium acetate medium. Chloroplasts were isolated from spinach leaves in 50 mM Tris·HCl buffer, pH 8.0, containing 175 mM NaCl, and were incubated in 150 mM sodium acetate (pH 6.7) plus 15 μM phenazine methosulfate (PMS). Chlorophyll concentration was 13 μg/ml; 90-degree light scattering (546 nm) and transmission (546 nm) levels of the chloroplasts in the dark were adjusted as 100% and then illuminated by actinic red light (600–700 nm) from the side by a tungsten lamp to induce chloroplast volume changes. [From S. Murakami and L. Packer, *J. Cell Biol.* 47, 332 (1970).]

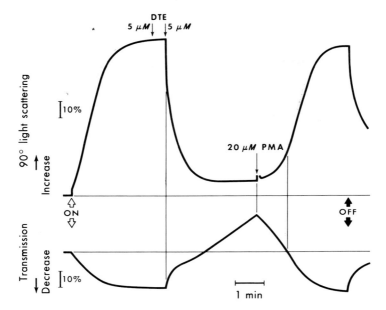

FIG. 3.   Simultaneous record of light-induced 90-degree light scattering (546 nm) and transmission (546 nm) changes of isolated spinach chloroplasts. Chloroplasts were isolated in 50 mM Tris·HCl, pH 8.0, plus 175 mM NaCl and incubated in the same medium to which 15 μM phenazine methosulfate (PMS) and 20 μM phenyl mercuric acetate (PMA) were added. Chloroplast suspension was illuminated by red light. Where indicated, 10 μM dithioerythritol (DTE) and 20 μM PMA were added to the incubation while illuminated. [From S. Murakami and L. Packer, *J. Cell Biol.* **47**, 332 (1970).]

and 3. Here chloroplasts were suspended in the incubation medium in a cuvette in a Brice-Phoenix Light Scattering photometer modified for recording. The temperature of the reaction mixture was controlled during experiments at 25° by circulating water through the jacket around the cuvette. Light scattering and transmission were measured, respectively, at 90 and 180 degrees to the direction of measuring light at 546 nm from high-pressure mercury lamp. Also both the scattered and transmitted light were filtered at the same wavelength (546 nm) through an interference filter, and the intensity in the dark before illumination was adjusted to read 100% on chart paper, using the minimum intensity of 546 nm incident light. Since light in this region is near the minimum of the photochemical action spectrum, this procedure minimized the possibility of electron flow being activated by the incident beam. When necessary, the chloroplast suspension was illuminated by a broad band of red actinic light (600–700 nm) from the opposite side of the light-scattering measurement. Chloroplasts containing less than 15 μg of chlorophyll per milliliter of incubation mixture should be used to prevent multiscattering of light by pigmented particles (chloroplasts) at higher concentration. Within this range of chlorophyll concentration, light scattering and transmission are proportional to the pigment content.

In Fig. 2 is presented an example of simultaneous recordings of light scattering and transmission changes of chloroplasts incubated in a weak acid anion (sodium acetate) solution in the presence of PMS, conditions under which shrinkage occurs upon illumination. The kinetics of light scattering are closely correlated with transmission changes. Increases in light scattering and decreases in transmission upon illumination parallel one another in time and, moreover, when the light is extinguished the relaxation or reversal of the changes to the initial condition are likewise well correlated. Thus, there is a general correlation between light scattering changes and transmission changes when chloroplasts undergo extensive photoshrinkage *in vitro*.

However, such agreement is not always found; selection of appropriate experimental conditions reveals discrepancies in the kinetics of light scattering and transmission changes. Such a case is demonstrated in Fig. 3. Here, chloroplasts are suspended in sodium chloride solution in the presence of a small quantity of PMS and phenylmercuric acetate (PMA). Under these conditions it has been established[15,16] that chloroplasts contract extensively upon illumination. This contraction can be reversed either by removing light or by removing PMA. The latter can be accomplished by adding a stoichiometric quantity of a dithio reagent, such as dithioerythritol (DTE). When this is done under conditions of continuous illumination, it can be seen, as in Fig. 3, that a rapid decay in the light scattering occurs. However, transmission changes under these conditions show a biphasic response. There is an initial rapid increase of transmission which does correlate with the light-scattering change, but this is then followed by a slower and much larger increase in transmission. The biphasic character of these responses can be further illustrated by adding a quantity of PMA larger than the amount of DTE present. Under these conditions the initial kinetics of light scattering and transmission changes show biphasic responses. It has been concluded that the initial response of light scattering and transmission changes under these conditions reflect the changes in the internal structure of the membranes themselves, and the secondary larger but slower changes reflect changes in the spacing of thylakoid membranes and in a volume change of chloroplasts as a whole.[17] Thus, transmission changes or ab-

[15]P. A. Siegenthaler, *Physiol. Plant.* **19,** 437 (1966).
[16]S. Murakami and L. Packer, *Biochim. Biophys. Acta* **180,** 420 (1969).
[17]S. Murakami and L. Packer, *J. Cell Biol.* **47,** 332 (1970).

sorbancy changes, under all circumstances thus far investigated with isolated chloroplasts, reflect changes in the chloroplast volume whereas light-scattering changes represent a mixture of two effects—a change in inner membrane structure, which is then superimposed and superseded by larger but slower volume changes that reflect the flattening of the thylakoid membrane.

Direct examination, by electron microscope, of individual chloroplast profiles under conditions of different metabolic states, i.e., light intensities and ionic environments, establishes the extent to which changes in chloroplast volume and/or configuration, i.e., shape of membranes, occurs upon illumination under various conditions. Such studies do not provide dynamic information, and therefore they are not required for studies of volume changes; however, they are of considerable importance for understanding the shape and/or conformational changes that may occur in the chloroplast membrane.[17]

Chloroplasts *in Vivo*

Only three methods have been useful, in general, for the study of chloroplast volume and/or configurational and conformational changes for chloroplasts in the living state. There are light and electron microscopy, absorbance, and light-scattering changes. Electron microscopy has been used successfully to demonstrate the changes in gross morphology, volume changes, and configurations of chloroplast membrane system,[18,19] and quite recently, from analysis of microphotodensitometric recording of high-resolution electron micrographs of chloroplasts in different metabolic states, has been successfully employed to demonstrate thickness changes of grana thylakoid membranes.[16,17] The dynamic techniques that have been applied to changes in chloroplast structure *in vivo* are photometric techniques, and these provide useful information as to changes in chloroplast volume and structure, especially as compared with electron microscope analysis.

Two general approaches have been employed to demonstrate changes in chloroplast structure *in vivo* upon illumination. One is to suspend leaves or the thalli of marine algae in aqueous media in the presence of various test solutions. In the case of leaves, the best results are obtained with vacuum-infiltrated leaves[19] when externally added test solutions are to be evaluated. Thalli of marine algae, which are one and two cell layers thick, pose no such problem because the cells face the aqueous medium in which they are suspended. The second technique is to examine the changes in transmission that occur in dry leaves or leaves en-

[18]H. Kushida, M. Itoh, S. Izawa, and K. Shibata, *Biochim. Biophys. Acta* 79, 203 (1964).
[19]L. Packer, A. C. Barnard, and D. W. Deamer, *Plant Physiol.* 42, 283 (1967).

closed in chambers where they are flushed under moist conditions in the presence of various gases to present conditions that permit the occurrence of photosynthetic activities. Studies have been primarily performed by Heber.[20]

An example of an apparatus suitable for the measurement of transmission (and/or light scattering not shown) changes in intact leaf tissue is shown in Fig. 4. The vacuum infiltrated leaf tissue or algal thallus is placed between two black Bakelite frames of identical size with 1 cm $\times$ 2 cm rectangular holes cut out in the center to permit light to strike the tissue (upper right-hand corner, Fig. 4). The edges of the leaf are trimmed with scissors in order to expose only a portion of the leaf section to the light beams (arrow pointing to leaf tissue in the diagram). Two O rings hold the frames together. A space is machined in the lower portion of the frame to permit movement of a magnetic stirring bar when the frame is within the cuvette. Light from a tungsten source filtered at 546 nm (filter A) passes through a porthole in the thermostatic holder, then through the optical window of the cuvette, striking the leaf obliquely. A portion of the light is transmitted through the leaf thickness and into the photomultiplier. The photomultiplier is protected by an interference filter (filter B) having a peak transmission at 546 nm. In this arrangement the measurement of transmission changes with low intensity of light (546 nm) is at a region where absorption by pigments in the leaf is nearly minimal. Hence the measuring wavelength, being near the minimum of the photochemical action spectrum, cannot significantly activate photosynthetic processes. Light reactions are activated by illumination with a tungsten light source at 600–700 nm (filter C in diagram), and the leaf surface is exposed obliquely to broad band red actinic light. The orientation of the leaf in the cuvette is such that reflection of the red light from the leaf surface is not detectable. If it is desirable, the light scattered by the leaf tissue can be filtered through an interference filter (546 nm) and recorded at 90 degrees from the opposite side of actinic light source.

A comparison of light-induced transmission changes and light-induced scattering changes in thalli of the marine alga *Porphyra* is shown in Fig. 5. It is clear from these results that light-induced increases in light scattering and decreases in transmission correlate well with one another with respect to their formation, initiation of change upon illumination, and the decay or relaxation of the response upon return to darkness.[21] Moreover, both responses are sensitive to the inhibitor of oxygen evolu-

[20]U. Heber, *Biochim. Biophys. Acta* **180**, 302 (1969).
[21]S. Murakami and L. Packer, *Plant Physiol.* **45**, 289 (1970).

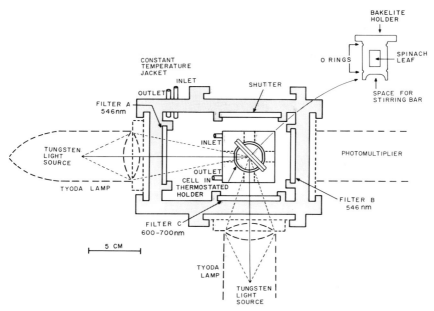

FIG. 4.   Apparatus for the measurement of transmission changes in leaf tissue. The block diagram of the spectrophotometric device used for leaf studies is drawn to scale. The main body of the apparatus itself, and the base plate upon which it rests (not shown), is machined from aluminum. Brackets in the outer walls accept aluminum holders which contain either photomultiplier tubes or light sources. Regulated power supplies are used for the tungsten light sources, and a high voltage supply is employed for the photomultiplier in conjunction with a circuit which permits the time recording of changes in transmission on an ink-writing recorder. The inner faces of the walls contain brackets for the location of filters or shutters. The optical cell used for angular scattering studies (Brice-Phoenix Precision Instrument Co., Philadelphia, Pennsylvania, type C-105) forms part of the walls of the temperature-jacketed holder, and the latter has four portholes through which the light beams may pass. The cuvette within the thermostatic holder rests on top of a magnetic stirring device. Leaf specimens are located within frames clamped together by O rings. The diagram shows the orientation used for the measurement of transmission changes at 546 nm in leaves or thalli under conditions where illumination with broad-band actinic light can be made. [From L. Packer, A. C. Barnard, and D. W. Deamer, *Plant Physiol.* **42**, 283 (1967).]

tion, 3-(3,4-dichlorophenyl)-1,1-dimethylurea (DCMU). These results suggest that conditions *in vivo* are such that weak acid anions dominate the ionic environment. Hence, light-induced shrinkage occurs. This is not unexpected because weak acid anions are the main products of photosynthetic activity. Hence, chloroplasts in the living cell are highly unlikely to be present in a starved condition, i.e., devoid of weak acid anions. Indeed, this seems to be the case in a wide variety of higher plant

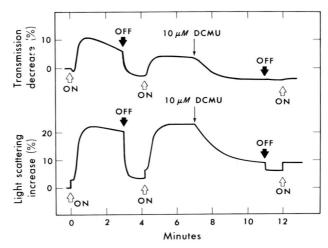

Fig. 5. Kinetics of light-induced 90-degree light scattering (546 nm) and transmission (546 nm) changes in thalli of *Porphyra* sp. Thalli which were sandwiched between two black Bakelite frames according to the procedure described by Packer [L. Packer, A. C. Barnard, and D. W. Deamer, *Plant Physiol.* **42**, 283 (1967)] were put into the cuvette containing 100 mM sodium acetate (pH 6.7) plus 2.5 mM potassium ferricyanide and then illuminated by red actinic light after 40 minutes' preincubation in the same medium in the dark at 0°. [From S. Murakami and L. Packer, *Plant Physiol.* **45**, 289 (1970).]

leaves and marine algae that have been investigated. Some results are summarized in Table III, where it can be observed that light-induced transmission changes occur on the average to the extent of about 10% or less with high intensity, broad-band illumination between 600 and 700 nm.[19]

Quite recently Heber[20] showed that illumination of leaves incubated in the absence or the presence of photosynthetic activity manifests fast absorbance changes that reflect the photoshrinkage of chloroplasts *in vivo*. The extent of the shrinkage is found to be strongly influenced by the quality and intensity of the exciting light, by the gas phases in which leaves are placed, and by the presence or the absence of the electron acceptors. The results suggest that shrinkage of chloroplasts may control electron flow and the regulation of the process of photosynthesis. The summary experiments taken from this pioneering study are illustrated in Fig. 6, where it is shown that the extent of the photoshrinkage induced in two types of higher plant leaves is strongly dependent upon the quality of light, i.e., red or far red, which activates photosystems II and I, respectively, in the presence or the absence of oxygen or carbon dioxide, which stimulate pseudocyclic electron flow and/or carbon assimitation. Similar results are obtained in a wide variety of leaves, but the extent of the photoshrinkage has not yet been determined by electron microscopy.

Chloroplasts Fixed in Their Metabolic and Structural States by Glutaraldehyde Treatment

A number of studies first initiated by Park et al.,[22] and later supplemented by Packer et al.,[23] West and Packer,[24] and Hallier and Park,[25] have demonstrated that chloroplasts treated with glutaraldehyde can be stabilized with respect both to their gross and morphological structure and to the retention, to some extent, of functional activities. The use of this technique enabled Park et al.[22] to fix chloroplasts *in vivo*,

TABLE III

PHOTOMETRIC EVIDENCE FOR RED-LIGHT INDUCED STRUCTURAL CHANGES IN CHLOROPLASTS OF LEAVES OF HIGHER PLANTS AND THALLI OF MARINE ALGAE[a]

Preparation	% Light-induced transmission change, 546 nm						Average	
Higher plants								
Spinach leaf	5.8	4.5	8.9	5.5			5.9	(4)
	3.4[b]	3.2[b]					3.3	(2)
Swiss chard leaf	1.5	2.6	1.3	1.0	1.6	2.5	1.8	(6)
Turkish tobacco leaf	1.3						1.3	(1)
Senecio leaf	7.8	8.9	9.9	9.3			9.0	(4)
Senecio stripped epidermis	1.2	1.1	0.2				0.8	(3)
Senecio stripped leaf	10.4	9.9					10.3	(2)
Marine algae								
Ulva lobata	8.0[c]	6.5[c]	15.6[d]	18.2[e]				
Enteromorpha sp.	0.3[f]						0.3	(1)
Porphyra sp.	10.8[g]	10.5[g]					10.7	(2)

[a]From L. Packer, A. C. Barnard, and D. W. Deamer, Plant Physiol. 42, 283 (1967).
[b]Percent of 90-degree light-scattering increase at 546 nm instead of transmission change.
[c]Experiment made in 0.5% seawater medium without vacuum infiltration. Thalli soaked 1 hour in seawater.
[d]Thalli soaked 1 hour in the dark at 2° in acetate (100 mM) + ferricyanide (6 mM) solution at pH 6.5.
[e]Thalli soaked 1 hour at 2° in acetate (100 mM) + PMS (20 μM) solution at pH 6.5.
[f]Ten layers in seawater.
[g]Thalli soaked 1 hour at 2° in sodium chloride (100 mM) + ferricyanide (6 mM) solution at pH 6.5.

[22]R. B. Park, J. Kelly, S. Drury, and K. Sauer, *Proc. Nat. Acad. Sci. U.S.* **55**, 1056 (1966).
[23]L. Packer, J. M. Allen, and M. Starks, *Arch. Biochem. Biophys.* **128**, 142 (1968).
[24]J. West and L. Packer, *Bioenergetics* **1**, 405 (1970).
[25]U. W. Hallier and R. B. Park, *Plant Physiol.* **44**, 544 (1969).

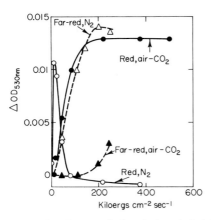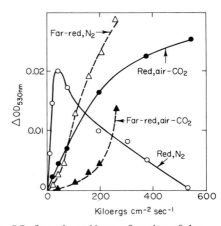

FIG. 6. Extent of photoinduced shrinkage in CO_2-free air or N_2 as a function of the intensity of exciting red or far-red light. In air, red light promotes shrinkage at high intensities, in N_2 only at low intensities. Far-red produces shrinkage readily in N_2; the far-red beam included a considerable proportion of light which is not absorbed by the pigment system of photosynthesis and the red beam contains also far-red light of a spectral composition similar to that of the far-red beam. For the measurement, a leaf was placed on a Lucite light pipe in a closed Lucite vessel directly above the phototube and was gassed with a stream of 99.997% pure moist N_2, with moist CO_2-free air, or with a mixture of both (approximately 0.8–1.2 liters per minute). Illumination was made by broad band of red (RG 2 filter, Schott, Jena) or far-red light (BG 8 filter) combined with additional filter (27 mm water and 1 mm Calflex). The phototube (Emi 9558B) was protected from exciting light by two No. 9782 Corning filters and by BG 18 (Schott, Jena). Transmittance changes were recorded on an oscillograph as the changes in the photocurrent produced by a measuring beam (usually at 530 nm, bandwidth 1 nm, intensity 9.4 ergs/cm² per second) which passed through the leaf. From U. Heber, *Biochim. Biophys. Acta* **180**, 302 (1969).

permitting their subsequent isolation and characterization in terms of volume. This procedure affords a novel experimental approach to stabilization of dynamic changes in structure brought about by a changeable metabolic state, thus permitting analyses of chloroplast volume and structure to be made "at leisure." For example, it is possible to analyze kinetics of light-induced changes in transmission and light scattering in chloroplasts *in vitro* under conditions where introduction of glutaraldehyde into the test system catches the structure during the time course of an experiment similar to that shown in Fig. 2. Examination of the changes in gross morphology made by electron microscopy under these circumstances substantiates the conclusion that transmission changes more nearly reflect the overall volume, and light-scattering changes more nearly reflect the internal membrane structure.[17,21]

Summary

In the appendix are summarized the various methods that have been applied in the main to the study of light-induced volume changes and configurational and conformational changes in chloroplasts *in vitro* and *in vivo*. This comparison of the various techniques illustrates the applicability of various methods, and the requirements for analysis of these parameters of chloroplast structure in relation to metabolic state.

Appendix

Tables AI and AII show changes in volume, shape, size, and internal structure of chloroplasts, *in vitro* and *in vivo*, as outlined below:

Table AI.   Isolated chloroplasts, *in vitro*
        A.   Light-induced shrinkage of isolated chloroplasts
        B.   Light-induced swelling of isolated chloroplasts
        C.   Volume and structural changes induced by various treatments
            1.   Effect of cations
            2.   pH-induced shrinkage
            3.   Effect of heating
            4.   Effect of fatty acids and lipids
            5.   Effect of solvents and detergents
        D.   Osmotic swelling
Table AII.   Chloroplasts, *in vivo*
        A.   Light-induced shrinkage

TABLE AI-A

LIGHT-INDUCED SHRINKAGE OF ISOLATED CHLOROPLASTS in Vitro

Materials	Reaction mixtures	Illumination	Methods and changes observed	References[a]
Spinach (Spinacia oleracea)	20 mM Tris·HCl, pH 7.5, + 35 mM NaCl + 5 mM MgCl₂ + 5 mM phosphate + 1 mM ADP + 3 mM NADP (28 μg Chl/ml), 25°	Red light (660 nm) 200 lumens	90° light scattering (LS) (546 nm); increased (15%)	(1, 2)
Spinach	20 mM Tris·HCl, pH 7.5, + 35 mM NaCl + 5 mM MgCl₂ + 4 mM phosphate + 5 mM K₃Fe(CN)₆ (40 μg Chl/ml), 25°	Red light (660 nm) 200 lumens	90° LS (546 nm); increased (18%)	(1, 2)
Spinach	40 mM phosphate, pH 7.2	White light 6000 lux	Coulter counter (CC), volume decreased; packed volume (PV), decreased (51–78%); turbidity (750 nm), increased; electron microscope (EM)	(3)
Spinach	40 mM phosphate buffer, pH 7.2 + 2 mM EDTA	435–800 nm 65–250 ergs/cm²/sec	CC, volume decreased	(4)
Spinach	40 mM phosphate buffer, pH 7.2, + 2 mM EDTA (+ 0.1–1.2 M sucrose or glucose)	White light 6000 lux	CC, volume decreased	(5)
Spinach	33 mM Tris·HCl, pH 6-7.1, + 35 mM NaCl + 1 mM TMQH₂ (25 μg Chl/ml)	640–670 nm	Absorbance (340, 510 nm), increased; light-dark difference spectrum 90° LS (546 nm), increased	(6)
Spinach	50 mM Tricine·NaOH, pH 7.4, + 50 mM KCl + 3 mM K₃Fe(CN)₆ + 60 mM methylamine·HCl or 100 μM atebrin	500 W projector lamp, λ>620 nm	PV, decreased; absorbance (580 nm), increased	(7)
Spinach	66 mM Tris·HCl, pH 6.2 (or 100 mM Tris·acetate, pH 7.0) + 0.13 mM TMQH₂	750 W projector lamp, red light 2.1 × 10⁵ ergs/cm²/sec	90° LS (546 nm), increased; absorbance, increased; CC, volume decreased	(8)

Spinach	5 mM Tris·HCl, pH 8, + 5 mM MgCl$_2$ + 2.5 mM ascorbate + 20 μM PMS + 2–10 μM PMA (10 μg Chl/ml) 25°	Red light	90° LS (546 nm), increased; CC, volume increased	(9)
Spinach	20 mM Tris·HCl, pH 8 + 35 mM NaCl + 5 mM MgCl$_2$; addition: various anions (40 mM) (20 μg Chl/ml) 25°	Red light	90° LS (546 nm), increased	(10)
Spinach	20 mM Tris·HCl, pH 8 + 35 mM NaCl + 5 mM MgCl$_2$ + 20 μM PMS + 10–50 μM PMA (15 μg Chl/ml) 25°	1200 lux	90° LS (546 nm), increased; CC, volume decreased	(11)
Spinach (Class I and II chloroplast)	100 mM NaCl + 20 μM PMS or 100 mM sodium acetate, pH 5.5–8.0 + 20 μM PMS (10–20 μg Chl/ml)	Red light (600–700 nm) 1000 ft-c	90° LS (546 nm), increased; CC, volume decreased; PV, decreased; EM	(12–14)
Spinach (Class I chloroplast)	50 mM MES, pH 6.1+330 mM sorbitol+ 20 mM NaCl + 2 mM NaNO$_3$ + 2 mM EDTA + 2 mM Na isoascorbate + 1 mM MnCl$_2$ + 1 mM MgCl$_2$ + 50 μM K$_2$HPO$_4$	Red light 10^5 ergs/ cm^2/sec	90° LS (520 nm), increased; EM (freeze-etching method)	(15)
(Class II chloroplast)	100 mM Tris·acetate, pH 7.0, + 1 mM K$_3$Fe(CN)$_6$ (20–25 μg Chl/ml)	Red light 10^5 ergs/ cm^2/sec	90° LS (520 nm), increased; EM (freeze-etching method)	(15)
Spinach (grana)	25 mM Tris·HCl, pH 7.5, + 100 mM sodium acetate + 20 μM PMS (5 μg Chl/ml)	Red light 650 nm, 1000 ft-c	90° LS (546 nm), increased; PV, decreased; EM	(16, 17)
Spinach (grana)	20 mM Tris·maleate, pH 7.0 + 5 mM MnCl$_2$ (5 μg Chl/ml)	Red light λ> 650 nm, 1000 ft-c	90° LS (546 nm), increased; PV, decreased; EM	(17)
Spinach	40 mM phosphate buffer, pH 7.2, + 10 μM DCPIP + 1 mM ascorbate (18 μg Chl/ml)	Red light λ>680 nm, 170 lux; far-red light, λ> 700 nm, 70 lux	Light-dark difference spectra absorbance (435 nm), decreased	(18)

TABLE AI-A (Continued)

LIGHT-INDUCED SHRINKAGE OF ISOLATED CHLOROPLASTS in Vitro

Materials	Reaction mixtures	Illumination	Methods and changes observed	References[a]
Spinach	50 mM Tris·HCl, pH 8.0 + 175 mM NaCl + 15 μM PMS + 20 μM PMA (18 μg Chl/ml) 25°	Red light, 600–700 nm	90° LS (546 nm), increased; transmission (546 nm), decreased; EM	(19)
Mnium undulatum	1/15 M phosphate buffer, pH 7.3, + 300 mM sucrose	Red light / Blue light (low int.) / Blue light (high int.)	Light microscope (surface area of chloroplast) Increased / Increased / Decreased	(20)
Euglena	20 mM Tris·HCl, pH 8, + 35 mM NaCl + 5 mM $MgCl_2$ + 2 mM ascorbate + 20 μM PMS (5 μg Chl/ml)	Red light (Wratten No. 26)	90° LS (546 nm), increased; PV, decreased	(21)
Rhodospirillum rubrum	20 mM Tris·HCl, pH 7.7, + 35 mM NaCl + 1 mM phosphate 25°		90° LS (546 nm), increased	(22)

[a]Key to references: (1) L. Packer, Biochem. Biophys. Res. Commun. 9, 355 (1962). (2) L. Packer, Biochem. Biophys. Acta 75, 12 (1963). (3) M. Itoh, S. Izawa, and K. Shibata, Biochim. Biophys. Acta 66, 319 (1963). (4) S. Izawa, M. Itoh, and K. Shibata, Biochim. Biophys. Acta 75, 349 (1963). (5) M. Itoh, Plant Cell Physiol. 6, 221 (1965). (6) R. A. Dilley and L. P. Vernon, Biochemistry 3, 817 (1964). (7) S. Izawa, Biochim. Biophys. Acta 102, 373 (1965). (8) R. A. Dilley and L. P. Vernon, Arch. Biochem. Biophys. 111, 365 (1965). (9) P. A. Siegenthaler and L. Packer, Plant Physiol. 40, 785 (1965). (10) L. Packer and P. A. Siegenthaler, Plant Physiol. 40, 1080 (1965). (11) P. A. Siegenthaler, Physiol. Planta. 19, 437 (1966). (12) D. W. Deamer, A. R. Crofts, and L. Packer, Biochim. Biophys. Acta 131, 81 (1967). (13) A. R. Crofts, D. W. Deamer, and L. Packer, Biochim. Biophys. Acta 131, 97 (1967). (14) D. W. Deamer and L. Packer, Arch. Biochem. Biophys. 119, 83 (1967). (15) R. A. Dilley, R. B. Park, and D. Branton, Photochem. Photobiol. 6, 407 (1967). (16) E. L. Gross and L. Packer, Biochem. Biophys. Res. Commun. 20, 715 (1965). (17) E. L. Gross and L. Packer, Arch. Biochem. Biophys. 122, 237 (1967). (18) K. Yamashita, M. Itoh, and K. Shibata, Biochim. Biophys. Acta 162, 610 (1968). (19) S. Murakami and L. Packer, Biochim. Biophys. Acta 180, 420 (1969). (20) J. Zurzycki, Acta Soc. Bot. Pol. 35, 281 (1966). (21) M. M. Belsky, P. A. Siegenthaler, and L. Packer, Plant Physiol. 40, 290 (1965). (22) L. Packer, R. H. Marchant, and Y. Mukohata, Biochim. Biophys. Acta 75, 23 (1963).

TABLE AI-B

LIGHT-INDUCED SWELLING OF ISOLATED CHLOROPLASTS in Vitro

Materials	Reaction mixtures	Illuminations	Methods and changes observed	References[a]
Spinach	20 mM Tris·HCl, pH 8.0, + 350 mM NaCl + 20 μM PMS (200 μg Chl/ml, 5 ng Chl/ml for CC) 25°	25,000 lux	PV, increased (100%); fresh packed weight (FPW), increased (60%); absorbance (546 nm), decreased; CC, volume increased	(1–3)
Spinach	75 mM Tris·HCl, pH 7.5, + 300 mM NaCl (15–20 μg Chl/ml)	30,000 lux	PV, increased; CC, increased	(4)
Spinach	100 mM Tris·HCl, pH 8.0, + 175 mM NaCl (1 mg Chl/ml) 20°	3.45×10^5 ergs/cm²/sec	PV, increased	(5)
Spinach	50 mM Tris·HCl, pH 7.9, + 175 mM NaCl (200 μg Chl/ml) 25°	30,000 lux (white light)	FPW, increased; EM	(6)
Spinach	50 mM Tricine-NaOH, pH 7.3, + 1 mM $K_3Fe(CN)_6$ + 4 mM NH_4Cl or methylamine-HCl	Red light, 600 nm	Absorbance (OD), increased; EM	(7)
Spinach	40 mM phosphate buffer, pH 7.2, + 66 mM methylamine + 60 μM $TMQH_2$	—	Absorbance (435 nm), increase; turbidity (530 nm), decrease; light-dark difference spectra	(8)
Elodea densa	0.03–0.1% nicotine solution; 20°	12,000 lux	Light and electron microscope	(9)

[a]Key to references: (1) L. Packer, P. A. Siegenthaler, and P. S. Nobel, J. Cell Biol. 26, 593 (1965). (2) P. A. Siegenthaler and L. Packer, Plant Physiol. 40, 785 (1965). (3) L. Packer, P. A. Siegenthaler, and P. S. Nobel, Biochem. Biophys. Res. Commun. 18, 474 (1965). (4) P. A. Siegenthaler, Bull. Soc. Bot. Suisse 78, 202 (1968). (5) P. A. Siegenthaler, Experientia 24, 1198 (1968). (6) S. Murakami and P. S. Nobel, Plant Cell Physiol. 8, 657 (1967). (7) S. Izawa and N. E. Good, Plant Physiol. 41, 544 (1966). (8) K. Yamashita, M. Itoh, and K. Shibata, Biochim. Biophys. Acta 162, 610 (1968). (9) H. Lindner, Protoplasma 51, 91, 507 (1960).

TABLE AI-C

Volume and Structural Changes Induced by Various Treatments in Vitro

Treatments	Changes	Materials	Reaction mixtures	Methods	References[a]
		1. Effect of cations			
10^{-2} M CaCl$_2$, BaCl$_2$, MgCl$_2$, Na$_2$SO$_4$, NaNO$_3$, CH$_3$COONa	Shrinkage	Spinach	0.5 M Sucrose	Turbidity (OD at 540 nm or 780 nm)	(1)
KCl	Swelling	Spinach	0.5 M Sucrose	Turbidity (OD at 540 nm or 780 nm)	(1)
NaCl	No effect	Spinach	0.5 M Sucrose	Turbidity (OD at 540 nm or 780 nm)	(1)
5 ~ 0.2 mM MnCl$_2$, ZnCl$_2$, CuCl$_2$, CoCl$_2$, NiCl$_2$	Volume decrease	Spinach	50 mM Tris·HCl + 100 mM sucrose, pH 8.5 (437 μg Chl/0.2 ml)	90° LS; PV	(2)
5 ~ 0.2 mM MgSO$_4$, MnSO$_4$, FeSO$_4$, CaCl$_2$, KCl	Volume decrease	Pisum (Sutton's variety 'Meteor')	250 mM sucrose + 27.5 mM Tris·HCl, pH 7.7	PV; PFW; 90° LS; fluorescence of chloroplast	(3)
10^{-2}–10^{-3} M NaCl, MgCl$_2$, CaCl$_2$	Volume decrease	Spinach (whole chloroplasts and fragments)	10 mM Tris·HCl, pH 7.8	PV; absorbance	(4)
5 mM MgCl$_2$, MnCl$_2$, MnSO$_4$, CaCl$_2$, CuSO$_4$, COCl$_2$, NiCl$_2$	Volume decrease	Spinach grana	20 mM Tris-maleate pH 7.0 (10 μg Chl/ml)	Absorbance (ratio 680:546 nm); PV; 90° LS; EM	(5)
		2. pH-Induced shrinkage			
0.05 N HCl	Shrinkage	Spinach	35 mM NaCl + 1 mM MgCl$_2$ + 100 μM K$_3$Fe(CN)$_6$	90° LS	(6)

Treatment	Organism	Effect	Medium	Method	Ref
0.05 N NaOH / 0.05 N KOH }	Spinach	Swelling	35 mM NaCl + 1 mM MgCl$_2$ + 100 μM K$_3$Fe(CN)$_6$	90° LS	(7)
0.05 N HCl, NaOH	Spinach	Shrinkage	100 mM NaCl + 20 μM PMS (10 μg Chl/ml)	90° LS EM	(8)
HCl	Spinach	Shrinkage	100 mM Sucrose (48.3 μg Chl/ml)	90° LS	(9)
0.05 N HCl, NaOH	Spinach	Shrinkage (contraction of thylakoid membrane)	100 mM NaCl (10–15 μg Chl/ml)	90° LS EM	(10)
3. Effects of heating					
40°	Spinach	Swelling	50 mM Tris·HCl, pH 7.8, + 175 mM NaCl	Absorbance (520 nm)	(11)
		Reversed by	ATP + MgCl$_2$ + serum albumin		
35°	Spinach	Swelling	50 mM Tris·HCl, pH 7.5, + 400 mM sucrose or 40 mM phosphate, pH 7.5, + 400 mM sucrose	PV; absorbance (520 nm); CC	(12)
		Reversed by	MgCl$_2$ + ATP + BSA		
4. Effect of fatty acids and lipids					
Sodium oleate (20 μM)	Spinach	Swelling	50 mM Tris·HCl, pH 7.8, + 175 mM NaCl (1 mg Chl/ml)	Absorbance (520 nm)	(13)
		Reversed by	ATP + MgCl$_2$ + BSA		
Sodium laurate or sodium oleate (300 μM)	Spinach	Dark: enhanced swelling	50 mM Tris·HCl, pH 7.9, + 175 mM NaCl (200 μg Chl/ml)	PFW:EM	(14)

TABLE AI-C (Continued)

VOLUME AND STRUCTURAL CHANGES INDUCED BY VARIOUS TREATMENTS in Vitro

Treatments	Changes	Materials	Reaction mixtures	Methods	References[a]
Lecithin (> 100 μM) or lysolecithin (> 100 μM)	Spinach	Light: inhibition of swelling Inhibition of light-induced swelling	50 mM Tris·HCl, pH 7.9, + 175 mM NaCl (200 μg Chl/ml)	PFW	(14)
5. Effect of solvents and detergents					
10^{-4} M Dodecylbenzene sulfonate	Spinach	Swelling (disintegration of lamellar structure)	350 mM NaCl + 40 mM phosphate buffer, pH 7.0	Absorbance (flattening effect); CC; EM	(15)
100–200 μM Triton X-100	Spinach	Swelling (disintegration of lamellar structure	100 mM NaCl or sodium acetate (15 μg Chl/ml)	CC; 90° LS; absorbance (546 nm); EM	(16)
Acetone or methanol	Spinach	Dark:<20%, swelling; >20%, inhibition of swelling Light: <10%, swelling; >10% inhibition of swelling	50 mM Tris·HCl, pH 7.9, + 175 mM NaCl	PFW; EM	(14)

[a]Key to references: (1) K. Nishida and K. Koshii, Physiol. Plant. 17, 846 (1964). (2) R. A. Dilley and A. Rothstein, Biochim. Biophys. Acta 135, 427 (1967). (3) A. P. Brown, Biochem. J. 102, 791 (1967). (4) N. Shavit and M. Avron, Biochim. Biophys. Acta 131, 516 (1967). (5) E. Gross and L. Packer, Arch. Biochem. Biophys. 121, 779 (1967). (6) Y. Mukohata, M. Mitsudo, and T. Isemura, Annu. Rep. Biol. Works, Fac. Sci., Osaka Univ. 14, 107 (1966). (7) Y. Mukohata, Annu. Rep. Biol. Works, Fac. Sci., Osaka Univ. 14, 121 (1966). (8) D. W. Deamer, A. R. Crofts, and L. Packer, Biochim. Biophys. Acta 131, 81 (1967). (9) R. A. Dilley and A. Rothstein, Biochim. Biophys. Acta 135, 427 (1967). (10) S. Murakami and L. Packer, J. Cell Biol. 47, 332 (1970). (11) Y. G. Molotkovsky and I. M. Zheskova, Biochem. Biophys. Res. Commun. 20, 411 (1965). (12) K. Nishida, N. Tamai, and T. Umemoto, Bot. Mag. Tokyo 79, 654 (1966). (13) Y. G. Molotkovsky and I. M. Zhestokova, Biochim. Biophys. Acta 112, 170 (1966). (14) S. Murakami and P. S. Nobel, Plant Cell Physiol. 8, 657 (1967). (15) M. Itoh, S. Izawa, and K. Shibata, Biochim. Biophys. Acta 69, 130 (1963). (16) D. W. Deamer and A. R. Crofts, J. Cell Biol. 33, 395 (1967).

TABLE AI-D

OSMOTIC SWELLING *in Vitro*

Materials	Reaction mixtures	Osmotic agents	Methods	References[a]
Spinach	50 mM Tris·HCl, pH 7.8, + 10 mM NaCl	0.05–0.4 M Sucrose	Phase contrast microscope; fluorescence microscope	(1)
Phytolacca americana	20 mM Tris·HCl, pH 7.5	0–0.5 M Sucrose	Gravimetry (weight); PV; absorbance (540:435 nm); light microscope	(2, 3)
Spinach		0.35 M NaCl, 0.5 M Sucrose	PV	(4)
Tomato, spinach, swiss chard, *Tetragonia*	25 mM Tris·HCl, pH 7.8	0–1.0 M Sucrose	PV; absorbance (OD 520 nm)	(5)
Spinach	SFDP	0.25–0.5 M Sucrose	Phase contrast microscope	(6)

[a]Key to references: (1) D. Spencer and S. G. Wildmena, *Aust. J. Biol. Sci.* **15**, 599 (1962). (2) K. Nishida, *Plant Cell Physiol.* **4**, 247 (1963). (3) K. Nishida and T. Hayashi, *Experientia* **21**, 705 (1965). (4) A. B. Tolberg and R. I. Macey, *Biochim. Biophys. Acta* **109**, 424 (1965). (5) S. Blumenthal-Goldschmidt and A. Poljakoff-Mayber, *Plant Cell Physiol.* **7**, 357 (1966). (6) T. Hongladarom and S. I. Honda, *Plant Physiol.* **41**, 1686 (1966).

TABLE AII-A

LIGHT-INDUCED SHRINKAGE OF CHLOROPLASTS *in Vivo*

Materials	Incubations	Illuminations	Methods	References[a]
Mnium undulatum	Tap water on microscope slide glass	240–340 nm 350–550 nm 650–850 nm	Light microscope (LM) Shape: round Surface area: decrease Surface area: increase (flattening)	(1)
Nitella flexilis	Water on slide glass	700 lux, 10 min	LM, chloroplast thickness decrease: 15–20%	(2)
Pisum sativum 'Blue bantam'	Leaf infiltrated with water	4000 lux, 10 min	LM, chloroplast thickness decrease ($2.68\ \mu \to 2.15\ \mu$)	(3)
Spinacia oleracea (leaf)	40 mM phosphate buffer, pH 7.2; fixation: 4% OsO_4	White light	EM, flattening of chloroplast $5.8 \times 3.4\ \mu \to 5.7 \times 2.2\ \mu$	(4)
Spinacia oleracea (leaf)	Infiltrated with 100 mM sodium acetate; fixation: 1% OsO_4	High intensity red light (600–700 nm)	EM, flattening of chloroplast and inner membrane system	(5)
Ulva sp.	100 mM sodium acetate, pH 6.7, + 2.5 mM $K_3Fe(CN)_6$; fixation: 1% glutaraldehyde and 1% OsO_4	Red light (600–700 nm)	EM; flattening of thylakoid and contraction of thylakoid membrane (25%)	(6)
Porphyra sp.	100 mM sodium acetate + 20 μM PMS; fixation: 1% glutaraldehyde and 1% OsO_4	Red light (600–700 nm)	EM; flattening of thylakoid and contraction of thylakoid membrane (23%)	(6)
Spinach leaf		Red light	Transmission decrease, 5.9%	(5)
Swiss chard leaf		Red light	Transmission decrease, 1.8%	(5)
Turkish tobacco leaf	20 mM Tris·HCl, pH 6.5, + 100 mM sodium acetate + 20 μM PMS	Red light	Transmission decrease, 1.3%	(5)
Senecio leaf		Red light	Transmission decrease, 9.0%	(5)
Senecio stripped epidermis		Red light	Transmission decrease, 0.8%	(5)
Senecio stripped leaf		Red light	Transmission decrease, 10.3%	(5)

Material	Medium/gas	Light	Measurement	Reference
Mimulus cardinalis Mimulus verbenaceus Pisum sativum Spinacia oleracea	CO₂-free air N₂ gas Low O₂ gas	Red light, high intensity Far-red light, low intensity Red or far-red light, high intensity	Absorbance (530 nm), increase	(7)
Ulva lobata	0.5% seawater	Red light	Transmission decrease, 8.0%	(5)
Ulva lobata	100 mM sodium acetate + 6 mM K₃Fe(CN)₆, pH 6.5	Red light	Transmission decrease, 15.6%	(5)
Ulva lobata	100 mM sodium acetate + 20 µM PMS, pH 6.5	Red light	Transmission decrease, 18.2%	(5)
Ulva sp.	100 mM sodium acetate + 2.5 mM K₃Fe(CN)₆	Red light	Transmission decrease, 10–15%	(6)
Enteromorpha sp.	Ten-layer thalli in seawater	Red light	Transmission decrease, 0.3%	(5)
Porphyra sp.	100 mM sodium chloride + 6 µM K₃Fe(CN)₆, pH 6.5	Red light	Transmission decrease, 10.7%	(5)
Porphyra sp.	100 mM sodium acetate + 20 µM PMS	Red light	Transmission decrease, 11%	(6)
Spinach leaf	Infiltrated with 20 mM Tris·HCl, pH 6.5, + 100 mM sodium acetate + 20 µM PMS	Red light	90° LS, increase 3.3%	(5)
Porphyra sp.	100 mM sodium acetate, pH 6.7, + 20 µM PMS	Red light	90° LS, increase 20%	(6)
Pisum sativum 'Lexton's Superb'	Leaf was illuminated in situ and then chloroplasts rapidly isolated were suspended in 200 mM sucrose plus 20 mM TES-NaOH buffer, pH 7.9	White light, 2000 lux	Coulter counter, $39\ \mu^3 \to 31\ \mu^3$	(8, 9)

^{a}Key to references: (1) J. Zurzycki, Protoplasma 58, 458 (1964). (2) H. Hilgenheger and W. Menke, Z. Naturforsch. B 20, 699 (1965). (3) P. S. Nobel, D. T. Chang, C. Wang, S. S. Smith, and D. E. Barcus, Plant Physiol. 44, 655 (1969). (4) H. Kushida, M. Itoh, S. Izawa, and E. Shibata, Biochim. Biophys. Acta 79, 201 (1964). (5) L. Packer, A. C. Barnard, and D. W. Deamer, Plant Physiol. 42, 283 (1967). (6) S. Murakami and L. Packer, Plant Physiol. 45, 289 (1970). (7) U. Heber, Biochim. Biophys. Acta 180, 302 (1969). (8) P. S. Nobel, Plant Physiol. 43, 781 (1968). (9) P. S. Nobel, Plant Cell Physiol. 9, 499 (1968).

[18] Circular Dichroism and Optical Rotatory Dispersion of Photosynthetic Organelles and Their Component Pigments

By Kenneth Sauer

The study of optical properties of photosynthetic materials has led to a wealth of information about their molecular composition and the functional relationships of the component molecules. Absorption spectra characterize the pigment molecules and reflect the heterogeneity of their sites or environments.[1] Transient changes in absorption produced by actinic light permit the detailed examination of reactive components and have helped to establish the manner in which they couple to one another. Fluorescence can be used as an indication of electronic excitation transfer, the state of activated sites, and the rates of depopulation or, in the case of delayed fluorescence, of repopulation of excited states. Activation spectra single out those components of the pigment array that are coupled to the photochemical reactions of photosynthesis.[2,3]

Within the past seven years much additional information has become available through the application to photosynthetic materials of the techniques of optical rotatory dispersion (ORD) and circular dichroism (CD). These spectroscopic measurements, which depend on the rotation of the plane of incident plane polarized light (ORD) or on the differential absorption of left- versus right-circularly polarized light (CD), date back to origins in the nineteenth century. Only recently, however, has instrumentation been available with sufficient sensitivity to measure the relatively small optical effects and render them useful for studies of photosynthetic materials.

Fundamental to the origin of the ORD or CD properties is the inherent (or induced) chirality or handedness of biological molecules. Molecules or structures which are superimposable with their mirror images do not exhibit these properties. There are three distinct types of asymmetries that result in optical activity. The first type arises from chirality intrinsic to the molecule itself. Most commonly this occurs be-

---

[1]"The Chlorophylls" (L. P. Vernon and G. R. Seely, eds.), especially Chapters 3, 4, 6, 11 12. Academic Press, New York, 1966.

[2]R. K. Clayton, "Molecular Physics in Photosynthesis" Blaisdell Publ., New York, 1965.

[3]"Progress in Photosynthesis Research" (H. Metzner, ed.), Vols. I–III. Tübingen, Germany, 1969.

cause of the presence in the molecule of carbon atoms which are bonded to four different substituents. Because of the tetrahedral nature of the bonding of saturated carbon, such structures cannot possess mirror symmetry. There exist, in addition, many molecules for which the symmetry test must be applied to the molecule as a whole. A simple example is hexahelicene, where the six fused benzene rings can form right- or left-handed structures that resemble lock washers.[4] Similar features occur in the urobilins[5] and may be present in the analogous chromophores of the phycobilins of marine algae[6,7] and in phytochrome.[8]

The second type of asymmetry occurs when a molecule, which may itself be symmetric, is placed in an asymmetric environment. Thus, an induced ORD or CD may result from the perturbation produced by an asymmetric solvent, by adsorption onto an asymmetric structure (e.g., protein or nucleic acid), or by the presence of an external magnetic field (MORD or MCD). The induced optical activity of the chromophore disappears, however, when it is removed from the asymmetric environment. Dyes such as acridine orange or proflavin adsorbed onto nucleic acids[9] or polypeptides[10] provide examples of this type of behavior.

A third type of asymmetry arises from the interaction of neighboring chromophores that are asymmetrically placed with respect to one another. This leads to degenerate or exciton ORD or CD features which are characterized by having multiple components for each electronic absorption band. This type of asymmetry is not independent of the other two, since the asymmetric arrangement arises only because of the presence of intrinsic asymmetry of the component molecules or an asymmetry of the host environment. Nevertheless, it is useful to treat it as a separate effect because the chromophore–chromophore interaction is the direct origin of the resulting optical properties, and because these components are frequently large in comparison with intrinsic components. The origin of the prominent ORD and CD bands of nucleic acids, proteins, and photosynthetic membranes lies primarily in this third area.

The theoretical origins of optical activity have been explored in terms

[4]A. Moscowitz, *Tetrahedron* **13**, 48 (1961).
[5]A. Moscowitz, W. C. Krueger, I. T. Kay, G. Skewes, and S. Bruckenstein, *Proc. Nat. Acad. Sci. U. S.* **52**, 1190 (1964).
[6]L. J. Boucher, H. L. Crespi, and J. J. Katz, *Biochemistry*, **5**, 3796 (1966).
[7]F. D. H. Macdowall, T. Bednar, and A. Rosenberg, *Proc. Nat. Acad. Sci. U. S.* **59**, 1356 (1968).
[8]H. H. Kroes, *Biochem. Biophys. Res. Commun.* **31**, 877 (1968).
[9]S. F. Mason and A. J. McCaffery, *Nature (London)* **204**, 468 (1964).
[10]L. Stryer and E. R. Blout, *J. Amer. Chem. Soc.* **83**, 1411 (1961).

of coupled oscillators and static field models,[11-14] and several recent reviews of the subject are available.[15-17] In addition, a number of excellent monographs describe both the experimental and theoretical aspects, as well as providing an extensive catalog of experimental data.[18-20] A review of Ke[21] covers some of the early applications to photosynthetic materials. Buckingham and Stephens have reviewed the subject of magnetic optical activity.[22]

Experimental

Most of the ORD and CD spectra published during the past five years have been obtained using commercial spectrometers, such as the Cary 60 (Applied Physics Corp., Monrovia, California) or the Durrum-JASCO (Durrum Instrument Corp., Palo Alto, California). These instruments are designed to achieve the high sensitivity and exceptional stability required for handling the small signals involved. Most current models have a serious shortcoming for photosynthetic materials, in that they are designed for optimum performance in the visible and ultraviolet region of the spectrum rather than the red or near infrared. They perform relatively poorly at wavelengths longer than 600 nm. For this reason, specially constructed instruments have been used for those measurements reported at the longest wavelengths.[23-25] Recently, the Durrum-JASCO J-20 spectropolarimeter has been made available with supplementary modifications which extend its useful range to 1000 nm.

The optical principles underlying an ORD or CD spectrometer are relatively simple, but the technical aspects of surmounting some of the

[11] W. Kuhn, *Annu. Rev. Phys. Chem.* **9,** 417 (1958).
[12] J. G. Kirkwood, *J. Chem. Phys.* **5,** 479 (1937).
[13] I. Tinoco, Jr. *Advan. Chem. Phys.* **4,** 113 (1962).
[14] J. A. Schellman, *Accounts Chem. Res.* **1,** 144 (1968).
[15] A. Moscowitz, *Proc. Roy. Soc. Ser. A,* **297,** 16 (1967).
[16] S. F. Mason, *Contemp. Phys.* **9,** 239 (1968).
[17] H. Eyring, H.-C. Liu, and D. Caldwell, *Chem. Rev.* **68,** 525 (1968).
[18] C. Djerassi, "Optical Rotatory Dispersion." McGraw-Hill, New York, 1960.
[19] P. Crabbé, "Optical Rotatory Dispersion and Circular Dichroism in Organic Chemistry," Holden-Day, San Francisco, 1965.
[20] L. Velluz, M. Legrand, and M. Grosjean, "Optical Circular Dichroism," Academic Press, New York, 1965.
[21] B. Ke, *in* "The Chlorophylls" (L. P. Vernon and G. R. Seely, eds.), p. 427. Academic Press, New York, 1966.
[22] A. D. Buckingham and P. J. Stephens, *Annu. Rev. Phys. Chem.* **17,** 399 (1966).
[23] E. A. Dratz, Ph.D. Thesis, University of California, Berkeley, 1966.
[24] M. Malley, G. Feher, and D. Mauzerall, *J. Mol. Spectrosc.* **26,** 320 (1968).
[25] B. Ke, R. H. Breeze, and M. Green, *Anal. Biochem.* **25,** 181 (1968).

problems encountered are formidable. The optical methodology is presented in some detail in the excellent monograph by Velluz et al.[20]

A simple way to outline the operation of an ORD or CD instrument is by comparison with a conventional absorption spectrophotometer. The light sources and scanning monochromators are essentially identical in the different types of instruments. The monochromatic beam is rendered plane polarized by passage through a suitable crystal polarizer of fixed (e.g., vertical) orientation. At this point the ORD and the CD instruments diverge. For ORD the plane polarized light is passed directly through the sample to be measured and then to a second crystal plane polarizer (the analyzer) essentially crossed to the first. If the sample does not rotate the plane of polarization, then no light will pass through the pair of crossed polarizers. If the sample does exhibit optical rotation, then the analyzer (or polarizer) must be rotated through an angle α in order to achieve the extinction condition. A plot of this angle versus wavelength constitutes on ORD spectrum. Since observed rotations are commonly of the order of millidegrees, some sophistication must be added in order to determine the extinction condition precisely. The light is measured using a suitable photomultiplier, amplifier, recorder arrangement. In order to increase sensitivity at the null point, a polarization modulation is superimposed on the optical beam after it has passed through the sample. This is done electrooptically using a Faraday cell in most modern instruments. The Faraday cell alternately rotates the plane-polarized light slightly to the right and then to the left at some convenient audiofrequency. This produces at the photomultiplier a modulated light intensity whose waveform is very sensitive to the position of the analyzer. The photomultiplier output is then used as an error signal to apply a fine correction to the analyzer orientation servomechanism.

For CD measurements, the monochromatic plane polarized light is passed into a quarter-wave plate, which produces either right- or left-circularly polarized light. In most automatic instruments these two polarizations are produced alternately using an electrooptical device (Fig. 1). Under these conditions the transmitted light, as monitored by a photomultiplier, appears as a full-wave rectified signal. The circular dichroism is proportional to the difference in amplitude between adjacent maxima. The problems associated with accurate measurement of CD are similar to those involved in absorption difference spectrophotometry, where the differences of interest are of the order of 10^{-3} compared with the absorbance in unpolarized light.

In general, the conditions required for precise CD and MCD measurements are similar to those for absorption measurements. Samples should

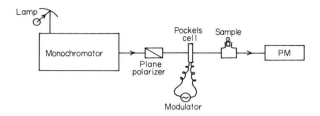

Fɪɢ. 1.   Schematic diagram of the optical components of a typical circular dichroism spectrometer. From E. A. Dratz, Ph.D. Thesis, University of California, Berkeley, 1966.

be relatively clear (although turbidities of the order of 0.3 appear not to interfere) and should have absorbances optimally in the range 0.5–1. For ORD measurements the requirements are more severe. Impurities which are optically active, but which have absorption bands outside the region of interest, may interfere because of the long Drude tails which extend in principle to infinite wavelength. The optical rotation of sucrose solutions in the visible region of the spectrum is a familiar example of this phenomenon. Likewise, any strains in the walls of the cuvettes result in a background signal which must be subtracted in careful work. Simply rubbing the outside surface of the cuvette windows with tissue can result in readily measurable changes in this optical artifact. These problems are much reduced in CD and MCD measurements. MORD is plagued by the additional problem that all cell windows and solvents, even if they are strain free and optically inactive at zero magnetic field, contribute significantly to the observed signal when the magnetic field is applied.

The current generation of ORD and CD spectrometers are all designed for single-beam operation. Accurate measurements require a high degree of instrument stability in order that suitable baseline corrections can be made. Because of the small signals involved, especially in CD measurements, it is becoming a common practice to use digital signal averaging techniques and multiple scans in order to improve signal-to-noise ratios. With these devices, the baseline correction can be applied conveniently by running the reference (solvent) spectrum in the subtract mode of the averaging device.

CD and MCD Spectra of Photosynthetic Materials

Chlorophylls and Related Molecules as Monomers in Solution

In order to illustrate the nature of CD and MCD spectra and their usefulness in interpreting the structure of photosynthetic materials, some examples from our laboratory are presented. In the presence of high magnetic fields the signals obtained from MCD measurements are typically much larger than those for the CD of the same solution. Examples of this phenomenon for seven chlorin free-base compounds were presented by Briat et al.[26] A comparison of the CD and MCD (in a field of 10 kgauss) for a solution of bacteriochlorophyll in ether is presented in Fig. 2. The CD and MCD features for isolated absorption bands (i.e., at 590 or 770 nm) have the same shape as the corresponding absorption bands, although there is the additional information residing in the sign of the circular dichroism phenomena. Where band overlap occurs, as in the 360–400 nm region, the CD and MCD effects for the individual components will combine arithmetically, with retention of their signs. This may result in a formidable problem of curve decomposition in order to "resolve" the individual components. The MCD spectrum contains less information relevant to the molecular structure, as it is not sensitive to the presence of molecular asymmetry.[23,27] The signs

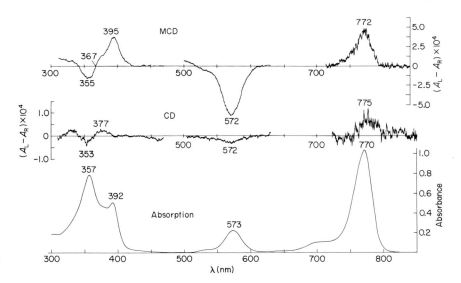

FIG. 2.   MCD (at 10 kgauss), CD, and absorption spectra of a solution of bacteriochlorophyll in ether. Concentration × pathlength is 1.1×10^{-5} mole-l^{-1}-cm. The MCD spectrum was obtained from 10 (blue), 2 (orange), and 4 (red) averaged scans; the CD spectrum from 40, 30, and 20 scans, respectively. Monochromator (Cary 14) slit width was fixed at 0.4 mm.

[26]B. Briat, D. A. Schooley, R. Records, E. Bunnenberg, and C. Djerassi, J. Amer. Chem. Soc. 89, 6170 (1967).
[27]C. Houssier and K. Sauer, J. Amer. Chem. Soc. 92, 779 (1970).

of the different components are determined largely by the vibronic mixing of a particular transition with other transitions of the same molecule; however, a completely satisfactory theoretical treatment for the origin of MCD in porphyrins has yet to be worked out.

The CD spectrum of bacteriochlorophyll occurs only because of the presence of asymmetrically substituted carbon atoms in the molecule. Of the seven such atoms in the molecule, five are closely associated with the porphyrin ring (two in each of the two saturated tetrapyrrole rings and one in the cyclopentenone ring) and two are in the phytyl esterifying group. The latter are relatively remote from the porphyrin chromophore and are thought not to contribute to the CD in the visible and near ultraviolet.[28] The former group of asymmetrically placed substituents, those directly adjacent to the porphyrin ring, interact with the electric dipole transition moments which give rise to visible and ultraviolet absorption and produce the observed CD bands (rotational strengths). In the chlorophyll-related molecules, the contributions from each of the asymmetric centers appear to be additive.[26-29] The molecule (or group) that is related in the sense of a mirror image to a given molecule (or group) contributes a component of equal magnitude, but of opposite sign. A relatively crude theoretical model based on the coupled oscillator theory of Kirkwood[12] appears to account for the signs of the CD components, but it greatly underestimates the magnitudes.[27] The use of point monopoles to represent the electronic transitions results in a considerable improvement in the agreement between experiment and theory.[28]

As a consequence of the relationship between the CD spectrum and molecular asymmetry, a useful application arises in assigning or confirming molecular structures. In this manner, CD and ORD spectra have been used to characterize an extensive series of chlorins that are enantiomeric at carbon atom C-10 (in the cyclopentenone ring).[28,29] ORD spectra have been used to confirm the assignment[30] of chlorophyll a' as the C-10 enantiomer of chlorophyll a.[27]

Chlorophyll Aggregates in Vitro

One of the most important features of the CD spectra of chromophores results from features that appear prominently as a consequence of molecular aggregation. Typically, for an aggregate containing N identical interacting molecules, the excited states will each be split into

---

[28]K. D. Philipson, S. C. Tsai, and K. Sauer, *J. Phys. Chem.* **75**, 1440 (1971).
[29]H. Wolf, H. Brockmann, Jr., I. Richter, C.-D. Mengler, and H. H. Inhoffen, *Justus Liebigs Ann. Chem.* **718**, 162 (1968).
[30]J. J. Katz, G. D. Norman, W. A. Svec, and H. H. Strain, *J. Amer. Chem. Soc.* **90**, 6841 (1968).

N components. These appear as individual bands (usually overlapping one another) in the absorption and CD spectra. Because of the sign differences involved in the CD components, these band splittings are much more prominent in the CD than in the absorption spectra.[31] This behavior is illustrated in Fig. 3 for bacteriochlorophyll dimers (present in carbon tetrachloride solution) in comparison with the corresponding monomer spectra. There is clear evidence in the CD spectrum of the dimers that each of the four major transitions observed in the absorption spectrum is associated with a double CD (here, both a positive and a negative component in each case). The band splitting also occurs in the absorption spectrum of the dimer solution, but it is unmistakable only for the longest wavelength band. The distinctive double CD features of the dimer should be symmetrical, in the sense that the positive and negative components for each band should have equal magnitudes and should cross the zero axis at the midpoint between the two components. The deviations from this behavior result from the simultaneous presence of the monomeric CD, somewhat enhanced by its altered environment in the dimer.[23] For the separated transitions near 580 and 770 nm in bacteriochlorophyll, it is evident that the distorting component in the dimer CD spectrum is of the same sign as the monomer component in each case.

The characteristic dimer CD results from the direct interaction between the transition moment oscillators of the two molecules. The inter-

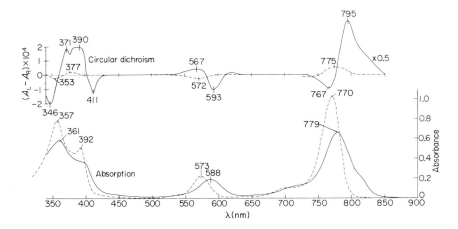

Fig. 3. CD and absorption spectra of bacteriochlorophyll as monomers in ether (dashed curves, 1.1×10^{-5} M, 1.0-cm path) and as dimers in carbon tetrachloride (solid curves, 1.1×10^{-4} M, 0.1-cm path).

[31]E. A. Dratz, A. J. Schultz, and K. Sauer, *Brookhaven Symp. Biol.* **19**, 303 (1966).

action has a vector character, such that magnitudes and the order of the signs depend intimately on the geometry of the dimer. Equations that apply to two interacting point transition dipoles have been presented by Tinoco.[32] The interaction decreases rapidly with increasing separation of the chromophores. For the chlorophyll solution dimers (Fig. 3), the distance between molecular centers is probably about 7 Å. In addition to the dimers, studies have been made of chlorophyll monolayers[33] and of microcrystalline chlorophyll.[31] It is clear from these studies that the physical state of the chlorophyll has a profound effect on the optical rotation spectra.

Photosynthetic Membrane Fragments

Pigment–protein–lipid complexes derived from active photosynthetic systems also possess distinctive ORD and CD spectra. Preparations of chloroplast lamellar fragments, capable of carrying out the photosynthetic electron transport reactions, exhibit a multiplicity of Cotton effects.[34,35] Some of these result from the multiplicity of pigments present; in particular, a mutant lacking chlorophyll *b* has a simpler CD spectrum, and those features attributable to chlorophyll *b* in the CD spectrum of the normal organism can be assigned.[31] Chromatophores from a variety of photosynthetic bacteria have been studied, and they show features attributable to carotenoids as well as to bacteriochlorophyll.[21,31] The latter exhibits multiple CD bands in the region from 800 to 900 nm, indicative of exciton interactions among these bacteriochlorophylls (Fig. 4). Of particular interest is a preparation of "reaction centers" from the purple bacterium *Rhodopseudomonas spheroides*. The reaction centers are obtained using detergents to remove selectively the "antenna pigments" that constitute the bulk of the normal bacteriochlorophyll absorption.[36] The resulting material retains the photoinduced absorption changes characteristic of this fraction of the bacteriochlorophyll in the intact chromatophores. In addition, the reaction centers exhibit characteristic changes in their CD spectra upon illumination or oxidation (Fig. 5).[37] We have interpreted the reduced or unilluminated spectrum in terms of a trimer of interacting bacteriochlorophylls, one of which bleaches upon illumination and thereby removes the interaction. The CD spectra of reaction centers of *Rhodopseudomonas viridis*, which contain

[32] I. Tinoco, Jr., *Radiat. Res.* **20**, 133 (1963).
[33] B. Ke and W. Sperling, *Brookhaven Symp. Biol.* **19**, 319 (1966).
[34] K. Sauer, *Proc. Nat. Acad. Sci. U. S.* **53**, 716 (1965).
[35] B. Ke, *Nature (London)* **208**, 573 (1965); *Arch. Biochem. Biophys.* **112**, 554 (1965).
[36] D. W. Reed and R. K. Clayton, *Biochem. Biophys. Res. Commun.* **30**, 471 (1968).
[37] K. Sauer, E. A. Dratz, and L. Coyne, *Proc. Nat. Acad. Sci. U. S.* **61**, 17 (1968).

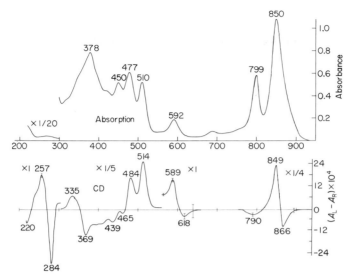

FIG. 4.   Absorption and CD spectra of chromatophores from *Rhodopseudomonas spher-oides*, wild type. Scale factors applying to different wavelength regions are shown above the CD curves.

a molecular variant of bacteriochlorophyll, have generally similar features.[38]

At least one example of a pigment–protein complex that serves an antenna or light-harvesting function, but has no measurable photo-chemical activity, has been studied. A bacteriochlorophyll-protein com-plex isolated from *Chloropseudomonas ethylica* by Olson[39] exhibits multi-ple ORD[21] and CD features. Both the orange and near infrared bands show strong evidence of pigment interactions (Fig. 6). The combination of low temperature CD and absorption spectra suggest the presence of 5 components, at 824, 815, 813, 805, and 792 nm, in the long wave-length band.[40] It is interesting to note that analytical and X-ray data show that the molecule contains 4 protein subunits each with 5 associated bacteriochlorophyll molecules. We conclude that because of their close proximity these molecules interact strongly with one another so as to split each excited state into 5 exciton components, but that the bacterio-chlorophylls of one subunit interact weakly or not at all with those of neighboring subunits.

[38]L. Coyne and K. Sauer, unpublished results.
[39]J. M. Olson, *in* "The Chlorophylls" (L. P. Vernon and G. R. Seely, eds.), p. 413. Academic Press, New York, (1966).
[40]K. D. Philipson and K. Sauer, *Biochemistry*, in press (1972).

Conclusion

The measurement of optical rotation, particularly CD, spectra can provide information relevant to a number of interesting molecular properties of chlorophylls. These include (1) determination of the absolute configurations at asymmetric carbon atoms of isolated chlorophylls; (2) characterization of the geometries of chlorophylls in dimers and higher aggregates; (3) evidence for the presence, number, and structures of associated chlorophylls in photosynthetic lamellae and information about changes in interaction upon activation by light; and (4) evidence bearing on the interactions of asymmetric environments, such as the protein matrix of the naturally occurring chlorophyll complexes.

In the future we may expect a better understanding of the theoreti-

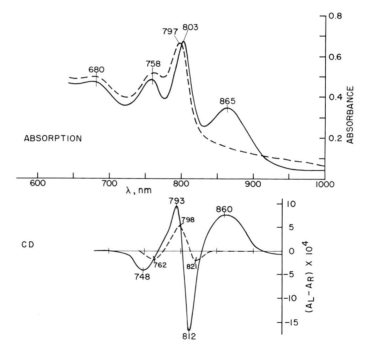

Fig. 5.   Absorption and CD spectra of reaction centers prepared from the blue-green (carotenoidless) mutant of *Rhodopseudomonas spheroides* (R-26), following the method of D. W. Reed and R. K. Clayton [*Biochem. Biophys. Res. Commun.*, **30**, 471 (1968)]. The solid curves correspond to the reduced or unilluminated material; the dashed curves were recorded in the presence of actinic light (550–600 nm). For details, see K. Sauer, E. A. Dratz, and L. Coyne, *Proc. Nat. Acad. Sci. U. S.* **61**, 17 (1968).

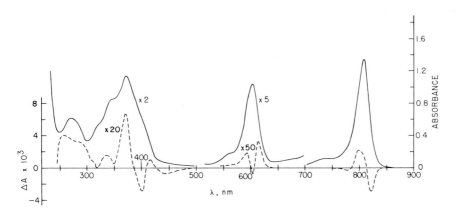

FIG. 6. Absorption (solid curves) and CD (dashed curves) of a bacteriochlorophyll-protein complex obtained from *Chloropseudomonas ethylica* [J. M. Olson, *in* "The Chlorophylls" (L. P. Vernon and G. R. Seely, eds.), p. 413. Academic Press, New York, 1966]. The author is grateful to Dr. John M. Olson for providing the sample and to Dr. E. A. Dratz for recording the CD spectra. The scale error referred to by Dr. Olson in his article [61] in Vol. XXIII of this series has been corrected in the figure shown here.

cal origin of the CD features of both individual and aggregated chlorophylls. Extension of the studies to a greater variety of photosynthetic materials will help to sort out those features common to reaction centers, to light-harvesting pigments, and to the association of chlorophylls with the electron transport components to which they are coupled. We are currently investigating the CD and MCD spectral changes associated with the conversion of protochlorophyll to chlorophyll following etiolation in higher plants.[41] Additionally, there are important changes that occur as chloroplasts synthesize more chlorophyll during maturation. While it is clear that much valuable correlative information is available from absorption, fluorescence, and structural measurements, the CD measurements provide a valuable tool in investigating the nature of pigment interactions in photosynthetic organelles.

Acknowledgments

The author wishes to acknowledge the very able assistance of Dr. Lelia Coyne, who obtained the CD and absorption spectra presented in this chapter.

The work described herein was supported, in part, by the U. S. Atomic Energy Commission.

[41] A. J. Schultz and K. Sauer, *Biochim. Biophys. Acta* **267**, 320 (1972).

[19] Estimation of Pool Sizes and Kinetic Constants[1]

By K. L. ZANKEL and B. KOK

In chloroplasts, electrons are driven from water to NADPH or other electron acceptors through a potential difference of >1.2 V by means of two series-connected photoacts. The photoproducts are processed in dark reactions which operate at both ends of the two photosystems and in the connecting chain. This essentially linear electron transport chain is linked to an ATP generating process which consumes part of the chemical potential generated by light. The electron transport chain contains a dozen or more components, each in very small concentration compared to the total chlorophyll (Chl). Observation of these components and that of the end products is done by a variety of direct or indirect methods, discussed in other parts of this volume. We will be concerned here with the nature and number of the electron transport components; particularly the interpretation of kinetic data concerning these intermediates. Any models presented are for illustrative purposes and are not an attempt to present a true and complete story.

Because of the ease with which chlorophyll can be determined, concentrations of the electron transport components are usually expressed in terms of total chlorophyll. One should realize that chlorophyll content varies with plant material and its pretreatment.

In strong, rate saturating light, green plants, algae, and well prepared spinach chloroplasts evolve up to 300–400 moles of O_2 per mole of chlorophyll per hour.[1a,2] This corresponds to 0.1 O_2/Chl·sec or approximately 0.5 electron equivalent/Chl·sec. Roughly then, assuming that 2 excitations are needed to move an electron through the chain (8 $h\nu/O_2$), we compute that at optimal rate each chlorophyll undergoes one useful excitation per second; i.e., at optimal rate the turnover time of the chlorophyll is 1 second.

Instead of giving continuous strong light, Emerson and Arnold[3,4] illuminated algae with repetitive short, bright flashes. They found that the individual flashes lost effectiveness in removing electrons from water if they were spaced closer than ~ 10 msec and that flashes, bright enough

[1]Preparation of this chapter was supported in part by the Atomic Energy Commission Contract No. AT(30-1)-3706 and National Aeronautics and Space Administration Contract ASW-1592.
[1a]S. Izawa and N. E. Good, *Plant Physiol.* **41,** 544 (1966).
[2]G. M. Cheniae and I. F. Martin, *Biochim. Biophys Acta* **153,** 819 (1968).
[3]R. Emerson and W. Arnold, *J. Gen. Physiol.* **15,** 391 (1932).
[4]R. Emerson and W. Arnold, *J. Gen. Physiol.* **16,** 191 (1932).

to saturate the photochemical events, could not process more than 1 electron per ~500 Chl. Optimal results in similar experiments[5] range to values of ≈5 msec and 1 eq/400 Chl, which is in rough agreement with the turnover time/Chl given in the preceding paragraph. The interpretation is that there is one electron transport system for each 400 Chl in the chloroplast. This is substantiated by the observation that certain important constituents, such as cytochrome f[6] and ferredoxin,[7] occur in an abundance of about 1/400 Chl. If we now assume that the total chlorophyll is equally divided between the two photosystems, the picture emerges (Fig. 1) of separate electron transport chains which each comprise 2 photosystems (I and II). Each photosystem contains some 200 light harvesting pigment molecules from which excitations flow to a trapping center where the photoconversion occurs. Arrival of a quantum in each reaction center results in a charge separation. In photosystem I an electron moves from the primary photooxidant P700 (a special chlorophyll complex of medium potential) to the primary photoreductant "X" which has a very low midpoint potential but is not further identified. In photosystem II the electron is transferred from an unidentified high potential oxidant "Z" to a medium potential reductant "Q", possibly a quinone. In subsequent dark steps Z^+ is reduced, ultimately by water, while O_2 is set free and X^- is reoxidized— ultimately in whole cells by CO_2. $P700^+$ is reduced by Q^- via the interconnecting reaction chain—probably including the site which generates ATP. The ATP is utilized in the fixation of CO_2. In isolated chloroplasts, one can arrange conditions so that only part of the chain is operative and suitable artificial electron donors or acceptors can replace the natural ones.

Of the 4 primary photoproducts, only P700 can be monitored directly. This can be done by spectroscopically viewing the loss or gain of absorption at 700 nm upon oxidation or reduction. The redox state of

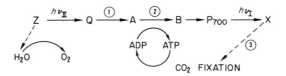

FIG. 1.   Scheme for the noncyclic electron flow from water to CO_2. Dashed lines indicate components omitted from the scheme.

[5]G. M. Cheniae and I. F. Martin, *Biochim. Biophys. Acta* **197**, 219 (1970).
[6]H. E. Davenport and R. Hill, *Proc. Roy. Soc. Ser. B* **139**, 327 (1952).
[7]H. E. Davenport, *in* "Non-Heme Iron Proteins" (A. San Pietro, ed.), p. 115. Antioch Press, Yellow Springs, Ohio, 1965.

Q can presumably be viewed by the fluorescence yield of photosystem II, an indirect but technically rather simple method. No direct observations have been made of Z, X, or most of the other components of the chain except for cytochromes and quinones which undergo adequate absorption changes upon oxidation or reduction. Thus, much of the kinetic information is obtained by indirect approaches in which one monitors products such as O_2, reduced acceptors, ATP, or pH.

We will discuss methods for determining the nature and number of components in the electron transport chain and the rates at which the equivalents are transferred from one component to another. First, we will discuss the primary photochemical donors and acceptors, then the pools of components in the chain, and finally some aspects of kinetics applied to photosynthetic systems.

Observation of P700 and Its Immediate Electron Donors

Light absorbed by photosystem I causes the photooxidation of P700, which is characterized by a loss of absorbance[8] and the formation of an EPR signal.[9] The photoconversion can take place in a single very brief flash and at temperatures as low as that of liquid helium.[10] This indicates a primary event, i.e., not a transfer of electrons in a thermal dark reaction. Under appropriate conditions one can slowly reduce P700 in dark and subsequently view the time course of its photooxidation in weak light. The kinetics of this conversion also indicate that the oxidation of P700 is a primary event: If the acceptor X is maintained in the active oxidized state, the rate of P700 photooxidation at any moment is proportional to the amount of reactive P700 (P700 still reduced) resulting in a first-order time course of this photooxidation. The rate of this oxidation is proportional to the number of incident photons per unit time so that the half-time of the photooxidation is inversely proportional to intensity. The observations are consistent with:

$$\frac{dP}{dt} = -\alpha_1 \, I\Phi_1 \, P/P_{tot,}$$

where Φ_1 is the quantum yield when all P is reduced, α_1 the fraction of photons absorbed by photosystem I, I the number of incident photons, P the amount of reduced P700 and P_{tot} is the total amount of P700. Assuming that the extinction coefficient of P700 is the same as that of the light-harvesting chlorophyll, one computes from the maximum change in absorbance upon photooxidation an abundance of the order of one P700/400 Chl.

[8]B. Kok, *Biochim. Biophys. Acta* **22**, 399 (1956).
[9]E. C. Weaver, *Annu. Rev. Plant Physiol.* **19**, 283 (1968).
[10]W. Arnold, personal communication.

In green cells, cytochrome f and plastocyanin act as electron donors to P700$^+$ and occur in a similar concentration, probably in a 1:1 ratio.[6,11] It is somewhat difficult to isolate the photochemical conversion of P700 since these donors react rapidly with P700$^+$, even at low temperatures.[12] For instance, at $-196°c$ (in liquid N$_2$) light photooxidizes both P700 and cytochrome f (denoted "C"), and even here the transfer can occur in a few milliseconds.[13] The ability of cytochrome to react rapidly with P700 at low temperature indicates that collision chemistry is not involved in the transfer and that the two components are arranged as pairs in close proximity to each other.

The Photoreductant of System I

The chemical nature of the primary electron acceptor in photosystem I is unknown ("X"). One *a priori* expects that its abundance is equal to that of P700 and that the two form a complex. X$^-$ is a strong reductant which can reduce many electron acceptors including low potential dyes like methylviologen. The semiquinone of viologen has a strong absorption band so that one can spectroscopically view its formation after a short light flash — and its subsequent disappearance due to reoxidation by O$_2$. As expected, the reaction $X^- + V + H^+ \rightarrow VH + X$ proceeds faster the more viologen (denoted V) is present. At a concentration of 10^{-3} M, the reaction is half completed in less than a millisecond. Indophenol dye reacts with even greater rapidity.[14] The rate of the natural reoxidation of X via ferredoxin and NADP reductase to NADP is not known; however, measurements with chloroplast fragments and exogenous donor indicate it is greater than 100/second.[15] For chloroplasts in the absence of an exogenous electron acceptor there remains a rather variable residual rate of electron transport in air (Mehler reaction),[16] indicating that X$^-$ reacts with O$_2$ in about 1 second.

The Photoreductant of Photosystem II

The chemical nature of the primary electron acceptor "Q" in photosystem II is also not known. It is speculated that it may be a plastoquinone. Therefore indirect means must also be used to determine the

[11]S. Katoh, I. Suga, I. Shiratori, and A. Takamiya, *Arch. Biochim. Biophys.* **94,** 136 (1961).
[12]B. Chance and W. D. Bonner, Jr., *in* "Photosynthetic Mechanisms in Green Plants," *NAS-NRC* **1145,** 66 (1963).
[13]B. Chance, R. Kihara, D. DeVault, W. Hildreth, M. Nishimura, and T. Hiyama, *in* "Progress in Photosynthesis Research" (H. Metzner, ed.), p. 1042. International Union of Biological Science, Tübingen, 1969.
[14]B. Kok, S. Malkin, O. Owens, and B. Forbush, *Brookhaven Symp. Biol.* **19,** 446 (1966).
[15]S. Katoh and A. San Pietro, *J. Biol. Chem.* **241,** 3575 (1966).
[16]A. H. Mehler, *Arch. Biochem. Biophys.* **34,** 339 (1951).

redox state of this component. These methods are based on the assumption that photochemistry can proceed only when the acceptor is in the oxidized, active state.

Fluorescence, being relatively easy to measure, is often used as a tool for monitoring the state of Q. The reasoning used is as follows: A quantum absorbed by one of the chlorophylls in a unit wanders through the pigment array until it encounters an active reaction center where it carries out a photoconversion. These events must take place in less than $\sim 10^{-8}$ second or else the excitation is degraded into heat or reemitted as (~ 700 nm) fluorescence. Blocking of the photochemistry is therefore accompanied by an increase in fluorescence; and fluorescence changes occur in a manner complementary to the photochemical rates.

Photochemistry in the two systems is blocked if any of the two donors or two acceptors are in the inactive state. However, no significant fluorescence changes have been observed which are conclusively attributed to the state of the photosystem I traps, a fact for which we have no ready explanation. Furthermore, fluorescence changes are usually attributed exclusively to the state of the photosystem II acceptor Q, since it is assumed that under most conditions the high potential photooxidant Z remains in its reduced, photoactive state.

The sum of the quantum yields for fluorescence f, degradation into heat H, and photochemistry r must be equal to unity ($f + H + r = 1$). The fluorescence yield then increases as the photosystem II traps close; i.e., $f \rightarrow 1 - H$ when $r \rightarrow 0$. Most fluorescence analyses assume that f is proportional to H. If this postulate is made, $r = 1 - F$; where $F = f/f_{max}$ and f_{max} (see Fig. 2) is the fluorescence for all traps closed, $r = 0$. With this postulate, changes in photochemical yields are equal to normalized negative changes in fluorescence yields. We can compare with the photochemical rate rather than the yield, since the quantum yield for

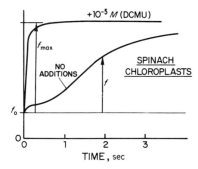

FIG. 2.   Fluorescence yield (arbitrary units) for isolated dark-adapted chloroplasts as a function of time in the absence of DCMU and in the presence of DCMU.

photochemistry is the rate per absorbed intensity; the latter is expressed in quanta per unit time. For a given intensity, the rate R is proportional to the quantum yield; i.e., $R =$ constant $\times r$. In addition to the variable fluorescence f described above, there is a constant fluorescence yield f_0, which is unaffected by the photochemistry. A problem encountered is the question whether f_0 is the fluorescence due to some small amount of disconnected pigment (dead fluorescence) or is due to fluorescence in competition with photochemistry. For either case $R/R_{max} = 1 - F$, where R_{max} is the maximum rate and f and f_{max} are measured from f_0. For the latter case (no dead fluorescence), f_0 must be included in f for computation of the quantum yield r.

The variations in fluorescence do tend to behave antiparallel to the rate of electron transport of photosystem II; i.e., variations of the rate of photochemistry as measured by the rate of oxygen evolution follow the properly normalized variable fluorescence yield.[17] With an artificial electron donor, such as hydroxylamine, replacing water, the fluorescence transient is not changed[18]; this supports the notion that one exclusively views the phenomena on the reducing side of photosystem II.

It is important to note that although the assumption that H is proportional to f does appear to be valid in many cases, and is used extensively in interpreting fluorescence data, the literature abounds with evidence that the situation is not so simple. One exception is seen in whole cells when considering long-term effects (~ 1 minute or longer).[19-21] Another exception to the simple behavior is observed at the onset of illumination after a dark period where F shows a peculiar rapid rise.[22] Still another problem encountered is the inability to induce the full fluorescence rise in less than about 50 msec, no matter how intense the light.[23] It should be mentioned also that we have considered only noncyclic flow through photosystem II. The O_2 evolution system is rather fragile, and Z^+ is a labile photoproduct that can oxidize substances other than water, probably including Q^-, so that a cyclic flow in photosystem II may occur. In spite of these reservations, fluorescence has been a very useful tool for the study of Q. Our application in the following will be for situations in which the deviations probably play little or no role.

[17]Y. de Kouchkovsky and P. Joliot, *Photochem. Photobiol.* **6**, 567 (1967).
[18]P. Bennoun and A. Joliot, in press.
[19]G. Parageorgiou and Govindjee, *Biophys. J.* **8**, 1316 (1968).
[20]T. T. Bannister and G. Rice, *Biochim. Biophys. Acta* **162**, 555 (1968).
[21]N. Murata, *Biochim. Biophys. Acta* **189**, 171 (1969).
[22]P. Joliot, *Biochim. Biophys. Acta* **162**, 243 (1968).
[23]P. Morin, *J. Chem. Phys.* **61**, 674 (1964).

Assuming that Z is always reduced, the rate of photosystem II photochemistry is proportional to the amount of Q in the active, oxidized state. This gives $Q/Q_{max} = R/R_{max} = 1 - F$. Actually the relation between Q and R (and therefore F) is not quite linear.[24] This nonlinearity is ascribed to a certain degree of cooperation between photosystem II pigment units—an absorbed quantum which finds the nearest center closed having a chance to travel to a neighboring unit. The reader is referred to the literature for this complicating phenomenon.[24,25] For simplicity we will adhere to the linear relation between Q, F and R.

The poison DCMU presumably prevents the normal rapid reoxidation of Q^- by its reaction partner A, discussed below. A slow dark oxidation remains, however, so that in darkness Q^- is returned to Q in about a second. If one now admits a light beam, the incoming quanta progressively convert Q to Q^-, and the fluorescence yield rises (Fig. 2). Since in the presence of DCMU Q^- is reoxidized slowly, reoxidation during this conversion can be neglected in all but the weakest intensities. Assuming that Z remains reduced, we expect that the photochemical rate is proportional to oxidized Q and the conversion should, as in the case for P, proceed in a first-order manner:

$$-\frac{dQ}{dt} = \alpha_2 \, I\Phi_2 \, Q = Q_{tot} \frac{dF}{dt}$$

Due to the excitation energy transfer between units mentioned above, this expectation proved only approximately fulfilled.[24]

In the presence of DCMU the conversion $Q \rightarrow Q^-$ as defined by fluorescence behaves as a purely photochemical act: (a) the rate is inversely proportional to light intensity; (b) most of the conversion can be carried out by a single 10 μsec light flash, provided it is bright enough to deliver sufficient quanta; (c) the phenomenon still occurs if the sample is dark restored at room temperature and then prior to illumination is cooled with dry ice.[26] Since the DCMU rise curve reflects the rate of photoconversion of the photosystem II traps in a sample, its half-time can be used as a rough measure of the effective photosystem II intensity of a light beam.

Pools: General Considerations

In speaking of pools we are concerned with the number of electrons that can be stored in either part or all of the chain between Q and P.

[24]A. Joliot and P. Joliot, *C. R. Acad. Sci.* **258**, 4622 (1964).
[25]R. K. Clayton, *J. Theoret. Biol.* **14**, 173 (1967).
[26]P. Joliot, personal communication.

Since we have not been able to measure all these storage components directly, indirect methods have been used.

In order to discuss these methods, we will first consider the problems involved in measuring the capacity of ordinary swimming pools. Suppose that we have, as in Fig. 3, two pools, A and B. We can measure their total capacity by emptying the pools and then measuring the total amount of water needed to fill the pools; alternatively we can fill the pools and measure the amount of water we can pump from them. It may be easier to measure the rate of water flowing through our pumps, in which case we can integrate this rate over the time necessary to fill or empty the pools: pool size = ∫ rate dt. Since our pools are leaky, we must pump fast enough so that we can neglect the water that leaks out. We could check this by doubling our pumping speed and seeing whether we get the same result.

If we wanted to measure the capacity of A alone, we must try to fill fast enough so that flow from A to B is negligible, and similarly to get the size of B we must empty B fast enough so that little water flows from A. Again we can check by varying the pumping speed.

We can now discuss photosynthetic systems by replacing our pools with electron transport components, the water by electron equivalents, the pumps by the photoacts and the connection between the pools by some (at times) rate-limiting step. We can measure the number of equivalents put into an emptied pool by the amount of either added or natural photosystem II donor oxidized during the filling. We can measure the number of equivalents removed from the pool from the behavior of photosystem I acceptors. We can also make pool size determinations from the rate of flow at either end.

Different pools can be studied by altering the light intensity (rate of pumping) and the rate-limiting steps. If we find that over a wide range

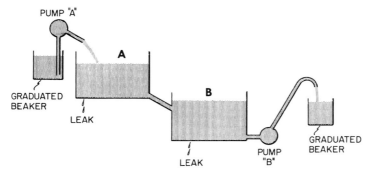

FIG. 3.   Model for pool size measurements.

of light intensities the pool size determinations are independent of light intensity, then we feel we are viewing a true pool of intermediates.

We can empty the pool with far-red light which is almost exclusively effective for photosystem I. (Care must be used to make this light not too strong, particularly without a photosystem I acceptor, or the small photosystem II component can become appreciable.) No light is exclusively effective for photosystem II, and we have to resort to rate limitations for filling.

Pool Size Determinations — Rate Limitations

Since rate limitations are used to separate pools, we will discuss these limitations before going on to specific examples of pool size determinations. The rate limitations depend very much on the system being studied. It is not always clear where the limitation occurs, and indeed this is one of the problems with pool size determinations; i.e., to know what pools are being measured.

The rate limitation in isolated chloroplasts without acceptor is in the transfer of electrons from $X \rightarrow O_2$ (Mehler reaction), i.e., at 3 in Fig. 1. This rate is on the order of 1 electron per second. Therefore, bright light can be used to fill (reduce) the entire chain in well washed chloroplasts.

This rate limitation is removed by the addition of an electron acceptor such as methylviologen. Now the rate limitation depends on the state of the ATP generating system. If the ATP apparatus is functional but "constipated" (a state which can be obtained by not supplying phosphorylation substrates such as ADP, P_i, etc.), then there is a limitation at the ATP site (2 in Fig. 1) of about 10 per second.[1,27] Therefore, for well prepared chloroplasts with an acceptor but no ADP, etc., bright light probably fills all pools up to an ATP site and empties all pools on the other side.

The limitation at the ATP site is relieved if the ATP process is short-circuited by an uncoupler. As will be discussed later, A possibly does not get fully reduced in uncoupled chloroplasts (with acceptor); i.e., A becomes reduced until the transfer time of $Q^- \rightarrow A^-$ equals the transfer time of $A^- \rightarrow B^-$. (The $Q^- \rightarrow A^-$ rate is proportional to A oxidized, while the $A^- \rightarrow B^-$ rate is proportional to A reduced.) The rate limitation is therefore possibly at A (1 in Fig. 1). The steady-state rate for such uncoupled chloroplasts is about 100 equivalents per second.[1] This is in agreement with the time for reduction of P700 after

[27]B. Kok, P. Joliot, and M. McGloin, *in* "Progress in Photosynthesis Research" (H. Metzner, ed.), p. 1042. International Union of Biological Sciences, Tübingen, 1969.

the cessation of bright light of ~10 msec and with the turnover time for a component of A, plastiquinone, which will be discussed later.

The situation in whole cells is not entirely clear. The rate in whole cells is comparable to that in uncoupled chloroplasts (with acceptor),[1] indicating that the cells behave almost like such chloroplasts. This is to be expected since both electrons from photosystem I and the ATP are being efficiently utilized. (Efficient utilization of ATP should give the same result as short circuiting the ATP system.) However, limitations might arise at the ATP site, in a way that balances the ATP production and CO_2 reducing processes.

Pools as Viewed from the Photosystem I Side

As was indicated above, with strong white light we can fill the pools from Q to the ATP site in isolated chloroplasts in the presence of a photosystem I acceptor (such as methylviologen). We can then use weak to moderate far-red light (effective mainly for photosystem I) to empty it, while measuring the rate of acceptor reduction. (The modulated polarograph[28] arranged to observe methylviologen reduction rate is best used for this.) The integral $\int Rdt$ is then a measure of the A pool. In the presence of an acceptor, the rate R_1 of photooxidation on photosystem I should be proportional to the amount of P700 in the reduced state. That the rate of methylviologen reduction does indeed parallel the behavior of reduced P700[27] is shown in Figs. 4 and 5. Initially P700 is fully reduced and the (modulated) photosystem I rate is maximal.

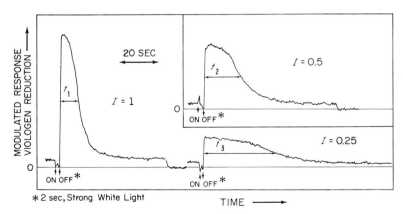

FIG. 4. Rate of viologen reduction as a function of time for three different intensities of far-red light for emptying the pool.

[28]P. Joliot and A. Joliot, *Biochim. Biophys. Acta* **153,** 625 (1968).

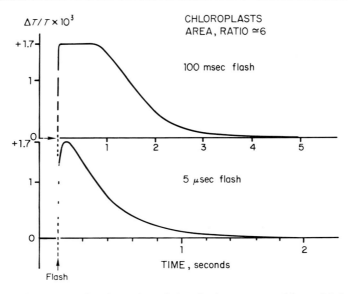

Fig. 5. Time course for absorption of P700 in the presence of far-red light after the pool was filled with a long flash (top) or partly filled by a brief flash (bottom).

At the end of the transient, P700 becomes largely oxidized and the rate very low because the reduced pool is depleted.

Since X is kept oxidized by an acceptor, $R_1 = \alpha_1 I \Phi_1 P/P_{tot}$ and $\int PIdt$ is proportional to the total number of equivalents. This integral is I times the area under the curve of P vs t given in Fig. 5. For a fixed far-red measuring beam the area is therefore proportional to the number of equivalents. If all the electrons are measured, one should get the same area for different light intensities if one plots P against $I \times t$ (i.e., photons) instead of t. As indicated previously, such a check of using different light intensities is necessary to ensure that all the electrons are counted and that the measurements are not limited by some dark reaction or leaking. The pools measured as in Figs. 4 and 5 are independent of light intensity over a large range of intensities. If the measurements are made in an intensity range such that dark reactions have no influence (are fast compared to the quantum flux), the kinetics of the P700 changes (the shape of the transient) should also be independent of intensity when plotted against $I \times t$. It is conceivable that dark reaction times alter the kinetics, but not the area or total equivalents counted. No such kinetic changes have been reported, perhaps owing to the fact that emptying times have been rather slow (seconds).

The areas shown above give a relative measure of pool size. It is sometimes difficult to determine absolute magnitudes of the pool. To be meaningful, these magnitudes must be compared to some standard, such as total chlorophyll, or to the amount of a particular molecule, such as P700. The areas above are referred to the number of P700 molecules in the following manner: Measurements[29] (which will not be discussed here) indicate that after about 20 minutes in the dark only P700 and another partner molecule are reduced. The area obtained when P700 is oxidized by far-red light is then two equivalents. All other areas can then be related to this "standard." Such measurements indicate a pool of ~10 equivalents/P700.

Care must be exercised with the use of artificial electron acceptors since these may entertain a cyclic electron flow. For example, a reduced photosystem I acceptor may donate its electron directly or indirectly back to $P700^+$ in competition with the flow from the pools. Viologens ($\rightarrow O_2$), ferricyanide and ferredoxin ($\rightarrow$ NADP) are commonly used acceptors which probably do not cycle.

Pools as Viewed from the Photosystem II Side

Pools can also be measured from the photosystem II side of the chain by giving light strong enough to fill (reduce) the chain up to some rate-limiting step and counting the number of equivalents put into the system to reach the steady state. The rate R of oxygen evolution can be measured while filling. The number of equivalents put into the system is then $4 \int_0^\infty R dt$. (The factor 4 comes from 4 equivalents per O_2 molecule.) The rate of filling must be much faster than the rate-limiting step, or subsequent flow negates the results. For example, the Mehler reaction may add oxidizing equivalents into the system if the measurement is too slow. The influence of this and of other dark reactions can be checked as for P700 by doing the experiments at different intensities and plotting the results in terms of $I \times t$.

Alternatively, since the rate of photosystem II is related to fluorescence yield, the integral $\int_0^\infty (1 - F) dt$ is also proportional to the number of equivalents put into the chain. This is the area above the rise curve for fluorescence (see Fig. 6). The area so obtained in the absence of an acceptor is about 20 times larger than that obtained in the presence of the poison DCMU.[30] Assuming that only Q is reduced in the presence of DCMU, whereas the whole chain is reduced in its absence, there are 20 equivalents in the chain for each Q. Again the

[29]T. Marsho and B. Kok, *Biochim. Biophys. Acta* **223**, 240 (1970).
[30]B. Forbush and B. Kok, *Biochim. Biophys. Acta* **162**, 243 (1968).

$I \times t$ law is fulfilled — be it now only over a restricted range of intensities where the dark reactions can keep pace with the quantum flux. Therefore it appears that a large pool of intermediates, made up largely of secondary reductants, is viewed. The complex time course of F and changes of kinetics with light intensity show that this pool is heterogenous.

In Fig. 6 is illustrated how the equivalents reduced during the rise were quantitated relative to the total chlorophyll.[31] If, after F_{max} was attained, a small, measured amount of acceptor (ferricyanide) was added, the fluorescence showed a temporary dip. The area bounded by this dip now reflected the number of added ferricyanide equivalents, all of which were reduced by the light. Therefore this added area could be used as a standard. The ratio of the area over the rise curve and the calibrated (standard) area yielded the size of the total internal reductant pool. With measurements of absorbed intensity, the quantum yield of the conversions could be estimated. It was found to be high (~ 1 eq/2 $h\nu$). Estimates of this type for the pool of internal reductants yielded values $\sim 1/35$ Chl, an order of magnitude greater than that of the trapping centers. This is in agreement with the pool size found (see above) using DCMU as a reference. The pool size of plastoquinone determined by absorption changes at 254 nm appears to be about 14 equivalents (7 molecules) per chain.[32]

Kinetics: General Principles

The behavior (redox state) of a component or pool X is governed by the flow of equivalents into it (f_{in}) and the flow out (f_{out}). The rate of change of X^- is the sum of these.[33] In kinetic analyses of X, one determines f_{in} and f_{out} and their functional dependences, so as to under-

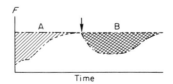

FIG. 6.   Fluorescence yield (arbitrary units), for isolated chloroplasts as a function of time. The shaded area A is obtained in the absence of an acceptor. An additional area B was obtained when the acceptor ferricyanide was added at time designated by arrow.

[31]S. Malkin and B. Kok, *Biochim. Biophys. Acta* **126**, 413 (1966).
[32]H. H. Stiehl and H. T. Witt, *Z. Naturforsch.* **246**, 1588 (1969).
[33]The capital letters will be used to denote concentrations of components and the superscript minus to designate the reduced state.

stand the mechanisms of the underlying processes. The flow rates depend on the redox state of X itself, or its electron donor (D) and of its electron acceptor (A). For example, the flow to a completely reduced component is zero. If the component is a primary photochemical acceptor or donor, the rate depends on light intensity (I). In addition the flow depends on other environmental factors, such as temperature, state of the membrane, pH. We have then

$$\frac{dX^-}{dt} = f_{in}\,(X^-, I, A, D) - f_{out}\,(X^-, I, A, D).$$

Techniques used in determining these functions involve a change of the system and a study of the subsequent time course of X and (when possible) other parameters. Isolation of each of the two functions (f_{in} and f_{out}) can often be achieved by blocking one of the transfers. For example, the flow from Q can be blocked chemically with DCMU, or flow from a component might be blocked by reducing its acceptor (A). Particular transfers may be lacking in mutants. If one of the rates (for example f_{in}) is light driven, the dark reaction (f_{out}) can be followed after removing the light.

To illustrate we will take a simple example in which light (I) reduces X and the electron is then transferred in a dark reaction to some component Y:

$$XY + h\nu \rightarrow X^-Y \xrightarrow{k'} XY^-.$$

Let us assume that f_{in} is proportional to X and I and that f_{out} is proportional to the number of donors X^- and acceptors Y. We further assume that Y is held constant (for example, by an excess of external oxidant). Then f_{out} is proportional to X^- alone and

$$\frac{dX^-}{dt} = \alpha\,\phi I\,(X_T - X^-) - kX^-$$

where (since Y is constant) $k = k'Y$ and the subscript T denotes total concentration of the component. For $X^- = X_0^-$ at $t = 0$, the solution to this equation yields:

$$X^- = X_F^- - (X_F^- - X_0^-)e^{-Kt},$$

where X_F^-, the steady-state level ($t \rightarrow \infty$) of X^-, is given by

$$X_F^- = \frac{\alpha\,\phi I}{k + \alpha\,\phi I}\,X_T$$

and $K = \alpha\,\phi I + k$. An illustration of the behavior of X^- at various light intensities for $X_0^- = 0$ is given in Fig. 7. The decay in the dark is given by $X^- = X_0^- e^{-kt}$. The dashed lines of Fig. 7 indicate the behavior when the light is extinguished after various levels of X^- are reached. There are

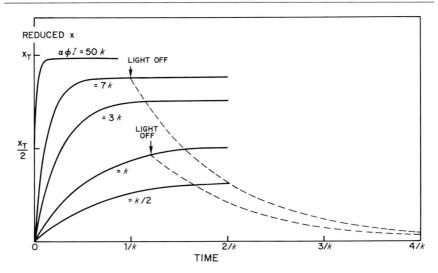

FIG. 7. X^- as a function of t for simple first-order behavior. Solid lines, light of given intensities turned on at $t = 0$ with $X_0^- = 0$. Dashed lines, light turned off at arrow.

several points to note. The steady-state level X_F^- asymptotically reaches X_T with increasing light intensity ($\alpha\,\phi I \gg k$). Since initially $X_0^- = 0$, the dark decay f_{out} is initially zero. The initial rate of f_{in} therefore is proportional to light intensity. The rise and decay are first order. One can demonstrate this for the rise by plotting $\log(X_F^- - X^-)$ vs time. This gives a straight line with slope $-K$. Similarly the decay gives a straight line with slope $-k$ when the $\log X^-$ is plotted as a function of time. Also, since the dark decay does not depend on how X^- is reached, the decays represented by the two dashed curves in Fig. 7 should be superimposable if one of the curves is shifted in time.

   Let us now turn the problem around, and assume that we are able to follow the rise and decay of X^- and want to determine f_{in} and f_{out}. To check whether these functions satisfy the first-order behavior described above, we can look at log plots and we can check whether we find the appropriate intensity dependence of the steady-state level and the initial rate.

   As an example of the above considerations, we will discuss some data on the redox state of plastoquinone in the presence of a system I acceptor and an uncoupler. Changes in optical density at about 254 nm have been attributed to changes in the redox state of plastoquinone.[34]

[34]H. H. Stiehl and H. T. Witt, Z. Naturforsch. **236,** 220 (1968).

These changes[35,36] appear rapidly (≤ 50 μsec) and have kinetics which suggest that pool A is plastoquinone. Although the data and interpretation may be obsolete in light of newer information,[32] they are presented because of pedagogical merits. The model assumes that at saturating light intensities all Q is reduced and that the time for the initial reduction of Q can be neglected. The rate between Q and A then is proportional to A. A, in turn, transfers its electrons to B which, because of rapid oxidation by the rest of the chain in the presence of acceptor and uncoupler, is always in the oxidized state. We then have

$$H_2O \xrightarrow{h\nu_{II}} Q \xrightarrow{k_Q} A \xrightarrow{k_A} B \xrightarrow{\text{fast}} P$$

The rate of reduction of A is given by:

$$\frac{dA^-}{dt} = k_Q (A_T - A^-) - k_A A^-,$$

where $k_A = k'_A \cdot B_T$, and A_T and B_T are the total amounts of A and B in a chain. With the appropriate substitution of symbols, this is the first-order equation given previously in the section. The solution to this equation with $A_0^- = 0$ at $t = 0$ is

$$A^- = A_F^-(1 - e^{-(k_Q + k_A)t}),$$

where $A_F^- = [k_Q/(k_Q + k_A)]A_T$ is the steady-state level of A^-. If the light is extinguished after A^- has reached a certain level, A^- should go oxidized with a rate $dA^-/dt = - k_A A^-$. The plastoquinone changes reported for these two situations follow this first-order behavior, with $(k_Q + k_A) = 46$/sec in the "light on" experiment; and $k_A = 11$/sec in the "light off" experiment. Therefore $k_Q = 35$/sec. The maximum amount of reduced A^- in saturating light is then $A_F^- = (35/46)A_T$, or about 80%. A comparison of absorption changes of plastoquinone and P700 (making assumptions concerning extinction coefficients) sets the number of A molecules (A_T) at 13 equivalents per P700 molecule or chain. Therefore about 10 A equivalents per chain are reduced in the steady state.

The model suggests that at the onset of illumination all Q's are reduced very rapidly (≤ 50 μsec) and further reduction of the system cannot take place until electrons go to A. The time for transfer from Q to A would be proportional to the amount of oxidized A (A^+). In the steady state, the rate is $k_Q A^+ = (35/\text{sec})(3) \simeq 100$/sec. Therefore the time is about 10 msec, again reminiscent of the time in the Emerson-Arnold experiment and consistent with the time for reduction of P when light is extinguished. Since in the steady state all rates must be the same, in

[35]P. Schmidt Mende and B. Rumberg, Z. Naturforsch. **236**, 225 (1968).
[36]P. Schmidt Mende and H. T. Witt, Z. Naturforsch. **236**, 288 (1968).

particular the rate of transfer from A $(k_A A^-)$ must be $\simeq 100/\text{sec}$. For A completely oxidized, the transfer rate $Q \rightarrow A$ is $(35/\text{sec})(13) \simeq 500/\text{sec}$ corresponding to a time of ~ 2 msec.

Since $e^{-0.69} = 0.5$, the half-time for the total plastoquinone change is given by $t_{1/2} = 0.69/(k_A + k_Q) \simeq 15$ msec. That is the time for 5 of the A equivalents to be reduced.

So far we have limited our discussion to first-order reactions. Other types of reactions such as second-order ones undoubtedly occur. We refer the reader to textbooks on kinetics for this. However, such discussions are of limited use for understanding more complex, interrelating processes. It should be emphasized that precise measurements over a large range are required to decide between a sequence of first-order reactions and a second-order or more complicated process.

For the case where the decay is not solely a function of X^-, other parameters must be varied in order to obtain the functional dependences of f_{in} and f_{out}. No longer can decays be superimposed. f_{in} can be determined by adding f_{out} to dX^-/dt at each point; i.e., f_{in} at some point is the slope of the rise curve at that point minus the magnitude of the initial slope of the decay curve when the light is extinguished at that point.

It is not always necessary to measure a component directly in order to determine its kinetic behavior. The next section gives an example of how the kinetic decay of an unidentified component which limits the formation of oxygen precursor is determined indirectly from oxygen evolution.

We have tacitly ignored differences between separate and interacting chains in the above kinetic discussions. This was possible because the behavior for a simple, first-order reaction is the same for both. This is not true in general. Some examples of separate chain behavior will be given in the last section.

O$_2$ Evolution

Intermediates of the oxygen evolution process, precursors of O_2, are unstable so that initially after a dark period precursors rather than oxygen are made.[37] Accordingly, if after a dark period a continuous light is given, the O_2 rate is initially zero and then rises to the steady-state value. Such rise curves of the rate follow the $I \times t$ relation; i.e., the oxygen deficit at the onset of illumination reflects a finite pool—a pool of O_2 precursors which are present during the steady state but absent after a dark period. Most of this deficit is removed by a single flash. In

---

[37] P. Joliot, G. Barbieri, and R. Chabaud, *Photochem. Photobiol.* in press.

other words, the deficit pool is of the order of 1 equivalent per trapping center.

The simplest kinetic interpretation[38] of the O_2 evolution process is a linear accumulation of four + charges by a catalyst S, connected to each individual trapping center,

$$S_0 \xrightarrow{h\nu} S_1 \xrightarrow{h\nu} S_2 \xrightarrow{h\nu} S_3 \xrightarrow{h\nu} S_4 \longrightarrow S_0 + O_2$$

where the subscript of S designates the amount of accumulated charge. The S_3 and S_2 states are unstable, being deactivated (reduced) to S_1, a stable state: $S_3 \longrightarrow S_2 \longrightarrow S_1$. Therefore after a long dark period the traps are either in the S_0 or S_1 state. Each of a series of brief flashes adds one oxidizing equivalent to each trapping center and brings S in the next higher oxidation state. After deactivation, the first two flashes yield no oxygen but bring the S_0 or S_1 states into S_2 or S_3 states. The third and fourth flashes now produce oxygen.

Photoact II cannot be repeated unless Q and Z are regenerated. If we assume a one-to-one correspondence between S, Q, and Z, the time between flash pairs is a measure of the slowest regeneration time in the sequence S → A. For example (Fig. 8), flash No. 2 lost half its effectiveness if spaced closer than 200 μsec to the first flash. The regeneration occurred in a second-order manner. The regeneration time necessary for flash No. 3 to be effective after flash No. 2 is somewhat longer. Thus with the assumption of a one-to-one correspondence, the recovery of both donor and acceptor for photosystem II (under certain conditions) takes place in ≤ 200 μsec. Under other conditions this recovery should be slower, reflecting the dependence of the transfer time from Q to A upon the redox state of A.

We could expect that absorbance changes associated with either a primary donor or acceptor to photosystem II would occur rapidly upon illumination. From the paragraph directly above, we would also expect a rapid decay of these changes. Fast changes in absorbance associated with photosystem II which may be related to the primary acceptor or donor have been observed at 682 nm[39,40] and 325 nm.[34] These changes were observed with the use of repetitive flashes of about 10–20 μsec duration in the presence of acceptor (benzylviologen) and uncoupler (NH_4Cl). The negative absorption change at 682 appears to be that of a chlorophyll. Because of its rapid return (~ 200 μsec), it can be separated kinetically from changes due to P700, which restore much more slowly.

[38]B. Kok, B. Forbush, and M. McGloin, *Photochem. Photobiol.* in press.
[39]G. Döring, J. L. Bailey, W. Kreutz, and H. T. Witt, *Naturwissenschaften* **55**, 220 (1965).
[40]G. Döring, H. H. Stiehl, and H. T. Witt, *Z. Naturforsch.* **226**, 639 (1967).

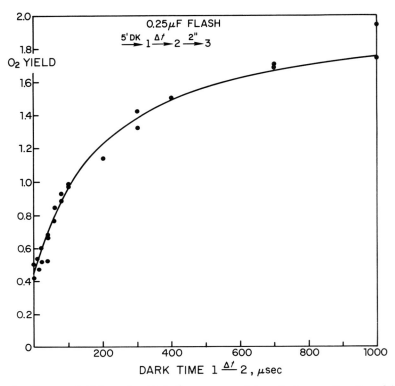

Fig. 8. Oxygen yield from the third of a sequence of three flashes as a function of time between the first two flashes.

We have observed (unpublished) a similar rapid return of the fluorescence yield, which was raised by a brief flash.

Separate Chains and Equilibria

The small temperature effect upon the $C \rightarrow P^+$ transfer shows that we are not dealing with conventional reactions in solution where all reactants can meet each other, but that the components are built together in a rigid matrix—so that each P700 communicates with only one cytochrome donor molecule. It is believed that the other components in the chain behave at least partly in this manner; i.e., to a large degree each chain operates independently of the others. The behavior of such systems is different enough from that of molecules in solution that a few comments are in order.

One feature of independent chains is that the various reactions will show first-order kinetics rather than second- or higher-order behavior;

i.e., in the reaction between P700 and cytochrome the rate will not be proportional to the product of the two concentrations (P$^+$·C). For each chain in the P$^+$C state the electron transfer from C $\rightarrow$ P$^+$ has a characteristic probability of occurrence $k(\sec^{-1})$, and viewing the overall system the rate will be the product of k and the number of P$^+$C states in the sample.

Another feature of independent chains is that normal equilibrium considerations may not be applicable. An example of this is obtained from the comparison of the steady-state reduction levels of P700 and cytochrome f in weak light of various wavelengths. In weak light, quanta arrive in the traps so infrequently that the dark reactions have ample time to proceed and equilibrium conditions are expected:

$$C + P^+ \underset{k'}{\overset{k}{\rightleftarrows}} C^+ + P \; ; \; \frac{C^+}{C} \cdot \frac{P}{P^+} = k/k' = K$$

One now observes, for example,[29] that at a wavelength where cytochrome f is $\sim 50\%$ reduced, P700 is $\sim 80\%$ reduced, and computes a value

$$K = \frac{0.5}{0.5} \cdot \frac{0.8}{0.2} = 4$$

In darkness, however, with the same chloroplasts, one meets the situation that virtually all P700 is reduced and all cytochrome f is oxidized, which suggests a very much higher equilibrium constant.

There are three possible explanations for this apparent change of K. The first is that the same entities are not measured in the two experiments. Spectral measurements indicate that they are. The second is that the equilibrium constant changes, perhaps owing to some change in membrane structure. Attempts to see changes caused by the additions of uncouplers have failed, indicating that the high-energy intermediate associated with ATP formation and associated phenomena is probably not involved. The third is that in at least one of the experiments equilibrium is not obtained. The rapidity of the reaction between C and P is such that equilibrium should have been obtained. In addition the results were independent of light intensity over a wide range.

The concept of "independent chains" not in equilibrium with one another can account for such an apparent change of K: At any moment the individual chains in the sample can have four possible states P$^+$C$^+$, P$^+$C, PC$^+$, PC. If we assume a high inherent K value, state P$^+$C with a lifetime of 1 msec does not accumulate, so that we have a distribution between only three states with either 0, 1, or 2 electrons in the complex (P$^+$C$^+$, PC$^+$, PC). While the high inherent K; observed in darkness, does not allow the occurrence of P$^+$C chains, the lack of electron transfer between chains allows a simultaneous accumulation of states P$^+$C$^+$

and PC. Since the entire system, not individual chains, is viewed, the presence of both P^+ and C at the same time gives the appearance of a low K. Malkin,[41] who treated this matter in detail, computes that for a 2-component system, like the one discussed, an irreversible step ($K = \infty$) would appear to have a varying K which, in the range of easy measurement, would appear to be as low as 4. He further calculated that this distortion rapidly decreases with an increase of the number of steps and intermediates in the chain.

A similar low apparent equilibrium constant in the light has been observed between Q and P. In the steady state, the rate of input to photosystem II must equal the output of photosystem I and $R_2 = R_1 = \alpha_2\phi_2 IQ = \alpha_1\phi_1 IP$ in which R_1 and R_2 are the rates of the two photosystems. Thus, considering the overall system (the total of all reaction chains), we will find a wavelength-dependent ratio of open traps in the two systems; $Q/P = \alpha_1\phi_1/\alpha_2\phi_2$. If, in addition, we assume equilibrium, the added constant fixes the levels of P and Q; i.e., both P and Q are independent of light intensity. These relationships are observed experimentally for light intensities weak enough so that neither the rate of transfer between Q and P nor the rate of transfer at the ends of the chain is limiting. Therefore the redox state of the two traps in continuous light of various wavelengths should reveal the value of K_{QP}. One can use one of the several procedures discussed above to determine the states P and Q. The results of such measurements[42] suggested an unexpectedly low value for K_{QP} (≤ 10). These low computed K values are clearly in conflict with estimates of K_{QP} in darkness and the respective midpoint potentials (the difference in midpoint potentials ΔE_m at room temperature is given by

$$\Delta E_m = RT \log K_{QP} = 60 \log K_{QP}$$

where the potential is expressed in millivolts. In this instance, owing to the large number of components between Q and P, the discrepancy does not appear to be due to nonequilibrium between chains.

[20] Photosensitization Reactions of Chlorophyll

By G. R. SEELY

A photosensitized oxidation-reduction reaction is one which goes only in the presence of, or is accelerated by, light absorbed by a sensitizer, which itself suffers no or relatively little net change in the process. Since

[41]S. Malkin, *Biophys. J.* **9**, 489 (1969).
[42]P. Joliot, A. Joliot, and B. Kok, *Biochim. Biophys. Acta* **153**, 635 (1968).

photosynthesis is a natural example of such a process with chlorophyll as the sensitizer, there is much interest in the reproduction *in vitro* of reactions like those presumed to be sensitized by chlorophyll *in vivo*.

It is now clear that there are three principal mechanisms whereby chlorophyll may sensitize oxidation–reduction reactions, distinguished by whether the photoexcited pigment is oxidized, reduced, or deactivated by energy transfer.[1] The first mechanism is responsible for most of the reactions reported in the literature, and probably for photosynthesis as well. The second mechanism is closely related to the Krasnovskii reaction, the reduction of chlorophyll to a pink dihydrochlorophyll product. The third mechanism is represented by the sensitized oxidation of substrates by O_2.

The experimental conditions under which reactions by these mechanisms can be realized, distinguished, and studied quantitatively will be discussed, with emphasis on reactions sensitized by the first mechanism. It will be assumed that the triplet state of chlorophyll, Chl', is the active one, which is true for most situations encountered *in vitro*. Study of reactions of the singlet excited state (Chl*) introduces special experimental difficulties and is beyond the scope of this section.

Sensitization through Reaction of Chl' with the Oxidant

Materials

Sensitizer. Chlorophyll and its derivatives should be carefully purified by a procedure like those published.[2-4] It is important to remove carotenoids, like lutein, which might intercept the energy of excited chlorophyll. Most of the lutein is removed by extracting a solution of chlorophyll in isooctane or petroleum ether repeatedly with 85% methanol–water.

Oxidants. Classes of compounds commonly used as oxidants include many of the azo dyes, quinones, flavins, Fe^{3+} complexes including those of porphyrins, nitroaromatics, azine and thiazine dyes. These should be purified by recrystallization and chromatography, especially the dyestuffs, which are often grossly impure. Most of the oxidants of these classes which have been investigated react rapidly with Chl', i.e., with a second-order rate constant approximately $10^9 \ M^{-1} \ sec^{-1}$.

Reductants. Ascorbic acid and phenylhydrazine are the two most

[1]G. R. Seely, *in* "The Chlorophylls" (L. P. Vernon and G. R. Seely, eds.), Chapter 17. Academic Press, New York, 1966.
[2]H. H. Strain and W. A. Svec, *in* "The Chlorophylls" (L. P. Vernon and G. R. Seely, eds.), Chapter 2. Academic Press, New York, 1966.
[3]A. F. H. Anderson and M. Calvin, *Nature (London)* 194, 285 (1962).
[4]H. J. Perkins and D. W. A. Roberts, *Biochim. Biophys. Acta* 58, 486 (1962).

commonly used reductants. Both reduce oxidized chlorophyll (Chl^+) rapidly, but also reduce Chl' slowly, with second-order rate constants[5] of the order of 10^4 to 10^5 M sec^{-1}. Hydrazobenzene reduces Chl^+ but not Chl' in dry, nonacidic solution. Sulfhydryl compounds react with Chl^+ and Chl', but often do so sluggishly, and give rise to complicated kinetics or side reactions.

The oxidized products of these compounds react slowly or not at all with Chl'. Many more compounds would be in common use as reductants if they were not extremely sensitive to air, and if their oxidized forms did not react rapidly with Chl'. This is true of hydroquinones and the leuco forms of flavins and many dyes. Some of these, the hydroquinones in particular, are useful in the study of photochemical transients.

It is important to note that ascorbic acid and phenylhydrazine can support oxidation–reduction reactions sensitized by either the first or the second mechanism. For determination of the mechanism, it is therefore advisable to measure the reaction rate over a range of concentration of reductant; reaction by the first mechanism persists at reductant concentrations much lower than reaction by the second.[6] Unfortunately, at reductant concentrations often used, 0.1–1 M or greater, it is not possible to decide without detailed examination whether the reductant is reacting with Chl', Chl^+, or, for that matter, with Chl^*. Hydrazobenzene reacts only with Chl^+, and thus discriminates between the two mechanisms.

Solvents. Particular attention should be paid to the purity of the solvent, and to assuring its freedom from oxidizing and reducing impurities which could react with chlorophyll or intermediates in the oxidation–reduction reaction. Even the best-grade solvents commercially available can be grossly impure for a photochemical reaction. An impurity present at only 1 ppm, and with a molecular weight of 100, would have a concentration of 10^{-5} M, sufficient to react with much or all of the chlorophyll sensitizer.

Among the more commonly used solvents for photosensitized reactions are ethanol, methanol, pyridine, acetone, dioxane, and their mixtures with each other and with water. Standard methods for purification of these solvents exist, and may or may not be adequate. The best procedure is probably to try solvents from different sources, and to purify them by preparative gas chromatography. A reaction exhibiting dubious kinetics should be checked in repurified solvent.

[5]R. Livingston and P. J. McCartin, *J. Amer. Chem. Soc.* **85**, 1571 (1963).
[6]G. R. Seely, *J. Phys. Chem.* **69**, 821 (1965).

Apparatus

A common practice is to run the reaction in a Thunberg tube, the body of which fits into a standard spectrophotometer cuvette holder, and the side arms of which contain one or more of the reagents before expulsion of oxygen.

For quantum yield measurements, the author prefers cylindrical cuvettes, 30 mm in diameter and 5-25 mm thick, which fit both in a constant temperature bath and in a spectrophotometer cuvette holder. The cuvette is sealed by a stopcock, through which N_2 and reagent solutions can be admitted by syringe. The stopcock is protected by a serum cap, but a serum cap alone is not proof against influx of detectable amounts of oxygen.

Deoxygenation

The O_2 content of solvents at room temperatures is of the order of 10^{-3} M. It must be reduced to the order of 10^{-7} M to prevent perversion of the intended reaction. The classical method of removing O_2 (and other gases) has been to freeze the contents of the reaction cell in liquid N_2, pump off uncondensed vapors, and thaw. This must be repeated many times for effectual removal of O_2.[7]

Passage through the reaction solution of purified N_2, argon, or helium is a simpler and popular alternative. Even the purest gases available commercially at reasonable cost contain traces of O_2 (prepurified grade N_2: 8 ppm O_2) which must be removed for best results. If any O_2 remains, reduction of the intended oxidant is delayed until all the O_2 is consumed, as shown rather dramatically in the early work of Ghosh and Sen Gupta with methyl red.[8] Really efficient removal of O_2 requires the action of free radical reducing agents, as in the sodium benzophenone ketyl solution of Fieser.[9] This solution is inconvenient in use because of its liability to deposit copious precipitates. A more convenient solution for scavenging O_2 is prepared as follows.

Ethylene glycol monomethyl ether (400 g) was treated with approximately 1 g of $NaBH_4$ in a 1-liter round-bottom flask and warmed gently under cover of N_2 until evolution of H_2 subsided. It was then distilled in a fractionating column at 5:1 reflux ratio, and the middle cut was taken. An excess of iron wire, as pure as possible, and 200 ml of the ether were put into a gas washing bottle and flushed with prepurified

[7]G. Tollin and G. Green, *Biochim. Biophys. Acta* 60, 524 (1962).

[8]J. C. Ghosh and S. B. Sen Gupta, *J. Indian Chem. Soc.* 11, 65 (1934).

[9]L. F. Fieser, "Experiments in Organic Chemistry," 2nd ed., p. 396. Heath, New York, 1941.

grade N_2 for 10 minutes. Then 0.25 g of methylviologen chloride was added, and N_2 flushing was continued for 15 minutes. If purification of the ether from carbonyl compounds has been effective, the solution begins to turn blue (reduced viologen cation radical), about 10 minutes after the viologen is added and becomes deep blue overnight. An H_2SO_4 bubbler and a KOH drying tube following the washing bottle remove any volatile impurities derived from the iron.

Dr. J. S. Connolly of the author's laboratory has compared lifetimes of bacteriochlorophyll transients in solutions flushed by N_2 purified as above, with those in solutions degassed by the much more arduous evacuation procedure, and found them to be identical.

Reaction Rate

If the products of the reaction are stable, as in the reduction of azo dyes or nitro compounds, the reaction may be followed by the decline of an absorption band of the oxidant, or, if the reductant is hydrazobenzene, by increase in absorption at 470 nm by azobenzene. However, the rate of net disappearance of a reactant or net appearance of a product is at best only an order of magnitude approximation to the rate of bimolecular reaction between Chl' and the oxidant.

If the products of the reaction back-react slowly, with lifetimes of seconds or more, the reaction can be followed by continuously recording spectrophotometry, as in the examination of photosensitized displacements of the equilibrium between N-methylphenazinium methyl sulfate (PMS) and ubihydroquinones by Vernon et al.[10] If the products back-react rapidly, or if it is desired to follow the primary photochemical steps of the reaction, the techniques of flash spectrophotometry must be employed.[11]

Radical species may be detected by their electron spin resonance (ESR) signals generated in the light. This is an especially convenient technique for quinones and flavins.[7,12,13]

Quantum Yield

For a proper understanding of sensitized oxidation–reduction reactions, it is necessary to estimate the amount of light absorbed by the sensitizer and relate the extent of the subsequent reaction thereto. It is quite difficult to do this in flash spectrophotometry; but it is straightforward when the products are stable.

[10]L. P. Vernon, W. S. Zaugg, and E. Shaw, *in* "Photosynthetic Mechanisms in Green Plants," p. 509. *Nat. Acad. Sci.–Nat. Res. Counc. Publ.* 1145, (1963).
[11]B. Ke, L. P. Vernon, and E. R. Shaw, *Biochemistry* 4, 137 (1965).
[12]G. Tollin and G. Green, *Biochim. Biophys. Acta* 66, 308 (1963).
[13]G. Tollin, K. K. Chatterjee, and G. Green, *Photochem. Photobiol.* 4, 593 (1965).

Light is measured quantitatively by a thermopile. A properly constructed thermopile converts the energy of light, regardless of frequency, into heat which sets up a current in the thermopile circuit. A procedure for calibrating a thermopile against a standard lamp is available from the National Bureau of Standards. Light may also be measured by a thermistor bolometer, most conveniently by one whose probe fits into a standard spectrophotometer cuvette holder.[14]

It is necessary to convert the light absorbed by the sample from watts cm^{-3} sec^{-1} to einsteins $liter^{-1}$ sec^{-1}, in order to calculate the quantum yield from the rate of a reaction expressed in moles $liter^{-1}$ sec^{-1}:

$$I_a \text{ (einsteins liter}^{-1}\text{ sec}^{-1}) = \frac{10^{10}\bar{\lambda}}{hcN} I_a \text{ (watt cm}^{-3}\text{ sec}^{-1}) \tag{1}$$

$$= 83.6 \, \bar{\lambda} \, I_a \text{ (watt cm}^{-3}\text{ sec}^{-1})$$

I_a is the light absorbed, in the units given; h, c, and N are Planck's constant, the speed of light, and Avogadro's number; and $\bar{\lambda}$ is a suitable weighted average wavelength in centimeters.

In order to get an accurate value of $\bar{\lambda}$, it is necessary to irradiate with a discrete line source, or with an incandescent lamp through a filter which confines the radiation absorbed to a narrow region of the spectrum. High intensity sources of lines absorbed strongly by chlorophyll are lacking, except for the Hg 436 nm line. Since it is usually desired that chlorophyll should absorb light in the red band, a tungsten lamp with a 650–670 nm interference filter is the preferred source. The value of $\bar{\lambda}$ can be estimated with sufficient accuracy, or calculated from Eq. (2).

$$\bar{\lambda} = \frac{\int \lambda T(\lambda) \, (1 - 10^{-\delta(\lambda)}) \, d\lambda}{\int T(\lambda) \, (1 - 10^{-\delta(\lambda)}) \, d\lambda} \tag{2}$$

where $T(\lambda)$ is the transmittance of the filter, and $\delta(\lambda)$ the absorbance of the sample.

Sensitization through Reaction of Chlorophyll with the Reductant

Krasnovskii Reaction

The photochemistry of chlorophyll was considerably advanced by the discovery[15] of its photoreduction to a pink dihydrochlorophyll ($ChlH_2$), by ascorbic acid or phenylhydrazine in aqueous or alcoholic pyridine solution. The pink product reduces added oxidants,[16] and in

[14]R. W. Treharne and H. W. Trolander, J. Sci. Instr. 42, 699 (1965).
[15]A. A. Krasnovskii, Dokl. Akad. Nauk SSSR 60, 421 (1948).
[16]A. A. Krasnovskii and G. P. Brin, Dokl. Akad. Nauk SSSR 73, 1239 (1950).

their absence reverts to chlorophyll by reduction of dehydroascorbic acid, or to pheophytin by displacement of Mg and oxidation. It was later recognized that reductions sensitized by this mechanism (e.g., of NAD) are accomplished by the one-electron reduced radical (ChlH·) rather than by the pink dihydrochlorophyll.[17]

The course and the quantum yield of photoreduction of chlorophyll are very sensitive to the nature of the medium.[18-20] Reduction with ascorbic acid requires the presence of an organic base, such as pyridine, and the presence of a nonbasic component which can also complex with the Mg of the chlorophyll, such as water or ethanol. The function of the base appears to be to form with ascorbic acid an ion pair, which is the actual reductant. The quantum yield is somewhat greater in aqueous pyridine than in ethanolic pyridine, but the rate of loss of Mg from $ChlH_2$ is also greater. The rate of reversion of $ChlH_2$ to Chl increases with the pyridine content of the solvent. Optimal conditions for formation of $ChlH_2$, which represent a comprise between conditions for high quantum yield, low rate of reversion, and low rate of Mg displacement, are attained in about 30% pyridine, 70% ethanol solution.[19] Phenylhydrazine reduces chlorophyll most efficiently in nonpolar media such as ethyl ether or toluene.[18] Rather high concentrations of reductant (0.01-1 M) are required, because reaction of Chl' with the reductants is slow.

Sometimes products other than $ChlH_2$ and the corresponding reduced pheophytin are obtained. Phenylhydrazine in EPA gives products with bands at 585 nm and 615 nm as well as $ChlH_2$ (520 nm).[21] Phenylhydrazine in nonpolar solvents containing benzylamine, allylamine, or ethanolamine gives a dihydrochlorophyll absorbing at 610 nm,[22] which may be identical with the 615 nm compound. Ascorbic acid, in pyridine containing ethanol and diazabicyclooctane, reduced chlorophyll to a dihydromesochlorophyll absorbing at 632 nm.[23] Structures for these two compounds have been proposed.[22]

Reduction of NAD and NADP

These compounds are not reduced directly by reaction with Chl'. They are reduced by reaction with ChlH· in aqueous pyridine, especially

[17]V. B. Evstigneev and V. A. Gavrilova, *Dokl. Akad. Nauk SSSR* **98**, 1017 (1954).
[18]T. T. Bannister, *in* "Photochemistry in the Liquid and Solid States" (F. Daniels, ed.), p. 110. Wiley, New York, 1960.
[19]G. R. Seely and A. Folkmanis, *J. Amer. Chem. Soc.* **86**, 2763 (1964).
[20]V. B. Evstigneev, *Photochem. Photobiol.* **4**, 171 (1965).
[21]S. S. Brody, *J. Amer. Chem. Soc.* **82**, 1570 (1960).
[22]W. Hendrich, *Biochim. Biophys. Acta* **162**, 265 (1968).
[23]G. R. Seely, *J. Amer. Chem. Soc.* **88**, 3417 (1966).

if a stronger base such as NH_3 is present to catalyze the reoxidation of $ChlH_2$.

NADP reduction by ascorbic acid can also be sensitized in aqueous solutions by chlorophyllin in the presence of ferredoxin–NADP reductase.[24] Good photoreduction rates were obtained in a solution of the following composition: chlorophyllin, $3–8 \times 10^{-6}$ M; sodium ascorbate, 5.5×10^{-3} M; NADP, 7.5×10^{-4} M; glutathione (reduced), 2×10^{-3} M; ferredoxin–NADP reductase, 0.1 mg; Na deoxycholate, 0.6%; Tris·HCl buffer (pH 7.8), 1.8×10^{-2} M.

For highest activity, the chlorophyllin was dialyzed against 10^{-2} M phosphate buffer (pH 8). The dialyzed chlorophyllin is partly aggregated, and deoxycholate apparently serves to loosen the aggregates. Ferredoxin did not stimulate the reaction. PMS, TMPD, and DPIP depressed the rate somewhat. The ferredoxin–NADP reductase was prepared by the method of Keister and San Pietro.[25] The extent of reaction was measured by the increase in absorbance at 340 nm (NADPH).

Sensitization by Energy Transfer

By this mechanism we do not mean the energy transfer that occurs in the photosynthetic unit (e.g., phycocyanin-sensitized photosynthesis), but transfer from the sensitizer to the oxidant or reductant, which then reacts as though it had been photoexcited directly. The best-documented examples of this reaction for chlorophyll are sensitized oxidations by O_2. For many of these it is now clear that they go through the agency of O_2 in its singlet excited state.[26] The biological manifestation of this type of reaction is called photodynamic action; it is sometimes caused by porphyrins related to or derived from heme and chlorophyll.

Oxidations by O_2, sensitized by chlorophyll and its immediate derivatives, have been investigated for many years, originally in the hope of finding a relationship to the sensitized evolution of O_2 in photosynthesis. Today, chlorophyll and its derivatives remain popular sensitizers, because their strong absorption bands in the red lie well beyond the range of absorption of most substrates and their oxidation products. However, in the absence of an easily oxidized substrate, chlorophyll itself is attacked by single excited O_2. Pheophytin is somewhat more resistant.

For quantitative information, it is usually necessary to determine

[24]L. P. Vernon, A. San Pietro, and D. A. Limbach, *Arch. Biochem. Biophys.* 109, 92 (1965).
[25]D. L. Keister and A. San Pietro, Vol. 6, p. 434.
[26]C. S. Foote, *Accounts Chem. Res.* 1, 104 (1968).

the rate of consumption of O_2. A classical method is to follow O_2 uptake manometrically with a Warburg apparatus. This apparatus is commercially available and does not require description here. The development of the O_2 electrode has provided another method. Pickett and French describe an electrode designed for rapid change of sample.[27] Fiala has described the use of polarography to follow photosensitized oxidations. Not only O_2, but other reagents and products can be quantitatively measured at the dropping Hg electrode.[28]

The method of choice obviously depends on the intended use. Electrode methods are better adapted for measuring departures of the O_2 concentration from an initial level, whereas the manometric method measures O_2 uptake rates at constant concentration.

Acknowledgment

The preparation of this section was supported in part by National Science Foundation Grant No. GB-7893.

[27]J. M. Pickett and C. S. French, *Proc. Nat. Acad. Sci. U. S.* **57**, 1587 (1967).
[28]S. Fiala, *Biochem. Z.* **320**, 10 (1949).

[21] Measurement of Photorespiration

By DAVID T. CANVIN and H. FOCK

The term photorespiration will be used only to refer to the evolution of CO_2 from a green leaf in the light. It is expected, because of the re-fixation of CO_2 in photosynthesis, that only a portion of the total CO_2 production in the leaf will diffuse to the surrounding atmosphere.[1-3] There is at present much debate about the relationship between CO_2 production by the leaf and CO_2 evolution from the leaf, and there is no easy solution to the problem. However, since total CO_2 production by the leaf cells cannot be measured, the methods that will be described must of necessity be restricted to the measurement of CO_2 evolution from the leaf in the light.

The exact relationship between CO_2 evolution and O_2 absorption by

[1]B. A. Bravdo, *Plant Physiol.* **43**, 479 (1968).
[2]J. V. Lake, *Aust. J. Biol. Sci.* **20**, 487 (1967).
[3]Y. Samish, and D. Koller, *Plant Physiol.* **43**, 1129 (1968).

the leaf in the light is unknown, but certainly oxygen absorption occurs.[4-7a] The measurement of the simultaneous fluxes of O_2 (i.e., oxygen evolution and oxygen absorption) is possible only through the use of the heavy stable isotopes of oxygen and the mass spectrometer.[8] Carbon dioxide fluxes too can be measured by this method, and the measurement of both fluxes have been reported.[4] Because of the author's lack of competence in this method and because of space limitations, the method will not be described here and readers are referred to the original literature.[9,10] A new development that should be mentioned, however, is the measurement of the net rate of O_2 exchange in a flowing gas stream by the use of the Oxygor oxygen analyzer.[11]

Zelitch[12] has proposed that the rate of photorespiration can be determined from the amount of $^{14}CO_2$ released by the leaf into a CO_2-free gas stream after a period of photosynthesis in $^{14}CO_2$. It should be emphasized that total CO_2 production cannot be determined by this method unless the specific activity of the $^{14}CO_2$ is known. Zelitch assumed that the specific activity of the $^{14}CO_2$ released in the CO_2-free gas stream was similar with time and that it was the same under conditions of illumination or darkness.

Although there may be occasions when this is so, most of the evidence indicates that in most conditions these assumptions are not justified (Fig. 1).[13-15] At best, then, the method would be of limited use, and for this reason it is not described here.

Gas Analysis Systems

The components for measuring CO_2 and $^{14}CO_2$ fluxes in leaves can be arranged to form several different types of gas circuits. Four of these basic circuits will be described, since each can be used to estimate or

[4]J. L. Ozbun, R. J. Volk, and W. A. Jackson, *Crop Sci.* **5**, 497 (1965).
[5]J. L. Ozbun, R. J. Volk, and W. A. Jackson, *Plant Physiol.* **39**, 523 (1964).
[6]W. A. Jackson and R. J. Volk, *Nature (London)* **222**, 269 (1969).
[7]H. Fock, H. Schaub, and W. Hilgenberg, *Z. Pflanzenphysiol.* **60**, 56 (1968).
[7a]H. Fock, H. Schaub, W. Hilgenberg, and K. Egle, *Planta* **86**, 77 (1969).
Schaub, W. Hilgenberg, and K. Egle, *Planta* **86**, 77 (1969).
[8]A. San Pietro, Vol. IV, 473.
[9]R. J. Volk and W. A. Jackson, *Crop Sci.* **4**, 45 (1964).
[10]A. H. Brown, A. O. C. Nier, and R. W. Van Norman, *Plant Physiol.* **27**, 320 (1952).
[11]H. Schaub, W. Hilgenberg, and H. Fock, *Z. Pflanzenphysiol.* **60**, 64 (1968).
[12]I. Zelitch, *Plant Physiol.* **43**, 1829 (1968).
[13]A. Goldsworthy, *Phytochemistry* **5**, 1013 (1966).
[14]J. L. Ludwig, Ph.D. thesis, Queen's University, Kingston, Ontario, 1968.
[15]A. D'Aoust, Ph.D. thesis, Queen's University, Kingston, Ontario, 1969.

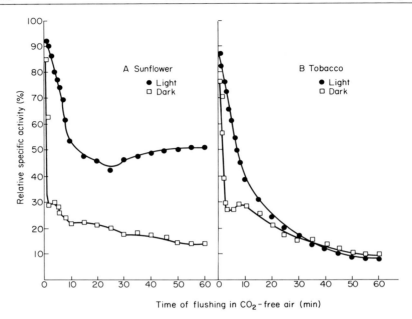

FIG. 1. Relative specific activity of $^{14}CO_2$ evolved into CO_2-free air in light or darkness from a sunflower and tobacco leaf immediately after a period of photosynthesis in $^{14}CO_2$. Relative specific activity is the specific activity of the $^{14}CO_2$ evolved into CO_2-free air as a percentage of the specific activity of the $^{14}CO_2$ supplied during photosynthesis (19.2 μCi $^{14}CO_2$/mg of CO_2). Measurements were made using an open gas analysis circuit equipped to measure CO_2 and $^{14}CO_2$. (A) A sunflower leaf was flushed with CO_2-free air in the light (3500 ft-c, 25°) or dark (25°) after 120 minutes photosynthesis (27 mg of CO_2 $dm^{-2}hr^{-1}$) in air containing 255 ppm CO_2 plus $^{14}CO_2$ (3500 ft-c, 25°). (B) A tobacco leaf was flushed with CO_2-free air under the same conditions after 15 minutes of photosynthesis (20 mg of CO_2 $dm^{-2}hr^{-1}$) in air containing 260 ppm of CO_2 plus $^{14}CO_2$ (3500 ft-c, 25°). Data from A. D'Aoust, Ph.D. thesis, Queen's University, Kingston, Ontario, 1969.

measure CO_2 evolution from green leaves in the light. The compensated semiclosed system[16] or semiopen system[17] possess some advantages when used for the measurements of the rate of apparent photosynthesis but offer no advantage for the measurement of the rate of CO_2 evolution.

In Fig. 2 are shown all the components of four types of gas circuits with differing capability. In assembling any one of the circuits, the unnecessary components and the connecting tubing would be omitted.

[16.]R. B. Musgrave and D. N. Moss, *Crop Sci.* **1**, 37 (1961).
[17]D. Koller and Y. Samish, *Bot. Gaz.* **125**, 81 (1964).

Components of Gas Analysis Circuits

Leaf Chamber (Cuvette). The leaf chamber is normally constructed of Plexiglas, dichloroethane being used to effect the sealing and attachment of joints. The dimensions of the leaf chamber and the shape are determined to some extent by these same properties of the plant material on which measurements are desired. It is impossible to describe all possible leaf chambers, and only one which we use routinely will be described in detail.

Small Plexiglas chambers[14] (Fig. 3) of the leaf cup design[18] are constructed in two halves; one half encloses a circular area on the upper surface of the attached leaf, and the other half encloses a similar area on the lower surface. Sponge rubber gaskets are used between the chamber and leaf surface to prevent damage to the leaf. A thin coat of lanolin

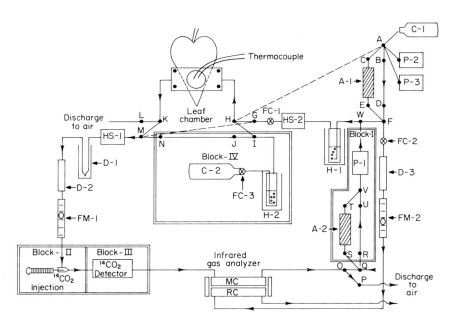

Fig. 2.   Schematic diagram of the components for several gas analysis circuits (not to scale): C-1, CO_2 gas mixture in pressurized cylinder; C-2, $^{14}CO_2$ gas mixture in pressurized cylinder; P-1 and P-2, gas pumps; P-3, precision gas mixing pumps; A-1 and A-2, CO_2 absorbers; FC-1, FC-2, FC-3, and H-2, humidifiers; D-1, low temperature drier; D-2 and D-3, $Mg(ClO_4)_2$ driers; FM-1 and FM-2, flow rate meters; MC, measuring cell; RC, reference cell.

[18]O. V. S. Heath and F. L. Milthorpe, *J. Exp. Bot.* **1,** 227 (1950).

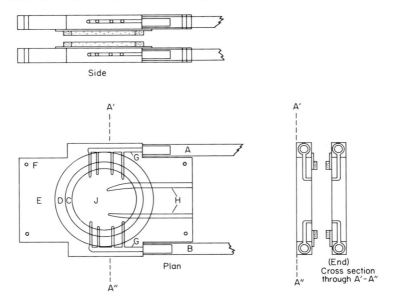

Side

Plan

(End)
Cross section
through A'–A'

FIG. 3.   Split-leaf chamber for measuring CO_2 evolution from a leaf. Side view shows the two halves in the position in which they would be clamped to a leaf. Plan view of one half of the leaf chamber and cross section through the center of the leaf chamber: A, inlet gas tubing; B, outlet gas tubing; C, sponge-rubber gasket; D, Lucite ring, 2.0 mm thick; E, 0.25-inch Lucite body; F, holes for screw clamps; G, inlet and outlet manifolds; H, thermocouples; J, leaf chamber.

is placed on the gasket before the leaf is inserted so that only a very light pressure from the screw clamps which hold together the two halves of the chamber is required to effect an air-tight seal. Each half of the chamber has its own multiple inlet and outlet systems, so that air passes evenly over both sides of the leaf.

The body of the cuvette is 0.25-inch Plexiglas and the inlet and outlet ports are 0.125 inch in diameter. A circular area 4.5 cm in diameter is enclosed by attaching 4 mm × 4 mm thick sponge rubber. When the leaf chamber is positioned on a leaf, 15–16 cm² of leaf is enclosed and the volume of the cuvette is 7–8 ml.

Air flow through the cuvette is 524 ml min⁻¹, and under these conditions the laminar resistance[19,20] of the leaf is 0.5–0.6 sec cm⁻¹.

Air and leaf temperature is measured with copper-constantan thermocouples (0.005 inch) sealed into the leaf chamber.

[19]P. Gaastra, *Med. Land. Wag. Ned.* **59**, 1 (1959).
[20]K. J. Parkinson, *J. Exp. Bot.* **19**, 840 (1968).

In the more general case, the flow rate of air through the leaf chamber exerts an effect by renewing the atmosphere in the cuvette and by affecting the laminar resistance next to the leaf surface and in both cases influencing the diffusion of CO_2 into or out of the leaf.[20-22] With any new leaf chamber and saturating light intensity, flow rate should be increased until no further increase in photosynthesis rate occurs. Flow rates of gas in the plateau region should then be used for all measurements as the rate of atmosphere renewal is now adequate and no further effects of flow rate on laminar resistance will be observed. In leaf chambers of the type described above, where the air stream is confined to an area near the leaf surface, the flow rate of air through the leaf chamber is sufficient to result also in a small laminar resistance. The laminar resistance in any new leaf chamber should also be measured, and if this value is not 0.5 sec cm^{-1} or less a redesign of the leaf chamber may be necessary or a fan must be placed in the leaf chamber to provide additional mixing or turbulence. Thus, to evaluate a leaf chamber it is not only necessary to provide data on the dimensions and rate of gas flow, but also to provide some estimate of the laminar resistance of the leaf in the chamber.

Leaf Temperature Control. The temperature of the leaf and air in the leaf chamber can be controlled by placing the leaf chamber and plant in a controlled-environment cabinet.[14] If such an arrangement is not available, then a second compartment must be sealed to the outside surface of the leaf chamber and water from a controlled temperature bath is then circulated through this second chamber.

Infrared Gas Analyzer (IRGA). In our laboratory we have used the UNOR 2 (H. Maihak A. G., 2000 Hamburg 39, Semperstrasse 38, Germany) and the Beckman IR-215 (Beckman Instruments, Inc. 2500 Harbour Blvd., Fullerton, California 91634), but other reliable instruments are available.

To be more versatile and provide greater sensitivity, both reference and measuring cells of the IRGA should be of the flow-through type with adequate diameter of inlet and outlet ports to allow the maximum required flow rate. The instrument should be capable of a calibration from 0 to 600 ppm of CO_2 or a full-scale differential calibration of 50 ppm of CO_2 anywhere in this range.

The analyzer may be equipped with filters to make it nonresponsive to water vapor, but sensitivity will usually be decreased. If it is not

[21]J. P. Decker, *Plant Physiol,* **22,** 561 (1947).
[22]D. J. Avery, *J. Exp. Bot.* **17,** 655 (1966).

equipped with filters, the air must be thoroughly dried or equal water vapor contents maintained in the sample and reference gas streams.

Calibration of IRGA. The instrument is calibrated for the appropriate range using gases of known CO_2 concentration before and/or after the experiments.[23]

Humidity Sensors. The water vapor content of the air streams entering and exiting from the leaf chamber is measured by psychrometers of the type described by Slayter and Bierhuizen.[24] The psychrometers are immersed in a constant-temperature bath at 35°, and the passage of the air through a 3 ft $\times$ 0.25 inch coiled copper tube immersed in the same bath prior to the psychrometer ensures that the air has equilibrated at that temperature. Other types of humidity sensors are available and may be preferable to the above psychrometers.

The measurement of the water vapor content allows the calculation of the transpiration rate and the total resistance for CO_2 diffusion from the leaf chamber atmosphere to the intracellular spaces of the leaf.[19] The measurement also provides a continuous record of the degree of stomatal opening.

Humidifiers. Gas diffusers are immersed in water maintained at 10° by a constant-temperature bath. If the temperature of the incoming air and leaf chamber is 25°, the relative humidity of the entering air will be approximately 40%. Humidification of the entering air is necessary to prevent excessive water loss and dehydration of the experimental material.

Driers. The major portion of water vapor is removed from the air stream by passing the air through a water bath maintained at 0° (D-1 in Fig. 2). Further drying is performed with anhydrous $Mg(ClO_4)_2$ (D-2 and D-3, Fig. 2) packed in glass tubes.

CO_2 Absorbers. CO_2 is absorbed when required with Indicarb (Fisher Scientific Co.) 10–20 mesh, packed in glass tubes.

Flowmeters: To suit the flow rates required flowmeters with capacity of 10–100 liters hr^{-1} or 50–650 liters hr^{-1} (Rota, 7867 Oeflingen, Baden, Germany; Manostat Corporation 20 N Moore St., New York, N.Y. 10013) are used.

Air Pumps. Bulk gas circulation pumps (P-1 and P-2, Fig. 2) of Type 406G (Reciprotor, Krogshojvej 47, Bagsvaerd, Copenhagen, Denmark) or small diaphragm pumps (e.g., Dyna-Vac, Model 2, Fisher Scientific Co.) are used in the systems.

Precision Gas Mixing Pumps. With these pumps (P-3, Fig. 2) many

[23]G. C. Bate, A. D'Aoust, and D. T. Canvin, *Plant Physiol.* **44,** 1122 (1969).
[24]R. O. Slayter and J. F. Bierhuizen, *Plant Physiol.* **39,** 1051 (1964).

different gas mixtures can be generated.[23,25] The composition of the gas mixtures is sufficiently accurate to be used for the calibration of infrared gas analyzers. Three Type NA27/3A and one Type NA18/3A precision gas mixing pumps (H. Wösthoff o. H.G., 463 Bochum, Germany) are arranged in series to generate the desired gas mixture.

$^{14}CO_2$ *Detection.* An ion chamber of 13 ml (1.75 cm i.d. × 5 cm i.d.) was machined from solid brass and fitted with the collecting electrode from a Nuclear Chicago (Model DCF500) ion chamber. The signal from the ion chamber is monitored with a Nuclear-Chicago (Model 6000) Dynacon (Nuclear-Chicago Corp., 333 E. Howard Ave., Des Plaines, Illinois 60018).

As an alternative the $^{14}CO_2$ may be monitored with a Geiger-Müller counter sealed into the gas circuit.[26]

$^{14}CO_2$ *Injection.* If the $^{14}CO_2$ is generated in another vessel, known volumes of the gas may be injected through a small piece of Tygon tubing in the circuit. The injection port may also be constructed of glass or metal fitted with rubber septums. Alternatively the generating system may be included in the gas circuit and the $^{14}CO_2$ generated from $Ba^{14}CO_3$ or $Na_2^{14}CO_3$ by injection of acid.[26]

Flow-Control Valves. Precision metering valves with the required flow capacity are used in the circuit.

Compressed Gas Containing $^{14}CO_2$. Such gas mixtures are commercially available, or they may be prepared in the laboratory.[14]

Connecting Tubing in the Gas Circuit. The various components of the gas exchange system are connected with glass tubing preferably, but Tygon tubing may be used, or a combination of these materials. In order to keep the total size of the system to a minimum, the connections should be as short as possible.

Recorders. The signals from the Dynacon and the infrared gas analyzer are recorded using a multiple-range dual-pen potentimetric recorder (Model DSRG, E. H. Sargent and Co.).

Closed Gas Circuit

The components shown in Blocks II, III, and IV are *not* required, but the components of Block I are inserted in the circuit (Fig. 2). If the infrared CO_2 analyzer is equipped with a sealed reference cell containing nitrogen, the portion of the circuit supplying a gas stream to the reference cell from C-1 or P-2 is not required, and F to W is not con-

[25]H. Fock and K. Egle, *Beitr. Biol. Pflanz.* **42**, 213 (1966).
[26]G. R. Lister, G. Krotkov, and C. D. Nelson, *Can. J. Bot.* **39**, 581 (1961).

nected. If the reference cell of the analyzer is not sealed but of the flow-through type, F to W is not connected and nitrogen or CO_2-free air may be supplied from C-1 via ABDF to the reference cell. Alternatively air may be pumped by P-2 via ACEF, and by passage through the CO_2 absorber (A-1), CO_2-free air is supplied to the reference cell. If the latter method is used as a source of reference gas, it may be advantageous to place a drier of type D-1 and also D-3 prior to A-1 as the absorption of water by Indicarb in A-1 may cause caking and interfere with gas flow. The flow rate of gas to the reference cell is controlled by the flow control valve (FC-2) and measured by the flowmeter (FM-2).

Initially the circuit is open so that the effluent from the measuring cell of the infrared CO_2 analyzer will exit via OP and the circuit at Q will be open to the atmosphere. Air would then be pumped by P-1 via QRUVWGH to the leaf chamber and travel to the infrared CO_2 analyzer via KM. The leaf is placed in the leaf chamber and illuminated. If a gas of composition other than ambient air (e.g., higher CO_2 concentration, different O_2 concentration) is required, it may be supplied from a cylinder or by precision gas-mixing pumps (P-3) connected at Q. When steady rates of photosynthesis have been obtained by the leaf the gas supply at Q is disconnected, OP is disconnected and the system is closed by connecting OQ. The gas stream is circulated in the closed system and the decrease in CO_2 concentration is measured by the infrared CO_2 analyzer.

As an alternative to the procedure described above, CO_2 replenishment or higher than ambient CO_2 concentrations in the closed circuit may be obtained by including the apparatus of Block II in the circuit and injecting CO_2 with a syringe.

A CO_2-free air stream may be circulated over the leaf by inserting the CO_2 absorber (A-2) into the circuit via QSTV. The flow rate of gas in the circuit is controlled by the flow control valve (FC-1) and measured by the flowmeter (FM-1). The gas is humidified at H-1, and the water vapor content of the entering air is measured at HS-2 and that of the exiting air at HS-1. The gas stream is dried by D-1 and D-2 before entering the measuring cell of the infrared gas analyzer.

The volume of the total closed circuit may be determined by filling the circuit with N_2 or CO_2-free air and injecting known quantities of pure CO_2 with a gas-tight syringe. The CO_2 concentration is determined by the calibrated infrared gas analyzer and the volume of the system determined from the dilution. If the system has a calibrated $^{14}CO_2$ detector (below), the volume of the system may be determined in the same manner using $^{14}CO_2$.

Closed Gas Circuit with $^{14}CO_2$ Measuring Capability

The circuit and procedure are the same as for the closed gas circuit described above except that Block II and III (Fig. 2) are inserted in the gas circuit.

After the pretreatment of the leaf and any time after the closure of the gas circuit, a quantity of $^{14}CO_2$ may be injected into the circuit or generated in the circuit at the position of Block II.

The $^{14}CO_2$ changes and CO_2 changes are then continuously measured. An improved procedure for this method (the circuit is shown partly in dashed lines in the diagram) is to supply the nonradioactive pretreatment gas to the illuminated leaf in the leaf chamber by a second pump (e.g., P-2) and direct connection of AH and have this air exit to the atmosphere via KL. GH is not connected and GM is connected — thus the closed gas circuit now excludes the leaf chamber. After $^{14}CO_2$ injection into the closed circuit, the gas stream is circulated until the $^{14}CO_2$ concentration is uniform. At that time the pump (P-1) is stopped, AH and GM are disconnected, GH and KM are connected, and the pump is started.

The closed gas circuits have the disadvantage of not being able to measure either evolution of CO_2 from leaves in the light or photosynthesis under steady-rate (state?) conditions. Since in all cases (except CO_2 at the compensation point) the CO_2 and $^{14}CO_2$ concentration is changing measurements at any one point are transient. The sensitivity of the system is usually also less than that of the open system because the infrared gas analyzer and the $^{14}CO_2$ measuring system must usually be calibrated for the full range of concentration change that can be expected. In addition, when the system is used for $^{14}CO_2$ measurements, changes in the rate of $^{14}CO_2$ concentration immediately after the release of $^{14}CO_2$ into the system are difficult to measure because it is difficult to obtain rapid uniform mixing of the $^{14}CO_2$ into the air stream. This disadvantage is less important if the system is modified to allow the pre-equilibration of the $^{14}CO_2$ before its introduction into the leaf chamber. The system must also be gas tight, as any leakage can lead to a considerable error in the measurements.

Open Gas Circuit

The open gas circuits have the advantage that the gas stream can be supplied to the leaf at any predetermined CO_2, $^{14}CO_2$, O_2, or water vapor concentration, and steady rates of CO_2 or $^{14}CO_2$ exchange and transpiration can be measured. Sensitivity of measuring instruments must be high, however, since marked differentials between entering and exiting gas streams are not desirable because the leaf environment is then no longer homogeneous.

The components of Blocks I, II, III, and IV are not required, and F to W is connected. For the measurement of apparent photosynthesis the gas stream is supplied to the leaf chamber from either C-1, P-2, or P-3 via ABDFWGH and is passed from the leaf chamber to the measuring cell of the IRGA via KM. The exit gas from the measuring cell is discharged via OP to the atmosphere. For open gas circuits, the infrared CO_2 analyzer is usually equipped with a flow-through reference cell and the same gas that enters the leaf chamber is passed through the reference cell of the IRGA. The flow rate to the reference cell is measured by flowmeter FM-2 and controlled by FC-2, and that to the leaf chamber and measuring cell by FM-1 and FC-1. The leaf is placed in the leaf chamber and illuminated. The difference between the CO_2 content of the gas entering the leaf chamber and that exiting from the leaf chamber is measured by the IRGA.

A CO_2-free gas stream may be supplied from either C-1, P-2, or P-3 via the absorber A-1 and gas circuit ACEF.

Open Gas Circuit with $^{14}CO_2$ Measuring Capability

The circuit is the same as the open gas circuit described above except that the components of Blocks III and IV are now necessary. A gas stream identical in composition to that in cylinder C-2, but containing no $^{14}CO_2$, is supplied from C-1, P-2, or P-3 to the leaf chamber, and the rate of apparent photosynthesis of the leaf is determined as outlined for the open gas circuit. After a steady rate of photosynthesis has been achieved, the gas stream is diverted to air via KL and the connection KM is broken. The $^{14}CO_2$-containing gas stream is now supplied from C-2 via IJNM to the measuring part of the gas circuit, and the initial $^{14}CO_2$ and CO_2 content of the gas is determined. To supply the $^{14}CO_2$ gas stream to the leaf chamber, GH and KL are disconnected, then MN is disconnected, MK is connected and IJ is disconnected, and IH is connected. A continuous trace of the changes in $^{14}CO_2$ and CO_2 content of the gas stream is measured.

The flow rate of gas from C-2 as controlled by FC-3 is identical to that which was supplied from C-1 and controlled at FC-1.

When using the open circuit with $^{14}CO_2$, a CO_2 absorber is connected to the exit line of the measuring cell to prevent the discharge of $^{14}CO_2$ into the atmosphere.

After $^{14}CO_2$ has been supplied to the leaf the leaf may be flushed with a CO_2-free gas stream and the evolution of $^{14}CO_2$ and CO_2 from the leaf can be measured. To do this a CO_2-free gas stream is supplied from C-1, P-2, or P-3 to the reference cell of the IRGA and to point G, which now exits to the atmosphere. At the required time, the gas supply from C-2 is discontinued, IH is disconnected, and GH is connected.

Estimation of CO_2 Evolution by Extrapolation of the Rate of Apparent Photosynthesis to a CO_2 Concentration of Zero

This method of estimating but not measuring the CO_2 evolution from a green leaf in the light was originally described by Decker.[27]

The closed gas circuit is usually used, and the IRGA is calibrated to measure CO_2 concentrations from 0 to maximum CO_2 content of the closed system. The decrease in the CO_2 concentration of the system is measured until the CO_2 compensation concentration is reached (Fig. 4A). The change in the CO_2 concentration in the closed system (CO_2) is determined from the trace for a small period of time (e.g., 0.25–1.0 min) and the total volume of the system is known. The rate of apparent photosynthesis is calculated as follows:

(i) CO_2 (ppm) min^{-1} × 1.8 = μg CO_2 liter^{-1}min^{-1}

(ii) μg CO_2 liter^{-1}min^{-1} × volume (liters) = μg CO_2 min^{-1}

(iii) $\dfrac{\mu\text{g } CO_2 \text{ min}^{-1}}{\text{leaf area (dm}^2)}$ × 0.06 = mg CO_2 dm^{-2}hr^{-1}

The rate of apparent photosynthesis is plotted vs the average CO_2 concentration of the system, and the linear portion of the line is extrapolated to zero CO_2 concentration (Fig. 4B). The intersection on the ordinate represents a minimum estimate of the rate of CO_2 evolution from the leaf. The method gives estimates of CO_2 evolution similar to

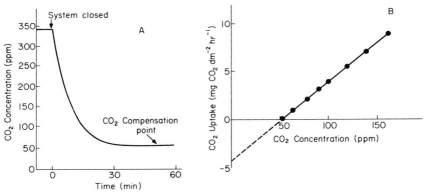

FIG. 4.   (A) Recorder trace of CO_2 concentration as obtained with a closed system. (B) Rate of apparent photosynthesis of a sunflower leaf vs CO_2 concentration. Point of intersection of extrapolated rate is an estimate of CO_2 evolution. Light intensity 3500 ft-c, 25°. Data from J. L. Ludwig.

[27]J. P. Decker, *J. Solar Energy Sci. Eng.* **1**, 30 (1959).

that measured by other methods. However, the method has been modified to approximate more closely the rate of apparent photosynthesis to the CO_2 concentration at the site of photosynthesis, and then higher rates of CO_2 evolution are estimated.[28]

CO_2 *Evolution into a CO_2-Free Gas Stream.* Using either a closed system with absorber or an open system, the rate of CO_2 evolution in the light is determined from the amount of CO_2 evolved into a CO_2-free air stream.[7,25,29,30] The infrared gas analyzer is calibrated from zero CO_2 to 50 ppm of CO_2.

The difference between CO_2 concentration of the gas stream entering and exiting (ΔCO_2) from the leaf chamber is measured, and the rate of CO_2 evolution is calculated as follows:

(i) ΔCO_2 (ppm) $\times 1.8 \times$ flow rate (liters min^{-1}) $= \mu g\ CO_2\ min^{-1}$

(ii) $\dfrac{\mu g\ CO_2\ min^{-1}}{\text{leaf area (dm}^2)} \times 0.06 = mg\ CO_2\ dm^{-2}hr^{-1}$

The rate of CO_2 evolution from the leaf is not constant; it is highest in the first 5 minutes after photosynthesis and slowly declines upon extended periods of flushing.

$^{14}CO_2$ and CO_2 Uptake in a Closed System. This method was used by Weigl et al.[31] and by more recent workers.[26] The method requires a closed gas circuit with $^{14}CO_2$ measuring capability and a second pump to supply air to the leaf chamber. Then the $^{14}CO_2$ can be equilibrated in the closed circuit, and the experiment is commenced by connecting the closed circuit containing $^{14}CO_2$ to the leaf chamber. Initial rates of $^{14}CO_2$ and CO_2 uptake are measured (both traces will be similar to the initial part of the trace in Fig. 4A), e.g., from 30 to 45 sec after connection to the leaf chamber.

A. True Photosynthesis (TPS) is calculated.

(i) $\dfrac{\mu C\ ^{14}CO_2\ min^{-1}}{\text{specific activity }(\mu C\ ^{14}CO_2 \text{ per mg } CO_2)} = mg\ CO_2\ min^{-1}$

(ii) $\dfrac{mg\ CO_2 min^{-1}}{\text{leaf area (dm}^2)} \times 60 = mg\ CO_2 dm^{-2}hr^{-1}$

B. Apparent photosynthesis (APS) is calculated from the CO_2 tracing as outlined in the previous section.

C. CO_2 evolution $=$ TPS $-$ APS.

[28]Y. Samish and D. Koller, *Ann. Bot.* **32**, 687 (1968).
[29]M. El-sharkawy and J. Hesketh, *Crop Sci.* **5**, 517 (1965).
[30]D. N. Moss, *Crop Sci.* **6**, 351 (1966).
[31]J. W. Weigl, P. M. Warrington, and M. Calvin, *J. Amer. Chem. Soc.* **73**, 5058 (1951).

Most data indicate that $^{14}CO_2$ is evolved from the leaf within 1 minute of the time that $^{14}CO_2$ is supplied to the leaf. The net effect of this evolution is an underestimation of the rate of photosynthesis calculated from the $^{14}CO_2$ uptake data and thus an underestimation of the rate of CO_2 evolution. Measurements must thus be obtained as soon as possible after $^{14}CO_2$ is supplied to the leaf.

$^{14}CO_2$ and CO_2 Uptake in an Open System. The method was devised by Ludwig[14] and requires the use of the open system with $^{14}CO_2$ measuring capability. The system has the advantage that it allows the measurement of CO_2 evolution during steady-rate photosynthesis. When the $^{14}CO_2$ gas stream from C-2 is connected to the leaf chamber, a chart tracing as shown in Fig. 5 is obtained. The initial large decrease in the $^{14}CO_2$ trace is caused by the passage of the nonradioactive gas that was in the leaf chamber through the $^{14}CO_2$ measuring system. The first reliable measurement with the system is obtained after 0.3 minute.

A. True photosynthesis (TPS) is calculated from the maximum rate of $^{14}CO_2$ uptake ($\Delta^{14}CO_2$ at 0.3 minute).

(i) $\Delta^{14}CO_2$ (μCi liter^{-1}) $\times$ flow rate (liters min^{-1}) = μCi $^{14}CO_2$ min^{-1}

(ii) $\dfrac{\mu\text{Ci } ^{14}CO_2 \text{ min}^{-1}}{\text{specific activity (}\mu\text{Ci } ^{14}CO_2 \text{ per mg } CO_2)}$ = mg CO_2 min^{-1}

The specific activity of the $^{14}CO_2$ used in this calculation is the average specific activity in the chamber calculated from the specific activity of the entering gas stream and the specific activity of the existing gas stream.

(iii) $\dfrac{\text{mg } CO_2 \text{ min}^{-1}}{\text{leaf area (dm}^2)} \times 60$ = mg CO_2 dm^{-2}hr^{-1}

B. Apparent photosynthesis (APS) is calculated from the rate of CO_2 uptake (ΔCO_2).

(i) $\Delta CO_2 \times 1.8 \times$ flow rate (liters min^{-1}) = μg CO_2 min^{-1}

(ii) $\dfrac{\mu\text{g } CO_2 \text{ min}^{-1}}{\text{leaf area (dm}^2)} \times 0.06$ = mg CO_2 dm^{-2}hr^{-1}

C. CO_2 Evolution = TPS $-$ APS

In general, if the methods are used with the same plant material and under the same conditions of light, temperature, and experimental apparatus, similar rates of CO_2 evolution will be obtained[32] (Table I).

[32]C. S. Hew, G. Krotkov, and D. T. Canvin, *Plant Physiol.* **44,** 662 (1969).

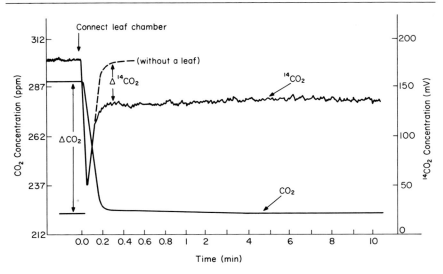

FIG. 5. Typical CO_2 and $^{14}CO_2$ recorder tracing showing the uptake of CO_2 by a sunflower leaf in the light (290 ppm of CO_2, 21% O_2, 3500 ft-c, 25°). CO_2 measurement was with a Beckman 215 infrared gas analyzer and $^{14}CO_2$ measurement was with a Nuclear-Chicago Dynacon equipped with a 13-ml ionization chamber. Data from A. D'Aoust, personal communication.

CO_2 Evolution from a Sunflower Leaf in Air in the Dark and in the Light Measured by Several Different Methods[a,b]

Method	CO_2 evolution (mg CO_2 dm^{-2} hr^{-1})
Closed system, extrapolation to zero CO_2 concentration, light	4.7 ± 0.2[c] ($n = 22$)
Closed system, $^{14}CO_2$ injection at CO_2 compensation point, light (rate calculated at 15 sec after $^{14}CO_2$ injection)	5.6 ± 0.2 ($n = 22$)
Open system, CO_2-free air, light (rate measured in first 3 min after photosynthesis)	5.4 ± 0.3
Open system, CO_2-free air, light (rate measured 60 min after photosynthesis)	4.2 ± 0.3
Open system, $^{14}CO_2$ during photosynthesis, light (rate calculated 0.5 min after connection of leaf chamber, $^{14}CO_2$ measured with 33 ml ionization chamber)	5.7 ± 0.3
Open system, CO_2-free air, dark (rate measured 60 min after photosynthesis)	1.6 ± 0.1

[a]Data of J. L. Ludwig, Ph.D. thesis, Queen's University, Kingston, Ontario, 1969.
[b]Light conditions, 3500 ft-c, 25°; dark conditions, 25°.
[c]Mean ± standard error.

[22] Measurement of the Intermediates of the Photosynthetic Carbon Reduction Cycle, Using Enzymatic Methods

By Erwin Latzko and Martin Gibbs

Preparation of the Extract

For intact chloroplasts[1] and unicellular algae,[2] rapid mixing with 20% $HClO_4$ at 0° and additional full power sonic oscillation for 30 seconds is recommended. After removal of the precipitate by centrifugation for 5 minutes at 7000 g, the supernatant fluid is carefully titrated potentiometrically with 5 M K_2CO_3 to pH 5.5. The precipitated $KClO_4$ is removed by centrifugation for 3 minutes at 3000 g. In the clear extract, glyceraldehyde 3-P, dihydroxyacetone-P, ATP, ADP, and AMP should be measured within 2 hours; ribose 5-P, ribulose 5-P, xylulose 5-P, ribulose 1,5-P_2, glucose 1-P, glucose 6-P, fructose 6-P, fructose 1,6-P_2, sedoheptulose 1,7-P_2, glycerate 3-P, and glycolate are stable for at least 2–3 days if the extract is stored at −20°. Whole leaves are immersed in liquid nitrogen and pulverized with a mortar.[1,3] Thawing has to be avoided until the powder is suspended in 5% $HClO_4$. The brei is sonicated for 30 seconds and further treated as described for chloroplasts.

Reagents
  I. $HClO_4$, 20% w/v
  II. K_2CO_3, 5 M

D-Ribulose 1,5-P_2, D-Ribulose 5-P, D-Ribose 5-P, and D-Xylulose 5-P

Assay

Principle.[4] Since the concentration of ribulose 1,5-P_2 is extremely low in most plant preparations,[1] a quantitative enzymatic-optical assay is not feasible. We have used an isotopic assay involving the incorporation

[1] E. Latzko and M. Gibbs, *Plant Physiol.* **44**, 396 (1969).
[2] E. Latzko and M. Gibbs, *in* "Methoden der enzymatischen Analyse" (H. U. Bergmeyer, ed.), 2nd ed. Verlag Chemie, Weinheim, 1970. , 1970.
[3] U. Heber, K. A. Santarius, M. A. Hudson, and U. W. Hallier, *Z. Naturforsch. B* **22**, 1189 (1967).
[4] E. Latzko and M. Gibbs, *in* "Methoden der enzymatischen Analyse" (H. U. Bergmeyer, ed.), 2nd ed. Verlag Chemie, Weinheim, 1970. 1970.

of $^{14}CO_2$ by ribulose 1,5-P_2 carboxylase (Eq. 1). The pentose monophosphates are converted into ribulose 1,5-P_2 by purified enzyme-reactions (Eqs. 2–4) and measured via Eq. 1. With a standard curve, the lower limit of the determination is < 5 nmoles of pentose phosphates per assay.

$$\text{Ribulose 1,5-}P_2 + {}^{14}CO_2 + H_2O \xrightarrow[\text{carboxylase}]{\text{ribulose 1,5-}P_2} 2 \text{ glycerate 3-P-}{}^{14}C \qquad (1)$$

$$\text{Ribulose 5-P} + \text{ATP} \xrightarrow[\text{kinase}]{\text{ribulose 5-P}} \text{ribulose 1,5-}P_2 + \text{ADP} \qquad (2)$$

$$\text{Ribose 5-P} \xrightarrow[\text{isomerase}]{\text{ribose 5-P}} \text{ribulose 5-P} \qquad (3)$$

$$\text{Xylulose 5-P} \xrightarrow[\text{3-epimerase}]{\text{xylulose 5-P}} \text{ribulose 5-P} \qquad (4)$$

Reagents

   III. Triethanolamine-HCl buffer, 0.4 M, in 0.2 M EDTA, pH 7.6 (tra-EDTA buffer)

   IV. $MgCl_2$, 0.5 M

    V. $NaH^{14}CO_3$, 0.5 M (5 mCi ^{14}C/mmole of CO_2)

   VI. HCl, 1.0 M

  VII. $KClO_4$ solution saturated at 0°

 VIII. Reduced glutathione, 0.1 M

   IX. ATP, 0.05 M ($\simeq$ pH 5.5)

    X. Ribulose 1,5-P_2, 0.001 M

   XI. Ribose 5-P, 0.01 M

  XII. Ribulose-1,5-P_2 carboxylase[5] ($\simeq$ 20 units/ml) free of phosphoenolpyruvate carboxylase, transketolase, and transaldolase

 XIII. Ribulose-5-P kinase[6] ($\simeq$ 40 units/ml)

 XIV. Ribose-5-P isomerase[7] ($\simeq$ 50 units/ml)

  XV. Xylulose-5-phosphate 3-epimerase[8] ($\simeq$ 50 units/ml) (XIII–XV free of transketolase and transaldolase; prepared according to the references; diluted from stock solutions with 0.05 M Tris buffer, pH 7.6, before use)

 XVI. Tris buffer, 0.05 M, pH 7.6

Procedure for Ribulose 1,5-P_2. The following solutions are mixed in the described order in a clinical centrifuge tube:

   A. 0.25 ml of extract, 0.02 ml of HCl* (* = after addition of this solution the assay tubes have to be carefully and vigorously

[5] E. Racker, Vol. V, p. 266.
[6] J. Hurwitz, Vol. V, p. 258.
[7] E. Racker, Vol. V, p. 280.
[8] M. Tabachnick, P. A. Srere, J. Cooper, and E. Racker, *Arch. Biochem. Biophys.* **74,** 315 (1958).

shaken), 0.075 ml of tra-EDTA buffer*, 0.01 ml of $MgCl_2$, 0.1 ml of H_2O, 0.02 ml of ribulose-1,5-P_2 carboxylase*, 0.02 ml of $NaH^{14}CO_3$.

B. In the blank control the extract is replaced by 0.25 ml of solution VII.

C. The standard curve is prepared using 0.001–0.05 μmole of ribulose 1,5-P_2 per assay.

The assay mixture is incubated for 30 minutes at 20°. The reaction is stopped by the addition of 0.1 ml of 20% $HClO_4$.

The precipitate is centrifuged off. An aliquot of the supernatant fluid is distributed homogeneously on a planchet, which is covered with one layer of lens tissue, dried under an infrared lamp, and assayed for radioactivity.

Procedure for D-Ribose 5-P, D-Ribulose 5-P, and D-Xylulose 5-P. Since the content of pentose monophosphates has to be corrected for ribulose 1,5-P_2, both assays, ribulose 1,5-P_2 and pentose monophosphates, are carried out simultaneously and with the same solutions. Analogous to the ribulose 1,5-P_2 assay, the following solutions are mixed in the described order:

D. 0.25 ml of extract, 0.02 ml of HCl* (* after addition of this solution the assay tubes have to be shaken), 0.075 ml of tra-EDTA buffer*, 0.01 ml of $MgCl_2$, 0.02 ml of ribulose-1,5-P_2 carboxylase*, 0.02 ml of $NaH^{14}CO_3$*, 0.01 ml of ATP, 0.02 ml of reduced glutathione, 0.02 ml of ribulose-5-P kinase*, 0.02 ml of ribose-5-P isomerase*, 0.02 ml of xylulose-5-P epimerase*.

E. In the blank control the extract is replaced by 0.25 ml of solution VII.

F. The standard curve is measured using 0.01–0.1 μmole of ribose 5-P per assay. Further treatments as described for ribulose 1,5-P_2.

Definition of Unit and Calculations. One unit of enzyme converts 1 μmole of substrate per minute. Calculations for ribulose 1,5-P_2 content per 0.25 ml extract: cpm for A − (cpm for B) = cpm for ribulose 1,5-P_2. Calculations for pentose monophosphates per 0.25 ml extract: cpm for D - (cpm for E + cpm for the ribulose 1,5-P_2 content) = cpm for pentose monophosphates.

D-Glyceraldehyde-3-P, Dihydroxyacetone-P, D-Fructose-1,6-P₂ and D-Sedoheptulose-1,7-P₂

Assay

Principle.[9] The combined enzymatic-optical assay is based on the dehydrogenation of NADH by the glycerol-1-P dehydrogenase reaction (Eq. 5). Glyceraldehyde 3-P, fructose 1,6-P₂ + sedoheptulose 1,7-P₂ are transferred into dihydroxyacetone-P in sequential steps (Eqs. 6–8).

$$\text{Dihydroxyacetone-P} + \text{NADH} \xrightarrow[\text{dehydrogenase}]{\text{glycerol-1-P}} \text{glycerol 1-P} + \text{NAD} \tag{5}$$

$$\text{Glyceraldehyde 3-P} \xrightarrow[\text{isomerase}]{\text{triose-P}} \text{dihydroxyacetone-P} \tag{6}$$

$$\text{Fructose 1,6-P}_2 \xrightarrow{\text{aldolase}} \text{glyceraldehyde 3-P} + \text{dihydroxyacetone-P} \tag{7}$$

$$\text{Sedoheptulose 1,7-P}_2 \xrightarrow{\text{aldolase}} \text{dihydroxyacetone-P} + \text{erythrose 4-P} \tag{8}$$

Reagents
  XVII. NADH, 0.005 M
  XVIII. Glycerol-1-P dehydrogenase ($\simeq$ 50 units/ml)
   XIX. Triose-P isomerase ($\simeq$ 50 units/ml)
    XX. Fructose-1,6-P₂ aldolase ($\simeq$ 50 units/ml) (XVIII–XX commercial preparations are diluted with 0.05 M Tris, pH 7.6 before use.)

Procedure. In a final volume of 1 ml, the following solutions are mixed in a microquartz cell with a 10-mm light path: 0.05 ml of tra-EDTA buffer (solution III), extract, 0.02 ml of NADH, 0.02 ml of glycerol-1-P dehydrogenase (ΔA_{340} = dihydroxyacetone-P); 0.02 ml of triose-P isomerase, (ΔA_{340} = glyceraldehyde 3-P); 0.02 ml fructose-1,6-P₂ aldolase (ΔA_{340} = dihydroxyacetone-P + glyceraldehyde 3-P from fructose 1,6-P₂ and dihydroxyacetone-P from sedoheptulose 1,7-P₂).

The reference cuvette contains instead of the extract a suitable volume of solution VII. The ΔA at 340 nm of the reference cuvette is subtracted from the ΔA of the assay cuvette.

Calculations. (ΔA_{340} nm)/6.2 = μmole per volume of extract used for the assay.

[9]T. Bucher and H. J. Hohorst, *in* "Methoden der enzymatischen Analyse" (H. U. Bergmeyer, ed.), 1st ed., p. 246. Verlag Chemie, Weinheim, 1962.

D-Glucose 6-P, D-Glucose 1-P, D-Fructose 6-P, and D-Fructose 1,6-P$_2$ or ATP

Assay

Principle.[10,11] The combined enzymatic-optical assay is based on the reduction of NADP by the glucose-6-P dehydrogenase reaction (Eq. 9). Glucose 1-P, fructose 6-P, and fructose 1,6-P$_2$ or ATP (Eqs. 10–13) are assayed with glucose 6-P in sequential steps.

$$\text{Glucose 6-P} + \text{NADP} \xrightarrow[\text{dehydrogenase}]{\text{glucose-6-P}} \text{6-P-gluconate} + \text{NADPH} \tag{9}$$

$$\text{Glucose 1-P} \xrightarrow{\text{phosphoglucomutase}} \text{glucose 6-P} \tag{10}$$

$$\text{Fructose 6-P} \xrightarrow{\text{hexose-P isomerase}} \text{glucose 6-P} \tag{11}$$

$$\text{Fructose 1,6-P}_2 \xrightarrow[\text{diphosphatase}]{\text{C-1-fructose}} \text{fructose 6-P} + \text{P}_i \tag{12}$$

$$\text{ATP} + \text{glucose} \xrightarrow{\text{hexokinase}} \text{ADP} + \text{glucose 6-P} \tag{13}$$

C-1-Fructose diphosphatase isolated from spinach leaves[12] does not react with sedoheptulose 1,7-P$_2$. The K_m for fructose 1,6-P$_2$ is about 0.3 mM which is lower than the K_m of rabbit muscle aldolase for fructose 1,6-P$_2$ ($\simeq$ 0.75 mM). For extremely low concentrations of fructose 1,6-P$_2$, assay by means of Eq. (12) rather than Eq. (7) is recommended.

Reagents
  XXI. Tris·HCl buffer, 1.0 M, pH 8.5
  XXII. NADP, 0.005 M
  XXIII. Glucose, 0.1 M
  XXIV. EDTA, 0.1 M, pH 8.5
  XXV. Glucose-6-P dehydrogenase ($\simeq$ 50 units/ml)
  XXVI. Phosphoglucomutase ($\simeq$ 50 units/ml)
  XXVII. Hexose-6-P isomerase ($\simeq$ 50 units/ml)
  XXVIII. Hexokinase ($\simeq$ 50 units/ml)
  XXIX. C-1-fructose diphosphatase[12] (50 units/ml) free of trans-
       aldolase and aldolase
  XXX. Tris·HCl buffer, 1.0 M, pH 7.6 (XXV–XXIX are diluted before use with 0.05 M Tris·HCl buffer, pH 7.6 XXV–XXVIII are commercial preparations.)

[10]H. J. Hohorst, *in* "Methoden der enzymatischen Analyse" (H. U. Bergmeyer, ed.), 1st ed., p. 134. Verlag Chemie, Weinheim, 1962.
[11]E. Racker, *in* "Methoden der enzymatischen Analyse" (H. U. Bergmeyer, ed.), 1st ed., p. 160. Verlag Chemie, Weinheim, 1962.
[12]E. Racker, Vol. V, p. 272.

Procedure for Hexose Monophosphates and Fructose 1,6-P_2. In a final volume of 1.0 ml, the following solutions are mixed in a microquartz cell with 10-mm light path: 0.1 ml of Tris-buffer (sol. XXI), extract, 0.05 ml of NADP, 0.02 ml of MgCl$_2$ (sol. IV), 0.02 ml of glucose-6-P dehydrogenase (ΔA_{340} = glucose 6-P), 0.02 ml of hexose-6-P isomerase (ΔA_{340} = fructose 6-P), 0.02 ml of P-glucomutase (ΔA_{340} = glucose 1-P), 0.02 ml of EDTA, 0.02 ml of C-1-fructose diphosphatase (ΔA_{340} = fructose 1,6-P_2).

In the blank cuvette the extract is replaced by an equal volume of solution VII.

Procedure for ATP. ATP may be determined instead of fructose 1,6-P_2 if EDTA and C-1-fructose diphosphatase are omitted from the above-mentioned procedure and 0.05 ml of glucose and 0.02 ml of hexokinase are added instead (ΔA_{340} = ATP). If solely ATP is assayed, the following solutions are added per milliliter: 0.1 ml of Tris-buffer (sol. XXX), extract, 0.05 ml of NADP, 0.02 ml of MgCl$_2$ (sol. IV), 0.02 ml of hexokinase (ΔA_{340} = ATP).

In the blank cuvette the extract is replaced by an equal volume of solution VII.

D-Glycerate 3-P

Assay

Principle.[13] Two glycolytic back reactions (Eqs. 14 and 15) are coupled to reduce glycerate 3-P to glyceraldehyde 3-P consuming stoichiometric amounts of ATP and NADH.

$$\text{Glycerate 3-P} + \text{ATP} \xrightarrow[\text{kinase}]{\text{3-P-glycerate}} \text{glycerate 1,3-}P_2 + \text{ADP} \qquad (14)$$

$$\text{Glycerate 1,3-}P_2 + \text{NADH} \xrightarrow[\text{dehydrogenase}]{\text{glyceraldehyde 3-P}} \text{glyceraldehyde 3-P} + \text{NAD} \quad (15)$$

Reagents

XXXI. Glycerate 3-P-kinase ($\simeq$ 50 units/ml)

XXXII. Glyceraldehyde 3-P-dehydrogenase ($\simeq$ 50 units/ml)

These enzymes (XXXI, XXXII) are diluted from commercial preparations with 0.05 M Tris-buffer, pH 7.6.

In a final volume of 1.0 ml, the following solutions are added into a microquartz cell with a 10-mm light path: 0.1 ml of tra-EDTA buffer (sol. III), extract, 0.02 ml of NADH (sol. XVII), 0.1 ml of ATP (sol. IX), 0.02 ml of MgCl$_2$ (sol. IV), 0.02 ml of glyceraldehyde 3-P dehydrogenase, 0.02 ml of glycerate 3-P kinase.

In the blank cuvette the extract is replaced by solution VII.

---

[13] T. Bucher, Vol. I, p. 415.

Adenosine Diphosphate and Adenosine Monophosphate

Assay

Principle.[14] The determination of ADP and AMP is coupled with NADH oxidation by the lactate dehydrogenase reaction (Eq. 17). In this series of reactions (Eqs. 16–18), 1 mole of oxidized NADH equals 1 mole of ADP or 0.5 mole of AMP.

$$\text{ADP} + \text{P-enolpyruvate} \xrightarrow[\text{kinase}]{\text{pyruvate}} \text{pyruvate} + \text{ATP} \tag{16}$$

$$\text{Pyruvate} + \text{NADH} \xrightarrow{\text{lactate dehydrogenase}} \text{lactate} + \text{NAD} \tag{17}$$

$$\text{AMP} + \text{ATP} \xrightarrow{\text{myokinase}} 2 \text{ ADP} \tag{18}$$

Reagents

XXXIII. P-enolpyruvate, 0.05 M, pH 5.5
XXXIV. ATP, 0.05 M (pH 5.5, free of ADP + UDP + AMP)
XXXV. KCl, 1.0 M
XXXVI. Lactate dehydrogenase ($\simeq$ 100 units/ml)
XXXVII. Pyruvate kinase ($\simeq$ 50 units/ml)
XXXVIII. Myokinase ($\simeq$ 50 units/ml)

(XXXVI–XXXVIII are commercial preparations and are diluted with 0.05 M Tris-buffer, pH 7.6.)

Procedure. The following solutions are mixed in a microquartz cell to a final volume of 1.0 ml: 0.1 ml of tra-EDTA buffer (sol. III), extract, 0.02 ml of NADH (sol. XVII), 0.02 ml of lactate dehydrogenase (ΔA_{340} = pyruvate), 0.02 ml of P-enolpyruvate, 0.05 ml of KCl, 0.02 ml of MgCl$_2$ (sol. IV), 0.02 ml of pyruvate kinase (ΔA_{340} = ADP), 0.01 ml of ATP (sol. XXXIV), 0.02 ml of myokinase ($\Delta A_{340}/2$ = AMP).

In the reference cuvette the extract is replaced by solution VII.

Glycolate

Assay

Principle. The assay depends on the reduction of 2,6-dichlorophenol-indophenol (DPIP) as catalyzed by glycolate oxidase (Eq. 19). Since this enzyme can also react with glyoxylate, the sensitivity of the method is increased if the glycolate concentration exceeds 0.5 mM. Therefore a standard curve rather than a coefficient factor is required in this assay.

$$\text{CH}_2\text{OH—COOH} + \text{DPIP} \xrightarrow[\text{oxidase}]{\text{glycolate}} \text{CHO—COOH} + \text{DPIPH}_2 \tag{19}$$

---

[14]H. Adam, *in* "Methoden der enzymatischen Analyse" (H. U. Bergmeyer, ed.), 1st ed., p. 573. Verlag Chemie, Weinheim, 1962.

Reagents

XXXIX. Tris·HCl buffer, 1.0 M, pH 8.0
   XL. DPIP, 0.01%
  XLI. KCN, 0.1 M, in 0.01 M NH$_4$OH
 XLII. FMN, 0.05 M (flavin mononucleotide)
XLIII. Glycolate oxidase[15] diluted with 0.01 M phosphate buffer, pH 8.0, to $\simeq$ 50 units/ml
 XLIV. Sodium glycolate, 0.01 M

Procedure. In a final volume of 1.0 ml, the following solutions are placed in a microquartz cell with a 10-mm light path: 0.1 ml Tris-buffer, extract, 0.1 ml of DPIP, 0.01 ml FMN, 0.003 ml KCN, 0.04 ml glycolate oxidase.

The blank cuvette contains solution VII instead of the extract. The reaction is followed at 587 nm, T = 20°. The standard curve is prepared using 0.025–0.25 μmole of glycolate. Since glycolate oxidase can react with glycolate (K_m = 0.38 mM) or with glyoxylate (K_m = 2.5 mM), the standard curve is not linear below 0.15 μmole of glycolate per milliliter. Ascorbate or other reducing compounds which could react with DPIP must be omitted from the reaction or extraction medium.

[15] I. Zelitch, Vol. I, p. 528.

[23] Light Intensity Measurement; Characteristics of Light Sources; Filters and Monochromators

By Richard W. Treharne

Light Intensity Measurement

Types of Light Measurement

The term "light" strictly can be applied only to the visible portion of the electromagnetic spectrum. Since radiation outside of the visible range also is of importance for photoreaction studies, we shall broaden the concept of light measurements to include all measurements of radiant power in the ultraviolet, visible, and infrared portions of the electromagnetic spectrum. The words "light" and "radiant power" thus will be used interchangeably.

There are three basic types of radiant power measurements and, as a further complication, each measurement can be expressed in two distinct types of units — radiometric or photometric units. The three types of measurements and corresponding radiometric or photometric terms are: (1) total radiant power emitted from a source in all directions (radiant or luminous intensity); (2) radiant power emitted from a surface at that surface (radiance or luminance); (3) radiant power incident on a surface at some distance from the source (irradiance or illuminance).

For application in photobiology or photochemistry, measurements of irradiance generally are of the most interest. Most of the discussion on light measurements will be concerned with the practical aspects of the measurement of irradiance.

Radiometric and Photometric Units

Light intensity measurements have been made using a variety of techniques that often are applied to unrelated disciplines. For example, light measurements sometimes are expressed in photometric units which generally are applicable only to the visible portion of the electromagnetic spectrum. For photometric type measurements, a detector with a response closely matched to the response of the average human eye is most desirable. Typical photometric terms, such as footcandles and lux, occasionally are used in the literature of photoreaction studies—often without regard to the spectral response sensitivity of the reaction under study.

Another type of light measurement employs a detector with a response independent of wavelength. Wavelength-independent measurements using radiometric units of radiant power are more significant than photometric units for most applications in photochemistry and photobiology. Typical wavelength-independent units of radiant power are microwatts per square centimeter (mW/cm^2) and ergs per square centimeter per second ($ergs/cm^2$-sec).

The result of such diverse approaches to measurements of radiant power has been a proliferation of units and terminology which often tend to be confusing. Basically, however, the accuracy of all light measurements is dependent upon the spectral response characteristics and sensitivity of the detector. Thus, to be meaningful, the spectral sensitivity limitations of the detector must be considered in all radiant power measurements. The conversion from photometric to radiometric units can be accomplished only if one specifies the wavelength at which the photometric measurement was made. Typical conversion factors from photometric to radiometric units at a specified wavelength are shown in Table I.

Types of Detectors

Many detectors of radiant power are photoemissive or photoconductive devices which inherently are wavelength dependent. Photovoltaic

TABLE I
TYPICAL CONVERSION FACTORS FOR PHOTOMETRIC TO RADIOMETRIC UNITS[a]

Photometric units	Radiometric units
1 lumen	1.471×10^{-3} W at 555 nm
1 footcandle	1.583 μW/cm^2 or 15.83 ergs/cm^2-sec at 555 nm
1 lux	0.1471 μW/cm^2 or 1.471 ergs/cm^2-sec at 555 nm

[a]Conversion factors are valid at 555 nm *only*.

cells, phototubes, and photomultipliers, for example, offer excellent sensitivity over certain wavelength ranges but, generally, cannot provide even relative measurements of radiant power at different wavelengths because of their nonuniform spectral response characteristics. Spectral response curves of phototubes and photomultipliers are available from manufacturers (e.g., RCA, EMI, Whitaker, Dumont, etc.). For light intensity measurements of less than 10^{-8} W/cm^2, photomultipliers are the only electronic detectors with adequate sensitivity, at present. Thus, for low level fluorescence, phosphorescence, and chemiluminescence measurements, a photomultiplier calibrated at higher light intensities against a wavelength-independent detector is required. An excellent discussion of the calibration of photomultiplier tubes for absolute spectral measurements has been presented by Lee and Seliger.[1] Primarily, we shall concentrate our discussion of light measurements on detectors that can be considered wavelength independent or radiometric-type detectors.

All detectors purporting to be wavelength independent are based on the principle of first converting the incident radiant power to heat and then measuring the temperature rise by some appropriate means. The conversion from radiant power to heat usually is accomplished by some form of "black body" coating material that acts, ideally, as a total absorber of all radiation independent of wavelength. Although no uniform absorber of radiant power for all wavelengths exists, materials such as finely divided carbon black, gold black, or zinc black commonly are used. Also some of the newer carbon black pigmented emulsions, if carefully prepared, can exhibit satisfactorily uniform response over wide portions of the ultraviolet, visible, and infrared range.[2] The black

[1]J. Lee and H. H. Seliger, *Photochem. Photobiol.* 4, 1015 (1965).
[2]Examples of emulsions which absorb visible radiation almost totally are: Parsons Optical Black (Parsons & Sons, Ltd.), Krylon Black No. 1602 (Krylon, Inc.), and Black Velvet Coating (3M Co.). For spectral absorption curves, contact manufacturer. Also see U. S. Dept. Commerce, *Weather Bur. Meteorol. Satellite Report No. 31* (1964).

body coating may be applied directly to the temperature sensor, or in some cases, to a heat absorber such as a metal disc with an embedded sensor.

Many different forms of radiant power detectors have been constructed. To name a few, appropriately coated thermocouples, thermopiles, thermistors, bolometers, and heat-sensitive pneumatic detectors all have been used with varying degrees of success. More recently, a renewed interest in pyroelectric detectors has been evident.

Thermocouples and Thermopiles. Some of the earliest devices for measuring radiant power were thermocouples or thermopiles. Dissimilar metals with high thermoelectric power coefficients, such as bismuth-antimony, bismuth-silver, iron-constantan, and copper-constantan have been used.[3] The earliest thermopiles normally were covered with finely divided carbon black, usually benzene lamp soot, to convert the incident radiant power to heat.

Today, the thermopile design has been developed to the stage where thermopiles can serve as satisfactory secondary radiant power measuring standards. A commercially available (Eppley Lab, Inc.) sixteen-junction bismuth-silver thermopile, for example, can provide sensitivities of approximately 1 μV per 80 ergs/cm²-sec. Moreover, the spectral response of this type of thermopile can be essentially flat over any portion of the range from 250 nm to more than 40 μ, depending, in part, upon the type of window material used to shield the thermopile from air currents. A quartz crystal window most commonly is used in the range from 250 nm to approximately 3.25 μ. An infrared transmitting material, such as Irtran-2 (Eastman Kodak Company), can be used above 3 μ. Spectral response characteristics of typical window materials and a black-body absorbent coating called Krylon Black No. 1602 (Krylon, Inc.) are shown in Fig. 1.

A microvolt bridge, or similarly sensitive device, is required to measure the output of a thermopile. This need for an ultrasensitive readout device, plus the need for conversion factors to convert from microvolts to radiant power units, has complicated the use of thermopiles as a general-purpose laboratory detector for the measurement of radiant power, particularly at low light levels. Nevertheless, the thermopile is, and probably will continue to be, one of the most accurate means for measuring radiant power. Commercial thermopile calibrations accurate to within ± 2% are claimed (Eppley Labs, Inc.). At least one manufacturer (Cintra) has facilitated the use of thermopiles by coupling a thermopile with a stable amplifier presenting a digital readout directly

---

[3] W. E. Forsythe, "Measurement of Radiant Energy." McGraw-Hill, New York, 1937.

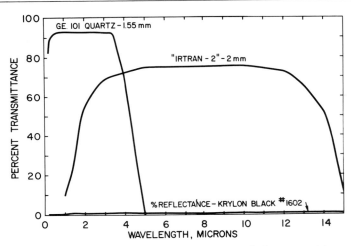

FIG. 1. Spectral response characteristics of typical window materials and "black body" absorbent coating used in radiometer probes.

in units of radiant power. Using a vacuum thermopile detector, this manufacturer claims a detectable limit of 2×10^{-7} W/cm^2 for radiant power measurements.

Thermistors and Bolometers. Thermistors and bolometer detectors are similar in that each exhibits a change in resistance when heated. They differ, basically, in that a bolometer usually is constructed of very thin metal strips, whereas a thermistor is composed of a semiconductor type material, such as a mixture of manganese and nickel oxides. The resistance of a thermistor varies as an exponential-type function of temperature whereas the resistance of a bolometer varies comparatively uniformly with temperature.

The change in resistance with temperature of a thermistor or bolometer can be detected by placing the element in one leg of an electrical bridge that initially is balanced. For radiometric applications, however, a more useful practice is to use two nearly identical thermistors or bolometers in adjacent legs of an electrical bridge. One of the thermistors or bolometer elements is blackened to absorb the incident radiant power; the other is shielded from the radiation and acts as a "reference." By this procedure, changes in ambient temperature tend to be offset by equal changes in the detector and reference legs; thus providing a more stable detector for radiant power measurements.

A typical bolometer design used as early as 1880 by Langley[4] consisted of two nearly identical strips of platinum wire which form two arms of a

[4] W. Langley, *Proc. Amer. Acad. Arts Sci.* **16**, 342 (1881).

Wheatstone bridge. The strips are blackened on one side. One strip is hidden and the other is exposed to the radiation which is to be measured. For fast response and good sensitivity, these strips should be very thin or preferably plated. To minimize thermal drift, the entire assembly should be insulated in a nearly constant temperature chamber. Vacuum mounting of the bolometer is most desirable, if the effect of air currents is to be eliminated. Cryogenic cooling of the bolometer also increases the sensitivity of a bolometer by several orders of magnitude, but few investigators can justify the expense and complication of cryogenic cooling for radiant power measurements. Because the voltage output of a bolometer bridge per degree rise in temperature inherently is not as large as that of a thermistor bridge, the bolometer has not been used as extensively as the thermistor for radiant power measurements.

A compact thermistor probe design is shown in Fig. 2. Two thermistors are mounted back to back, one shielded from the incident radiant power and the other blackened and exposed to the radiant energy to be measured. The dimensions of this probe are of particular interest since the probe was designed specifically to fit into a standard spectrophotometer cuvette holder. This probe design thus is particularly useful for photoreaction studies where physical size is a limiting factor.

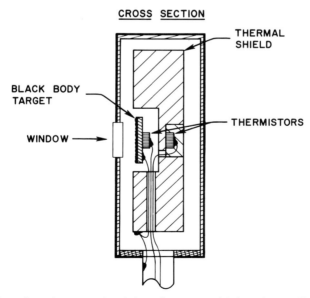

CROSS SECTION

FIG. 2.   An enlarged cross-sectional view of a commercial thermistor radiometer probe (Yellow Springs Instrument Co., Inc.). The thermistor detector is mounted in a "black body" target. Actual probe dimensions are less than 1 cm × 1 cm × 4 cm.

The basic circuit schematic used with this thermistor probe is shown in Fig. 3. In this circuit, the dc voltage output of the electrical bridge is converted to an ac voltage by "chopping" the dc voltage with a vibrating contact. Since ac voltage amplifiers are much easier to construct than dc voltage amplifiers, this chopping technique is commonly used when the need arises to amplify low level dc voltages. The useful sensitivity range of this particular thermistor radiometer (Yellow Springs Instrument Company) extends from 2500 ergs/cm²-sec for full-scale reading on the lowest range to 2.5×10^6 ergs/cm²-sec on the highest range.[5,6] This range compares favorably with the range of most thermopiles and is quite adequate for most actinic light intensity measurements. For measuring the radiant power output of low-intensity monochromators, however, a radiometer with more sensitivity is required.

Pneumatic Detectors. An ultrasensitive detector for radiant power in the range from 200 nm to the millimeter region of the electromagnetic spectrum was invented by Golay.[7] The detector operates on the pneumatic principle with radiant energy causing expansion of gas in a chamber which distends a flexible mirror. A detector light beam reflected off the flexible mirror into a light-sensitive cell measures the expansion

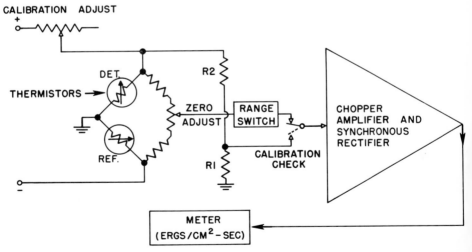

FIG. 3.   A block diagram of a commercial thermistor radiometer circuit (Yellow Springs Instrument Co., Inc.). Detector and reference thermistors, mounted as shown in Fig. 1, are connected in a differential bridge arrangement. See text for details.

[5] R. W. Treharne and H. W. Trolander, *J. Sci. Instrum.* **42**, 699 (1965).
[6] G. D. Milam and R. W. Treharne, *Inst. Soc. Amer. Trans.* **6**, 94 (1967).
[7] M. J. E. Golay, *Rev. Sci. Instrum.* **18**, 346 (1947).

produced by incident radiant power. The instrument was designed primarily for the infrared range, but it is adaptable with appropriate coating and window material for use at shorter wavelengths. A commercial (Eppley Labs, Inc.) version of this detector claims a radiant power range from 5×10^{-8} W (0.5 erg/sec) to 10^{-6} W (10 ergs/sec) with a linearity of 1%. Neutral density screens or similar types of radiant power attenuators could be used to extend the range of the detector for higher levels of radiant power.

Although the sensitivity of the pneumatic detector still is several orders of magnitude below detectors such as photomultiplier tubes, the pneumatic detector has provided some of the most sensitive, wavelength-independent measurements of total radiant power made to date. The sensitivity of the pneumatic detector, however, currently is being challenged by recently developed pyroelectric detectors.

Pyroelectric Detectors. Certain types of crystals, when heated, generate a voltage due to reorientation of the polarized crystalline structure. This is known as the pyroelectric effect. There is a marked similarity between the better known piezoelectric (pressure sensitive) effect and the pyroelectric effect. In fact, all classes of crystals which exhibit the pyroelectric effect also are piezoelectric. However, the converse statement is not true.

According to Lang,[8,9] the pyroelectric effect has been known since 1703, but the principle found little application until the relatively recent development of materials with high pyroelectric coefficients. Pyroelectric materials are characterized by internal polarization properties that occur within certain temperature limits known as Curie points. The direction of polarization is determined by application of a strong electric field applied to an appropriate crystalline structure as it is cooled through its Curie point. Examples of crystalline structures which exhibit good pyroelectric properties are barium titanate, lead zirconium titanate, and lithium sulfate monohydrate. Pyroelectric coefficients of hundreds or even thousands of volts per degree centigrade can be obtained. However, since the pyroelectric crystal itself behaves as a very small capacitor, only a minute current can be drawn from the crystal. Moreover, the charge generated in a pyroelectric crystal by incident radiant power is dissipated or neutralized in a relatively short time by factors such as leakage and charge redistribution within the crystal. For this reason, it is not feasible to make steady-state measurements of the charge induced in the pyroelectric crystal. In actual

[8]S. B. Lang, "Temperature—Its Measurement and Control in Science and Industry," Vol. 3, Part 2. Reinhold, New York, 1962.
[9]S. B. Lang, *Nat. Bur. Stand. U. S. Rep.* **9289**, (1967).

practice, some form of radiant energy chopping means in front of the crystal is desirable to convert the direct current output to an alternating current signal.

Most applications of the pyroelectric effect to date have been for measurements of high intensity radiant power — mostly in the infrared range.[10-16] At our laboratory, we have shown that the basic principles of pyroelectricity can be extended to provide radiant power measurements of less than 1 μW/cm² through the near UV, visible, and infrared range.

One form of the pyroelectric detector probe which we have developed is shown in Fig. 4. A lead zirconium titanate crystal (Clevite Corporation) is the sensing element. Dimensions of the active area of the crystal are 10 × 5 mm, so that the entire probe, including a chopping vane driven by a meter movement, is contained in the volume of a standard 1 × 1 × 4 cm cuvette. Evaporated silver electrodes are applied to the two faces of the crystal, and leads are attached with a conducting silver epoxy.

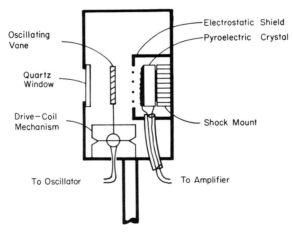

FIG. 4. Cross-sectional view of a radiometer probe used to measure radiant power levels as low as 1 μW/cm². A pyroelectric crystal coated with a "black body" absorbing material serves as a radiant power to electrical signal transducer.

[10]A. G. Chynoweth, *J. Appl. Phys.* **27**, 78 (1956).

[11]T. A. Perls, T. J. Diesel, and W. I. Dobrov. *J. Appl. Phys.* **29**, 1297 (1958).

[12]B. L. Mattes and T. A. Perls, *Rev. Sci. Instrum.* **32**, 332 (1961).

[13]J. Cooper, *J. Sci. Instrum.* **39**, 467 (1962).

[14]J. Cooper, *Rev. Sci. Instrum.* **33**, 92 (1962).

[15]S. B. Lang and F. Steckel, *Rev. Sci. Instrum.* **36**, 1817 (1965).

[16]J. H. Ludlow, W. H. Mitchell, E. H. Putley, and N. Shaw, *J. Sci. Instrum.* **44**, 694 (1967).

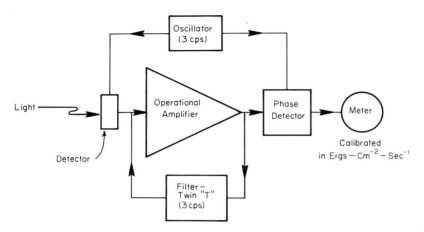

FIG. 5. Block diagram of radiometer circuit used with the ultrasensitive radiometer probe shown in Fig. 4. Incident radiant power is chopped at 3 cps and phase detected to provide low light level measurements. See text for details.

One surface of the pyroelectric element is then covered with an optically flat black coating. We found that Krylon Black No. 1602 (Krylon, Inc.) gave satisfactory results over the range from 250 nm to more than 3.3 μ, depending upon the type of window material used in the probe.

A block diagram of the pyroelectric radiometer is shown in Fig. 5. The incident radiant power to be measured is chopped at a low frequency (typically 3 cps) by an oscillating vane mechanism driven by an oscillator. The output of the pyroelectric detector is amplified and phase detected to minimize sources of noise such as undesirable piezo-electric-type voltages generated within the crystal. The meter readout provides direct measurements of radiant power from 100 ergs/cm²-sec for full-scale reading on the lowest range to 250,000 ergs/cm²-sec on the highest range.[17] The sensitivity and wide-range linearity of pyro-electric crystal type radiometers thus appear to be ideally suited for radiant power measurements.

Pulsed-Light Detectors. The preceding discussion has been concerned mainly with steady-state radiant power measurements. The response time of all the above wavelength-independent detectors of radiant power is, at best, measured in milliseconds. Faster response detectors are required for measurements of light pulses. There are several methods applicable to pulsed light measurements. Photomultipliers and photodiodes calibrated at specific wavelengths have been used

[17]R. W. Treharne and C. K. McKibben, U.S. Patent 3,398,381.

for radiant power measurements down to the nanosecond range.[18] Chemical quantum counters and actinometers also can be used over selected portions of the UV, visible, and infrared range.

The quantum counter, introduced by Bower,[19] is a concentrated solution of fluorescent substance which converts all incident radiation above a certain energy into fluorescence at specific wavelengths. The wavelengths are determined by the type of fluorescent substance used. Bowen used esculin and quinine as quantum counters for the ultraviolet region. Melhuish used rhodamine B down to 590 nm and also determined the fluorescence efficiency of a number of other substances.[20-22] A photomultiplier calibrated at the specific wavelengths of the fluorescing material is used as the detector readout in quantum counters. G. R. Seely has produced an excellent survey of this field.[23]

An interesting alternate approach to quantum measurements, first suggested by Dr. H. H. Seliger to this writer, is to use a photodiode with an S-14 spectral-sensitivity response. Close examination of the spectral sensitivity characteristic of a photojunction cell having an S-14 response shows that in the range from 450 nm to 1.3 μ the relative sensitivity increases linearly with wavelength.[24] Over this wavelength range, therefore, the output of a photodiode with an S-14 response can be calibrated directly in number of quanta per second incident on the photodiode. We have found, for example, that a carefully selected germanium photodiode can be employed as a quantum counter with accuracies approaching ±10% over the above specified wavelength range.

As in the case of steady state measurement of radiant power, the pyroelectric detector offers certain advantages for the measurement of the total energy in light pulses. One commercial unit (Barnes Engineering Company) claims a sensitivity of better than 1 mJ over the wavelength range from 0.3 to 40 μ. This energy meter was designed primarily for laser pulse measurements, but also is applicable to high intensity light pulse measurements, such as commonly are employed in actinic beams.

---

[18]A typical pulsed-light radiometer is the "lite-mike" of E. G. & G., Boston, Massachusetts.

[19]E. J. Bowen, *Proc. Roy. Soc. Ser. A* **154**, 349 (1936).

[20]W. H. Melhuish, *N. Z. J. Sci. Technol. Sect. B* **37.2**, 142 (1955).

[21]W. H. Melhuish, *J. Phys. Chem.* **65**, 229 (1961).

[22]W. H. Melhuish, *J. Opt. Soc. Amer.* **54**, 183 (1964).

[23]G. R. Seely, *in* "Photophysiology" (A. C. Giese, ed.), Vol. 3. Academic Press, New York, 1968.

[24]For S-14 response curve see, for example, RCA Electron Tube Handbook HB3 (Volume on Phototubes and Photomultipliers).

Characteristics of Light Sources

The requirements for light sources in photosynthesis and nitrogen fixation studies are dependent upon whether the light source is to be used as an analytical light source for spectroscopy measurements or as an actinic light source for photoreaction studies. Since the intensity requirements for actinic light sources are generally at least an order of magnitude greater than the requirements for analytical spectroscopy light sources, this discussion will center on intense actinic light sources. Similar types of light sources can and have been used as spectroscopic light sources.

Actinic Light Source Requirements

Radiant energy in relatively narrow wavelength bands generally is preferred for most photoreaction studies. The radiant power, ideally, should supply adequate energy in the absorption band or bands of the reactants and negligible energy outside of these bands.

The conventional approach in searching for an ideal light source to drive a particular photoreaction has been first to select a lamp with adequate radiation in the wavelength range of interest. The radiation from the selected lamp then usually is passed through filters and/or monochromators in an attempt to eliminate all except the desired wavelength bands. Within the past decade, however, an alternate approach to obtaining nearly pure monochromatic light at any desired power level has been made possible with the invention of the laser primarily accredited to T. H. Maiman[25–28] but predicted by Schawlow and Townes.[29] The laser promises to be the ultimate monochromatic light source and probably will replace more conventional radiant power sources for many photoreaction studies in the near future.

Lasers

The laser is a source of intense coherent radiation. The radiation is very nearly monochromatic, collimated, in phase, and polarized. Continuous-wave gas lasers, such as He-Ne lasers,[30] and solid-state diode

[25] T. H. Maiman, *Nature (London)* **187**, 493 (1960).
[26] T. H. Maiman, *Brit. Commun. Electron.* **7**, 674 (1960).
[27] T. H. Maiman, *Phys. Rev.* **123**, 1145 (1961).
[28] T. H. Maiman, R. H. Haskins, I. J. D'Haenens, C. K. Asawa, and V. Evtukov, *Phys. Rev.* **123**, 1151 (1961).
[29] A. L. Schawlow and C. H. Townes, *Phys. Rev.* **112**, 1940 (1958).
[30] R. J. Collins, D. F. Nelson, A. L. Schawlow, W. Bond, C. G. B. Garrett, and W. Kaiser, *Phys. Rev. Lett.* **5**, 303 (1960).

lasers[31] can provide relatively low radiation power levels in the milliwatt per square centimeter to watt per square centimeter range and are quite suitable for many photoreaction studies. Pulse-type lasers, such as the ruby rod and CO_2 lasers, can radiate at power levels up to the multimegawatt range and, therefore, are considerably excessive for most photochemical investigations. Nevertheless, for practically every application, there now exists a laser which presents the closest approach to an ideal radiant power source for many photoreaction studies.

The unique properties of the laser are produced by creating an electron energy population inversion in certain atoms by some form of external stimuli, such as an external activating light source or an electrical stimulus.[32] When the atoms of certain materials are excited to a high energy state, they subsequently drop back to their normal state, ejecting the excess energy as photons of light. Since the first ruby rod laser developed by T. H. Maiman in 1960,[25] a deluge of different types of lasers have been described in the literature. To name a few: solid state lasers, gas lasers, liquid lasers, metallic vapor lasers, injection lasers, glass rod lasers, plastic rod lasers all have been reported using many different materials in a variety of configurations (e.g., see footnotes 33–41 for a partial bibliography). To make a complete list of lasers presently available is a formidable task, and no doubt the list would be hopelessly out of date in a short time. However, a partial sampling of lasers presently available commercially is shown in Table II.

Aside from economic considerations, which are rapidly becoming secondary considerations by virtue of more economical designs and increasing competition, there have been at least two technical considerations which have limited the use of lasers in photoreaction and

[31]A typical example of a "laser" diode is a diffused planar gallium arsenide diode such as the infrared emitter M series by Monsanto, Cupertino, California.
[32]One of many fine text books describing laser principles is that of B. A. Lengyel, "Lasers — Generation of Light by Stimulated Emission," Wiley, New York, 1963.
[33]A. L. Schawlow, "Quantum Electronics-Infra-red and Optical Masers" (C. H. Townes, ed.). Columbia Univ. Press, New York, 1960.
[34]A. L. Schawlow, *Sci. Amer.* June 1961.
[35]G. B. Bryd and J. P. Gordon, *Bell Syst. Tech. J.* **40**, 489 (1961).
[36]A. D. White and J. D. Rigden, *Proc. IRE* **50**, 1697 (1962).
[37]W. R. Bennett, Jr., *Phys. Rev.* **126**, 580 (1962).
[38]A. Javan, W. R. Bennett, Jr., and D. R. Herriott, *Phys. Rev. Lett.* **6**, 106 (1961).
[39]A. Javan, "Advances in Quantum Electronics." Columbia Univ. Press, New York, 1961.
[40]P. D. Maker, R. W. Terhune, M. Nisenoff, and C. M. Savage, *Phys. Rev. Lett.* **8**, 21 (1962).
[41]W. Groner, *Electron. World,* September, 1965.

TABLE II
Typical Lasers[a]

Laser material	Type	Emission wavelengths
CO_2-N_2-He_2	Gas	10.6 μ
Cs-He	Gas	7.18 μ
Xe	Gas	3.5 μ
Nd:YAG	Solid	1.06 μ
GaAs	Diode	8450 Å
Ruby	Solid	6943 Å
He-Ne	Gas	6328 Å, 1.15 μ, 2.026 μ 3.39 μ, 3.5 μ
Hg^+	Gas	5677 Å, 6150 Å 7346 Å, 1.0583 μ
Ne	Gas	5401 Å
Nd:YAG	Solid (frequency doubled)	5300 Å
Ar	Gas	4880 Å
Ar-Kr	Gas	4880 Å, 5145 Å 5682 Å, 6471 Å
N_2	Gas	3371 Å
He-Cd^{2+}	Gas	3250 Å, 4416 Å
Nd:YAG	Solid (frequency quadrupled)	2650 Å

[a]A partial listing of lasers commercially available as of September, 1969. Completeness of list is not claimed. The table is intended to indicate only the wide range of wavelengths available.

spectroscopy studies to date. First, lasers have not been available at all wavelengths. Lasers in the visible and near-infrared range have been relatively plentiful. However, not until very recently have commercial lasers been available for the ultraviolet range. One breakthrough in producing an ultraviolet laser was achieved by using a nonlinear optical crystal to achieve frequency doubling.[42] The basic principle of frequency doubling and quadrupling presents the possibility of generation of intense radiant power at any wavelength by proper choice of laser. One is no longer restricted to the fundamental laser frequencies, since now it is also possible to generate radiant power at the harmonic frequencies. This technique, combined with the wide array of laser fundamental frequencies now available, promises to provide laser radiant power at virtually any desired wavelength in the ultraviolet, visible, or infrared range.

A second, and even more recent, breakthrough in ultraviolet laser design is the discovery that certain metallic vapors, such as Cd, Sn,

[42]R. W. Terhune, *Int. Sci. Technol.* August, 1964.

and Zn, have fundamental laser resonances in the ultraviolet range.[43] A helium-cadmium ion laser very recently has been marketed (Spectra-Physics) which introduces a new class of lasers for the UV range. This particular laser produces 10 mW continuous output at 4416 Å and 3250 Å.

Another limitation of the laser has been that it is not tunable to any desired wavelength. This limitation also is yielding to continued research and development. Experimental lasers have been described which are tunable over narrow wavelength ranges.[44] The tunable feature was accomplished by controlling the temperature of the laser crystal to alter slightly the crystal lattice structure and thereby shift the electron energy levels within the crystal. Extension of techniques such as this eventually may eliminate the last technical advantages which more conventional light sources can claim over lasers for photoreaction studies. At present, however, lasers cannot compete economically with conventional lamp sources for most applications in photochemistry and spectroscopy.

Lamps

The most logical approach in attempting a survey of light sources available is to attempt to classify the lamps according to the type of radiation emitted by the particular lamp. With only a few exceptions, however, it is difficult to categorize most lamps neatly under such headings as ultraviolet, or visible, or infrared lamps. Most lamps produce radiation which overlap into two, if not three, of these regions. For example, many lamps which emit strong ultraviolet radiation also emit significant, if not predominant, amounts of visible radiation. One notable exception is the low-pressure mercury arc lamp, sometimes called a mercury "resonance" or a germicidal lamp, which emits 80–90% of its radiation at 2537 Å. Higher-pressure mercury arc lamps capable of producing higher levels of ultraviolet energy, however, also produce intense radiation in the visible and near infrared region. Conversely, tungsten-filament lamps, operated at temperatures less than 2500°K, generally are considered to be visible and/or infrared sources only. At higher temperatures, nevertheless, a tungsten-filament lamp also can emit significant quantities of ultraviolet radiation, as will be discussed later.

[43]W. J. Silfast, Bell Telephone Lab, Murray Hill, New Jersey. Reported in *EDN-Design News*, July 1, 1969, p. 22; also see *Electronics*, August 4, 1969, p. 177.
[44]S. E. Harris, Stanford Univ., Stanford, California. Reported in *Electronic News*, May 15, 1967; also see review by S. L. Norman, *Res. Devel.* **20**, 36 (1969).

The classification of lamps, therefore, is somewhat arbitrary and, in this abbreviated review of the subject, only the more common, commercially available types of lamps will be included.

ULTRAVIOLET REGION

Most photochemical reactions of interest are produced by radiation in the ultraviolet range (200–400 nm). Consequently lamps that provide radiation in this range are of particular interest to photochemists. Two of the more common types of ultraviolet lamps are mercury and xenon lamps.

Mercury Arc Lamps. One of the most widely used radiant power sources for the ultraviolet region is the low pressure (<0.001 atm) mercury arc lamp which emits a predominantly strong resonant line of mercury at 2537 Å. Several other weaker lines of mercury, including a line at 1849 Å, also are generated by a mercury arc under low pressure. The 1849 Å line generally is absorbed by the quartz envelope of the lamp so that only a few relatively weak lines in the near UV and visible portion are radiated in addition to the predominant 2537 Å line. As mentioned above, up to 90% of the energy emitted from a low-pressure mercury arc is radiated at 2537 Å. Consequently the spectral purity of low-pressure mercury lamps generally exceeds that of most other types of lamps.

Low-pressure mercury lamps are available in a wide variety of envelope configurations and with different types of electrode structures in order to provide flexibility in coupling the light source to a photo-reaction vessel. One manufacturer (Ultra-violet Products Inc.), for example, provides coil-type, grid-type, and immersible-type low-pressure mercury lamps with radiation intensities at 2537 Å as high as 20 W.

Although the radiation from low-pressure mercury lamps is predominantly at 2537 Å, the radiation from higher-pressure (> 10 atmospheres) mercury lamps is shifted more toward the visible lines in the mercury spectrum. Self-absorption or "reversed" radiation of the resonance line at 2537 Å appears as reradiated energy at the longer-wavelength resonance lines of the mercury spectrum. These lines, in turn, are superimposed on a continuum background radiation. Typical spectra of a high-pressure mercury arc lamp are included in Fig. 6.

Xenon-Arc Lamps. Xenon-arc lamps are a particularly useful source of ultraviolet radiation since the radiation from a xenon arc, under proper conditions, can approximate solar radiation in the ultraviolet range. One of the most thorough investigations of xenon arcs has been con-

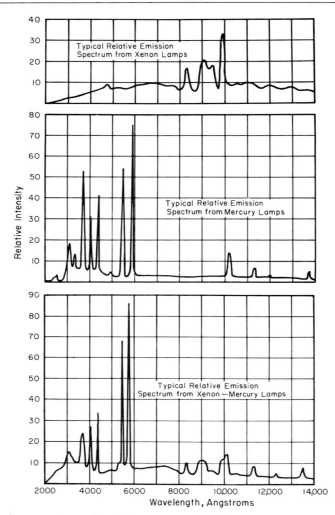

Fig. 6. A comparison of emission spectra of high-pressure xenon and mercury lamps with a high-pressure xenon–mercury lamp. Data obtained from Oriel Optics Corp., Stamford, Connecticut, the manufacturer of this light source.

ducted by Searle *et al.*[45–48] Some of the data of these authors, reproduced by permission of N. Z. Searle, are shown in Fig. 7. The graph compares the radiated intensity of a 150-W Osram (XBO 150 W), a 900-W Hanovia

[45]N. Z. Searle, P. Gieseke, R. Kinmonth, and R. C. Hirt, *Appl. Opt.* **3**, 923 (1964).

[46]R. C. Hirt and N. Z. Searle, *Appl. Polymer Symp.* **4**, 61 (1963).

[47]N. Z. Searle and R. C. Hirt, *J. Opt. Soc. Amer.* **55**, 1413 (1965).

[48]R. C. Hirt, R. G. Schmitt, N. Z. Searle, and A. P. Sullivan, *J. Opt. Soc. Amer.* **50**, 706 (1960).

(538 C9), and a 6000-W Osram (XBF 6000 W) xenon-arc lamp compared to solar radiation above and below the earth's atmosphere on a clear day in July in Stamford, Connecticut. The 150-W and 900-W xenon lamps are air-cooled arcs, and the irradiance measurements were made at a distance of 25 cm from these lamps. The 6000-W xenon lamp is a water-cooled arc lamp; in this case, the irradiance measurements were made at 48 cm from the source. Measurements were made at 30-Å bandwidths using a calibrated ultraviolet photomultiplier standardized against a ferrioxalate actinometer. The original papers should be consulted for further details. In summary, these investigators have found that a 6000-W xenon arc closely approximates the ultraviolet radiation from the extraterrestrial sun. A xenon-arc lamp, therefore, can be an excellent source of ultraviolet radiation for the range from 2200 Å to 4000 Å. At the same time, however, a xenon arc provides characteristic intense xenon band spectra superimposed on continuum radiation in the visible spectrum, as can be seen in Fig. 6. Some form of filtering is required with xenon-arc lamps if the radiation is to be restricted to the ultraviolet region.

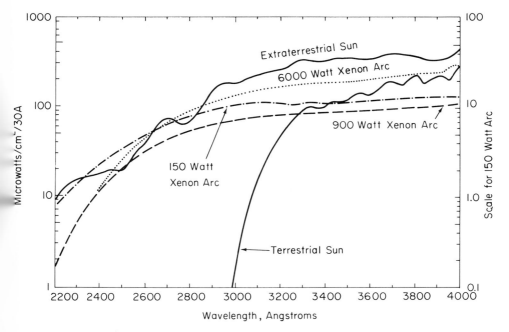

FIG. 7.   Absolute spectral distribution of irradiance received from high-pressure xenon lamps in the ultraviolet range compared to solar energy distribution. Reprinted by permission from N. Z. Searle, P. Gieseke, R. Kinmouth, and R. C. Hirt, *Appl. Opt.* 3, 923 (1964).

Other UV Lamps. Arc lamps containing combined gas vapors often provide more nearly uniform radiation than can be obtained from the individual gases separately. For example, a mercury–xenon-arc lamp produces a better distribution through the UV visible range than either mercury or xenon alone. The spectra of mercury, xenon, and mercury–xenon arcs are compared in Fig. 6.

Numerous other types of gaseous discharge lamps have been used in the UV region. Constricted arc discharge lamps of argon, in particular, exhibit strong continuum in the ultraviolet and visible.[49] Both low- and high-pressure hydrogen- and deuterium-arc lamps have found extensive use in UV spectroscopy. Carbon-arc lamps also have been used, but the necessity for periodic replenishment of the carbon electrodes makes the carbon arc too inconvenient for most investigators.

Low-pressure helium-, argon-, and nitrogen-arc lamps have been used into the far ultraviolet range below 2000 Å. An excellent survey of all types of lamps for the UV region is presented by Calvert and Pitts.[50]

VISIBLE AND NEAR-INFRARED REGION

Incandescent light sources generally are preferred when continuous radiation is desired in the visible and/or near-infrared range. The spectral distribution of the light from incandescent sources approximates the radiation from a black body radiator at some temperature. This temperature, known as the "color temperature" of the lamp, serves as an indication of the spectral distribution of radiation from that particular lamp. Higher color temperatures indicate not only more radiant energy at all wavelengths, but also a shift in the energy maximum toward the blue end of the spectrum. Color temperature ratings commonly are applied to tungsten lamps.

Tungsten Lamps. Tungsten-filament lamps provide a fair approximation to black body radiation if allowances are made for the spectral emissivity of tungsten, which may vary from 0.4 to 0.6 that of a black body (e.g., see footnote 3 reference). Projector type lamps normally operate in the color temperature range 2800°–3200°K. Tungsten lamps containing iodine vapor to minimize blackening of the bulb can be operated at color temperatures as high as 3400°K.[51] With proper

---

[49]An example of a constricted arc discharge lamp is the Vortex Stabilized Radiation Source (VSRS) of Giannini Scientific Corp.

[50]J. C. Calvert and J. N. Pitts, Jr., "Photochemistry." Wiley, New York, 1966.

[51]Tungsten-iodide lamps are available from Sylvania under the trade name "Sun-Gun" lamps. Similar lamps also are available from other manufacturers, e.g., Westinghouse and General Electric.

optical focusing, a tungsten–iodide lamp can provide an intense source of radiation spanning the ultraviolet, visible, and infrared range. One commercial light source (Cintra) claims to provide the highest spectral irradiance possible from a tungsten source. This light source employs a tungsten–halide lamp mounted at one focal point of an optically polished nickel ellipsoid. The lamp filament thus is imaged at the other focal point of the ellipsoid to provide intense illumination of any sample placed at this focal point. A plot of the spectral output of this tungsten–halide light source as provided by the manufacturer is shown in Fig. 8. The output approximates black body radiation although noticeable deviation from characteristic black body radiation is evident, particularly in the range above 3 μ.

Other Visible:Infrared Lamps. Although tungsten lamps are the most widely used for the visible and near infrared range, the radiation of tungsten lamps is somewhat limited in the blue region by the maximum operating temperature of tungsten, which melts near 3643°K. Higher color temperature sources are required in order to obtain more radiation in the shorter-wavelength range. Color temperature designations strictly are applicable only to continuous spectrum sources such as

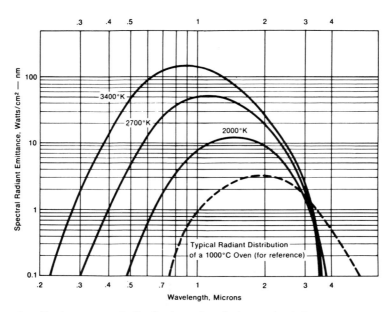

Fig. 8. Absolute spectral distribution of radiation emitted from a high-intensity tungsten–iodide light source as supplied by manufacturer (Cintra, Mountain View, California).

incandescent lamps. Nevertheless, combination line spectra and continuous spectra sources such as xenon-arc lamps are sometimes referred to in terms of "effective" color temperature. High-pressure xenon light sources, for example, can be operated at effective color temperatures of 6000°K, or solar temperature, as discussed above.

Other sources with effective high color temperatures are carbon arcs and enclosed metallic arcs such as zirconium arcs, which commonly are employed as point light sources.[52]

At the other end of the visible spectrum, typical lamps which exhibit strong infrared emission accompanying low visible radiation are Nernst Glower and Globar lamps. These lamps are particularly useful in infrared spectroscopy.

Filters and Monochromators

Filters and/or monochromators are used to select wavelength bands of light. Each method has certain advantages and limitations. If large areas and/or intense monochromatic light are required, filters are generally preferred although high intensity monochromators for photochemical applications have been described.[53-58] If high resolution or variable wavelength is desired, such as in spectroscopy, monochromators are required. Monochromators, in turn, can be divided into two basic types: refraction or prism monochromators and diffraction or grating monochromators.

Monochromators

Grating Monochromators. Grating monochromators are characterized by linear dispersion, so that the wavelength output is a linear function of the angular position of the grating. As a result, the bandwidth at the output slit of a grating monochromator is nearly independent of wavelength. Grating monochromators have the additional advantage of offering better resolution and aperture per dollar investment than prism monochromators. However, grating monochromators present one serious pitfall that warrants concern. Several overlapping orders of spectrum are an inherent characteristic of gratings, so that the spectral purity of the output of a simple grating monochromator must be cri-

[52] Typical examples of metallic arc lamps and point light sources are the "Metalarc" Lamps of Sylvania and zirconium concentrated arc lamps, such as those from Spectroline.
[53] C. S. French, G. S. Rabideau, and A. S. Holt, *Rev. Sci. Instrum.* **18**, 11 (1947).
[54] D. S. Villars, *J. Amer. Chem. Soc.* **49**, 326 (1927).
[55] L. J. Heidt and F. Daniels, *J. Amer. Chem. Soc.* **54**, 2384 (1932).
[56] G. S. Forbes, *J. Phys. Chem.* **32**, 482 (1928).
[57] F. Benford, *J. Opt. Soc. Amer.* **26**, 99 (1936).
[58] G. R. Harriman, *Rev. Sci. Instrum.* **5**, 149 (1934).

tically appraised. For example, it is not uncommon to set a second-order grating monochromator on the 2537 Å line from a high-pressure mercury lamp and visually observe at the output slit a green line produced by the first-order spectrum from the grating. In addition, scattered, nondispersed light reflected from the grating, acting as a partial mirror, also can degrade the spectral purity of the monochromator output.

There are several methods of minimizing the problems associated with grating monochromators. Double-pass, single-grating, or double-grating monochromators exhibit markedly improved spectral purity over simple-grating monochromators. Two separate monochromators or monochromator–filter combinations also increase spectral purity.

Prism Monochromators. Unlike gratings, prism monochromators provide only a single refraction spectrum, and thus prisms may be preferable if spectral purity is desired. Fewer possibilities of spectral errors exist with prism monochromators. A disadvantage of prism monochromators, however, is that the dispersion is nonlinear with wavelength. The wavelength readout of a prism monochromator not only is nonlinear but also the bandwidth, for a fixed slit width, varies with wavelength. Resolution in the red end of the spectrum thus is inherently inferior to resolution at shorter wavelengths. This can be compensated, partially, by decreased slit widths at longer wavelengths.

Combination prism-grating monochromators attempt to combine the advantages of both prism and grating monochromators and, to a certain extent, approach this objective.[59] This requires a rather elaborate, and costly, mechanical linkage system between the prism and grating.

Monochromator Optics. The method of collecting and focusing light in either a prism or grating monochromator is as important as the choice of the light-dispersing element itself. Many different types of optical configurations have been used in an attempt to compromise and/or minimize such problems as spherical aberration and astigmatism while maintaining optimum resolution, spectral purity, and aperature at reasonable cost. The subject is too vast to be covered in even a superficial manner. For more details the reader should consult texts on optics or spectroscopy,[60] where a few of the more popular types of monochromator mounts are listed under names such as Ebert, Littrow, Czerny-Turner, Pfund, Rowland, and Pashen-Range.

[59]An example of a prism-grating monochromator is the monochromator used in a Cary 14 spectrophotometer. In this instrument a 30-degree fused silica prism is connected by cam and bar linkages to a 600-line/mm echellete grating.
[60]The following are representative texts on monochromator optics: R. A. Sawyer "Experimental Spectroscopy," Prentice-Hall, Englewood Cliffs, New Jersey, 1946; F. A. Jenkins and H. E. White "Fundamentals of Optics," McGraw-Hill, New York, 1957.

As a typical example, a relatively new, commercial, double-grating monochromator optical configuration is shown in Fig. 9. This mono-chromator example is based on a modified Czerny-Turner mount employing two 1200 grooves per millimeter plane reflection gratings. A linear dispersion, or resolution of 15 Å per millimeter of slit width is claimed. The efficiency of the optical system at 2537 Å is stated as 60%, and the stray light at this wavelength is claimed to be less than 0.01% with a high-pressure mercury-arc lamp source. The wavelength range of this particular instrument is 190 nm to 825 nm. The statistics are not necessarily optimal, but they do provide a measure of the performance obtainable from a moderately priced monochromator. The double-grating system provides high dispersion in a short path length, and consequently, a compact monochromator design.

Filters

The most common types of filters for isolating a band or bands of wavelength are: (1) Wratten or gelatin filters, (2) colored glass or plastic filters, (3) chemical filters in either liquid or gas form, and (4) interference or dielectric filters. Another type of filter with unique, but relatively unused, potential is the Christiansen filter.

For most applications, the measure of a good filter is the ratio of transmission at the desired wavelength or wavelengths to the transmission at all other wavelengths. This ratio is dependent upon several

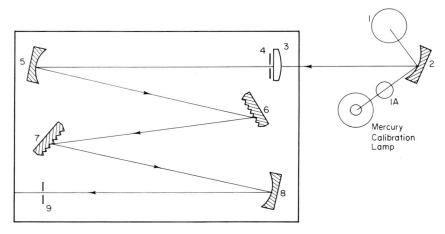

FIG. 9. A commercial double-grating monochromator (Bausch and Lomb, Rochester, New York). 1, (1A), light source; 2, condensing mirror; 3, collective lens; 4, entrance slit; 5 and 8, collimating mirrors; 6 and 7, reflectance gratings; and 9, exit slit. A modified Czerny-Turner mount is used.

basic filter parameters: peak transmission, bandwidth, shape of spectral response curve, and blocking effectiveness outside of the pass bands.

Wratten Filters. Wratten filters are made by coating gelatin containing a given weight of organic dye upon a prepared carrier glass plate. After drying, the film is stripped from the glass and remounted between optically flat glass cover plates. Wratten filters have been commercially available (Eastman Kodak Company) for several decades and have found extensive use, particularly in photographic work. Either narrow band-pass or wide-band cutoff filters are available for the visible or near-infrared range, but at present Wratten filters are not available for the UV range. In general, gelatin filters are the least expensive and yet most reproducible of the different filter types. Long-term stability of the organic dyes used in Wratten filters must be considered. Data concerning the degradation of a particular filter, particularly under prolonged UV irradiation, are available from the manufacturer.

Colored Glass and Plastic Filters. Glass or plastic plates and lenses, colored with metallic ions, such as nickel, cobalt, tungsten, in the case of glass and with organic dyes in the case of plastics, are effective, and moderately priced, filters. Glass filters are available in a wide variety of wavelengths and pass bands including the ultraviolet range down to 220 nm. In general, colored glass or plastic filters are intended primarily for wide bandwidth applications in the UV, visible, or infrared range. For example, sharp cutoff filters, which effectively absorb all radiation shorter than a certain wavelength, and pass all longer radiation into the near infrared, are available in glass filters (Corning). Relatively narrow-bandwidth characteristics also can be obtained, at some loss in peak transmission, by cascading several filters. The stability of most glass filters is excellent except under intense illumination, such as from a laser.

Chemical Filters. Liquid filters containing salts or dyes are inexpensive and effective filters, if properly prepared and periodically checked. Calvert and Pitts[50] and Seely[23] have described preparation procedures and transmission data for various salt solutions such as $CuSO_4$, $NiSO_4$, and $CoSO_4$. Many chemical filters for isolating portions of the ultraviolet and visible region have been described.[61-65] Some of

[61]R. E. Hunt and W. Davis, Jr., *J. Amer. Chem. Soc.* **69**, 1415 (1947).
[62]E. J. Bowen, "Chemical Aspects of Light." Oxford Univ. Press (Clarendon), London and New York, 1946.
[63]M. Kasha, *J. Opt. Soc. Amer.* **38**, 929 (1948).
[64]W. A. Noyes, Jr. and P. A. Leighton, "Photochemistry of Gases." Reinhold, New York, 1941.
[65]C. R. Masson, V. Boekelheide, and W. A. Noyes, Jr., "Photochemical Reactions, Technique of Organic Chemistry" (A. Weissberger, ed.), Vol. II. Wiley (Interscience), New York, 1956.

the more elaborate filter combinations, such as those described by Calvert and Pitts,[50] have included gases such as Cl_2 as one section of a composite filter. Most chemical filters experience some deterioration with time and irradiation, so that periodic rechecking of the transmission spectrum is necessary.

Interference Filters. Interference filters, sometimes called dielectric filters, probably offer the best narrow-bandwidth characteristics of any commercially available filters. An interference filter consists of multiple, partially reflecting, surfaces sandwiched between alternate thin layers of transparent dielectric materials of selective refractive indices. The layers normally are vacuum deposited on a substrate of glass or quartz. The interference filter works on the principle of selective reinforcement between direct and reflected beams of light at certain wavelengths dependent upon the thickness of the layers. Interference filters are commercially available at all wavelengths including the ultraviolet range down to 210 nm, as listed by Seely.[23]

Since the effective pathlength of a layer of an interference filter is dependent upon the angle at which the incident light beam traverses the filter, optical alignment of interference filters is important. Temperature also is a factor that may contribute to an effective change in pathlength, and thus wavelength characteristics, of an interference filter.

Within the past few years, rapid advances in the techniques of vacuum deposition has made the interference filter less costly to construct and consequently more readily available to the photochemist.

Christiansen Filter. As a final type of filter for consideration in this abbreviated survey, the Christiansen filter[66,67] is worth mentioning because its unique features are only recently being fully appreciated and explored in detail. A Christiansen filter consists of small transparent particles, having some dispersive power, immersed in a liquid of different dispersive power. The filter will transmit only that wavelength where the refractive indices of the two materials are exactly equal and will scatter all other wavelengths to some degree. In theory, a Christiansen filter of 5-Å bandwidth with transmission approaching 100% is possible; in practice, filters of 50–100 Å bandwidth with 70% transmission readily can be made. Matovich has constructed a unique pressure-cell version of a Christiansen filter in which the wavelength of the filter can be shifted to any wavelength throughout the visible and

[66]C. Christiansen, *Ann. Phys. Chem.* **23**, 298 (1884).
[67]E. D. McCalister, *Smithsonian Misc. Coll.* **93**, 7 (1935).

near-infrared range by varying the pressure on the cell.[68] To a certain degree, therefore, the Christiansen filter could combine some of the tunable advantages of a monochromator with the larger aperture potential of a filter.

[68]E. Matovich, *ISA J.* December, 1965.

[24] Enhancement

By G. Ben Hayyim and M. Avron

The quantum efficiency of photosynthesis in green plants is relatively independent of the sensitizing light wavelength within the range of 580–680 nm, but is strongly decreased above 680 nm.[1] This low quantum efficiency of photosynthesis at the long wavelength region, known as the red-drop phenomenon, can be improved when light of a shorter wavelength (580–680 nm) is simultaneously provided. The effect, known as the Emerson enhancement, is usually experimentally demonstrated by showing that the rate of the process under simultaneous illumination is greater than the sum of the rates obtained with each light beam separately. It provides one of the most important pieces of evidence for the concept of two photoacts in photosynthesis.

The enhancement effect (E) can be quantitatively expressed in several ways, the most sensitive and commonly employed[2] is

$$E = \frac{R_{12} - R_2}{R_1}$$

where R_1 and R_2 are the rates of the process during far-red and shorter wavelength illumination, respectively, and R_{12} the rate during simultaneous illumination with both light beams.

Two hypotheses are currently under discussion as models for the enhancement effect, named "spillover" and "separate package."[2,3] Theoretical analyses were suggested in an attempt to determine which one of these models is compatible with the data in the literature.[2,4]

[1]R. Emerson and C. M. Lewis, *Amer. J. Bot.* **30**, 165 (1943).
[2]J. Myers, *in* "Photosynthetic Mechanisms of Green Plants," *Nat. Acad. Sci.—Nat. Res. Counc. Publ.* 1145, p. 301 (1963).
[3]G. Hoch, *Rec. Chem. Prog.* **25**, 165 (1964).
[4]S. Malkin, *Biophys. J.* **7**, 629 (1967).

The Emerson enhancement effect is most commonly observed in algae for the complete photosynthetic process. Recently, it was observed also in isolated chloroplasts during the reduction of certain oxidants, such as NADP. These observations relate the effect to the electron transport pathway itself. Measurement of the enhancement effect is one of the few available methods to determine whether the photoreduction of an electron acceptor requires both photosystems or can proceed by excitation of only one or the other photosystem.

The observation of the enhancement effect for NADP photoreduction has been confirmed in several laboratories, although some controversy still exists. More controversial results appear in the literature concerning the enhancement effect for ferricyanide photoreduction.

Method

Reagents

Tricine, 0.3 M, pH 7.4
MgCl$_2$, 0.8 M
CH$_3$NH$_3$Cl, 0.5 M
NADP, 0.01 M
Cytochrome c, 0.001 M
Ferricyanide, 0.01 M
Ferredoxin purified from green plants as described in Vol. 23 [39]. Chloroplasts, isolated from spinach or lettuce as described in Vol. 23 [18].

Materials

Monochromatic light can be provided either through a monochromator or by light from a commercial projector, filtered through sharp cutoff interference filters blocked to infinity, having half bandwidth of 5–30 nm.

The intensity of the light can be varied through the use of neutral filters or metal screens, and must be measured by an instrument which measures light energy independent of wavelength, such as the Yellow Springs Instrument Radiometer (Yellow Springs, Ohio).

Procedure

Four procedures are currently employed in the measurement of enhancement.

Rate of Change in O_2 Concentration—Stationary Platinum Electrode. This electrode is not commercially available, but several detailed descriptions of its construction have appeared.[5-7] It has the great advantage, in the determination of enhancement values, of measuring directly the *rate* of oxygen evolution or uptake, rather than the *concentration* of oxygen in solution. However, since a constant response demands a sustained linear rate for relatively long periods of time, it has been employed almost exclusively in studies of whole algal cells, and its usefulness in studies of isolated chloroplasts has been very limited.

In operation, a drop of algal suspension is allowed to settle on the platinum surface, forming a layer 5–30 μ thick. The liquid is covered with a piece of dialysis membrane over which a constant flow of well aerated medium is provided. The nutrient solution employed for the growth of the algae is usually also used in the flowing medium. The cells are illuminated from above, through the flowing solution.

O_2 Concentration—Platinum Electrode. The more usually employed and commercially available (an appropriate one is Model 4004, manufactured by Yellow Springs Instruments, Yellow Springs, Ohio) membrane-covered combination electrode can be utilized to follow the oxygen concentration of the medium with preparations of both whole algae and chloroplasts. The electrode is placed inside a flat-bottom tube whose diameter is only slightly (1 mm) wider than that of the electrode employed. Rapid, homogeneous stirring is provided by a magnetic stirrer, with a distance of 3–10 mm between the stirring magnet and the bottom of the electrode. Illumination is provided from the sides. Several arrangements of this sort have been described.[8] Reaction medium consists of the nutrient medium in the case of whole algae, or the appropriate reaction mixture (see below) for chloroplast preparations.

Direct Monitoring of the Rate of Reduction in a Spectrophotometer. Many spectrophotometers may be adapted for the purpose. A possibility to illuminate the reaction cuvette through appropriate filters, and to place a protective filter in front of the photomultiplier tube must be provided. Both of the actinic beams can enter the cuvette through a half-silvered mirror placed at an angle of 45 degrees to the cuvette; 50% of one of the beams passes through the mirror into the cuvette, and 50% of the other is reflected into the cuvette. When red (630–660 nm) and far-red (> 700 nm) light is used for excitation, appropriate wavelengths and protective filters for the photomultiplier tube are as tabulated:

[5] J. Myers and J. Graham, *Plant Physiol.* **38**, 1 (1963).
[6] G. Gingras, *Physiol. Veg.* **4**, 1 (1965).
[7] D. C. Fork, *Plant Physiol.* **38**, 323 (1963).
[8] Y. de Kouchkovsky, *Physiol. Veg.* **1**, 15 (1963).

Compound measured	Wavelength (nm)	Protective filter
Ferricyanide	420	Corning 4-96
Cytochrome c	550	Corning 4-96
NADP	350	1 cm of saturated $CuSO_4$ + Corning 7-60

Assay of Reduced Reaction Components. In this procedure 2 cuvettes are placed inside a small black box fitted on both sides to receive 2 × 2 inch interference filters. Illumination is provided from the two sides for a period not exceeding 2 minutes. The reaction is stopped by turning the light off, and the reaction mixture is assayed for its content of the reduced product in question. Sufficiently sensitive methods for assaying NADP[9] or ferrocyanide[10] have been published. This method has the advantage that two different reaction mixtures may be analyzed simultaneously for enhancement for comparative purposes.

The following reaction medium is appropriate for studies of enhancement in chloroplast preparations: tricine, 0.15 ml; $MgCl_2$, 0.1 ml; CH_3NH_3Cl, 0.05 ml; and chloroplasts containing 20–90 μg of chlorophyll in a total volume of 3.0 ml. One of the following electron acceptor systems can be employed: NADP, 0.1 ml, plus a saturating amount of ferredoxin; or 0.1 ml of cytochrome c, plus a saturating amount of ferredoxin; or 0.15 ml of ferricyanide.

Properties

All theoretical and experimental analyses of the enhancement effect show that the degree of enhancement increases linearly with the ratio of R_2 to R_1. Significant values of E are normally obtained when this ratio exceeds 4. Maximal enhancement values, as high as 3–4, were obtained when this ratio exceeded 10. High ratios of R_2 to R_1 can be achieved by either increasing R_2 or decreasing R_1. The former is limited by the requirement that any approach to saturation should be avoided (one tries to stay within the linear range of the rate vs intensity plot, since only within this range may addition or subtraction of rates reasonably be attempted). The latter possibility is limited by the sensitivity of the assay. This may be overcome to a certain extent by either prolonging the illumination period or increasing the chlorophyll concentration. However, such devices should be employed with caution,

[9]G. Ben Hayyim, Z. Gromet-Elhanan, and M. Avron, *Anal. Biochem.* **28**, 6 (1969).
[10]M. Avron and N. Shavit, *Anal. Biochem.* **6**, 549 (1963).

since in longer periods of illumination the rate of the reaction may not be linear with time. An increase in the chlorophyll concentration to more than 40–50% absorption of the actinic light may also give rise to inaccurate measurements (see below).

High values of the enhancement effect are also dependent on the wavelength of light employed. For the far-red light region, the longer the wavelength, the better is the chance to improve its efficiency. With chloroplasts from higher plants or green algae, the best region lies between 715 and 730 nm, where the quantum yield of the process is rather poor, but the activity is not too low to be measured. For the shorter wavelength, it was shown that the region of the accessory pigments, namely chlorophyll b (650 nm) for isolated chloroplasts or green plants, gives the highest values. If the shorter wavelength is chosen at this range, low chlorophyll concentration should be employed to avoid any error due to high absorption. If the absorption is too high, there is a large drop in the actual intensity of the actinic light beam along the light path in the cuvette, and the average light intensity seen by most of the chloroplasts is very much lower than that measured at the surface. This can mask the difference between R_2 and R_{12}, which is most important for the calculations.

Transients are often observed on turning a light on or off in methods that employ a continuous recording, particularly when whole cells are employed. These should be ignored and only the steady-state rate eventually reached be measured and used in the calculation of the results.

[25] Steady-State Relaxation Spectrophotometry

By G. E. Hoch

Many valuable data on the mechanism of electron transport in photosynthesis have been obtained by optical spectrophotometry. This form of "nondestructive" testing can give both qualitative and quantitative information about electron-transport carriers and their individual role in the overall mechanism. In general the information obtained thus far has been qualitative in a kinetic sense. To be sure of the role of a proposed carrier, one needs to know the electron flux through the carrier and compare this with the flux through the entire electron-transport chain. Is the carrier involved with all the electrons, or one-half, or one-tenth? Or does it just respond to the overall oxidation-reduction state of the system? This is a difficult problem. An attempt to deal with it is presented below.

General

Spectrophotometry of photosynthetic material is beset by a number of problems which include: (1) the measurement by light of a light-induced absorption change; (2) the strong absorption of the photosynthetic pigments which limit the absorption changes of electron transport components to about 1% transmission or less. (3) Transmission changes which do not result from absorption changes, principally changes in scattered light.

In general the design of a spectrophotometric system is concerned with maximizing the information-to-noise ratio, by increasing the information, minimizing misinformation, and minimizing noise. Misinformation includes stray light within the monochromator, changes in light scatter, changes in fluorescence yield, effect of the measuring light on the reaction, artifacts from the actinic light, etc. Electrical noise may be reduced by using differential measurements, by causing the signal to appear at a fixed frequency and "narrow banding" the detector system about this frequency to reject noise, and by making this frequency as high as possible to minimize low frequency noise. Great care should be taken to minimize mechanical vibrations.

For kinetic experiments, the information desired is the amplitude of absorption changes and the kinetic parameters governing the formation and relaxation of intermediates in the photosynthetic process. Information about the amplitude of absorption changes, but few kinetic data, can be obtained from steady-state "light minus dark" difference spectra. Differential methods and narrow-banded ac systems can be put to good use in this method.[1,2] In general, the response is too slow to follow photosynthetic relaxations (the overall relaxation time of photosynthesis is about 10–50 msec.[3] At the other extreme is the "flash photolysis" method, in which a single, brief flash of light causes the absorption change to occur and permits the relaxation to be observed during a long dark time. Hence amplitude and relaxation (and perhaps formation) kinetics can be observed. This system appears to give maximum information. Unfortunately, it also yields the greatest noise, due to the wide bandwidth necessary in the detecting system, thus restricting it to fairly large absorption changes. Another unfortunate feature is the presence of induction phenomena in photosynthetic systems which makes comparison to "steady-state" results hazardous (see Witt et al.[4]).

[1] B. Chance, Vol. IV, p. 273.
[2] L. N. M. Duysens and H. Gaffron, "Research in Photosynthesis," p. 59. Wiley (Interscience), New York, 1957.
[3] R. Emerson and W. Arnold, J. Gen. Physiol. 16, 191 (1932).
[4] H. T. Witt, B. Suerra, and J. Vater, in "Currents in Photosynthesis" (J. B. Thomas and J. C. Goedheer, eds.), p. 273. Ad Donker, Rotterdam, 1966.

A natural extension of this technique is to employ periodic flashes and average the results of many flashes to reduce noise. At this point the simple kinetic beauty of the infinitely long dark period starts to disappear. How frequently can the flashes be given and a faithful replica of the long dark time result still be obtained? Even neglecting the difference between induction effects and "steady state," this is a difficult question, and obviously it depends upon the relaxation kinetics. If the flashes are too closely spaced, the observed decay will give an erroneous kinetic result and the signal amplitude will be different from that of the single flash. The experimenter's thinking must switch from the time domain to the frequency domain. Instead of simply asking what is the signal amplitude and decay as a function of time, it is necessary to ask what these are as functions of time and frequency. If this is going to be done, there is another way to do it which is considerably easier.

Steady-State Relaxation of Sinusoidally Driven Systems

In this method the photosynthetic apparatus is driven by actinic light, which is modulated to give 50% on and 50% off time of variable frequency. The wave form of the actinic light is preferably sinusoidal (square waves may be used, for the fundamental of a square wave is a sinusoid with zero phase shift). This forces the electron carriers to respond in a sinusoidal response at the frequency of the actinic light. The amplitude and phase relationship of the carrier's response will be governed by the appropriate kinetic parameters. The method is nearly identical to that used in some fluorescence lifetime experiments.[5] It allows noise reduction through extreme narrow banding but still preserves kinetic information.

Kinetic Theory

Methods for determining linear forced responses to periodic forcing functions may be found in any elementary textbook on electrical network analysis (see, for example, Hayt and Kemmerly[6]). An extensive treatment has been given by Eigen and DeMaeyer.[7] The following is an outline in trigonometric notation.

[5]W. L. Butler and K. H. Norris, *Biochim. Biophys. Acta* **66,** 72 (1963).
[6]W. H. Hayt, Jr. and J. E. Kemmerly, "Engineering Circuit Analysis." McGraw-Hill, New York, 1962.
[7]M. Eigen and L. DeMaeyer, *in* "Technique of Organic Chemistry," (S. L. Friess, E. S. Lewis, and A. Weissberger, eds.), Vol. 8, part 2, p. 895. Wiley (Interscience), New York, 1963.

For the photochemical reaction

$$h\nu + A \xrightarrow{\hspace{1cm}} A* \xrightarrow{k(1/\tau)} A \tag{1}$$

the rate equation is

$$\frac{dA*}{dt} = I \cdot \phi \, (A_{\text{total}} - A*) - A*(1/\tau) \tag{2}$$

where I is the absorbed light intensity and ϕ the intrinsic quantum yield. Even this simple system is nonlinear owing to the second-order nature of the formation reaction. The solution will be given below, but if the concentration of A* may be neglected in comparison to A_{total} (a condition which holds at low intensities—i.e., when $I\phi \ll k(1/\tau)$) the equation is linearized. Alternatively, if the excitation is sensitized (for example, by bulk chlorophylls) and excitation energy has a choice of many traps, then the conversion of some A to A* does not seriously affect the probability of a successful encounter and the equation is linear over this range. In these cases

$$\frac{dA*}{dt} = I\phi A_{\text{total}} - A*(1/\tau) \tag{3}$$

If I is completely modulated,

$$I = I_0 \, (1 + \sin \omega t) \tag{4}$$

the forced response can be

$$A* = A_0^* + A_\omega^* \sin(\omega t - \theta) \tag{5}$$

where A_0^* is the time average value of A* and A_ω^* is one-half the peak-to-peak amplitude of A* at the angular frequency of the light, ω. After substituting this solution into the differential equation (Eq. 3) and after equating the zero-order and first-order coefficients, one obtains

$$\tan \theta = \frac{\omega}{k} = \omega\tau \qquad A_\omega^* = A_0^* \cos \theta = \frac{A_0^*}{(1 + \omega^2\tau^2)^{1/2}}$$

As $\theta \longrightarrow 90°$, $A_\omega^* \longrightarrow 0$. For this mechanism the time average, steady-state velocity or flux may be computed:

$$v_0 = A_0 k = \frac{A_\omega}{\cos \theta} \cdot \frac{\omega}{\tan \theta} = \frac{A_\omega \omega}{\sin \theta}$$

If A* is returned to the dark state by component B and if A and B are localized in noncommunicating chains, the bimolecular reaction can still be characterized by a first-order time constant. Thus

$$h\nu + A \longrightarrow A*$$

$$A* + B \xrightarrow{k_1(1/\tau_1)} B* + A$$

$$B* \xrightarrow{k_2(1/\tau_2)} B$$

The same results as above are obtained for the primary component A, and extension of the kinetic analysis for the second component yields,

$$\tau_2 = \frac{\tan\theta_B}{\omega\left(1 + \dfrac{A_\omega}{B_\omega}\dfrac{\cos\theta_A}{\cos\theta_B}\right)}$$

The velocity equation changes similarly.

For this exceptionally fortunate linear relationship of carriers, it is therefore theoretically possible to obtain the sequence and individual reaction times.

For the "second-order" forward reaction, rearrangement of the original differential equation gives

$$\frac{dA*}{dt} = I\phi A_{total} - (k + I\phi)\,A*$$

An approximate solution is easily obtained by neglecting the nonlinear term (ignoring the time dependency of I in the right-hand term), in which case the time average value of $I_0\phi$ must be added to k. The true response of A* in this system can be described by a Fourier series and, assuming such a solution (closed after three harmonics), results in

$$\tan\theta = \frac{\omega}{I_0\phi + k} + \frac{I_0\phi}{I_0\phi + k} \cdot \frac{A^*_{2\omega}\cos\theta_{2\omega}}{2A^*_\omega\cos\theta}$$

for the fundamental frequency. The range of intensities over which the right-hand term (containing cross products) may be neglected depends upon the ratio of ω to k. Machine calculations show that with $\omega = k$, the error in k becomes 17% as I is increased to k. At lower frequencies the range is less; at higher frequencies it is greater.

Instrumentation

A diagram of an instrument useful for these measurements is given in Fig. 1. The instrument includes three main parts:

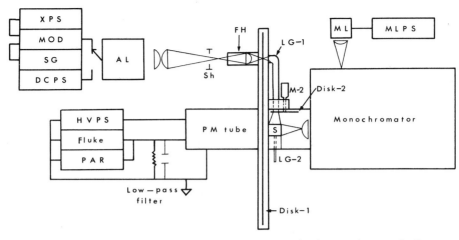

FIG. 1.   Diagram of spectrophotometer. MLPS, measuring lamp power supply (Sorenson, Q Nobatron, Model QB28-8); ML, measuring lamp (Sylvania Tungsten Halogen Lamp, 200 W, 6.6 Å; XPS, xenon power supply; MOD, current modulator for xenon lamp; SG, signal generator (Hewlett-Packard, audio oscillator, Model 2001); DCPS, dc power supply for tungsten halogen lamp; AL, 500-W xenon lamp (Hanovia, 959C); PM, photomultiplier (EMI, 9558); HVPS, high-voltage power supply (Keithley, Model 240A); Fluke, vacuum tube voltmeter (Model 845A); PAR, "lock-in amplifier" (Princeton Applied Research, Model HR-8); M-2, motor (Insco Corp.); Sh, shutter; FH, filter holder; LG, light guide; S, sample. Not shown in diagram: synchronous motor to drive Disk-1 (Globe, Type FC).

(1) a source of monochromatic light and suitable detecting system; (2) a source of strong modulated actinic light; (3) a rotating sector disk which causes the sample to be alternately illuminated by the measuring and actinic lights.

In the present instrument a Bausch and Lomb grating monochromator (500 mm focal length, f 4.5) was employed with the sample cuvette (1 cm × 1 cm) placed at the grating image. The monochromator source was a quartz-iodine (6.6 A, 250 W) lamp driven by a regulated power supply. This lamp proved to be very stable, and sufficient light was obtained at a lamp current of 6 A, which permitted long lamp life. Light intensity at the cuvette varied between 5 and 50 μW. The transmission of the sample was monitored by photomultipliers having either S-11 or S-20 photocathodes. The voltage developed by the anode current passing through a 1-megohm resistor was measured by an electrometer for time average (dc) amplitude, and by a "lock-in amplifier" for ac amplitude and phase measurements.

Modulated actinic light was obtained by either mechanically interrupting a light beam from a quartz–iodine lamp or by modulating the current to a xenon arc. The latter method provides great flexibility in modulation frequency (90% modulation up to ~ 20 kc) but has instabilities in intensity due to rather unpredictable arc wandering. In addition, magnetic fields develop as a result of the large current changes, which cause spurious signals to appear in the detector at the modulation frequency. Circuits for control and modulation of xenon arcs can be obtained commercially or built after DeSa and Gibson.[8]

Mechanical chopping is most satisfactory for low frequencies. Synchronous motors with variable gear drives are the most constant (Insco. Corp., Groton, Massachusetts). Feedback motors provide greater speed range, but their stability is marginal for the HR-8 type amplifiers. Modulation may also be obtained by inserting two polarizers in the light beam and rotating one of them. The transmission of the polarizers, however, is only about 30%.

Good separation of the actinic light from the measuring beam is required in this method. Color separation may be employed. For example, red actinic and blue measuring lights might be used, with a blue filter at the photomultiplier to eliminate actinic light and any chlorophyll fluorescence excited by either beam. S-11 photomultipliers are a good choice here. However, the experimenter is apt to feel cramped with this arrangement in a short time. Time separation of the actinic light from the measuring light provides a great deal more flexibility in wavelength of observation, but this greater flexibility of wavelength is obtained at the expense of response time. In the instrument depicted, a disk containing 18 sectors is turned at 60 rps. Observations (and flashes from the actinic light) occur at the rate of 1080 per second, and the time between center of flash and center of observation is somewhat less than 0.5 msec. An absorption change ($\geq 1 \times 10^{-5}$ $\Delta I/I$) must be detectable at this time to be recorded. This arrangement has advantages other than convenience; the instrument can also function as a phosphoroscope, or one may measure the changes in fluorescence yield from a weak monochromator beam as influenced by the actinic light.

[8]R. J. DeSa and Q. H. Gibson, *Rev. Sci. Instrum.* **37**, 900 (1966).

[26] Contact Shifts and Magnetic Susceptibilities in Iron-Sulfur Proteins as Determined from Nuclear Magnetic Resonance Spectra

By W. D. PHILLIPS and MARTIN POE

The remarkable electronic, magnetic, and geometrical properties exhibited by the iron-sulfur proteins have stimulated a variety of physical studies.[1] These physical approaches, a number of which are discussed elsewhere in this volume, include X-ray diffraction analysis, electron spin resonance (ESR), and Mössbauer, optical, and dichroic spectroscopies, as well as magnetic susceptibility determinations. Largely as a result of improvements in sensitivity and resolving power, nuclear magnetic resonance (NMR) spectroscopy has recently been found to be valuable in the elucidation of protein structure in solution.[2] These instrumental advances involve the use of computer averaging of multiple spectral scans to improve signal-to-noise ratios and the introduction of high-frequency PMR spectrometers based on superconducting solenoids. The primary purpose of this section is to illustrate the use of PMR in the determination of the geometrical structures and the electronic and magnetic properties of the iron/sulfur/cysteine centers in the iron-sulfur proteins.

Iron-sulfur proteins exhibit multiple oxidation states and generally possess low redox potentials.[3] The redox centers of these proteins appear to be associated with the iron-sulfur moiety. At least one oxidation state of each of the iron-sulfur proteins studied to date has been found to be paramagnetic, and this has permitted a number of important and revealing ESR and magnetic susceptibility studies.[4,5] The existence of such paramagnetic centers in these proteins makes the PMR approach to their electronic and magnetic structures, as well as to their geometrical conformations, particularly appropriate.

Molecular paramagnetism can produce profound effects on the characteristics of NMR. In favorable situations, analysis of such effects leads to determination of magnetic susceptibilities and provides insight

[1] J. C. M. Tsibris and R. W. Woody, "Structural Studies of Iron-Sulfur Proteins." Coordination Chemistry Reviews, Elsevier, Amsterdam, 1970.

[2] C. C. McDonald and W. D. Phillips, *in* "Proton Magnetic Resonance Spectroscopy of Proteins" (S. N. Timasheff and G. D. Fassman, eds.), "Biological Macromolecules," Vol. 4. Dekker, New York, 1970.

[3] D. O. Hall and M. C. W. Evans, *Nature (London)* **223**, 1342 (1969).

[4] W. H. Orme-Johnson and H. Beinert, *J. Biol. Chem.* **244**, 6143 (1969).

[5] T. H. Moss, D. Petering, and G. Palmer, *J. Biol. Chem.* **244**, 2272 (1969).

into the distributions of unpaired electrons in molecules. We will present briefly relevant background considerations and illustrate with applications to the iron-sulfur proteins rubredoxin and ferredoxin from *Clostridium pasteurianum* and the high potential iron protein (HiPIP) from *Chromatium.*

Magnetic Susceptibilities

A number of excellent monographs are available that describe the basic nuclear magnetic resonance phenomenon and applications of the technique to the study of molecules in solution.[6,7] Here we will outline only those concepts and relations needed to follow the subsequent discussion of applications of NMR to iron-sulfur proteins. The reader is referred to the cited references for detailed descriptions.

A nucleus of magnetogyric ratio γ subjected to magnetic field H_0 will absorb energy of a frequency given by the Larmor relation

$$f = \gamma H_0 \tag{1}$$

Extranuclear electrons shield nuclei in molecules so that a given nucleus actually does not experience the field H_0 but $H_0 (1 - \sigma)$, where σ is a shielding constant that depends on the particular bonding situation of the nucleus in the molecule. The appropriate Larmor relation is

$$f_i = \gamma H_0 (1 - \sigma_i), \tag{2}$$

where f_i is the resonance frequency of the ith nucleus in the molecular bonding situation characterized by the shielding constant σ_i. The shielding constants are responsible for the chemical shift effects of NMR that make NMR so powerful for the elucidation of molecular structure.

The bulk magnetic susceptibilies of samples on which NMR experiments are performed can additionally affect the magnetic fields experienced by nuclei that comprise the sample. Dickinson[8] computed classically the Lorentz cavity field produced on a uniform distribution of magnetizable material surrounding a paramagnetic species. He showed that the resonance field shift, ΔH, induced in the diamagnetic medium by the paramagnetic solute is given by

$$\frac{\Delta H}{H_0} = \left(\frac{4\pi}{3} - \alpha\right) \Delta K \tag{3}$$

[6]J. W. Emsley, J. Feeney, and L. H. Sutcliffe, "High Resolution Nuclear Magnetic Resonance Spectroscopy." Pergamon, London, 1965.
[7]J. A. People, W. G. Schneider, and H. J. Bernstein, "High Resolution Nuclear Magnetic Resonance" McGraw-Hill, New York, 1959.
[8]W. C. Dickinson, *Phys. Rev.* 81, 717 (1951).

where α is the demagnetization factor and ΔK is the change in volume susceptibility of the sample caused by introduction of the solute. We will return later to a discussion of α. Equation (3) can be recast in the form

$$\chi = - \frac{\Delta f}{\left(\frac{4\pi}{3} - \alpha \right) fm} + \chi_0 + \frac{\chi_0 \, (d_0 - d_s)}{m} \tag{4}$$

where χ is the mass susceptibility of the dissolved substance, f is the resonance frequency of the medium, or, as is more frequently the case, of a susceptibility "probe" dissolved in the medium, Δf is the change in this resonance frequency caused by introduction of the paramagnetic solute, χ_0 is the mass susceptibility of the solvent, m is the mass concentration of solute, and d_0 and d_s are, respectively, densities of solvent and solution.[9-11]

The actual procedure utilized in a NMR magnetic susceptibility determination is as follows. An NMR sample tube, typically 5 mm o.d. and 4 mm i.d., is filled with the solution containing solvent and the solute whose susceptibility is to be determined. To this is added a small amount of a third component, the susceptibility "probe" mentioned above, whose resonance will be the one actually observed for the susceptibility determination. A variety of materials may be used as "probe," e.g., tetramethylsilane for nonaqueous systems and sodium 2,2-dimethyl-2-silapentanesulfonate and tetramethylammonium chloride for aqueous solutions. Care must be exercised in the choice of "probe" molecule to ensure that significant interaction does not occur between "probe" and solute, otherwise spurious effects can be introduced.[12] A separate capillary that contains only solvent and "probe" is also prepared. This capillary then is introduced into the NMR sample tube discussed above, and Δf, the difference of resonance frequencies of the "probe" species in the sample tube and capillary, is carefully measured. For many determinations the third term of the right-hand side of Eq. (4) can be ignored since d_0 and d_s for dilute solutions often are very similar; obviously caution must be exercised in this approximation.

The demagnetization factor, α, in Eqs. (3) and (4) depends upon sample geometry and upon the relative orientations of the principal geometrical axes of the sample tube and the polarizing field H_0. Values

[9]D. R. Eaton and W. D. Phillips, *Advan. Magn. Resonance* 1, 103 (1965).
[10]D. F. Evans, *J. Chem. Soc.* p. 2003 (1959).
[11]K. D. Bartle, D. W. Jones, and S. Maricic, *Croat. Chem. Acta* 40, 227 (1968).
[12]M. Epstein and G. Navon, *Biochem. Biophys. Res. Commun.* 36, 126 (1969).

of α and the factor $[(4\pi/3) - \alpha]$ for commonly encountered sample tube geometries and magnetic field orientations are tabulated in the table. It is apparent from the table that there is no susceptibility shift for a spherical sample tube. Thus, susceptibility effects are eliminated by utilization of this geometry, but, conversely, a magnetic susceptibility determination is not possible.

For conventional NMR spectrometers in which the polarizing field is provided by electromagnets or permanent magnets, the cylindrical sample tube is positioned so that its long axis is perpendicular to H_0. This is the transverse cylinder of the table. On the other hand, for cylindrical samples and superconducting solenoids, the long axis of the sample tube is parallel to H_0, the longitudinal cylinder geometry. From the table and Eq. (4), it is seen that, for a given magnetic susceptibility difference between solute and solvent, the susceptibility shift is twice as large and opposite in sign for a coaxial, solenoidal field in comparison to the shift in a conventional magnet of equivalent field. In addition, the susceptibility shift is linearly dependent on the magnitude of the polarizing field. A principal virtue of the superconducting solenoid over conventional magnets is that, for purposes of high-resolution NMR, higher values of H_0 are practically achievable. Thus, a superconducting solenoid possesses major advantages over electromagnets and permanent magnets for the determination of magnetic susceptibilities, since larger susceptibility shifts are observed by reason of sample geometry and magnetic field strength.

The magnetic susceptibilities of rubredoxin and ferredoxin from *Clostridium pasteurianum* have been determined using the above NMR method.[13,14] The results on ferredoxin illustrate the remarkable spin-ordering features that appear to exist throughout the class of iron-sulfur proteins that contain two or more iron atoms per molecule. Rubredoxin with but a single iron atom exhibits "normal" magnetic properties.

DEMAGNETIZATION FACTORS FOR VARIOUS SAMPLE GEOMETRIES

Sample container geometry	α	$(4\pi/3)$-α
Sphere	$4\pi/3$	0
Transverse cylinder	2π	$-2\pi/3$
Longitudinal cylinder	0	$4\pi/3$

[13]M. Poe, W. D. Phillips, C. C. McDonald, and W. Lovenberg, *Proc. Nat. Acad. Sci. U. S.* **65**, 797 (1970).
[14]W. D. Phillips, M. Poe, J. F. Weiher, C. C. McDonald, and W. Lovenberg, *Nature (London)* **227**, 5258 (1970).

Rubredoxin undergoes a reversible, one-electron oxidation and reduction with $E_0' = -0.057$ V, and both redox forms are stable.[15] It is the
magnetic properties of the redox states that we wish to investigate. The
primary sequence of clostridial rubredoxin is partially known; the four
component cysteine residues are found in pairs, separated in each instance by two amino acids.[16] The X-ray crystallographic study to 2.5 Å
has revealed that the iron atom is bound to the polypeptide chain
through coordinative binding to the sulfur atoms of the four cysteine
residues. The sulfur ligands are arrayed approximately tetrahedrally
about iron.[17]

The temperature dependences of the paramagnetic components of
the molar magnetic susceptibilities of the oxidized and reduced forms
of rubredoxin from *C. pasteurianum* are plotted in Fig. 1. The results
are for an aqueous solution and extend over the temperature range
6°–78°. χ was derived experimentally as discussed above, and χ^p, the
paramagnetic component of the susceptibility, was obtained from

$$\chi = \chi^p + \chi^d \tag{5}$$

χ^d is the diamagnetic component of the magnetic susceptibility and was

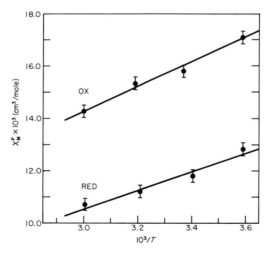

Fig. 1.   Temperature dependence of the paramagnetic contribution to the molar magnetic susceptibility of oxidized ("ox") and reduced ("red") *Clostridium pasteurianum* rubredoxin. From W. D. Phillips, M. Poe, J. F. Weiher, C. C. McDonald, and W. Lovenberg,
Nature (London) **227**, 5258 (1970).

[15]W. Lovenberg and W. M. Williams, *Biochemistry* **8**, 141 (1969).
[16]K. McCarthy and W. Lovenberg, personal communication (1970).
[17]J. R. Herriott, L. C. Sieker, L. H. Jensen, and W. Lovenberg, *J. Mol. Biol.* **50**, 391 (1970).

obtained from a separate determination of the magnetic susceptibility of aporubredoxin.

χ_m^p, the molar paramagnetic susceptibility, is related to the effective magnetic moment of the paramagnetic center, μ_{eff}, by the expression

$$\mu_{eff}^2 = \frac{3kT}{N\beta^2}\chi_m^p \tag{6}$$

In Eq. (6), N, β, k, and T are, respectively, the Avogadro number, Bohr magneton, Boltzmann constant, and absolute temperature. The effective magnetic moment is related to the spin quantum number, S, of the paramagnetic site

$$\mu_{eff} = g\sqrt{S(S+1)} \tag{7}$$

where g is the g value of the paramagnetic center.

Over the temperature range examined, the paramagnetic susceptibilities of both oxidized and reduced forms of rubredoxin were proportional to $1/T$, i.e., exhibited Curie law behavior (Fig. 1). For oxidized rubredoxin, μ_{eff} equals 5.85 Bohr magnetons, while reduced rubredoxin exhibits a μ_{eff} of 5.05 Bohr magnetons. These susceptibility results indicate that the iron constitutively associated with oxidized rubredoxin is high-spin Fe(III) (S = 5/2). Similarly, the iron of reduced rubredoxin is high-spin Fe(II) (S = 2). These conclusions are compatible with the results of Mössbauer studies on rubredoxin.[14]

The magnetic characteristics of the iron-sulfur proteins that contain two or more iron atoms are in significant contrast to the approximately normal behavior encountered in the single-iron rubredoxin. This is well illustrated by the temperature dependence of the magnetic susceptibility of the oxidized form of the eight-iron ferredoxin from *C. pasteurianum* (Fig. 2). The temperature dependence is decidedly non-Curie law in that χ_m^p increases with temperature rather than falling off as $1/T$. The temperature dependence of μ_{eff} per iron atom, calculated from Eq. (6), is plotted in Fig. 3. Two features are of particular interest. The first is that μ_{eff}, rather than being invariant, increases with temperature, further reflecting deviation from Curie law behavior. The second is the absolute value of μ_{eff}, ranging from 1.0 to 1.15 Bohr magnetons over the interval 5° to 65°. These values of μ_{eff} are considerably less than the value of about 1.73 Bohr magnetons to be expected if each of the eight iron atoms of clostridial ferredoxin possessed only a single unpaired electron. If the iron in clostridial ferredoxin were either high-spin Fe(III) or Fe(II) as in rubredoxin, μ_{eff} per iron atom would be expected to be about 5.8 or 5.0 Bohr magnetons, respectively.

The reason for the anomalously low value of μ_{eff} for oxidized clos-

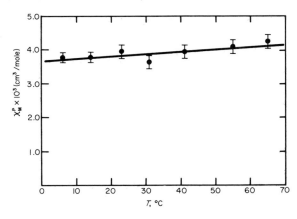

FIG. 2. Temperature dependence of the paramagnetic component of the molar magnetic susceptibility of oxidized *Clostridium pasteurianum* ferredoxin. From M. Poe, W. D. Phillips, C. C. McDonald, and W. Lovenberg, *Proc. Nat. Acad. Sci. U. S.* **65**, 797 (1970).

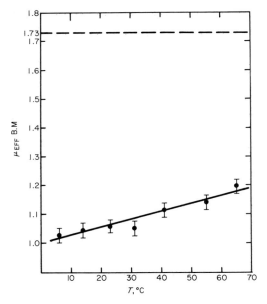

FIG. 3. Temperature dependence of μ_{eff} per iron atom, for oxidized *Clostridium pasteurianum* ferredoxin. B. M. = Bohr magnetons. From M. Poe, W. D. Phillips, C. C. McDonald, and W. Lovenberg, *Proc. Nat. Acad. Sci. U.S.* **65**, 797 (1970).

tridial ferredoxin, as for the other multiple iron iron-sulfur proteins, appears to be the existence of extensive antiferromagnetic exchange coupling between iron atoms. This leads to a ground magnetic state in which the net paramagnetic contribution to the magnetic susceptibility of the iron-sulfur moiety vanishes or becomes small at low temperature. With increasing temperature, upper, more paramagnetic states are thermally populated and χ_m^p and μ_{eff} increase.

Contact Shifts

The existence of molecular paramagnetism can be reflected in the characteristics of NMR absorption in two principal ways in addition to the already discussed magnetic susceptibility effects. One is the general effect of paramagnetic centers on nuclear relaxation times. The topic is complex and will be dealt with here only in a highly superficial fashion. The effect has been developed into an extremely powerful approach to the elucidation of structure and interactions around paramagnetic centers in proteins, and the reader is referred to an excellent recent review of the topic.[18]

Nuclear spin systems are characterized by two relaxation parameters: T_1, the longitudinal relaxation time, and T_2, the transverse relaxation time. T_1 reflects the time for a nuclear spin system to reestablish thermal equilibrium after saturation. Paramagnetic centers tend to reduce T_1 through electron–nucleus dipolar interaction. A practical consequence is that higher levels of radiofrequency power often can be employed, with resulting enhancement of spectral sensitivity for nuclei with shortened T_1 values. T_2 largely characterizes NMR line widths; long T_2's are associated with sharp nuclear resonances and short T_2's with broad nuclear resonances. As with T_1, paramagnetic centers can profoundly reduce T_2's and, consequently, broaden nuclear resonances, sometimes beyond observability. In this case, molecular paramagnetism has highly undesirable effects on the characteristics of nuclear resonances.

An invaluable feature of molecular paramagnetism, however, is that its presence often can give rise to contact shifts in NMR absorption that can provide deep insight into electronic and geometrical structures of molecules in general, and iron-sulfur proteins in particular.[19,20] There are two kinds of contact shifts.[9] The first is the pseudocontact interaction, which reflects the coupling of the nuclear spin under observa-

---

[18]A. S. Mildvan and M. Cohn, personal communication.
[19]T. Yamane, K. Wüthrich, R. G. Shulman, and S. Ogawa, *J. Mol. Biol.* 49, 197 (1970).
[20]D. G. Davis, N. L. Mock, V. R. Laman, and C. Ho, *J. Mol. Biol.* 40, 311 (1969).

tion with the dipolar field associated with the paramagnetic center. This interaction falls off as the third power of the distance, d, separating the paramagnetic center and resonating nucleus. For a paramagnetic center that is in a ligand field environment of tetragonal symmetry, the pseudocontact shift, ΔH_p, in solution is given by

$$\Delta H_p = -(3 \cos^2 \zeta - 1) (g_{\parallel} - g_{\perp}) (g_{\parallel} + 2g_{\perp}) \frac{\beta^2 H_o S (S+1)}{27kTd^3} \qquad (8)$$

Here ζ is the angle between the distance vector and the tetragonal axis of the ligand field, $g_{\parallel}$ and $g_{\perp}$ are the components of the g-tensor parallel and perpendicular, respectively, to the principal symmetry axis of the ligand field; the other symbols already have been defined. It is seen from Eq. (8) that for a highly symmetrical ligand field environment about the paramagnetic center, e.g., a tetrahedral or octahedral environment where $g_{\parallel} = g_{\perp}$, the pseudocontact field vanishes. This probably is the situation for many of the iron-sulfur proteins with the iron residing in ligand field environments largely dominated by sulfur atoms arrayed about the iron in approximately tetrahedral configurations. To the extent that this is the case, the contact shifts observed in the iron-sulfur proteins arise from the second kind of contact shifts, those due to isotropic hyperfine contact interactions.

Isotropic hyperfine contact shifts in NMR originate from the same electron–nucleus interaction that gives rise to hyperfine splitting in ESR. For a paramagnetic system that exhibits Curie law behavior, isotropic hyperfine contact shifts, ΔH_i, are given by

$$\Delta H_i = \frac{A\gamma_e}{\gamma_n} \frac{g\beta H_o S (S+1)}{3kT} \qquad (9)$$

Here, γ_e and γ_n are, respectively, electron and nuclear magnetogyric ratios, and, with the exception of A, the other symbols are as defined earlier.

A is the isotropic hyperfine coupling constant and may be calculated from Eq. (9) and experimental values of ΔH_i. In favorable instances, values of A derived from observed isotropic hyperfine contact shifts have proved to be of great value in mapping spin-density distributions in paramagnetic coordination compounds.[9] To illustrate the origin of contact shifts and the manner in which they can be employed to obtain spin densities, consider the paramagnetic fragment CH_3-$C^{.}$. The unpaired electron indicated by the dot is considered to be centered on a $p\pi$-orbital of the carbon atom to which the methyl group is attached. The hydrogen atoms of the methyl group can interact with the unpaired electron by a hyperconjugative mechanism that can lead to an isotropic hyperfine contact shift in PMR for these methyl protons. The value of A derived from the contact shift is related to the unpaired electron density on the paramagnetic carbon atom, ρ_c, by the expression

$$A = Q\rho_c \tag{10}$$

where Q is a constant of proportionality characteristic of this particular geometry and mode of electron-nucleus hyperfine interaction.

To illustrate the above, we present and analyze the contact-shift spectrum of an iron-sulfur protein. Among the proteins that can be isolated from the photosynthetic bacterium *Chromatium* is an electron-transfer agent with unusually high, positive redox potential known as HiPIP (high-potential iron protein). HiPIP has a molecular weight of 10,074, contains 4 atoms of iron and "labile" sulfide and 4 cysteine residues per mole of protein.[21] This protein has been the subject of a number of physical studies, including X-ray diffraction analysis, ESR, Mössbauer, optical, and dichroic spectroscopy, as well as a magnetic susceptibility determination.[22] Mössbauer spectroscopy suggests that the four iron atoms of the protein are equivalent in each of the redox forms.[23] The magnetic susceptibility of oxidized HiPIP corresponds to one unpaired electron per molecule of protein, while reduced HiPIP appears diamagnetic.[5] The four component iron atoms by X-ray diffraction are in a single unresolved cluster at a resolution of 4Å.[24]

The PMR spectrum of the oxidized form of *Chromatium* HiPIP is shown in Fig. 4. On this chemical shift scale diamagnetic proteins nor-

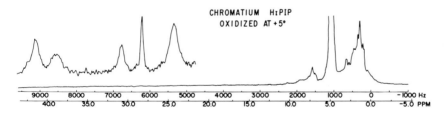

FIG. 4. Paramagnetic resonance (PMR) spectrum of oxidized *Chromatium* high-potential iron protein (HiPIP). The lower portion of the figure is a single-pass spectrum; the inset at upper left is a 100-pass, computer-averaged spectrum. From W. D. Phillips, M. Poe, C. C. McDonald, and R. G. Bartsch, *Proc. Nat. Acad. Sci. U.S.* 7, 682 (1970).

[21]R. G. Bartsch, *in* "Bacterial Photosynthesis" (H. Gest, A. San Pietro, and L. P. Vernon, eds.), p. 315. Antioch Press, Yellow Springs, Ohio, 1963.

[22]W. D. Phillips, M. Poe, C. C. McDonald, and R. G. Bartsch, *Proc. Nat. Acad. Sci. U. S.* **67**, 682 (1970).

[23]T. H. Moss, A. J. Bearden, R. G. Bartsch, M. A. Cusanovich, and A. San Pietro, *Biochemistry* 7, 1591 (1968).

[24]G. Strahs and J. Kraut, *J. Mol. Biol.* **35**, 503 (1968).

mally exhibit proton resonance absorption only in the region −2 to +10 ppm. The intense peak at about 5.5 ppm arises from residual HDO in the system. Resonances in the 6.5 to 10 ppm region can be assigned to protons of component aromatic amino acid residues and nonexchanged N*H* protons. Resonances of the −2 to +4 ppm region of resonance absorption derive from saturated C*H* protons of the amino acid residues.

A good deal of information concerning the secondary and tertiary structures of proteins has been derived from PMR spectra in the −2 to +10 ppm region of resonance absorption.[2,25] However, we are interested here in the magnetic properties and redox center of HiPIP, and, therefore, are more concerned with contact shifted resonances. These are to be found in the weak, broad resonances that appear in the 23 to 43 ppm region of the PMR spectrum. These resonances are displayed in the inset of Fig. 4 and can be observed only upon extensive accumulation on a computer of average transients.

In contrast to most nuclear resonances of diamagnetic proteins, the five low-field resonances of HiPIP to which we have assigned a contact shift origin exhibit striking temperature dependences (Fig. 5). This temperature dependence is in fact a principal discriminant for contact shift interaction since, in the absence of conformational changes,

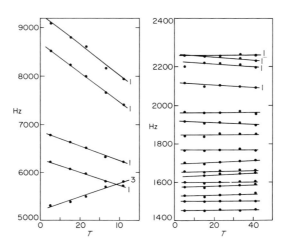

FIG. 5.   Temperature dependence of resonance positions for low field resonances in oxidized *Chromatium* high-potential iron protein. From W. D. Phillips, M. Poe, C. C. McDonald, and R. G. Barsch, *Proc. Nat. Acad. Sci. U. S.* **67**, 682 (1970).

[25]J. L. Markley, M. N. Williams, and O. Jardetzky, *Proc. Nat. Acad. Sci. U.S.* **65**, 645 (1970).

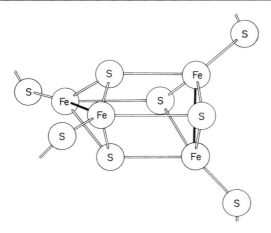

FIG. 6. Model for Fe/S/cysteine center in *Chromatium* high-potential iron protein. From W. D. Phillips, M. Poe, C. C. McDonald, and R. G. Bartsch, *Proc. Nat. Acad. Sci. U. S.* **67**, 682 (1970).

resonance positions in diamagnetic proteins are temperature indepen-dent. Further confirmation of the contact shift origin of the low-field resonances is the widths of the resonances, extending up to 250 Hz, that are indicative of electron–nucleus dipolar interaction and proximity to a paramagnetic center.

Using suitable comparison standards, the integral intensities of the contact-shifted resonances can be measured and the absolute number of protons contributing to each resonance determined. This has been done for oxidized HiPIP, and the numbers of protons per HiPIP mole-cule responsible for each of the five contact-shifted resonances are indicated in Fig. 5.

The reader is referred to the original literature for details of the PMR spectral analysis on HiPIP.[22] We only summarize here. The five contact-shifted resonances of oxidized HiPIP are assigned to the β-CH$_2$ protons of the four cysteine residues that are believed to bind the iron-sulfur moiety to the polypeptide chain. A model for this structure is presented in Fig. 6. It has been postulated from the PMR results that upon oxida-tion an electron is removed from a center consisting of two iron, two in-organic sulfur, and two cysteine sulfur atoms. The four β-CH$_2$ protons of these two cysteine residues exhibit the approximately Curie law chem-ical shifts of Fig. 5. The β-CH$_2$ protons of the other two cysteine residues removed from the environment exhibit the contact shift of relative in-tensity three (Fig. 5), whose temperature dependence appears to be com-patible with the underlying antiferromagnetic exchange coupling of the four atoms. That the resonance exhibits a relative intensity of three, rather than four to be expected from two β-CH$_2$ groups, is attributed tentatively to excessive dipolar broadening of a component of unit intensity.

We now turn to discussion of the PMR characteristics of reduced HiPIP from *Chromatium*. The magnetic susceptibility of reduced HiPIP to 200°K in frozen solution has been reported to be diamagnetic.[5] The extreme low-field region of resonance absorption at 5° and 40° of reduced HiPIP is, however, that shown in Fig. 7. The temperature dependences of the resonances of reduced HiPIP between 1400 and 3700 Hz are plotted in Fig. 8. It is clear that the three resonances between 2700 and 3700 Hz, each arising from a single proton, are being influenced by contact shift interaction. These resonances are attributed to three of the eight β-CH$_2$ protons of the four cysteine residues that presumably bind the iron-sulfur moiety to the polypeptide chain. It is assumed that the missing five resonances are to be found in the complex PMR absorption below 2000 Hz.

The increase of contact shifts of reduced HiPIP with increasing temperature is compatible with an antiferromagnetic exchange coupling between the component iron atoms. Whether the exchange coupling proceeds through direct iron–iron interactions, a super-exchange mechanism via the ligand sulfur atoms, or a combination of the two, is not presently clear. Similar contact shift manifestations of antiferromagnetic exchange couplings have been observed in the formally diamagnetic oxidized form of the two-iron ferredoxins from spinach,

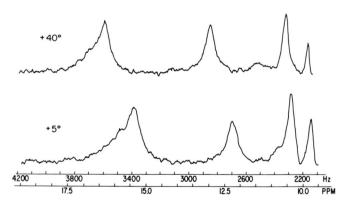

Fig. 7. Absorption in the 9.5–19 ppm region of the paramagnetic resonance spectrum of reduced *Chromatium* high-potential iron protein at +5° and +40°. From W. D. Phillips, M. Poe, C. C. McDonald, and R. G. Bartsch, *Proc. Nat. Acad. Sci. U. S.* **67**, 682 (1970).

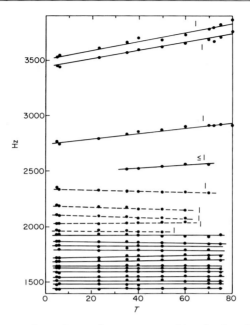

FIG. 8. Temperature dependence of resonance position for low field resonances in reduced *Chromatium* high-potential iron protein. From W. D. Phillips, M. Poe, C. C. McDonald, R. G. Bartsch, *Proc. Nat. Acad. Sci. U. S.* **67**, 682 (1970).

parsley, and alfalfa, and the two-iron putidaredoxin from *Pseudomonas putida*.

Much obviously remains to be done before the geometrical, electronic, and magnetic properties of the iron–sulfur proteins are understood in detail. Many physical approaches will contribute to this understanding. However, it appears that NMR will be prominent among these.

[27] Exchange Reactions (Plant)
$^{32}P_i \rightleftarrows$ ATP, ATP $\rightleftarrows$ H^{18}OH, and P$_i \rightleftarrows$ H^{18}OH

By Noun Shavit

An ATP $\rightleftarrows$ HOH exchange reaction is catalyzed by isolated chloroplasts in the light.[1] P$_i \rightleftarrows$ ATP and P$_i \rightleftarrows$ HOH exchange reactions occur under net phosphorylation conditions in the light,[2] and in the dark with sensitized chloroplast preparations.[3,4] Studies of ^{18}O exchange reactions in photophosphorylation were reviewed in a recent publication.[5] The general principles involved in the measurement of these reactions with mitochondrial preparations as well as the analysis of the ^{18}O content of the P$_i$ isolated from the reaction mixture have been described in previous volumes.[6] The procedures for measuring these reactions with freshly isolated and sensitized chloroplast preparations are presented.

Chloroplast Preparations

Freshly isolated chloroplasts are prepared according to standard procedures.[7] They are resuspended in the homogenizing medium or in 10 mM Tris·HCl or tricine·NaOH, pH 7.8.

Sensitized chloroplasts are prepared by incubating freshly isolated chloroplasts in the presence of a thiol and illuminating them or subjecting them to an acid-base "bath" in the dark. These preparations have been shown to catalyze enhanced ATP hydrolysis, and P$_i \rightleftarrows$ ATP, ATP $\rightleftarrows$ HOH, and P$_i \rightleftarrows$ HOH exchange reactions in the dark.

$^{32}P_i \rightleftarrows$ ATP Exchange

Activation by Light. The final reaction mixture contains in a volume of 1.0 ml at pH 7.8: 20 mM tricine·NaOH; 20 mM NaCl; 3 mM MgCl$_2$; 3 μM PMS[8]; 15 mM DTT[8]; 2 mM P$_i$ (containing about 10^6 cpm of $^{32}P_i$); 2 mM ATP and chloroplasts containing 60 μg of chlorophyll. Initially, 0.9 ml of an incubation mixture containing the appropriate amounts of tricine·NaOH, NaCl, MgCl$_2$, PMS, DTT, and chloroplasts

[1]M. Avron, V. Grisaro, and N. Sharon, *J. Biol. Chem.* **240**, 1381 (1965).
[2]N. Shavit, G. E. Skye, and P. D. Boyer, *J. Biol. Chem.* **242**, 5125 (1967).
[3]C. Carmeli, and M. Avron, *Eur. J. Biochem.* **2**, 318 (1967).
[4]G. E. Skye, N. Shavit, and P. D. Boyer, *Biochem. Biophys. Res. Commun.* **28**, 724 (1967).
[5]P. D. Boyer, *Curr. Top. Bioenerg.* **2**, 99 (1967).
[6]See Vol. VI, exchange reactions [32]; Vol. X, exchange reactions [9]; ^{18}O methods [10].
[7]See Vol. 23, [18].
[8]Phenazine methosulfate (PMS); dithiothreitol (DTT).

are incubated in test tubes (22°) and illuminated for 2 minutes at 70,000 lux. Immediately after illumination, 0.1 ml of a solution containing $^{32}P_i$ and ATP is added and the combined mixture is well mixed. After incubation in the dark for 10 minutes at 22°, the reaction is terminated by addition of 0.1 ml of 30% trichloroacetic acid.

Activation in the Dark by Acid-Base Incubation. Freshly prepared chloroplasts suspended in 10 mM Tris·HCl or tricine·NaOH are incubated in 80 mM DTT at pH 7.5 at 0° for 45 minutes.[9] An aliquot of this suspension (0.1 ml, containing 100–200 μg of chlorophyll) is added to a reaction mixture that, in a final volume of 1.0 ml, contains: chloroplasts, 10 mM succinate, pH 3.5, and 30 μM DCMU.[10] The final pH at this stage should be 4.0. This mixture is incubated at 0° for 20 seconds, then injected with a syringe into a reaction mixture of 1.0 ml containing 100 mM tricine·NaOH, pH 8.5; 2 mM ATP; 5 mM MgCl$_2$ and 1 mM P$_i$ (containing 10^6 cpm of $^{32}P_i$). The final pH in the combined reaction mixture should be about 7.8. The reaction is allowed to proceed for 45 seconds at 0° and then terminated by the addition of 0.2 ml of 30% trichloroacetic acid.

Measurement of $^{32}P_i$ ⇄ *ATP.* Denatured chloroplasts are removed by centrifugation, and aliquots are analyzed for [^{32}P]ATP formed after removal of the $^{32}P_i$ by the isobutanol–benzene extraction of the phosphomolybdate complex.[11] Before addition of ammonium molybdate and extraction, the thiol present is oxidized by addition of Br$_2$-H$_2$O until the solution becomes pale yellow. The amount of $^{32}P_i$ incorporated into ATP is calculated by dividing the *total* radioactivity of the aqueous phase by the specific radioactivity of the P$_i$ in the original reaction mixture.

Comments. It is assumed that the radioactivity in the aqueous phase represents [^{32}P]ATP. It should be remembered that phosphate esters, both organic and inorganic, remain in this phase after extraction of the phosphomolybdate complex. Thus, apparent effects on the P$_i$ ⇄ ATP exchange reaction may arise. These can be mostly overcome by determining the specific radioactivity of ATP.[3,6]

Activation of chloroplasts by illumination requires the presence of DTT whereas the activation by acid-base incubation reveals most of the exchange activity in the absence of DTT.[9] A weak exchange activity is observed with chloroplasts in the light without any sensitization by exposure to thiols or acid-base incubation.[2]

[9] J. H. Kaplan and A. T. Jagendorf, *J. Biol. Chem.* 243, 972 (1968).
[10] 3(3,4-Dichlorophenyl)-1,1-dimethylurea (DCMU); it is added to inhibit light-dependent reactions. When the incubation is performed in darkness, DCMU addition is omitted.
[11] M. Avron, *Biochim. Biophys. Acta* 40, 257 (1960).

ATP ⇌ HOH Exchange

Exchange under Continuous Illumination. In a final volume of 3.0 ml at pH 7.8, the reaction mixture contains: 20 mM NaCl, 4 mM MgCl$_2$; 0.033 mM PMS; 1.33 mM ATP; 10 mM ascorbate; and freshly isolated chloroplasts containing 100–150 μg of chlorophyll. Water in the reaction mixture contains about 1 atom % excess ^{18}O. After addition of chloroplasts, the suspension is illuminated for 15 minutes at 160,000 lux at 20°. At the end of the illumination period, 0.15 ml of 70% perchloric acid is added, and the sample is chilled. Protein is removed by centrifugation at 0°. The supernatant fluid is decanted, ATP is adsorbed on charcoal, and KH$_2$PO$_4$-^{18}O is isolated and analyzed as described by Boyer and Bryan.[6]

Acid-Base Incubation. Chloroplasts, 1.5 ml, suspended in 10 mM tricine·NaOH, pH 7.8, containing 1.5–2.0 mg of chlorophyll, are mixed with 13.5 ml of a chilled solution containing 150 μmoles of succinate, pH 3.5. The final pH of this mixture should be about 4.0. After incubation at 0° for 20 seconds this suspension is mixed with 15 ml of a reaction mixture at 0° containing 100 mM Tris·HCl or tricine·NaOH buffer, pH 8.5; 5 mM MgCl$_2$ and 0.2 mM ATP. The final pH at this stage should be about 7.8. Water in both reaction mixtures contains about 1 atom % excess ^{18}O. After 60 seconds in the dark at 0°, the reaction is terminated by addition of 1.5 ml of 70% perchloric acid. Denatured chloroplasts are removed by centrifugation, and the sample is treated as mentioned above for isolation of KH$_2$PO$_4$ from ATP for ^{18}O analysis.

P$_i$ ⇌ HOH and ATP ⇌ HOH Exchanges

The P$_i$ ⇌ HOH exchange under continuous illumination, with freshly isolated chloroplasts that have not been sensitized, is rather weak. Therefore, we shall consider the procedure to assay this exchange as well as the ATP ⇌ HOH in the dark, with chloroplasts sensitized in the presence of thiols and light. The incubation conditions are identical to those described for the assay of P$_i$ ⇌ ATP exchange, except that the concentrations of P$_i$ and ATP in the reaction mixture are raised to 4 mM. The isolation of P$_i$ and of P$_i$ from ATP are performed as described.[6] The presence of P$_i$ and ATP in the reaction mixture stimulates, respectively, the ATP ⇌ HOH and the P$_i$ ⇌ HOH exchange reactions.[4] Thus to measure both exchanges in the dark with sensitized chloroplasts, it is best to test these exchange reactions in the presence of ATP and P$_i$. A correction has to be made due to hydrolysis of ATP under these conditions. This can be achieved by introducing the ^{32}P as P$_i$ or ATP in separate incubation controls. The net ATP hydrolysis can be calculated from the release of ^{32}P from ATP which reflects both exchange ($P_i \rightleftarrows$ ATP) and ATPase, and from the incorporation of ^{32}P into ATP, which gives a measure of exchange.

Negligible contamination by other sources of oxygen was found in samples containing 3–5 μmoles of P_i. The precision of measurement of the ^{18}O exchanges is limited by the amount of isotope present in the water of the incubation mixture. Using highly ^{18}O labeled P_i or ATP enables measurement of the exchanges by loss of ^{18}O from P_i and ATP, and allows the use of smaller samples, to which carrier P_i can be added.

[28] A Green Safelight for the Study of Chloroplast Development and Other Photomorphogenetic Phenomena

By JEROME A. SCHIFF

The excellent safelight of Withrow and Price[1] has been extremely useful in many laboratories, including our own,[2] for studying photomorphogenetic phenomena that are not promoted by green light. Since this safelight employs filters made from dye solutions in gelatin, they are somewhat tedious to prepare and do not last indefinitely. Faced with the problem of constructing many safelights for permanent installation, we have explored the possibility of using colored plexiglass in place of the gelatin filter.

We now routinely use ordinary industrial fluorescent light fixtures in which we replace the diffusing honeycomb with Roehm and Haas plexiglass filters. The fixture is then taped around with lightproof tape to contain any light that does not pass through the filters. The filter consists of one sheet of deep blue plexiglass (No. 2424), and one sheet of green plexiglass (No. 2092). These transparent sheets of plastic can be obtained from local dealers since Roehm and Haas does not sell directly. We obtain ours from Commercial Plastics, 352 McGrath Highway, Somerville, Massachusetts. The thickness of the plexiglass does not matter since the amount of dye the manufacturer adds is prorated to the thickness. Therefore, all thicknesses have the same transmission. We find 1/8 inch thickness to be convenient. The lamps we use are yellow

[1]R. B. Withrow and L. Price, *Plant Physiol.* **32**, 244 (1957).
[2]H. Lyman, H. T. Epstein, and J. A. Schiff, *Biochim. Biophys. Acta* **50**, 301 (1961).

or golden fluorescent tubes taped at each end with about 1 inch of black electrician's tape, where the phosphor becomes unreliable. The reason for selecting the yellow tubes is that they do not emit any radiation below 500 nm, since they do not emit the blue and near-UV mercury lines. Thus, there is a sharp cutoff on the short-wavelength side in the lamp. The green and blue filters take care of the long-wavelength side and also catch any stray blue light. The result is a green emission with a fairly sharp peak in the trough of most morphogenetic phenomena in plants but near the peak of visual acuity for the human eye. With a hand spectroscope, the emission band of the safelight extends roughly from 515 to 560 nm with a peak in the region of the 546 nm mercury line emitted by the yellow tubes. After dark adaptation of our eyes, we find that these lamps give adequate light at a distance of about 4–6 feet above the working space and that they are completely safe even if organisms are exposed to them continuously for several days. These lamps have proved to be safe in not promoting the protochlorophyll(ide) to chlorophyll conversion and the photoreactivation of UV inactivation in *Euglena* or the protochlorophyllide to chlorophyll transformation and phytochrome-controlled morphogenetic phenomena in beans.

Acknowledgment

I wish to thank Dr. John Olson, Brookhaven National Laboratory, who first called my attention to the availability of green plexiglass. I am also indebted to Mr. Jack Hilbert, Biology Department, Brandeis University, for assembling the prototype safelight and its descendants. Dr. A. O. Klein provided the observations on beans. This work was supported by GM 14595 from the National Institutes of Health. The technical assistance of Miss Nancy O'Donoghue is appreciated.

[29] Principles of Differential Spectrophotometry with Special Reference to the Dual Wavelength Method

by Britton Chance

Introduction

Any significant level of understanding of a chemical, physical, or biological system requires the measurement of its intermediates. The properties of these intermediates are reported by the behavior of either intrinsic or extrinsic probes. The most common and sensitive intrinsic probes of the biological systems are their very constituents: cytochromes, flavoproteins (Fp), quinones, and pyridine nucleotides

(PN).[1] The more readily measurable signals for these constituents are the electronic and nuclear transitions associated with their oxidation and reduction which can be recorded by a number of physical methods: absorbance spectrophotometry, fluorometry, electron spin resonance, nuclear magnetic resonance, etc. Oldest and historically of greatest importance is the absorption spectrophotometry. Fluorescence techniques which are feasible for flavoproteins and are of high signal to noise ratio for the pyridine nucleotides[1] "come into their own" when membrane proteins are labeled with high fluorescence-yield probes.[2,3] Electron spin resonance is applicable for the measurements of iron–sulphur- and copper-containing components. Optical methods can be also used for this purpose but temperatures as low as 4°K may be necessary for significant sensitivity and selectivity. The application of nuclear magnetic resonance to membrane systems is just in its infancy but presumably will come into much greater usefulness as labeled carbon components can be inserted into the membrane.

The Nature of the Optical Data. The optical methods record either the reduced (cytochromes, pyridine nucleotides) or the oxidized state (flavoprotein). The visible bands of the cytochromes in the oxidized state are not readily distinguishable. Nevertheless the peak of the oxidized Soret band (390–420 nm) can be measured, and the kinetic studies of the transition from the reduced to oxidized state may furnish information on the intermediate steps in the reactions of the cytochromes.

Use of Inhibitors. The greatest aid in spectroscopic identification of components of the photosynthetic and respiratory system have been specific inhibitors. The inhibitors serve, on the one hand, to identify the location of the components within the mitochondrial or photosynthetic membranes, e.g., cyanide, antimycin A, and rotenone in the case of mitochondria; DCMU, cyanide, and antimycin A in the case of chloroplasts or chromatophores, on the other. Some of them (the so-called heme-linked inhibitors as cyanide, azide, sulfhide and carbon monoxide) alter the spectral properties of the respiratory chain components and thus furnish an independent tool for their studies.

One of the widely recognized advantages of the absorption spectrophotometry lies in the fact that it can be easily combined with other techniques for which it provides the basic type of readout. Absorption spectroscopy can be supplemented with the low-temperature studies, with fast kinetic techniques, and finally with the oxidation–reduction poises.

[1]B. Chance, A. Azzi, I-Y. Lee, C. P. Lee, and L. Mela, *FEBS Symp.* 17, 233 (1969).
[2]A. Azzi, B. Chance, G. K. Radda, and C. P. Lee, *Proc. Nal. Acad. Sci. U.S.* 62, 612 (1969).
[3]G. K. Radda, *Current Topics Bioenergetics* 4, 81 (1970).

Temperature as a Parameter. Low-temperature studies offer a significant advantage since the optical transitions are better resolved at low temperatures and the competing reactions are slowed down so that the reaction of interest may come within the measurable time range. This has been long recognized in the case of photosynthetic systems where the rates of electron transfer decrease down to 120°K with an apparent activation energy of 3.5 Kcal.[4] It has been only recently in our studies that the rates of mitochondrial electron transfer have been measured at subzero temperatures, and clear kinetic resolution of the oxidation times of cytochrome a_3 and a has recently been accomplished.

Kinetics as a Parameter. Optical spectroscopy has been widely used in combination with the fast kinetic technique. Systems which are recalcitrant to inhibitors can be resolved on a kinetic basis by light pulses (for the photosynthetic systems) or oxygen pulses (for the mitochondrial systems). Pulsing with reductants is possible in system II–system I interaction but is difficult in chromatophores and mitochondria. Some progress has been made using hydroquinones as fast reductant of the ubiquinone–cytochrome b region of the respiratory chain[5] and with pulses of hydrated electrons obtained from linear accelerators.[6] The latter method, though still in its infancy, may provide an answer to time-resolved spectroscopy of electron transport system in the direction of reduction.

Oxidation–Reduction Poise. The segregation of the components of the mitochondrial multienzyme system has recently received a new spur of experimental activity due to introduction of the oxidation–reduction poises combined with optical spectroscopy, fluorometry, or electron spin resonance.[7] The segregation of the carriers can be accomplished either by the use of ATP, ADP + Pi under conditions where the phosphate potential causes redox potential shifts of components involved in the energy coupling, or in the systems equilibrated with the redox buffers of different potentials. It is possible to scan through the redox range so that low-potential components may remain oxidized and vice versa. Of even greater value is the possibility that the system may be poised at a particular potential in the dark in the case of photosynthetic systems or under anaerobic conditions in the case of the respiratory systems and respectively pulsed with a flash of light or a pulse of oxygen.

[4]D. DeVault and B. Chance, *Biophys. J.* **6**, 825 (1966).
[5]A. Boveris, R. Oshino, M. Erecinska, and B. Chance, *Biochim. Biophys. Acta* **245**, 1 (1971).
[6]I. Pecht, *FEBS Letters* **13**, 221 (1971).
[7]D. F. Wilson and P. L. Dutton, *in* "Electron and Coupled Energy Transfer in Biological Systems" (T. E. King and M. Klingenberg, eds.), Vol. 1, p. 221. Marcel Dekker, New York, 1971.

Only those components whose potential lies above that of the poise will respond. This technique not only affords an identification of such components from the redox potential standpoint, but as well may determine the extent to which the components of lower potential alter the kinetic properties of those at higher potential.

In summary, the spectroscopic technique supplemented with an appropriate array of inhibitors, suitable and adequately controlled redox poises, and variation of temperature over the widest possible range can provide us with detailed and versatile information on the system investigated.

Principles of the Dual Wavelength Method. The dual wavelength technique supplements the information obtainable from the single wavelength technique with important signals from a neutral (or active) reference wavelength which is either continuously or intermittently subtracted from that of the measuring wavelength. The information from the reference wavelength may be of low-frequency character such as light source fluctuations or scattering changes in the sample, or it may be of high-frequency character such as artifacts from actinic illumination or a portion of the signal itself which is either redundant or undesirable.

As an illustration of the dual wavelength technique in the interference filter version,[8] Fig. 1 shows a single tungsten–halogen lamp illuminating $f/2$ lenses, mechanical shutters, a pair of optical filters, a vibrating mirror, and an output lens. This diagram illustrates the general features of the technique, the dual light paths of a wide optical aperture, the adjustment of the relative intensity light paths, and the projection of an image of the apertures along coincident axes towards the sample. The proximity of the detector and the sample is essential for operation with highly scattering material such as chloroplasts and mitochondria.

Time Resolution. In systems which continuously subtract the reference wavelength from the measure wavelength (i.e., do not time-share reference and measure light beams) the time resolution is equal to that of the single beam method (see this Volume, Chap. [2]) (examples of this are dual wavelength instruments which employ dual detectors and monochromating elements such as filters[9] or which employ orthogonal light beams and dual detectors[10,11]).

[8]B. Chance, D. Mayer, and V. Legallais, *Anal. Biochem.* 42, 494 (1971).
[9]B. Chance, *Rev. Sci. Instr.* 13, 158 (1942).
[10]H. Schleyer, personal communication.
[11]A. Crofts, personal communication.

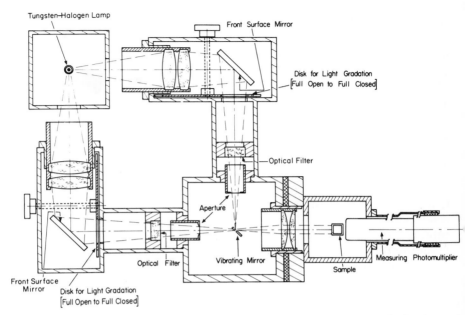

FIG. 1.   A configuration of the dual wavelength spectrophotometer employing interference filters and wide aperture optics. The components are identified in the text.

An example of a fast dual wavelength spectrophotometer employing interference filters is shown in Fig. 2.[12] This design, based upon the differential colorimeter on which most of the early work on peroxidase has been carried out,[6] now affords an extremely versatile and powerful instrument for fast spectrophotometry of oxygen or light activated electron transport. The 48-W tungsten–iodine lamp, rated at 6.7 A, can be pulsed to 10 A during the interval of the measurement. In some cases a 1-kW water-cooled mercury arc has been found to be suitable. Initial filtering of the light is accomplished by an interference filter. The observation chamber is of adequately large size to accomodate the $f/1.5$ optical system, and the photomultipliers are of equal window area. The light is split in the two paths, one for the measuring wavelength and one for the reference wavelength. The high voltage applied to the dynodes of these two photomultipliers is adjusted so that the signals across their load resistors are equal. The distance between the photomultipliers and the sample may be adjusted slightly so that they receive equal amounts of the actinic source artifact; in this way transients may be appropriately cancelled. As indicated generally here and described in detail below,

[12]B. Chance, N. Graham, J. Sorge, and V. Legallais, *Rev. Sci. Instr.* in press.

synchronization signals are provided for the Biomation computer and the monitor oscilloscope. Such an apparatus is suitable for operation in very short time ranges and has been shown to be optimal for rejection of the laser artifacts. One objection to the apparatus for photosynthetic systems is the fact that the majority of the measuring light over a significant spectral interval passes through the sample. This may be avoided by a very simple flow cell which replaces the material in the observation chamber a few milliseconds prior to the laser flash and thus presents the observation system with essentially "dark" material.

Usually the time resolution of the dual wavelength apparatus is limited to the time-sharing frequency, but recently the difference between the single and dual wavelength technique has been substantially diminished. Single beam operation of the dual wavelength method can be obtained during one measuring light flash, and dual wavelength operation is obtained after a complete cycle of the time sharing.[12,13] Then, reference wavelength cancels out the light source fluctuations.

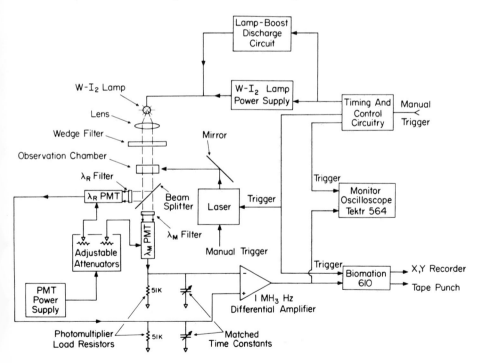

FIG. 2. An example of a fast dual wavelength spectrophotometer employing interference filters.

[13]B. Chance and M. Erecinska, *Arch. Biochem. Biophys.* **143,** 675 (1971).

Figure 3 illustrates how the dual wavelength spectrophotometer may be employed for fast single beam recording.[12,13] The optical configuration is similar to that already identified in Fig. 1 where interference filters or indeed "straight through" monochromators may be employed for the monochromating elements. The beam is split to compensate for fast fluctuations of the light source (particularly useful where xenon or mercury arcs are employed) which occur during the interval of the flash of the measuring light. In this case, however, only 5–10% of the beam is split off to the compensating photomultiplier, and the beam splitter occurs before rather than after the sample. As in Fig. 2 the outputs of the two photomultipliers are rendered equal by appropriate dynode voltages, and their output is fed into a differential amplifier as before.

The fast or single beam output is fed directly to the computer or a fast sweep storage scope; to all intents and purposes the operation occurs as if a single beam spectrophotometer were employed. However, after the transition of the vibrating mirror to the reference wavelength,

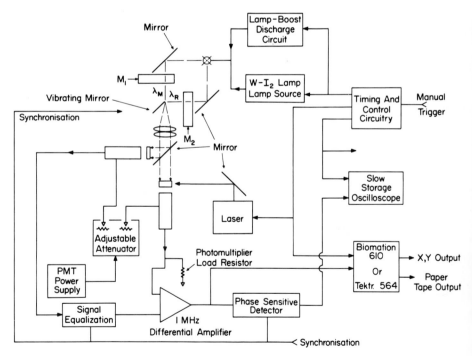

FIG. 3.   A modification of the conventional dual wavelength spectrophotometer permitting rapid readout in the interval of the measuring light pulse by selection of this cycle to appropriate time relation to the flash of the laser pulse.

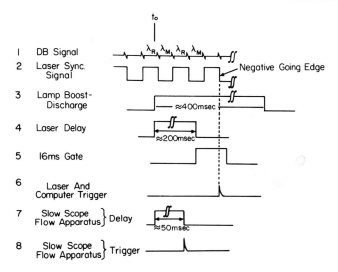

Fig. 4. The timing diagram of the fast dual wavelength spectrophotometer, illustrating how a particular cycle of the flash of the measure wavelength is selected and affords a trigger for flash photolysis by the laser. Details are described in the text.

the phase sensitive detector operates and the usual dual wavelength output can be obtained.

Since synchronization of the measuring light flash to the laser flash is essential for this technique, an appropriate timing waveform diagram is shown in Fig. 4. The top line (1) represents the double beam (DB) signal. The system can be triggered at any time (t_0) which starts three gates, lines 3, 4, and 7, lamp boost, laser and computer, slow scope, and flow apparatus triggers. The lamp boost pulse is the longest and covers the entire recording interval. In order to obtain a laser trigger which is synchronized with the measuring light pulse of the spectrophotometer, the negative going portion of line 2 coincides with the beginning of this interval. If desired the laser trigger can be delayed after the computer trigger to give a convenient "start point" on the digitized output. The time gap afforded by the reference interval can be filled in by a second laser trigger 180° delayed with respect to the first trigger. The sum of the two sets of data is complete. Thus, this method affords the great advantages of the dual wavelength technique for long-term high-stability recording and supplements it with a compensated single beam technique that is as fast as could be obtained with any of the configurations specialized to this point.

Components Employed in the Dual Wavelength Spectrophotometer

Monochromating Elements. The requirement for two adjustable wavelengths has been satisfied by the use of two identical monochromators (see this Volume, Chap. [2] and Refs. 14–16) (for geometric reasons one may be illuminated in the inverse direction from the other, a topic which does have some impact on the stray light performance) or by a split grating in one monochromator.[17] The two portions of the grating may be independently movable in order to obtain appropriately adjustable wavelengths. The spectrum from a single monochromating element may be imaged upon multiple exit slits, thereby securing the appropriate separated wavelengths. One particular version of this employs a rotating exit slit. In one position, the emitted light is highly monochromatic; at 90° a number of wavelengths are obtained which can be selected appropriately by mechanical means. This apparatus is useful for "triple beam spectrophotometers"[18,19] which are used where light scattering changes are extremely large, as in the case of stimulated muscle. Another alternative is to scan the spectrum very rapidly across the exit slit so that a whole series of wavelengths are obtained, a part of which may be selected for the dual wavelength readout.[20]

Monochromators which have proved useful in the construction of various dual wavelength spectrophotometers are the Bausch and Lomb 250- and 500-mm focus grating monochromators, Hilger D330, and Zeiss single M4Q III or double MM12 prism monochromators.[21] The advantages of grating over prism monochromators are discussed elsewhere.[22] The grating monochromators used by Aminco, Phoenix, and Hitachi have generally similar characteristics to that of the 250-mm focus Bausch and Lomb.[16,17] On the more economical side, the Heathkit EU700/701 monochromators have proved useful as have the Hilger D-252 grating monochromators. In addition to monochromators, interference filters (see this Volume, Chap. [2]) are now to be seriously

[14]B. Chance, *Rev. Sci. Instr.* 22, 627 (1951).

[15]B. Chance, *in* "Methods in Enzymology" (S. P. Colowick and N. O. Kaplan, eds.), Vol. IV, p. 273. Academic, New York, 1957.

[16]Hitachi–Perkin-Elmer, *U.V. Technical Memo #1 on Model 356 Instrument*, Perkin Elmer, Norwalk, Connecticut.

[17]*Aminco Bulletin, No. 2383*, American Instrument Co., Bethesda, Maryland.

[18]F. F. Jobsis, personal communication.

[19]B. Chance, unpublished observation.

[20]D. W. Lübbers and R. Wodick, *Appl. Opt.* 8, 1055 (1969).

[21]B. Hess, H. Kleinhaus, and H. Schlüster, *Hoppe Seyler's Z. Physiol. Chem.* 351, 515 (1970).

[22]R. Rikmenspoel, *Appl. Opt.* 3, 351 (1964).

considered for use alone or in conjunction with monochromators. The high transmission of interference filters together with their large areas permits wide aperture optics to be employed with consequently significant improvements of signal to noise ratio. A number of apparatuses for time sharing employing interference filters have been developed.[23] Infrared transmission of interference filters, as well as microholes, should be checked.

Appropriate specifications for a monochromator are summarized:

Resolution: A spectroscopic resolution down to 1 nm is usually considered necessary and is obtainable with most grating units.

Stray light: Stray light should be less than 1%, especially when dealing with turbid suspensions. The use of color or interference filters to check stray light errors is always necessary.

Intensity: Wide aperture optics consistent with the needs for a spectral purity and stray light are essential. Precise mechanical adjustment of slits is required.

Light Switches. The simplest light switch is a slotted disk which may or may not have reflecting surfaces.[17,21] Such disks have the advantage of a chopping speed that is a multiple of the rotation speed, but they may exhibit a light modulation at the rotation speed due to small differences in the transmission or reflectance of the slots or mirrors. A commercial instrument employs 180° sectors in a rotating disk at 60 Hz and more recently a multisector at 250 and 1000 Hz.[17] Multisector disks are used with light pipes at 1-k Hz chopping frequency[21] as illustrated in Fig. 5. In this apparatus[21] two light sources are employed and the output of the two monochromators is chopped by slotted disks A and B which in turn illuminate two light pipes so that a common optical axis is obtained for illumination of a cuvette and the photomultiplier. Recently Schott has indicated their ability to make quartz light guides so that the unit will be suitable for the ultraviolet region.

The vibrating mirror has been used as an inexpensive and robust light modulator.[15,23] For example, the 60-Hz Brown converter can be fitted with a mirror which vibrates in a low amplitude mode (1–2 mm) or in a large amplitude mode (~4 mm). In the latter configuration, the contact of the reed with the pole pieces may set up high-frequency oscillations in the light output.[13,23] These are usually not objectionable in the reference wavelength. Since the oscillations are of high frequency, they do not interfere with the frequency response from the demodu-

[23]D. Mayer, B. Chance, and V. Legallais, *in* "Probes of Structure and Function of Macromolecules and Membranes" (B. Chance, C. P. Lee, and J. K. Blasie, eds.), Vol. 1, p. 527. Academic, New York, 1971.

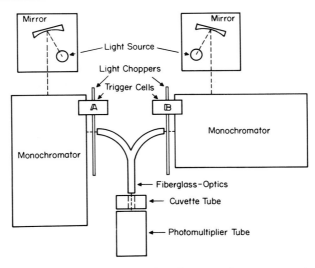

FIG. 5.   A configuration of the dual wavelength spectrophotometer employing rotating sectors for light chopping and light pipes for combining the two beams.[21]

lated output of the system. Bulova (American Time Products) manufactures a 200- and 400-Hz light modulator. The 1-kHz model has been employed.[24] A 75-Hz light chopper is available.[25]

In wavelength scanning devices[24] a 1-kHz torsional light modulator is substituted for the above-mentioned light chopper. The most rapid torsion scanner available is the 12–25-kHz galvanometer driven vibrating mirror.[20,26] In combination with a 50-Hz mirror for wavelength scanning, it is possible to obtain the absorbancy difference at any two wavelengths by a suitable synchronized gating circuit.[20,27]

A rotating mirror driven by an air turbine[12] has been employed for chopping at ~4 KHz (Fig. 6) and gives paired pulses of reference (λ_R pulse = 12 μsec) and measure (λ_M pulse = 20 μsec) wavelengths with a dark interval of 56 μsec. Flash activation occurs conveniently during the 56-μsec interval between the light pulse pairs. The turbine-driven mirror affords an economical device for increasing the time range of the dual wavelength technique.

Faster Light Modulators. The whole problem of fast light modulation has received tremendous impetus from the development of the Q-

[24]P. Garland, unpublished observation.
[25]R. Pye-ling, Herts, England.
[26]W. Niesel, D. W. Lübbers, D. Schneewolf, J. Richter, and W. Botticher, *Rev. Sci. Instr.* **35**, 578 (1964).
[27]D. W. Lübbers and R. Wodick, *XXV Internat. Cong. Physiol. Sci. Munich, 1971.*

switched lasers, developments which have obvious applications for light modulated spectroscopy, for fluorescence lifetime measurements, etc. Pockel cells (this Volume, Chap. [2]) afford high-speed switching of collimated light beams. There are further developments in mechanical light modulators, for example, a magnetically suspended rotating ball similar to that used for the measurement of the velocity of light could be constructed.

Light Sources. The light source is of greatest significance, and tungsten–iodine, xenon, or mercury arcs or even xenon flash lamps are currently in use. All these sources should be pulsed or boosted so as to give maximum intensity over the interval of the spectroscopic recordings.[28,29]

A novel dual wavelength spectrophotometer in which modulated xenon arcs are employed not only to obtain high-intensity illumination of the monochromators but also to provide time sharing is indicated in Fig. 7.[30] Pulses 180° out of phase are applied to the two monochromators which switch the xenon arc from low to high intensity and hence modulate the light through to PM-1 and PM-2. Frequencies of several kilohertz are obtained in this method. The instability of the xenon arcs necessitates a compensating photomultiplier which can be employed in the circuit configuration of Fig. 3.

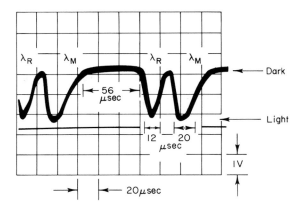

FIG. 6. Oscilloscope photographs of the light pulses generated by the turbine driven rotating mirror. The top "dark" baseline corresponds to no light transmitted, and the pulses shown are 12 and 20 μsec in duration. Other details are described in the text.

[28]D. Devault, *in* "Rapid Mixing and Sampling Techniques in Biochem" (B. Chance, R. H. Eisenhardt, Q. H. Gibson, and K. K. Longberg-Holm, eds.), p. 165. Academic, New York, 1969.
[29]B. W. Hodgson and J. P. Keene, *Rev. Sci. Instr.* in press.
[30]R. J. DeSa and Q. H. Gibson, *Rev. Sci. Instr.* 37, 900 (1966).

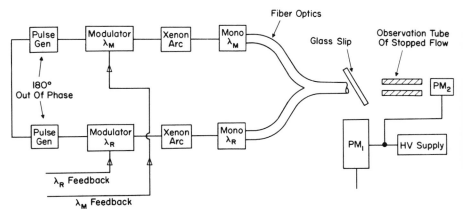

FIG. 7. Dual wavelength modulation of the light sources. The xenon arcs are alternately modulated 180° out of phase to provide time-sharing function of the dual wavelength method.[28] The light beams are combined with fiber optics.

Eventually lasers may supplant the gas discharge sources. The 400-nsec pulse of the liquid dye laser is already suitable for some applications.

The fast fluctuations in these sources can be minimized with a compensation circuit[31] which will operate as well with the dual wavelength technique as with single wavelength techniques.

Cuvette Dimensions. Of critical importance is the dimensions of the optical cell (see this Volume, Chap. [2]), particularly in the light of photosynthetic systems where the intensity of the measuring light can be appropriately diminished by increasing the area of the sample. Highly turbid materials can be used in sufficiently thin layers that self-absorption phenomena may be minimized, except in the case of macromolecular aggregates.

Detectors. Photoemission surfaces (some types are described in this Volume, Chap. [2]) are superior to solid state (PIN diodes) in the visible and ultraviolet regions, but the latter offer significantly better performance in the infrared region. Electron multiplication is usually employed, although for very rapid recordings at high light fluxes only a few stages are required and perhaps even none may eventually be optimal as lasers are used as measuring light sources (an objection to electron multiplication is its saturation feature which may lead to unexpected nonlinearities at high light flux). The recovery characteristics of the photoelectric devices to light-induced overloads needs special

[31] B. Chance, D. Mayer, N. Graham, and V. Legallais, *Rev. Sci. Instr.* **41**, 111 (1970).

attention. The surface area of the detector should be appropriate to the area of the optical cuvette.

Electronic Circuits. While it is not the purpose of this chapter to describe electronic circuits, it is appropriate to point out the need for skillful design of electronic coupling between the photodetector and the data display or readout unit. Poor overload characteristics in any part of this link may lead to artifactual results which may be difficult to identify.

In fast readout with the dual wavelength method it is necessary to compensate for fast light fluctuations in the interval of the measuring light flash. Essentially the electronic circuits subtract a portion of the light prior to the optical cuvette from that following the cuvette. In order that equal quantities be subtracted a gain modulation circuit is employed.[31]

Signal Averaging. This is often used to improve the signal to noise ratio, provided the noise is of equal energy at all frequencies, i.e., it is a "white-noise." However, the responses of the biological materials may not be exactly repeatable, leading to errors in signal averaging.

Synchronization Problems. Chapter [2] of this volume describes actinic flash spectrophotometry and its associated synchronization problems. Time sharing of the dual wavelength system is synchronized with the flash activation for two reasons: first, in order to render the circuits minimally sensitive to overload by the actinic flashes, and secondly, to display on an oscilloscope the measuring light flash corresponding to the actinic flash. The circuit affords oscilloscope and solenoid triggers, a fast sweep trigger, and an actinic flash trigger. The latter two are time selected so that the sweep begins with a pulse of the measuring light. For example, with 60-Hz modulation, the fast scope trace has an 8-msec duration and the actinic light flash occurs 1 msec of the beginning of the sweep. The demodulation switch closes about 1 msec thereafter. In this way, the demodulator senses the change of the measuring light signal as soon as the flash artifact has abated.

Wavelength Scanning. While the dual wavelength technique is in principle primarily used for measurement at fixed pairs of wavelengths, the possibility of scanning through a variety of wavelengths on each side of the reference wavelength has been realized by the use of a compensation circuit[31,32] which allows scanning over 40 nm at rates of 10 nm/sec. Such high-sensitivity wavelength scanning is readily alternated with the static measurements at appropriate pairs of wavelengths in the scan region.

[32]B. Chance and N. Graham, *Rev. Sci. Instr.* **42**, 941 (1971).

[30] Acid–Base Indicator for the Measurement of Rapid Changes in Hydrogen Ion Concentration

By B. CHANCE and A. SCARPA

Sensitive measurements of pH changes can be carried out with glass electrodes or by spectrophotometric or fluorometric measurements using appropriate indicators. In spite of recent technical improvements in reducing the response time of glass electrodes, the use of pH indicators and appropriate spectrophotometric techniques are the only ways extremely sensitive measurements of rapid pH changes can be followed. In the case of photosynthetic systems, glass electrodes are seldom equal to the task of providing an adequately rapid response for relating H^+ changes to electron transport in chromatophores and in chloroplasts. With these systems, indicators like bromocresol purple (BCP), bromophenol red (BPR), and bromothymol blue (BTB) have already been successfully employed for the measurements of rapid pH changes.[1-3] For this reason, this brief report focuses its attention on spectrophotometric methods based on changes in absorbance by these dyes.

Criteria for the Selection of an Indicator

In principle, any substance which changes color when the hydrogen ion concentration varies can be used as a pH indicator. However, a number of criteria must be met by the dye to be employed as an indicator for fast H^+ concentration changes in biological materials.

Binding of indicators. Preferably the indicator should be free in solution surrounding the vescicles of chromatophores or chloroplasts. If it is bound, it should be readily influenced by external pH changes and should not be affected by the state of the membrane. Through studies of centrifugation and detection of indicators in the pellet and in the supernatant fractions, it has been found that only 15% of BPR was recovered in chromatophores; under the same conditions 19% of BCP and 80% BTB were bound to the chromatophores.[2] Therefore BCP and BPR seem to meet the requirements of accessibility to external pH changes. BTB, on the other hand, binds the membrane tightly and is

[1] B. Chance and L. Mela, *J. Biol. Chem.* **241**, 4588 (1967).
[2] J. B. Jackson and A. R. Crofts, *Eur. J. Biochem.* **10**, 226 (1969).
[3] B. Chance, A. R. Crofts, M. Nishimura, and B. Price, *Eur. J. Biochem.* **13**, 364 (1970).

relatively insensitive to pH changes external to some types of vescicles.[1,2,4]

Response Time. Color indicators equilibrating directly with hydrogen ions have a very fast response. Field jump studies by Ilgenfritz[5] show that BCP measures pH changes at diffusion-limited reaction rates. Presumably, similar fast response times can be obtained with other indicators. The glass electrode, however, may have a response speed in the order of milliseconds, due to an intrinsic limitation set by the electrode capacitance.

Sensitivity. The sensitivity of the indicator used with spectrophotometric techniques depends upon the change in the number of photons transmitted or emitted per unit time per pH unit. Using BCP or BPR in suspension of chromatophores containing 10–20 μM of chlorophyll, sensitivities greater than 0.1 μM of H^+ at a response speed of 40 μsec were obtained.[3]

pK_a. The pK_a should be appropriate for the biological system under study (see Table I). No shift of apparent pK_a values for the indicator should be observed upon addition of biological material. This is the case for BCP and BPR when the color change was titrated against pH in the presence of or in the absence of chromatophores. On the other hand, with BTB, there was a shift in pK_a upon binding of chromatophores (about 0.3–0.8 pH units in the direction of increased pH).[2,6]

Interferences. Interferences from carotenoids or cytochromes can be minimized through the wide range of wavelengths obtainable by these indicators. In the case of BCP and BPR, mutants such as blue-green *R. rubrum* chromatophores have been employed to abolish changes due to chlorophyll at 585 nm.[2]

TABLE I

Indicator	Color change	pH of color change	pK_a	ϵ, basic form [mM cm]$^{-1}$
Bromcresol purple (BCP) 5,5'-Dibromo-*o*-cresosulfonphthalein	yellow–purple	5.2–6.8	6.47	58 (at 585 nm)
Bromphenol red (BPR) 3'-3"-Dibromophenolsulfonphthalein	yellow–red	5.2–7.0	6.59	87 (at 552 nm)
Bromthymol blue (BTB) 3,3'-Dibromothymolsulfonphthalein	yellow–blue	6.0–7.6	7.3	39 (at 618 nm)

[4]M. Nishimura, K. Kadota, and B. Chance, *Arch. Biochem. Biophys.* **125**, 308 (1968).
[5]G. Ilgenfritz, Chemical relaxation in strong electric fields. Dissertation, Univ. Gottingen, 1966.
[6]B. Chance and V. Legallais, *Rev. Sci. Instr.* **22**, 627 (1951).

Detailed Characteristics

The absorption difference spectra of BCP and BPR in acidic and basic media are shown in Figs. 1A and 1B, respectively. The difference spectra were obtained by raising the pH in the measuring cuvette from 4 to 8 through the addition of KOH. It is apparent that there is a wide range of wavelengths over which appropriately high absorbancy changes can be obtained.

The apparent millimolar extinction coefficients of these indicators are given in Table I, and the titration curve of BCP in the presence and in the absence of blue-green mutant *R. rubrum* chromatophores is shown in Fig. 2.

Some comparison between BCP and pH electrode methods are given in Fig. 3, wherein the pH changes of the medium were followed by the two methods after illumination of blue-green mutant *R. rubrum* chromatophores. In the absence of BCP (Fig. 3A), there was a small change in absorbance which was negligible compared with the response in the presence of the indicator (Fig. 3B). In the presence of valinomycin and KCl (Fig. 3C), the initial rate of external pH change on illumination of chromatophores was greatly enhanced. It is evident that even in these relatively slow reactions, the response of the electrode is limiting.

The rapidity in recording of pH changes with BCP in a flash-illuminated suspension of chromatophores is illustrated in Fig. 4. The reaction is completed in less than 1 msec and has a half-time of about 400 μsec. The great sensitivity in measuring H^+ concentration is indicated

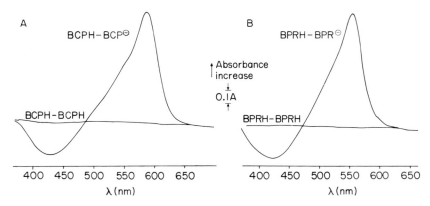

FIG. 1.   Difference spectra of bromocresol purple (BCP) (A) and bromophenol red (BPR) (B) in the acid and basic form. The medium contained 0.1 M KCl, 2 mM Tris maleate and 15 μM BCP or 10 μM BPR. The difference spectra were obtained by raising the pH of the measuring cuvette from 4 to 8 by addition of KOH.

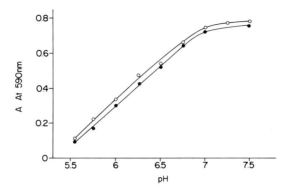

FIG. 2.   Apparent pK$_a$ value of bromocresol purple in the presence and in the absence of *R. rubrum* chromatophores. The medium contained 0.1 M KCl, 2 mM Tris maleate at the pH indicated in the figure, and 10 μM BCP in the presence (●——●) and in the absence of chromatophores (0.07 mg of BCHl/ml of blue-green mutants).

by the calibrations of the top traces which are illustrated at a 10-fold reduction of the absorbance scale and a 17,000-fold reduction in time scale. With time proceeding from left to right, one observes the typical response of chromatophores to steady state illumination. The cuvette is stirred in order to mix the contents and 6 μM NaOH is added and back titrated by two additions of HCl totaling 6.7 μM. Thus, the calibration for the low sensitivity is 2 μM H$^+$ per division. It can be seen from the noise level employed in the lower trace that 0.03 μM H$^+$ is detectable, and from the slope of the trace the response could follow a change of less than 40 μsec.

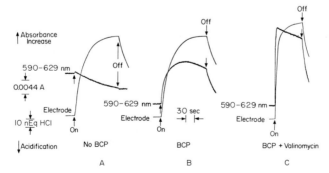

FIG. 3.   Double beam spectrophotometer and glass electrode traces of blue-green mutant chromatophores: 0.1 ml of chromatophore suspension (0.19 mg of bacteriochlorophyll) in 6 ml of 0.1 M KCl, pH 6.2. (A) No added BCP; (B) 8.2 μM BCP; (C) 8.2 μM BCP and 0.20 μg/ml of valinomycin. [Redrawn from J. B. Jackson and A. R. Crofts, *Eur. J. Biochem.* **10**, 226 (1969).]

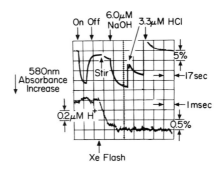

FIG. 4   Illustration of fast light-induced absorbance changes in BCP supplemented *Chromatium* chromatophores. The medium contained 0.1 *M* KCl, 67 *μM* BCP, 8.1 *μM* chlorophyll from *Chromatium* chromatophores. In the bottom trace the illumination was supplied by 720 nm xenon flash lamp for 10 *μ*sec.

Instrumentation

The detection of absorbancy changes of the indicator as a function of pH requires the use of a dual beam spectrophotometer to compensate for the interference by scattered light or by nonspecific broad-band absorbancy changes. A spectrophotometer similar to that described by Chance and Legallais[6] can be successfully employed (Fig. 5 and this volume [29]). One wavelength, the reference, is set in a region wherein the sample does not show specific absorbance changes, while the measure wavelength is set close to an absorbance peak of the indicator. Differences in the absorbance between the two beams can be amplified[7] and displayed on a storage oscilloscope. An attachment at 90° enables actinic illumination of the sample at right angles to the beams of the spectrophotometer. Kodak Wratten 88A gelatin filter (cut off below 720 nm) can be used to provide near-infrared irradiation from a lamp with a shutter device. An alternative light source can be provided by the rapid flash of a Q-switched ruby laser as described by DeVault[8] (light pulse duration 20 nsec and 694.3 nm wavelength). For such rapid events a single beam spectrophotometer or a "cycle selection" dual wavelength spectrophotometer using interference filters or monochromators may be required (see this volume [29]).

[7]B. Chance, this volume [29].
[8]D. DeVault, *in* "Rapid Mixing and Sampling Techniques in Biochemistry" (B. Chance, R. H. Eisenhardt, Q. H. Gibson, and K. K. Lambergholm, eds.), p. 165. Academic Press, New York, 1964.

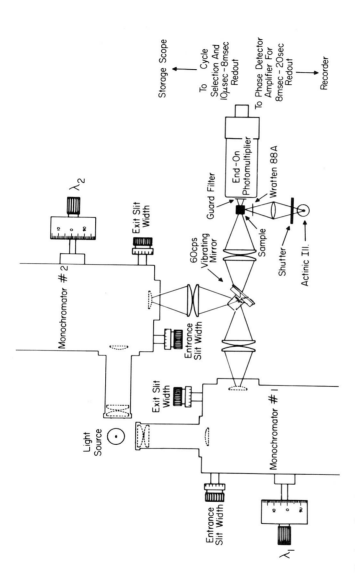

FIG. 5 Dual wavelength spectrophotometer with 90° actinic light attachment for fast kinetic measurements of H⁺ movements in photo-synthetic systems.

New Developments

Unpublished data from Crofts and co-workers[9,10] suggest the use of phenol red as an indicator of external pH changes in photosynthetic systems. According to the authors (A. R. Crofts, personal communication), the use of phenol red has some advantages over BCP and BRP. These are: (a) a lower binding of the indicator to the chromatophores (less than 5%); (b) a higher pK_a, 7.6, which does not shift upon addition of chromatophores; (c) a color which is not lost upon reduction with dithionite or oxidation with ferricyanide; (d) and a response time which is comparable with that of BCP. All these characteristics should render phenol red as the best indicator available so far for spectrophotometric measurements of external pH with photosynthetic and other biological systems.

Concluding Remarks

In summary, the glass electrode has the advantage of specificity, while the color indicator has the advantage of a faster time. In spite of sophisticated apparatuses recently introduced to reduce the response time, glass electrodes are at the present time inadequate techniques for routine measurements of fast pH changes. On the other hand, the color indicator technique suffers from possible interference due to unrelated absorbance changes by other pigments present in biological material. The use of acid–base indicators requires therefore that a series of controls should be carried out and a number of conditions satisfied. The most important of these are the selection of suitable wavelengths at which absorbance changes of the indicator can be obtained without interference by other absorbing pigments and the knowledge of evidence of lack of binding of the indicator to the material used. However, the above-mentioned indicators, used under well controlled experimental conditions, are the only tools so far available when fast kinetic measurements of H^+ movements have to be measured in photosynthetic and other biological systems.

[9] A. R. Crofts, R. J. Cogdell, and J. B. Jackson, in "Energy Transduction in Respiration and Photosynthesis," *Colloq. Bioenergetics.* Pugnochiuso, Italy, 1970, in press.
[10] A. R. Crofts, J. B. Jackson, E. H. Evans, and R. J. Cogdell, *2nd Intern. Congr. Photosynthesis.* Stresa, Italy, June 1971, in press.

[31] Spectrophotometric Measurement of Calcium by Murexide

By Antonio Scarpa

The importance of calcium in regulating the properties of biological systems and the activity of isolated enzymes has generated much interest in recent years on the mechanisms of Ca^{2+} transport and Ca^{2+} interactions in whole cells and in subcellular fractions.

Various procedures are available for the measurement of Ca^{2+} movements and binding in biological systems. Atomic absorption and radioactive ^{45}Ca are frequently employed; however, both techniques require long periods of time for sampling and separation of the material and are inapplicable to kinetic detection of Ca^{2+} movements. Another method of Ca^{2+} measurement is based on the use of cation-sensitive electrodes, but their use is limited by the low sensitivity of the electrodes and, even more so, by their slow response time ($\sim$ 5 sec). An alternative method is based on the spectrophotometric detection of the change in absorbance of the dye murexide, an indicator which is responsive to the Ca^{2+} concentration in the medium. The latter technique has the advantages of a fast response time, high sensitivity, and small volume of sample required. Furthermore, it has been successfully applied to the measurement of Ca^{2+} transport in sarcoplasmic reticulum,[1] mitochondria,[2] and intact cells.[3,4]

Principle

Murexide is a metallochromatic indicator which can form complexes with Ca^{2+}, other divalent cations, and rare earths. Murexide and the murexide–Ca^{2+} complex absorb light at different wavelengths such that when Ca^{2+} ions are added to a solution of murexide, there is an observable color change from purple to orange. The mechanism of the absorption change towards the ultraviolet region when the free monovalent murexide forms a complex with cations is not entirely understood, but some details can be obtained from the papers of Geier[5] and Schwarzenbach.[6] For practical reasons we can formulate the complexa-

[1] T. Ohnishi and S. Ebashi, *J. Biochem.* **55**, 599 (1964).
[2] L. Mela and B. Chance, *Biochemistry* **7**, 4059 (1968).
[3] F. F. Jobsis and M. G. O'Connor, *Biochem. Biophys. Res. Commun.* **25**, 246 (1966).
[4] A. Cittadini, A. Scarpa, and B. Chance, *FEBS Lett.* **18**, 98 (1971).
[5] G. Geier, *Helv. Chim. Acta* **51**, 94 (1968).
[6] G. Schwarzenbach and H. Gysling, *Helv. Chim. Acta* **32**, 1314 (1949).

tion as a simple equilibrium of the following type: $(Ca^{2+})(Mu)/(Ca^{2+} -$ Mu) $= K_D$. The value of K_D has some variability depending on the type of cation complexed and the characteristics of the medium, *e.g.*, pH, ionic strength, and temperature. The high values of the K_D for the murexide–Ca^{2+} complex have made possible quantitative determinations of pCa^{2+} values over a wide pH range in biological fluids.[7] The following intrinsic properties of murexide make this dye very useful for the kinetic measurement of Ca^{2+} binding or transport in the presence of biological membranes.

(1) The low affinity of murexide toward Ca^{2+} (Table I) allows the measurement of Ca^{2+} binding and transport in a biological system without appreciable disturbance of the Ca^{2+} distribution by murexide; for example, in the experiments shown, the free $[Ca^{2+}]$ is reduced by less than 1% by the addition of murexide.

(2) The high coefficient of absorption (~ 1–3×10^4 cm $mole^{-1}$) permits the use of very small amounts of murexide as indicator. Murexide concentrations routinely used (20–50 μM) have been demonstrated to be without effect on the metabolic properties and permeability characteristics of cells[4] and subcellular fractions.[2]

(3) The insolubility of murexide in nonaqueous solvents is the most likely reason for its lack of solubility in and permeability through biological membranes; murexide, when added in micromolar concentrations to a suspension of cells or mitochondria, is completely recovered in the external medium.[2,4]

(4) Murexide does not respond to Mg^{2+} which is often present in

TABLE I

MUREXIDE-CATION COMPLEXES[a]

Cation	K_D (mM)	λ_{max} (nm)	Relaxation time (μsec)	Relative $\Delta\epsilon$ (mM^{-1} cm^{-1})
Ca^{2+}	2.7	483	<5	0.96
Sr^{2+}	10.0	495	<5	0.58
Ba^{2+}	8.8	510	<5	0.61
Mn^{2+}	8.3	484	~10	0.54

[a]Dissociation constant (K_D), wavelengths of maximum of absorption (λ_{max}) were measured in reaction mixtures identical to those described in Fig. 1. The values of relaxation times for the complex dissociation are from Geier[5] and were obtained at pH 4 and 10°. Relative $\Delta\epsilon$ (mM^{-1} cm^{-1}) refer to differences in the change of extinction of murexide at 540–507 nm by addition of 420 μM cation in the same medium as in Fig. 1.

[7]W. T. G. M. Smeets and L. Seekles, *Nature* **169**, 802 (1952).

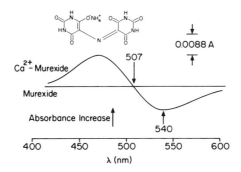

FIG. 1. A difference spectrum of Ca²⁺–murexide complex versus free murexide. The spectrum was obtained in a split-beam spectrophotometer after the addition of 420 μM Ca²⁺ to the measuring cuvette. Both cuvettes contained 125 mM KCl, 5 mM morpholino-propanesulphonate (MOPS) (pH 7.4), and 35 μM murexide. The temperature was 20°.

high concentrations or is required for certain functions of biological membranes.

(5) The rate constant for the reaction of murexide with Ca²⁺ is high enough to be used for very fast kinetic measurements of Ca²⁺ binding or transport (see Table I).

(6) The apparent disadvantage of the relatively broad absorbance spectrum of the murexide–Ca²⁺ complex can be used to advantage because this allows a choice in wavelengths in which unrelated changes in absorbance by other interfering compounds (e.g., cytochromes) can be minimized.

General Procedures

Figure 1 shows that the Ca²⁺–murexide complex, as compared with murexide alone, exhibits a lower light absorbance at 540 nm and a higher absorbance at 470 nm with an isosbestic point at 507 nm. The formation and the disappearance of the Ca²⁺–murexide complex can therefore be measured through changes in absorbance at 540 or 470 nm. Measurements in turbid solutions require the simultaneous de-tection of two close wavelengths and subtraction of the reference from the observation wavelength. This is necessary to minimize the inter-ference caused by the difference in absorbance due to changes in volume or refractive index of the system. The experiments with cells or subcellular fractions must therefore be performed using a dual wavelength double beam spectrophotometer as described in this volume, Chapter [29]. We suggest the use of 540 and 507 nm as mea-suring and reference wavelengths, respectively. Although a reference at 470 nm would improve our sensitivity by more than twofold, it was avoided in order to minimize the above-mentioned nonspecific light scattering changes. The experiments described in Figs. 2 and 3 were carried out with an Aminco-Chance dual wavelength spectrophotometer which recorded the change in absorbance at 540 minus the change in absorbance at 507 nm. The experiments described in Fig. 4 have been performed with a filtered double beam spectrophotometer built at the Johnson Research Foundation of the University of Pennsylvania.[8] Other instruments with similar characteristics can be successfully employed. Interference filters with a 2-nm bandwidth and more than 50% transmitance can be ordered from Omega Optical Co., Brattleboro, Vermont. Murexide (ammonium purpurate) is commercially available through most chemical companies. Because this compound is sparingly soluble in water and rather unstable, solutions were made fresh daily, and they did not exceed 1 mM.

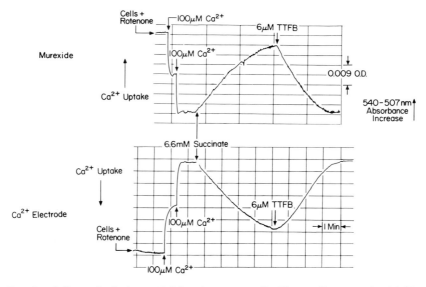

FIG. 2. Ca^{2+} uptake by intact *Erlich ascites* tumor cells. The medium contained 0.25 M sucrose, 10 mM MOPS (pH 7.4), 50 μM murexide, 6.7 μM rotenone, and 18 mg (dry wt) of *Ascites* tumor cells/ml. As a control in the lower figure, Ca^{2+} was measured with a Ca^{2+} electrode (from Orion, Cambridge, Massachusetts) in a stirred vessel, temperature 24°. Final volume was 3 ml. [From A. Cittadini, A. Scarpa, and B. Chance, *FEBS Lett.* **18**, 98 (1971)].

[8]B. Chance, *Rev. Sci. Instr.* **22**, 634 (1951).

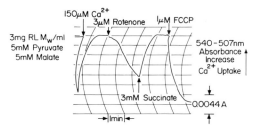

Fig. 3. Energy-linked transport of Ca²⁺ in rat liver mitochondria (RLM). The medium contained 125 mM KCl, 10 mM MOPS (pH 7.4), 3 mM Na phosphate, 5 mM pyruvate, 5 mM malate, and 3 mg of RLM/ml. Final volume was 2 ml; temperature, 24°.

Ca²⁺ Measurements in Intact Ascites Tumor Cells

The upper trace of Fig. 2 shows a typical experiment of Ca²⁺ uptake by intact *Ascites* tumor cells. The addition of Ca²⁺ to the cells incubated in the presence of rotenone induced a steep downward deflection which corresponds to a decrease in absorbance at 540 nm due to the formation of the Ca²⁺–murexide complex. Two successive additions of Ca²⁺ cause identical deflections, indicating that we are in a concentration range in which changes in absorbance of murexide respond linearly to the amounts of Ca²⁺. The trace thereafter remains linear indicating that the Ca²⁺ added is not disappearing from the medium. The addition of succinate induced an upward deflection of the trace, indicating that the Ca²⁺ is progressively taken up by the cells and is less available to form the Ca²⁺–murexide complex. The addition of the uncoupling agent TTFB gives rise to an opposite deflection, indicating that all the Ca²⁺ accumulated by the cells is being released. The bottom trace shows Ca²⁺ movements measured with a Ca²⁺-sensitive electrode under identical experimental conditions. With the exception of a slightly slower response time from the Ca²⁺ electrode, the two traces are indistinguishable. Although settling of cells can be a cause of artifacts, this was not the case for the first 10 minutes of this experiment. Those desiring measurements of longer periods of time with other systems in which precipitation of cells might occur should equip their cuvette with a stirring device.

Ca²⁺ Movements in Mitochondria

Kinetic studies of Ca²⁺ transport in mitochondria with the use of murexide are reported in the literature.[2,9,10] At a concentration of

---

[9]L. Mela, *Biochemistry* **8**, 2481 (1969).
[10]A. Scarpa and G. F. Azzone, *Europ. J. Biochem.* **12**, 399 (1970).

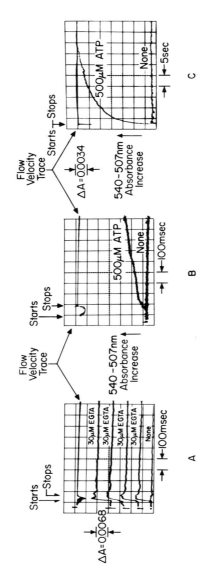

Fig. 4. Fast kinetics of ATP-dependent Ca²⁺ uptake by sarcoplasmic reticulum. The medium contained 20 mM Tris maleate (pH 6.8), 50 mM KCl, 140 μM CaCl₂, 10 mM MgCl₂, and 100 μM murexide. In B and C, 0.7 mg of fragmented sarcoplasmic reticulum from white muscle of rabbit hind leg was also present. In A, the reactions were started with the addition of 30 μM ethylene glycol (β-aminoethyl ether) N,N'-tetraacetic acid (EGTA). In B and C, the reaction was initiated by the addition of 500 μM Ca²⁺-free ATP. Note that the absorbance scale in A is half of that in B and C. More details are found in Inesi and Scarpa.[12]

20–40 μM, murexide has been demonstrated to be entirely localized externally to mitochondria[2] and not to affect mitochondrial respiratory control or other energy-linked functions.[2] In addition, the wavelengths of 540–507 (or 510) nm have been shown to be unaffected by the absorbancy changes of the respiratory carriers.[2] Figure 3 gives an example of the use of murexide for kinetic readout of energy-linked Ca²⁺ transport in rat liver mitochondria. As in Fig. 2, the downward deflections (decrease in absorbance at 540 nm) correspond to an increase in the [Ca²⁺] in the medium, while an upward deflection corresponds to the disappearance of Ca²⁺ from the medium due to mitochondrial uptake.

The addition of Ca²⁺ to mitochondria which are oxidizing pyruvate and malate causes an abrupt decrease in absorbance which is quickly followed by an increase in absorbance until all the externally added Ca²⁺ is accumulated inside the mitochondria. The addition of rotenone, which blocks the oxidation of both endogenous and exogenous substrates, induces a slow release of the Ca²⁺ into the medium. The oxidation of added succinate again provides energy, and Ca²⁺ is taken up at a rapid rate. The addition of FCCP, an uncoupler of oxidative phosphorylation, gives rise to a rapid release of Ca²⁺, the amount of which exceeds the amounts of Ca²⁺ added and can be explained as a release of endogenous Ca²⁺. The rapidity of the Ca²⁺ uptake by mitochondria does not permit a precise evaluation of the decrease in absorbance due to the addition of Ca²⁺. Such calibrations can be obtained in a parallel experiment in which mitochondria are incubated in the presence of a respiratory inhibitor which precludes the uptake of Ca²⁺ by mitochondria. We wish to emphasize that the Ca²⁺ movement described in this experiment is too fast to be followed with the Ca²⁺ electrode, and therefore murexide is the only tool so far available for the direct readout of the kinetics of Ca²⁺ uptake in mitochondria.

Fast Kinetics of Ca²⁺ Uptake by Sarcoplasmic Reticulum (SR)

The above-mentioned high rate constant for the formation of the Ca²⁺–murexide complex has made possible the use of the murexide technique for the detection of the first milliseconds of the Ca²⁺ uptake by SR.[1] For this reaction, the mixing time is the limiting factor and a fast mixing device is required. Therefore in the experiments reported in Fig. 4, changes in absorbance of murexide were followed in a double beam spectrophotometer, and ATP and other reagents were added to the reaction mixture using the stopped-flow apparatus described by

Chance *et al.*[11] This instrument has a mixing time of less than 12 msec, and the volume ratio of added reagents to the reaction mixture is 1:80; both characteristics render this or similar mixing devices mandatory for such measurements. The use of mixing devices with two syringes of identical volume has already been employed for similar studies,[1] but this can present technical difficulties and problems of interpretation of the data obtained.

Figure 4A shows a titration experiment in our system: the large syringe of the stopped flow contained the medium with murexide and Ca^{2+} but devoid of SR. EGTA was added through the small syringe, and mixing times and absorbancy changes were displayed in a storage oscilloscope. The figure shows that Ca^{2+} chelation by addition of EGTA produces an increase in light absorbance; this effect was completed within the mixing time and could be reproduced by further additions of EGTA. Figures 4B and 4C show an experiment of Ca^{2+} uptake by SR induced by the addition of ATP and obtained at a different time scale by using different oscilloscopes. In this case, the large syringe contained the same medium as in the above titration but supplemented with SR, and the small one contained the solution of ATP. The traces show a rapid but easily measurable increase of absorbancy due to the ATP-driven Ca^{2+} uptake by SR. The addition of ATP which can contain appreciable amounts of Ca^{2+} and is able to form complexes with the Ca^{2+} present in the medium can cause artifacts. Both causes of error can be avoided by the use of Ca^{2+}-free ATP and in the presence of Mg^{2+}, which effectively prevents the formation of Ca–ATP complex in sizable amounts. More details are found in the paper by Inesi and Scarpa.[12]

Concluding Remarks

The above-described experiments indicate that murexide, if used under well-controlled experimental conditions in which unspecific changes have been eliminated, is a useful tool for kinetic measurements of Ca^{2+} uptake by cellular and subcellular systems. Table I shows that this method can be successfully applied to the study of other divalent cations, whose interaction with biological membranes can be of physiological significance. By this method, the limit of detection of Ca^{2+} is $\sim 3 \mu M$ or $\sim 6 \mu M$ for Mn^{2+} and Sr^{2+}. In addition to the reported experimental systems, the use of murexide can be applied to the study of

[11]B. Chance, D. DeVault, V. Legallais, L. Mela, and T. Yonetani, *in* "Fast Reactions and Primary Processes in Chemical Kinetics" (S. Claesson, ed.), p. 437. Interscience, Stockholm, 1967.

[12]G. Inesi and A. Scarpa, *Biochemistry* 11, 356 (1972).

Ca^{2+} (or other divalent cations or rare earths) in different biological systems. In each case, the following experimental conditions should be satisfied: (a) suitable wavelengths at which a maximum in absorbance can be obtained without interference by scattering or by other absorbing components; (b) the lack of penetration or binding of murexide to the system; (c) absence of side effects of murexide on the functional properties of the biological system.

Section II
Inhibitors

[32] Inhibition of Photosynthetic Electron Transport and Photophosphorylation[1]

By S. IZAWA and N. E. GOOD

In this section we have been particularly concerned with results of inhibition studies in isolated chloroplasts and bacterial "chromatophores." Usually the chloroplasts employed have been naked lamellae from spinach (*Spinacia oleracea* L.), and the chromatophores have been from *Rhodospirillum rubrum* or, more rarely, *Rhodopseudomonas* and *Chromatium*. It is probable that most of the chloroplast inhibitions described are of a general nature affecting chloroplasts from all plants in a similar way. However, certain aspects of the biochemistry of photosynthesis varies so much among the bacteria that it can be dangerous to extrapolate from one organism to another. Consequently we have specified the bacteria used in all reports of inhibitions of bacterial photosynthesis.

Electron Transport Inhibition

Definition. Substances or treatments that inhibit electron transport, regardless of its nature, are thought to act directly on some oxidoreduction step in the electron transport chain. Figure 1a illustrates the mode of action of a typical electron transfer inhibitor, DCMU.[2] The basal nonphosphorylating electron transport, the phosphorylating electron transport, and phosphorylation are similarly inhibited.

Inhibition of Photosynthetic Electron Transport in Plants

In postulating the sites of action of the various inhibitors, we have assumed the reality of the electron transport pathways depicted in Fig. 2. This scheme or other very similar schemes have now gained wide acceptance. Nevertheless it should be realized that the basic model has not been rigorously verified and that many of the details are at best

---

[1]For the classification of inhibitors see E. C. Slater, Vol. X [8].

[2]Abbreviations used: CCCP, ketomalononitrile 3-chlorophenylhydrazone (carbonyl-cyanide 3-chlorophenylhydrazone); CMU, 3-(4-chlorophenyl)-1,1-dimethylurea; DCCD, N,N'-dicyclohexylcarbodiimide; DCIP, 2,6-dichlorophenolindophenol; DCMU, 3-(3,4-dichlorophenyl)-1,1-dimethylurea; FCCP, ketomalononitrile 4-trifluoromethoxyphenyl-hydrazone (carbonylcyanide 4-trifluoromethoxyphenylhydrazone); HQNO, 2-*n*-heptyl-4-hydroxyquinoline-*N*-oxide; P_i, inorganic orthophosphate, PMS, *N*-methylphenzonium methosulfate (phenazine methosulfate); PP_i, inorganic pyrophosphate; *R. rubrum, Rhodospirillum rubrum;* TFTB, 4,5,6,7-tetrabromo-2-trifluoromethylbenzimidazole.

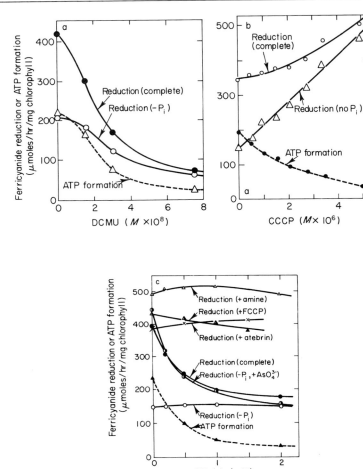

Fig. 1. (a) Effect of 3-(3,4-dichlorophenyl)-1,1-dimethylurea (DCMU), a typical electron transport inhibitor, on photophosphorylation, coupled electron transport, and the basal, nonphosphorylating electron transport in spinach chloroplasts. Note that these three processes are inhibited to approximately the same extent by increasing concentrations of the inhibitor. (b) Effect of carbonylcyanide 3-chlorophenylhydrazone (CCCP), a typical phosphorylation uncoupler. Note that the coupled and nonphosphorylating electron transport are markedly stimulated as the phosphorylation is suppressed. (c) Effect of phlorizin, a typical energy transfer inhibitor. Note that only the phosphorylation and that part of the electron transport stimulated by ADP plus P_i or ADP plus arsenate are inhibited.

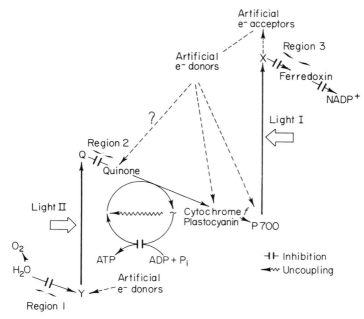

FIG. 2.   Model of electron transport pathways in chloroplasts used in assigning sites of actions to inhibitors and uncouplers. Solid arrows represent natural pathways. Broken arrows represent donation or acceptance of electrons by exogenous substances. Some donors to photosystem II are p-phenylenediamine, benzidine, hydrazines, hydroquinone, and high concentrations of ascorbate. Among the donors to photosystem I are reduced indophenol dyes (DCIPH$_2$), diaminodurene (DAD), N,N,N',N'-tetramethylphenylenediamine (TMPD), reduced phenazine methosulfate (PMSH$_2$), and reduced pyocyanine. Among the exogenous electron acceptors (Hill oxidants) are ferricyanide, flavins, viologens, and various dyes.

conjectural. Assignments of sites of action can be less accurate, but never more accurate, than the model used in making the assignments.[3]

Inhibition of Water Oxidation (Region 1)

The following inhibitors and treatments are effective in selectively blocking or destroying the water oxidation step. Chloroplasts treated thus are still capable of transferring electrons from various artificial donors to the standard chloroplast electron acceptors.

[3]The reader is referred to a review article by R. P. Levine [*Annu. Rev. Plant Physiol.* **20**, 523 (1969)] for a detailed description of the photosynthetic electron transport chain. For the details of current theories about the reaction steps around photosystem II, see (a) B. Kok and G. M. Cheniae, *Curr. Top. Bioenerg.* **1**, 1 (1966); (b) G. Hind and J. M. Olson, *Annu. Rev. Plant Physiol.* **19**, 249 (1968).

Ammonia, Methylamine and Hydroxylamine. The unprotonated base seems to be the active form, and therefore the inhibition is most pronounced at higher pH. At pH > 8.0, concentrations of ammonia $> 10^{-2}$ M and concentrations of methylamine $> 10^{-1}$ M inhibit water oxidation,[4] but at lower concentrations or lower pH levels the uncoupling effect of these typical amines predominates (see below). Hydroxylamine is a much weaker base and probably partly for this reason is effective at lower pH levels and lower concentrations (10^{-4} to 10^{-3} M).[4,5] A higher concentration ($> 10^{-2}$ M) of hydroxylamine can also serve as an artificial electron donor for photosystem II in isolated chloroplasts.[4]

Tris-Washing. Pretreatment of chloroplasts with a high concentration of Tris buffer also destroys the ability of chloroplasts to oxidize water. The Tris-washed chloroplasts retain their ability to phosphorylate when transferring electrons from artificial donors to NADP or methylviologen. See footnote 6 for procedures.

Chloride Removal. Chloride ion is involved in the water oxidation step.[4] Chloroplasts prepared from spinach can be rendered Cl$^-$-deficient by repeated washings in Cl$^-$-free media at alkaline pH.[4,7,8] The Hill reaction activity of the Cl$^-$-depleted chloroplasts is then severely inhibited (down to 10–20% of the control). This inhibition is readily reversed by the addition of 5×10^{-3} M Cl$^-$ or Br$^-$. See footnote 9 for procedures.

Heat Treatment. Mild heating (e.g., 3 minutes at 50° or 5 minutes at 45°) destroys the mechanism of water oxidation and oxygen production

---

[4] S. Izawa, R. L. Heath, and G. Hind, *Biochim. Biophys. Acta* **180**, 388 (1969).
[5] A. Joliot, *Biochim. Biophys. Acta* **126**, 587 (1966).
[6] T. Yamashita and W. L. Butler, *Plant Physiol.* **43**, 1978 (1968); *Plant Physiol.* **44**, 435 (1969). Preparation of Tris-washed chloroplasts: Spinach leaves are homogenized with a solution containing 0.4 M sucrose, 0.05 M Tris·HCl (pH 7.8), and 0.01 M NaCl, filtered, and centrifuged. A fraction of chloroplasts sedimented between 300 g (1 minute) and 600 g (7 minutes) is resuspended in Tris·HCl (0.8 M at pH 8.0 or 0.05 M at pH 8.8). After 10 minutes, the fraction sedimenting between 300 g (1 minute) and 1500 g (7 minutes) is collected and suspended in the sucrose·Tris·NaCl medium.
[7] J. M. Bové, C. Bové, F. R. Whatley, and D. I. Arnon, *Z. Naturforsch. B* **18**, 683 (1963).
[8] G. Hind, H. Y. Nakatani, and S. Izawa, *Biochim. Biophys. Acta* **172**, 277 (1969).
[9] Procedure for Cl$^-$ removal: For the preparation of Cl$^-$-deficient, *coupled* chloroplasts, see footnotes 7 and 8. More rigorous Cl$^-$ depletion is achieved with EDTA-washed (uncoupled) chloroplasts: Spinach leaves are homogenized with a medium containing 0.1 M Na$_2$SO$_4$, 0.03 M TES (see Chap. [3])-NaOH (pH 7.4), filtered, and centrifuged at 2500 g (5 minutes). The sedimented chloroplasts are washed 3 times (including a light centrifugation to remove debris) with a medium containing 0.1 M sucrose, 2 mM Tricine·NaOH (pH 7.8), and 0.5 mM Na-EDTA, and are finally taken up in a medium containing 0.1 M sucrose, 6 mM TES·NaOH (pH 7.4), and 5 mM MgSO$_4$. A brief exposure of suspensions to room temperature (5–10 minutes) during the washing procedure promotes the release of Cl$^-$ from the lamellae (footnote 4).

in chloroplasts.[10-12] Similar results can be obtained by heating the leaves before isolating the chloroplasts.[10] The heat treatment causes Mn to be lost from the chloroplast lamellae, but the readdition of Mn^{2+} does not restore oxygen production.[10]

Manganese Deficiency. Algae grown in the absence of Mn lose their ability to use water as an electron donor while retaining their ability to use hydrogen for photosynthesis.[13] Recent studies have located the primary site of Mn involvement on the O_2 evolving side of photosystem II. However, reversible inactivation and reactivation of water oxidation by the removal and restoration of Mn has been successful only in living cells. Moreover, light is required for the *in vivo* reactivation of the oxygen producing system.[14]

Miscellaneous. Ultraviolet irradiation destroys the mechanism of water oxidation[12] and at the same time partially destroys the plastoquinones of the chloroplasts.[15] Aging chloroplasts for 20 minutes at pH 4, or at pH 8.3 and room temperature for somewhat longer, also selectively destroys the mechanism of water oxidation.[4,11]

Inhibition on the Reducing Side of Photosystem II (Region 2)

Phenylureas. 3-(3,4-Dichlorophenyl)-1,1-dimethylurea (DCMU; MW 233) and 3-(4-chlorophenyl)-1,1-dimethylurea (CMU; MW 198) were developed by du Pont as herbicides and were later described as exceptionally potent and specific inhibitors of the Hill reaction.[16] Currently these are the most widely used inhibitors of photosynthetic electron transport. Both are very stable compounds only sparingly soluble in water (DCMU, 2×10^{-4} M; CMU, 10^{-3} M) and about 10 times more soluble in ethanol, methanol, or ethylene glycol. Concentration giving 50% inhibition of electron transport in isolated chloroplasts: DCMU, 5×10^{-8} to 10^{-7} M; CMU, 5×10^{-7} to 10^{-6} M. For intact algal cells, the concentrations required are about 5 times higher. The inhibition, which is completely reversible, is most pronounced when the light intensities are low enough to be rate-determining.[17,18] It is believed[19]

[10]P. Homann, *Biochem. Biophys. Res. Commun.* 33, 229 (1968).
[11]S. Katoh and A. San Pietro, *Arch. Biochem. Biophys.* 128, 378 (1968).
[12]T. Yamashita and W. L. Butler, *Plant Physiol.* 43, 2037 (1968).
[13]E. Kessler, *Arch. Biochem. Biophys.* 59, 527 (1955).
[14]G. M. Cheniae and I. F. Martin, *Brookhaven Symp. Biol.* 19, 406 (1966); *Plant Physiol.* 44, 351 (1969).
[15]K. E. Mantai and N. I. Bishop, *Biochim. Biophys. Acta* 131, 350 (1967).
[16]J. S. C. Wessels and R. van der Veen, *Biochim. Biophys. Acta* 19, 548 (1956).
[17]G. Gingras, C. Lemasson, and D. C. Fork, *Biochim. Biophys. Acta* 69, 438 (1963).
[18]S. Izawa and N. E. Good, *Biochim. Biophys. Acta* 102, 20 (1965).
[19]L. N. M. Duysens and J. Amesz, *Biochim. Biophys. Acta* 64, 243 (1962).

that the inhibition site is closely associated with the electron acceptor Q (see Fig. 2). The use of high concentrations of DCMU ($> 10^{-4}$ M) should be avoided because the inhibitor then produces a secondary effect — an inhibition of the photosystem I-independent cyclic photophosphorylation probably through an uncoupling action.[20] Many structurally related compounds such as chlorophenylcarbamates[21] and acylchloroanilides[22] are also potent inhibitors which probably act at the same site.

Symmetrical Aminotriazines. These were introduced by Geigy as herbicides. They also reversibly inhibit electron transport near Q.[23,24] Indeed it has not been possible to distinguish between triazine inhibition and phenylurea inhibition, and there seem to be equal numbers of sites in chloroplasts sensitive to inhibition by these two classes of inhibitor.[18] Atrazine (2-chloro-4-(2-propylamino)-6-ethylamino-*s*-triazine; MW 216) and Simazine (2-chloro-4,6-bis(ethylamino)-*s*-triazine; MW 202) are about as potent as CMU with 50% inhibition at 5×10^{-7} M. However, Simazine is rather difficult to use *in vitro* because of its very low solubility in water (about 3×10^{-5} M). Atrazine is about 10 times more soluble. Like the phenylureas, the triazine inhibitors are ordinarily very stable substances but, unlike the ureas, the triazines are destroyed by the tissues of certain plants, notably corn (*Zea mays*). A review covering these and other herbicidal inhibitors is available.[25]

2-Alkyl-4-hydroxyquinoline Oxides. These inhibitors of the respiratory electron transfer chain (see Vol. X [8]) also inhibit electron transport in chloroplasts.[26] The 2-heptyl compound (HQNO; MW 258) and 2-nonyl compound (NQNO; MW 286) are soluble in ethanol and methanol but only sparingly soluble in water (HQNO, 10^{-5} M at pH 7 and 2.5×10^{-4} M at pH 9; NQNO is about 1/10 as soluble). Molecular extinction coefficients in 10^{-3} M NaOH are $\epsilon_{253} = 24 \times 10^3$, $\epsilon_{346} = 9.5 \times 10^3$.[26a] In chloroplasts, 10^{-6} M NQNO or 10^{-5} M HQNO give approximately 50% inhibition. The inhibition characteristics are similar to those of DCMU,[27] although there is evidence that the exact site of inhibition may be different.[28]

[20]S. Izawa, *in* "Comparative Biochemistry and Biophysics of Photosynthesis" (K. Shibata, A. Takamiya, A. T. Jagendorf, and R. C. Fuller, eds.), p. 140. Univ. of Tokyo Press, Tokyo, 1968.
[21]D. E. Moreland and K. L. Hill, *J. Agr. Food Chem.* **7**, 832 (1959).
[22]N. E. Good, *Plant Physiol.* **36**, 788 (1961).
[23]D. E. Moreland, W. A. Gentner, and J. L. Hilton, *Plant Physiol.* **34**, 432 (1959).
[24]G. Zweig, I. Tamas, and E. Greenberg, *Biochim. Biophys. Acta* **66**, 196 (1963).
[25]D. E. Moreland, *Annu. Rev. Plant Physiol.* **18**, 365 (1967).
[26]M. Avron, *Biochem. J.* **78**, 735 (1961).
[26a]J. W. Cornforth and A. T. James, *Biochem. J.* **63**, 124 (1956).
[27]S. Izawa, T. N. Connolly, G. D. Winget, and N. E. Good, *Brookhaven Symp. Biol.* **19**, 169 (1966).
[28]G. Hind and J. M. Olson, *Brookhaven Symp. Biol.* **19**, 188 (1966).

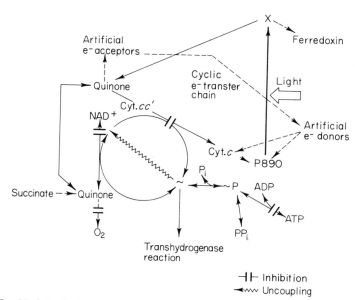

FIG. 3. Model of electron and energy transfer pathways in *Rhodospirillum rubrum* used in assigning sites of actions to inhibitors and uncouplers. The exact position of the inhibition site (antimycin block) in the cyclic chain relative to the site of coupling is not clear except that both the inhibition site and the coupling site are between cytochrome cc' and cytochrome c_2. The two sites could be identical. In such schemes cytochrome b is often included between cytochrome cc' and the coupling site (see, however, Olson, ref. b in text footnote 3), but it is omitted in this figure for simplicity. The cyclic electron transfer chain involves at least one more phosphorylation site near quinone in addition to the one shown, which is bypassed by artificial electron carriers such as PMS.

monas are rather resistant to HQNO but are still sensitive to antimycin.[43] The antimycin and HQNO blocks are almost completely bypassed by phenazine methosulfate (PMS)[40] and are partially bypassed by reduced indophenol dyes.[44]

Miscellaneous. 1,10-*Phenanthroline* (10^{-5} M) inhibits the photophosphorylation mediated by ascorbate in *R. rubrum* chromatophores, but PMS-mediated photophosphorylation is relatively insensitive.[45] It has been suggested that orthophenanthroline may complex the Fe^{2+} of a hypothetical nonheme iron protein which acts at a locus preceding cytochrome cc'.[45,46] *Phenylmercuric acetate* inhibits a variety of reactions in

[43]J.-H. Klemme, *Z. Naturforsch. B* **24**, 67 (1969). (A paper dealing with the inhibition study of the energy-linked NAD reduction in *Rhodopseudomonas capsulata*.)
[44]S. K. Bose and H. Gest, *Proc. Nat. Acad. Sci. U. S.* **49**, 337 (1963).
[45]T. Horio and J. Yamashita, *Biochim. Biophys. Acta* **88**, 237 (1964).
[46]H. Sato, K. Takahashi, and G. Kikuchi, *Biochim. Biophys. Acta* **112**, 8 (1966).

photosynthetic bacteria, especially photophosphorylation and NAD photoreduction.[47] These phenomena have been interpreted in terms of electron transport inhibition. However, the more recent observation[48] that the dark ATPase activity in *R. rubrum* chromatophores is sensitive to this compound suggests that the inhibition of NAD photoreduction may be an indirect effect of an inhibition of the energy coupling mechanism. At very high concentrations (10^{-4} to 10^{-3} M), *DCMU* and *Simazine* inhibit photophosphorylation and some oxidoreduction reactions in *R. rubrum* chromatophores.[49] It has been suggested that the inhibition site is around cytochrome *b*. The autoxidizable cytochrome *cc'* involved in the cyclic electron transport chain (*Rhodospirillum, Rhodopseudomonas, Chromatium*, etc.) binds *carbon monoxide* but not cyanide.[50] The CO-complex is readily photodissociated.

Inhibition of the Auxiliary ("Oxidative") Electron Transfer Chain

Amytal, Rotenone, and Piericidin A. These are known to inhibit mitochondrial electron transport, and they also inhibit the respiratory chain of the facultative phototrophs *Rhodospirillum*[51] and *Rhodopseudomonas*.[43] Effective concentrations; amytal, 10^{-3} M; rotenone, 10^{-5} M; piericidin, 10^{-7} M. They are specific inhibitors of the NADH dehydrogenase system and, as such, suppress NADH respiration and light-induced reactions involving NAD.[38,43]

Cyanide and Carbon Monoxide. Cyanide (10^{-4} M)[51,52] and CO[53] inhibit dark respiration by affecting the terminal oxidase (cytochrome *o* ?) of the respiratory chain. The CO inhibition is reversed by light. Aerobic photooxidations (e.g., photooxidation of cytochrome *c*) are rather resistant to cyanide[54] suggesting that in photooxidations the terminal oxidation probably occurs via cytochrome *cc'* (cf. Yamashita *et al.*[52]). These two poisons inhibit hydrogenase, a common enzyme among photosynthetic bacteria, thus suppressing all the reactions involving H_2.

In contrast to the photosynthetic chain, the respiratory chain of *R. rubrum* is relatively insensitive to antimycin and HQNO.[40,51]

---

[47]M. Nozaki, K. Tagawa, and D. I. Arnon, *Proc. Nat. Acad. Sci. U. S.* **47**, 1334 (1961).

[48]S. K. Bose and H. Gest, *Biochim. Biophys. Acta* **96**, 159 (1965).

[49]J. Yamashita and M. D. Kamen, *Biochim. Biophys. Acta* **153**, 848 (1968).

[50]S. Taniguchi and M. D. Kamen, *Biochim. Biophys. Acta* **74**, 438 (1963).

[51]D. M. Geller, *J. Biol. Chem.* **237**, 2947 (1962).

[52]J. Yamashita, S. Yoshimura, Y. Matuo, and T. Horio, *Biochim. Biophys. Acta* **143**, 154 (1967).

[53]T. Horio and C. P. S. Tayler, *J. Biol. Chem.* **240**, 1772 (1965).

[54]L. P. Vernon and M. D. Kamen, *J. Biol. Chem.* **211**, 643 (1954).

Uncoupling in Chloroplasts and Bacterial Chromatophores[55]

Definition. Uncoupling abolishes photophosphorylation without inhibiting electron transport. The electron transport, freed of restraints imposed by the coupling mechanism, is often greatly accelerated as the phosphorylation rate diminishes, and very high rates of electron flow may be attained in the absence of phosphorylation. Figure 1b illustrates the action of CCCP, a typical uncoupler. It is assumed that uncouplers either catalyze the breakdown of high energy intermediates or divert the energy made available through electron transport toward some work other than ATP formation, for example, ion translocation. See footnote 56 for uncoupling in bacteria.

Uncouplers and Uncoupling Treatments

Ammonia and Simple Aliphatic Amines. In isolated chloroplasts these uncouple electron transport from phosphorylation.[57,58] Ammonium salts reversibly uncouple at pH 8 and concentrations from 10^{-3} M to 10^{-2} M but become inhibitory at higher pH levels or higher concentrations (see above under inhibition of water oxidation). The uncoupling effect is approximately proportional to the amount of unprotonated ammonia or amine and therefore increases with increasing pH.[58] Methylamine hydrochloride (effective at 10^{-3} to 10^{-2} M) is often preferred to the more active ammonium salts because it seems to have less inhibitory effect. Amine-uncoupled electron transport is associated with an uptake of the amine together with other cations and available anions.[59] Consequently the chloroplast lamellae swell.[60] With high con-

---

[55]See the review articles: (a) N. E. Good, S. Izawa, and G. Hind, *Curr. Top. Bioenerg.* **1**, 75 (1966) (an article on the inhibitors and uncouplers of photophosphorylation); (b) M. Avron and J. Neumann, *Annu. Rev. Plant Physiol.* **19**, 137 (1968).

[56]In bacteria the main photosynthetic electron transfer pathway producing ATP is a cyclic system, so that there is no easy way of analyzing the effect of an inhibitor on the electron flow. Neither is there any noncyclic reaction that has been shown to be coupled tightly enough for kinetic analyses. There are, however, indications that many of the uncouplers of chloroplasts do *uncouple* (not block) bacterial photophosphorylation. For instance, CCCP, atebrin, and dinitrophenol have been shown to accelerate markedly the dark decay of light-induced cytochrome changes in *Chromatinum*, suggesting a speeding of electron flow (ref. b, footnote 42). It has also been found that gramicidin and atebrin in fact stimulate certain photoreduction reactions in *R. rubrum* chromatophores [J. W. Newton, *Biochim. Biophys. Acta* **109**, 302 (1965); O. K. Ash, W. S. Zaugg, and L. P. Vernon, *Acta Chem. Scand.* **15**, 1629 (1961)].

[57]D. W. Krogmann, A. T. Jagendorf, and M. Avron, *Plant Physiol.* **34**, 272 (1959).

[58]N. E. Good, *Biochim. Biophys. Acta* **40**, 502 (1960); G. Hind and C. P. Whittingham, *Biochim. Biophys. Acta* **75**, 194 (1963).

[59]A. R. Crofts, *Biochem. Biophys. Res. Commun.* **24**, 725 (1966).

[60]S. Izawa and N. E. Good, *Plant Physiol.* **41**, 533 (1966).

TABLE I
EFFECT OF UNCOUPLERS AND ENERGY TRANSFER INHIBITORS ON PARTIAL
REACTIONS OF PHOSPHORYLATION IN CHLOROPLASTS IN THE DARK

Compound	Type	Effective conc. (M)	$Ca^{2+}ATPase^a$	$Mg^{2+}ATPase^b$	$ATP-P_i^b$ exchange
Ammonia	Uncoupler	$10^{-3}-10^{-2}$	Inhibits slightly[c,d]	Not clear[e-i,d]	Inhibits[d,h,i]
Atebrin	Uncoupler	$10^{-6}-10^{-5}$	Little effect[d]	Inhibits[d,i,j]	Inhibits[h,i]
CCCP	Uncoupler	$10^{-6}-10^{-5}$	No effect[c,k]	Inhibits[g]	Inhibits[l]
Octyl-DNP[m]	Uncoupler[m]	$10^{-6}-10^{-5}$	—	Inhibits[l]	Inhibits[l]
Nigericin	Uncoupler	$10^{-7}-10^{-6}$	No effect[n]	Inhibits[n]	Inhibits[n]
Dio-9	Inhibitor	1–2 $\mu g/ml$	Inhibits[d]	Inhibits[d,l]	Inhibits[d,i,l]
Phlorizin	Inhibitor	$10^{-4}-10^{-3}$	Inhibits[k]	Inhibits[i,o]	Inhibits[h,i]
Antibody to CF$_1$	Inhibitor	—	Inhibits[d]	Inhibits[d]	Inhibits[d]
DCCD[p]	Inhibitor[p]	$10^{-5}-10^{-4}$	No effect[k,q]	Inhibits[q]	—

[a] Isolated coupling factor (CF$_1$) treated with trypsin or dithiothreitol.
[b] Light- or pH-transition-triggered, —SH dependent activity in chloroplasts.
[c] V. K. Vambutas and E. Racker, *J. Biol. Chem.* 240, 2660 (1965).
[d] R. E. McCarty and E. Racker, *J. Biol. Chem.* 243, 129 (1968).
[e] Reported results conflicting: ref. *f* (stimulation); refs. *g, h* (no effect); refs. *d, i* (inhibition).
[f] G. Hoch and I. Martin, *Biochem. Biophys. Res. Commun.* 12, 223 (1963).
[g] B. Petrack, A. Craston, F. Shappy, and F. Farron, *J. Biol. Chem.* 240, 906 (1965).
[h] J. H. Kaplan and A. T. Jagendorf, *J. Biol. Chem.* 243, 972 (1968).
[i] C. Carmeli and M. Avron, *Eur. J. Biochem.* 2, 318 (1967).
[j] A. Bennun and M. Avron, *Biochim. Biophys. Acta* 109, 117 (1965).
[k] A. E. Karu and E. N. Moudriankis, *Arch. Biochem. Biophys.* 129, 655 (1969).
[l] K. G. Rients, *Biochim. Biophys. Acta* 143, 595 (1967).
[m] 4-Isoctyl-2,6-dinitrophenol, an inhibitory uncoupler (see text).
[n] N. Shavit, R. A. Dilley, and A. San Pietro, *Biochemistry* 7, 2356 (1968).
[o] S. Izawa, T. N. Connolly, G. D. Winget, and N. E. Good, *Brookhaven Symp. Biol.* 19, 169 (1966).
[p] N,N'-Dicyclohexylcarbodiimide; probably a mixed type inhibitor (see text).
[q] R. E. McCarty and E. Racker, *J. Biol. Chem.* 242, 3435 (1967).

centrations of salts, the swelling is very great but the uncoupling is much diminished. Ammonia and simple amines have little effect on photophosphorylation in *Anabaena* lamellar fragments[61] and in very small fragments of chloroplast lamellae (subchloroplast particles).[62]

Ammonium sulfate at concentrations up to 10^{-2} M has no effect on photophosphorylation in *R. rubrum* chromatophores.[45]

Complex Amines (Atebrin, Chlorpromazine, Brucine, Chloroquine). Atebrin (quinacrin; MW 509 as hydrochloride) and chlorpromazine (MW 355

[61] W. C. Duane, Sr., M. C. Hohl, and D. W. Krogmann, *Biochim. Biophys. Acta* 109, 108 (1965).
[62] R. E. McCarty, *J. Biol. Chem.* 244, 4292 (1969).

TABLE II

EFFECT OF VARIOUS COMPOUNDS ON PARTIAL REACTIONS OF PHOSPHORYLATION AND ENERGY-LINKED TRANSHYDROGENASE REACTION IN *Rhodospirillum Rubrum* CHROMATOPHORES

Compound	Effective conc. (M)	Partial reactions (dark)			Transhydrogenase reaction[a]		
		ATPase	ATP-P$_i$ exchange	PP$_i$ase	Energy source		
					Light	ATP	PP$_i$
CCCP	10^{-6}–10^{-5}	Stimulates[b]	Inhibits[c]	—	Inhibits	Inhibits	Inhibits
Atebrin	10^{-6}–10^{-5}	Inhibits[d]	Inhibits[d]	—	Inhibits	Inhibits	Inhibits
Octyl-DNP[e]	10^{-6}–10^{-5}	Stimulates[f]	—	Stimulates[f]	—	—	—
Gramicidin D	10^{-7}–10^{-6}	Inhibits[f]	Inhibits[c] slightly	—	Inhibits	Inhibits	Inhibits
Desaspidin	10^{-7}–10^{-6}	Stimulates[f]	—	Stimulates[f]	Inhibits	Inhibits	Inhibits
Oligomycin	2–5 μg/ml	Inhibits[f,g]	Inhibits[g]	Stimulates[f]	No effect	Inhibits	Little effect
Antimycin	10^{-7}–10^{-6}	No effect[h]	No effect[c]	Little effects[f,g]	Inhibits	No effect	No effect

[a]D. L. Keister and N. J. Yike, *Biochemistry* 6, 3847 (1967).
[b]S. K. Bose and H. Gest, *Biochim. Biophys. Acta* 96, 159 (1965).
[c]W. S. Zaugg and L. P. Vernon, *Biochemistry* 5, 34 (1966).
[d]H. Baltscheffsky and M. Baltscheffsky, *Acta Chem. Scand.* 12, 1333 (1958).
[e]4-Isoctyl-2,6-dinitrophenol.
[f]M. Baltscheffsky, H. Baltscheffsky, and L.-V. von Stedingk, *Brookhaven Symp. Biol.* 19, 246 (1966).
[g]T. Horio, K. Nishikawa, Y. Horiuti, and T. Kakuno, *in* "Comparative Biochemistry and Biophysics of Photosynthesis" (K. Shibata, A. Takamiya, A. T. Jagendorf, and R. C. Fuller, eds.), p. 498. Univ. Park Press, State College, Pennsylvania, 1968.
[h]T. Horio, K. Nishikawa, M. Katsumata, and J. Yamashita, *Biochim. Biophys. Acta* 94, 371 (1965).

as hydrochloride) are known to be uncouplers of oxidative phosphorylation. They are also uncouplers of chloroplast electron transport. Uncoupling by atebrin (10^{-6} to 10^{-5} M) is only partially reversible.[32] Chlorpromazine is structurally related to atebrin and uncouples similarly.[32] Brucine[63] and chloroquine[8] are effective at 10^{-4} to 10^{-3} M. Chloroplasts uncoupled by any of these complex amines shrink markedly during electron transport.[60] The shrinking is associated with an extrusion of salts.[64]

Atebrin (10^{-6} to 10^{-5} M) also inhibits photophosphorylation in R. rubrum chromatophores. The suppressed photophosphorylation is stimulated by FAD (10^{-2} M) and weakly by FMN, but it is not clear that this represents a recovery of the impaired coupling mechanism since FAD also stimulates photophosphorylation in the control chromatophores.[65]

Carbonylcyanide Phenylhydrazones (CCCP and FCCP). These substances, better described as ketomalononitrile phenylhydrazones, are powerful uncouplers of both oxidative phosphorylation and photophosphorylation.[66] They are weak acids (pK_a for CCCP, 5.95), soluble in ethanol and methanol but only sparingly soluble in water ($< 10^{-4}$ M). They are more soluble in alkaline media but are unstable therein. In mitochondria CCCP-uncoupling is prevented by cysteine and other 1,2- or 1,3-aminothiols, but not by glutathione.[66] Uncoupling in chloroplasts can be relieved by removal of the uncoupler or by the addition of cysteine.[67] In chloroplasts the effective concentrations are 10^{-7} to 10^{-6} M for FCCP and 10^{-6} to 10^{-5} M for CCCP. Higher concentrations are strongly inhibitory.

Both photophosphorylation and oxidative phosphorylation in R. rubrum chromatophores are sensitive to CCCP.[68]

Phenols (Dinitrophenols, Dicoumarol, Pentachlorophenol). These are classical uncouplers of oxidative phosphorylation. However, 2,4-dinitrophenol (pK_a 4.1) even at 10^{-3} M has only mild effects on chloroplasts, uncoupling slightly at pH 6.0.[69] More lipophilic alkylated dinitrophenols such as 6-(1,2-dimethylbutyl)-2,4-dinitrophenol (pK_a 4.5) and 6-(1-methylpentyl)-2,4-dinitrophenol (pK_a 4.6) act as typical uncouplers at pH 7.6–8.0 (effective concentrations, 10^{-6} to 10^{-5} M), but they are potent

[63]S. Izawa and N. E. Good, *Biochim. Biophys. Acta* **162**, 380 (1969).
[64]R. A. Dilley and L. P. Vernon, *Arch. Biochem. Biophys.* **111**, 365 (1965).
[65]H. Baltscheffsky, *Biochim. Biophys. Acta* **40**, 1 (1960).
[66]P. G. Heytler, *Biochemistry* **2**, 357 (1963).
[67]D. Y. de Kiewiet, D. O. Hall, and E. L. Jenner, *Biochim. Biophys. Acta* **109**, 284 (1965).
[68]D. M. Geller, *in* "Bacterial Photosynthesis" (H. Gest, A. San Pietro, and L. P. Vernon, eds.), p. 161. Antioch Press, Yellow Springs, Ohio, 1963.
[69]J. Neumann and A. T. Jagendorf, *Biochem. Biophys. Res. Commun.* **16**, 562 (1964).

inhibitors at higher pH levels.[70] The better known 4-isoctyl-2,6-dinitrophenol (see Vol. X [8]) is similarly potent as an uncoupler but seems to be more inhibitory.[71] Dicoumarol and pentachlorophenol are rather strong electron transport inhibitors (effective concentrations above 10^{-5} M), but at pH 6.0 they can act as uncouplers.[72]

Photophosphorylation in R. rubrum chromatophores is rather insensitive to 2,4-dinitrophenol (50% inhibition at 10^{-3} M), but oxidative phosphorylation is slightly more sensitive.[51] 4-Isoctyl-2,6-dinitrophenol (10^{-5} M)[71] and dicoumarol (10^{-4} M)[73] strongly inhibit R. rubrum chromatophore photophosphorylation.

Desaspidin. This phlorobutyrophenone derivative (MW 446) is a potent uncoupler of oxidative phosphorylation. The cyclic photophosphorylation in chloroplasts mediated by PMS plus ascorbate is severely inhibited by the compound at 10^{-7} M whereas noncyclic photophosphorylation is unaffected until the uncoupler exceeds 10^{-6} M.[74] Originally this difference in sensitivity to desaspidin was interpreted as a difference in sensitivities of two different phosphorylation sites. However, there is strong evidence that desaspidin is rapidly photooxidized in the noncyclic systems.[75]

Desaspidin is a widely used and powerful inhibitor of chromatophore photophosphorylation. Effective concentration, 10^{-7} to 10^{-6} M. See Table II.

Phosphate Analogs (Arsenate and Thiophosphate). Arsenate[76] uncouples chloroplasts only in the presence of ADP. That is to say, arsenate alone does not stimulate the basal electron transport in chloroplasts. Arsenate and thiophosphate[32] are competitive with phosphate in inhibiting photophosphorylation. However, arsenate competes rather poorly with phosphate. In the ferricyanide reducing system the K_i for arsenate is 8.3×10^{-4} M while the K_m for phosphate is 5.4×10^{-4} M.[76] Thus with arsenate and phosphate in equal concentrations, the inhibition of phosphorylation is only 30–40%. Thiophosphate competes somewhat more favorably with phosphate, but it is slowly oxidized by ferricyanide.[32] The formation of ADP-arsenate and ADP-thiophosphate has not been detected; if these high-energy compounds are formed they must break down rapidly.

[70]K. S. Siow and A. M. Unrau, *Biochemistry* 7, 3507 (1968).
[71]H. Baltscheffsky, *Acta Chem. Scand.* 19, 1933 (1965).
[72]A. T. Jagendorf and J. Neumann, *J. Biol. Chem.* 240, 3210 (1965).
[73]H. Baltscheffsky and M. Baltscheffsky, *Acta Chem. Scand.* 14, 257 (1960).
[74]H. Baltscheffsky and D. Y. de Kiewiet, *Acta Chem. Scand.* 18, 2406 (1964); Z. Gromet-Elhanan and D. I. Arnon, *Plant Physiol.* 40, 1060 (1965).
[75]Z. Gromet-Elhanan and M. Avron, *Plant Physiol.* 41, 1231 (1966); G. Hind, *Plant Physiol.* 41, 1237 (1966).
[76]M. Avron and A. T. Jagendorf, *J. Biol. Chem.* 234, 967 (1959).

It has been reported that 4×10^{-2} M arsenate inhibits 50% of photo-phosphorylation by $R.$ $rubrum$ chromatophores when the phosphate concentration is 8×10^{-3} M.[45] Arsenate stimulates dark ATPase activity in the chromatophores but inhibits pyrophosphatase activity and the ATP–P_i exchange reaction.[77]

$EDTA$ $Treatment.$ Washing chloroplasts with EDTA (optimum 5×10^{-4} M, pH 7.8) in low salt media causes uncoupling.[78,79] This uncoupling results from the removal of a proteinacious coupling factor,[79] a latent Ca^{2+}-dependent ATPase[80] which can be activated by trypsin or dithiothreitol. Simple salts (NaCl 5×10^{-2} M, MgCl$_2$ 10^{-3} M) prevent EDTA uncoupling. The uncoupled chloroplasts can be recoupled when incubated with concentrated coupling factor in the presence of Mg^{2+}.

Inhibition of photophosphorylation in $R.$ $rubrum$ chromatophores by 10^{-2} M EDTA has been reported, but it is not clear whether this inhibition involves anything more complex than the chelation of the Mg^{2+} which is essential for phosphorylation.[45]

$Poly$-L-$lysine.$ This polyamine uncouples chloroplast electron transport but has little effect on the electron transport-associated pH rise of the medium.[81] Most other uncouplers do abolish this pH rise. Higher molecular weight polymers show stronger uncoupling action. Amounts of polylysine equivalent to between 10^{-5} and 10^{-4} M lysine uncouple, but the monomer lysine at these concentrations has little uncoupling effect.

$Salicylanilide.$ 5-Chloro-3-(3-chlorophenyl)-2′,4′,5′-trichlorosalicylanilide uncouples chloroplasts at concentrations between 10^{-7} and 10^{-6} M.[82] Like polylysine, such anilides affect the pH rise of the medium only weakly. They inhibit the light-triggered Mg^{2+}-dependent ATPase activity but do not inhibit the Ca^{2+}-dependent ATPase.[82]

Photophosphorylation in $R.$ $rubrum$ chromatophores is also very sensitive to the salicylanilides.[82]

$Indophenols.$ 2,6-Dichlorophenolindophenol (DCIP) and 2,6,3′-trichlorophenolindophenol (TCIP) are useful as electron acceptors with isolated chloroplasts. However, in their oxidized forms they are uncouplers. The transport of electrons to DCIP is very rapid and independent of phosphorylation when the concentration of this acceptor is

[77]T. Horio, K. Nishikawa, M. Katsumata, and J. Yamanaka, $Biochim.$ $Biophys.$ $Acta$ **94,** 371 (1965).
[78]A. T. Jagendorf and M. Smith, $Plant$ $Physiol.$ **37,** 135 (1962).
[79]M. Avron, $Biochim.$ $Biophys.$ $Acta$ **77,** 699 (1963).
[80]V. K. Vambutas and E. Racker, $J.$ $Biol.$ $Chem.$ **240,** 2660 (1965).
[81]R. A. Dilley and J. S. Platt, $Biochemistry$ **7,** 338 (1968).
[82]C. C. Black, Jr., $Biochim.$ $Biophys.$ $Acta$ **162,** 294 (1968).

greater than 5×10^{-5} M, but at lower concentrations the transport is partially coupled.[83] It is more probable that indophenols have a dual role, functioning independently as electron acceptors and uncouplers, than that they uncouple because they accept electrons before the phosphorylation site, thereby bypassing phosphorylation.

Sonication, Detergent Treatment, etc. Disruption or damaging of lamellar membranes by physical or chemical methods results in partial to complete uncoupling.[84] It has been suggested that these methods can cause "site specific" uncoupling in spinach chloroplasts (n-heptane treatment),[85] in *Anabaena* particles (sonication)[86] and in *R. rubrum* chromatophores (Triton X-100 treatment).[45]

Transport-Inducing Antibiotics ("Ionophores").[87] This group of antibiotics deserves special consideration because of their well-characterized function as ion carriers and their consequent importance as tools for the study of the role of ion transport in uncoupling. They can be classified as belonging to two types.[87,88] The *valinomycin type* (valinomycin, gramicidin D, nonactin, monactin, etc.)[89] forms lipophilic charged complexes with alkali ions and greatly stimulate the active (energy-linked) uptake of K^+ in mitochondria. All these antibiotics can uncouple mitochondria in the presence of specific alkali ions (see, however, Harris *et al.*[90]). Substances of the *nigericin type* (nigericin, dianemycin, monensin, etc.)[89] carry alkali ions as lipophilic zwitterionic complexes and hydrogen ion in protonated carboxyl groups. These antibiotics abolish the active K^+ uptake, release accumulated K^+ from membranes, and inhibit oxidation of certain substrates in mitochondria by K^+ depletion.

Only those antibiotics capable of carrying both alkali ions and H^+ (nigericin type and gramicidin D) are strong uncouplers of chloroplasts (Table III). However, in the presence of K^+, combinations of valinomy-

[83]Z. Gromet-Elhanan and M. Avron, *Biochem. Biophys. Res. Commun.* 10, 215 (1963); D. L. Keister, *J. Biol. Chem.* 238, PC2590 (1963).

[84]S. Izawa and N. E. Good, *Biochim. Biophys. Acta* 109, 372 (1965).

[85]L. J. Laber and C. C. Black, Jr., *J. Biol. Chem.* 244, 3463 (1969).

[86]S. S. Lee, M. Young, and D. W. Krogmann, *Biochim. Biophys. Acta* 180, 130 (1969).

[87]B. C. Pressman, E. J. Harris, W. S. Jagger, and J. H. Johnson, *Proc. Nat. Acad. Sci. U. S.* 58, 1949 (1967); B. C. Pressman, *Fed. Proc., Fed. Amer. Soc. Exp. Biol.* 27, 1283 (1968).

[88]H. A. Lardy, S. N. Graven, and E. Estrada-O, *Fed. Proc., Fed. Amer. Soc. Exp. Biol.* 26, 1355 (1967).

[89]*Valinomycin*, a cyclic depsipeptide (MW = 1110); *nonactin* and *monactin*, two of the four serial macrotetralide 'nactins (MW = 664 and 720, respectively); *gramicidin D*, N-formyl-pentadecapeptide ethanolamine (MW ≅ 1880), a mixture of six slightly different compounds (gramicidin S is a different compound and inactive as ionophore); *nigericin, dianemycin, monensin* are monocarboxylic cyclic polyethers (MW = 736, 956, and 670, respectively). All these antibiotics are applied as alcohol solutions.

[90]E. J. Harris, M. P. Höfer, and B. C. Pressman, *Biochemistry* 6, 1348 (1967).

TABLE III

Ion Transport-Inducing Antibiotics and Their Effects on Photophosphorylation and Light-Induced pH Rise in Chloroplasts and Bacterial Chromatophores

Antibiotic	Alkali ion specificity[a]	Ability[a] to carry H^+	Effect[b]			
			Chloroplasts		Chromatophores	
			Photophosphorylation	pH rise in medium	Photophosphorylation	pH rise in medium
Valinomycin	K^+, Rb^+, Cs^+ >> NH_4^+, Na^+, Li^+	No	Little effect[c,d]	Little effect[c,d]	Little effect[e]; partially inhibits[f-h]	Enhances rate, extent[e-v,i]
Nonactin (+ monactin)	K^+, Rb^+, Cs^+ > NH_4^+, Na^+ >> Li^+	No	Little effect[c]	Little effect[c]	Little effect[v]	Enhances rate extent[v]
Gramicidin D	K^+, Rb^+, Cs^+, NH_4^+, Na^+, Li^+	Limited	Inhibits[j,k] (uncouples)	Suppresses[l]	Inhibits[m]	Suppresses extent[f,i]
Nigericin	K^+, Rb^+ > Na^+ > Cs^+ > Li^+	Yes	Uncouples strongly[l]	Suppresses[l]	Little effect[e-v]	Suppresses[e-v] rate[f], extent[f,v]
Dianemycin	Na^+, K^+, Rb^+, Li^+, Cs^+	Yes	Uncouples[n]	—	Little effect[e-v]	Suppresses[e-v]; stimulates at low conc.[e]
Monensin	Na^+ > K^+, Li^+ > Rb^+ > Cs^+	Yes	—	—	Little effect[e,v]	Suppresses[e,v] extent[v]

[a] After P. J. F. Henderson, J. D. McGivan, and J. B. Chappel, *Biochem. J.* 111, 521 (1969).

[b] The concentration of antibiotics used, approximately 10^{-8} to 10^{-6} M; K[+], 0.02–0.1 M. In all the cases tested (refs. f,l) a K[+] efflux and H[+] influx was observed on addition of antibiotics to chloroplasts or chromatophores suspended in low K[+] media (in the dark).

[c] S. J. D. Karlish, N. Shavit, and M. Avron, *Eur. J. Biochem.* 9, 291 (1969).

[d] R. E. McCarty, *J. Biol. Chem.* 244, 4292 (1969).

[e] M. Nishimura and B. C. Pressman, *Biochemistry* 8, 1360 (1969).

[f] J. B. Jackson, A. R. Crofts, and L.-V. von Stedingk, *Eur. J. Biochem.* 6, 43 (1968).

[g] A. Thore, D. L. Keister, N. Shavit, and A. San Pietro, *Biochemistry* 7, 3499 (1968).

[h] H. Baltscheffsky and B. Arwidsson, *Biochim. Biophys. Acta* 65, 425 (1962). These authors observed "50% inhibition" of photophosphorylation and postulated that one of the two phosphorylation sites is valinomycin sensitive. The "50% inhibition" has not been confirmed (refs. f,g). The partial inhibition was shown to be K[+] independent (ref. f).

[i] L.-V. von Stedingk and H. Baltscheffsky, *Arch. Biochem. Biophys.* 117, 400 (1966).

[j] F. R. Whatley, M. B. Allen, and D. I. Arnon, *Biochim. Biophys. Acta* 32, 32 (1959).

[k] H. Baltscheffsky, *Acta Chem. Scand.* 14, 264 (1960). In these experiments (refs. j,k) electron flow was not measured. Cation dependency of uncoupling not known.

[l] N. Shavit, R. A. Dilley, and A. San Pietro, *Biochemistry* 7, 2356 (1968). Uncoupling and inhibition of H[+] change were absolutely K[+] dependent.

[m] H. Baltscheffsky and M. Baltscheffsky, *Acta Chem. Scand.* 14, 257 (1960).

[n] P. Plengvidhaya and R. H. Burris, *Plant Physiol.* 40, 997 (1965). Cation dependency of uncoupling not tested.

cin or nonactin (which induce K^+ permeability but not H^+ permeability) with 2,4-dinitrophenol or very low FCCP (which induce H^+ permeability but not K^+ permeability) can produce uncoupling in chloroplasts relatively insensitive to the same concentrations of these substances when used singly.[91] The same combinations abolish the electron transport-associated pH rise of the medium. Valinomycin alone enhances ammonia uncoupling, but not methylamine uncoupling, in the amine-insensitive subchloroplast particles.[62] See ammonia uncoupling above.

In *R. rubrum* chromatophores gramicidin strongly inhibits photophosphorylation but none of the other ionophores is very effective. Nevertheless, most of them have marked effects, either stimulatory or suppressive, on the pH change in the medium (see Table III). Some authors have found synergistic effects of the valinomycin and nigericin types antibiotics,[92,93] but some have not.[94] Valinomycin plus CCCP show no synergism.[93]

Miscellaneous. Long-chain fatty acids such as *linolenic* are strongly uncoupling (10^{-6} M) as well as inhibitory (10^{-5} M) in chloroplasts.[95] Many common *anions* such as citrate, phosphate, or sulfate can uncouple weakly or strongly depending on other ions present and on prior treatment of the chloroplast preparation.[96] This is an insidious effect which must be guarded against by careful attention to media used and method of chloroplast preparation. *Octylguanidine* (10^{-4} to 10^{-3} M) and *gramicidin S* (10^{-6} to 10^{-5} M) also uncouple chloroplasts.[32] The uncoupling potencies of *2-alkylbenzimidazoles* depend critically on the length of the alkyl side chain; compounds with C_8 to C_{14} side chains uncouple strongly at 10^{-5} M.[97] Finally, it should be reemphasized that whenever *amino compounds* are tested they are usually found to be uncouplers. The major exceptions are aromatic amines (which are always very weak bases) and highly polar amines, such as amino acids and amines with many hydroxyl groups.

Energy Transfer Inhibition in Chloroplasts and Bacterial Chromatophores

Definition. During respiration and photosynthetic electron transport a portion of the free energy of the oxidoreduction reactions is transferred

[91]S. J. D. Karlish, N. Shavit, and M. Avron, *Eur. J. Biochem.* **9**, 291 (1969).

[92]J. B. Jackson, A. R. Crofts, and L.-V. von Stedingk, *Eur. J. Biochem.* **6**, 43 (1968).

[93]A. Thore, D. L. Keister, N. Shavit, and A. San Pietro, *Biochemistry* **7**, 3499 (1968).

[94]M. Nishimura and B. C. Pressman, *Biochemistry* **8**, 1360 (1969).

[95]R. E. McCarty and A. T. Jagendorf, *Plant Physiol.* **40**, 725 (1965).

[96]N. E. Good, *Arch. Biochem. Biophys.* **96**, 653 (1962).

[97]A. Trebst, E. Pistorius, and K. H. Büchel, *Z. Naturforsch. B* **21**, 667 (1966).

through a coupling mechanism to phosphorylate ADP. Substances which inhibit this capture of energy in ATP by inhibiting any of the unknown steps in the phosphorylation reaction are known as energy transfer inhibitors or simply phosphorylation inhibitors. Such inhibitors prevent both phosphorylation and phosphorylating electron transport, but they do not affect nonphosphorylating electron transport. Thus any inhibition of electron transport caused by an energy transfer inhibitor is relieved by uncouplers which act on reaction steps preceding the inhibition site. See Fig. 1c.

Dio-9. This is an antibiotic of undetermined structure which is known to inhibit oxidative phosphorylation. Administered in alcohol solution (final concentration, 1–2 μg/ml), it inhibits phosphorylation and phosphorylating electron transport in chloroplasts, but the inhibition of electron transport is overcome by uncouplers.[98] It also inhibits the Mg^{2+}-dependent and Ca^{2+}-dependent ATPase and the ATP–P_i exchange activity of chloroplasts in the dark. See Table I.

Dio-9 at somewhat higher concentrations (10–20 μg/ml) inhibits photophosphorylation in *R. rubrum* chromatophores,[99,100] but its effect on energy-linked reactions are different from those of oligomycin, resembling rather the effects of uncouplers.[99] Dio-9 stimulates the basal (dark) ATPase as uncouplers do, but it also inhibits the FCCP-stimulated ATPase reaction.[100]

Phlorizin. This phloretin-2′-β-glucoside (MW 436) is a well known inhibitor of glucose transport across animal cell membranes. At 4×10^{-4} M it reversibly inhibits photophosphorylation in chloroplasts (cyclic or noncyclic) by about 50%, but at low phosphate concentrations the inhibition is appreciably stronger. Uncoupled electron transport is unaffected.[27] The synthetic analogs 4′-deoxyphlorizin and 4′-deoxyphloretin-2′-galactoside are equally specific inhibitors of phosphorylation but are almost 10 times more potent than phlorizin.[101] The aglucon phloretin is a potent but unspecific inhibitor of all kinds of electron transport. Phlorizin is only sparingly soluble in water. A stock solution 10^{-2} M in alkaline buffer (pH 8.2) is recommended, but this aqueous solution is not very stable. A concentrated solution in methanol:ethylene glycol (1:1) can also be used but the final concentration of the organic solvent must be kept below 2% if the specificity of the phlorizin effect is to be maintained.

Phlorizin (2×10^{-3} M) inhibits 50% of the photophosphorylation in *R. rubrum* chromatophores.[99]

[98] R. E. McCarty, R. J. Guillory, and E. Racker, *J. Biol. Chem.* 240, PC4822 (1965).
[99] Z. Gromet-Elhanan, *Arch. Biochem. Biophys.* 131, 299 (1969).
[100] R. R. Fisher and R. J. Guillory, *Biochim. Biophys. Acta* 143, 654 (1967).
[101] G. D. Winget, S. Izawa, and N. E. Good, *Biochemistry* 8, 2067 (1969).

Synthalin (decamethylenediguanidine; see Vol. X [8]). This compound inhibits noncyclic photophosphorylation in chloroplasts at concentrations between 10^{-4} and 10^{-3} M. Cyclic photophosphorylation mediated by PMS is slightly less sensitive. The inhibition of electron transport is relieved by uncouplers, including nigericin and gramicidin D.[102]

Mercurials. p-Chloromercuribenzoate (PCMB) at concentrations between 10^{-5} and 3×10^{-4} M acts as a typical energy transfer inhibitor in the ferricyanide-reducing system.[63,103] Other mercurials such as p-mercuribenzenesulfonate and phenylmercuric acetate exhibit similar inhibition characteristics but sometimes show slight uncoupling activity. The inhibition by PCMB is relieved by cysteine in the FMN-reducing system.[103] None of these inhibitors can be used in NADP reducing systems or any other systems involving ferredoxin since this iron protein is destroyed by mercurials.

Phenylmercuric acetate at 10^{-4} M suppresses photophosphorylation in *R. rubrum* chromatophores.[40] The dark ATPase activity is sensitive to either PCMB or phenylmercuric acetate at 10^{-4} M.[48]

Antibody against Coupling Factor. An antibody against chloroplast coupling factor has been prepared from mice immunized against the isolated protein. The serum and the purified antibody inhibit the ATPases, the ATP–P_i exchange reaction, photophosphorylation and "acid-bath" phosphorylation in chloroplasts but do not inhibit the electron transport-associated pH change in the medium.[104] Techniques for preparing similar antisera and purifying the active γ-globulins have been described in this series.[105]

N,N'-Dicyclohexylcarbodiimide (DCCD). This compound inhibits oxidative phosphorylation in mitochondria much as does oligomycin. It inhibits photophosphorylation and phosphorylating electron transport in chloroplasts at 10^{-5} M. The inhibition of electron transport is only partially relieved by ammonia and atebrin, which suggests that the inhibitor may have secondary effects on the electron transport chain. In chloroplasts partially uncoupled by EDTA, low concentrations of DCCD can stimulate residual phosphorylation and largely restore the pH rise in the medium. These observations have been taken to indicate that low

---

[102]E. Gross, N. Shavit, and A. San Pietro, *Arch. Biochem. Biophys.* **127,** 224 (1968).

[103]S. Izawa and N. E. Good, *in* "Progress in Photosynthesis Research" (H. Metzner, ed.), Vol. III, p. 1288. International Union of Biological Sciences, Tübingen, Germany, 1969.

[104]R. E. McCarty and E. Racker, *Brookhaven Symp. Biol.* **19,** 202 (1966).

[105]J. M. Fessenden and E. Racker, Vol. X [107].

DCCD prevents the breakdown of high-energy intermediates while high concentrations inhibit ATP formation.[106]

Oligomycin. This inhibitor of energy transfer reactions in mitochondria (Vol. X [8]) has no effect on photophosphorylation in chloroplasts at concentrations up to 10 μg/ml. Higher concentrations are weakly uncoupling.[32]

Photophosphorylation[73] and oxidative phosphorylation[68] in *R. rubrum* chromatophores are inhibited by oligomycin at 3 μg/ml as in mitochondrial oxidative phosphorylation. The dark ATPase activity, the ATP–P$_i$ exchange reaction and other reactions of chromatophores involving ATP are also sensitive to oligomycin. On the other hand, photosynthetic pyrophosphate formation,[107] pyrophosphatase activity, and other reactions involving pyrophosphate are either stimulated or unaffected by oligomycin (Table II). These observations locate the site of oligomycin inhibition near the terminal step in ATP synthesis.

Miscellaneous. At very low concentrations (10^{-7} M) *chlorotri-n-butyltin* inhibits phosphorylating electron transport and the basal, nonphosphorylating electron transport, but not amine-uncoupled electron transport in chloroplasts. The ATPases are not affected.[108] Further clarification of the role of this inhibitor is needed. *4,5,6,7-Tetrabromo-2-trifluoromethylbenzimidazole* (TFTB) inhibits transport of electrons from water (the Hill reaction) whether or not the transport is coupled to phosphorylation. The transport of electrons from reduced indophenols (DCIPH$_2$) to NADP is not inhibited, but the concurrent phosphorylation is abolished.[109] *Phenol, o-cresol,*[110] and other phenolic compounds, such as *salicylaldoxime*[110,111] and certain *chloroacylanilides,*[20] show similar effects. These compounds (TFTB, etc.) may represent a special type of energy transfer inhibitors acting very close to the level of electron transport.[55b] Alternatively they may represent examples of numerous deleterious compounds unselectively affecting two major weak points of the lamella-bound photosynthetic apparatus—photosystem II and the coupling mechanism.

---

[106]R. E. McCarty and E. Racker, *J. Biol. Chem.* 242, 3435 (1967); see also Table I.
[107]H. Baltscheffsky and L.-V. von Stedingk. *Biochem. Biophys. Res. Commun.* 22, 722 (1966).
[108]J. S. Kahn, *Biochim. Biophys. Acta* 153, 203 (1968).
[109]K. H. Büchel, W. Draber, A. Trebst, and E. Pistorius, *Z. Naturforsch. B* 21, 243 (1966).
[110]J. Neumann and Z. Drechsler, *Plant Physiol.* 42, 573 (1967).
[111]A. Trebst, *Z. Naturforsch. B* 18, 817 (1963).

Section III
Synthesizing Capabilities of the Photosynthetic Apparatus

[33] Synthesizing Capability of the Photosynthetic Apparatus: Proteins

By Robert M. Smillie

In photosynthetic cells, the chloroplasts are an important site of protein synthesis, and in some leaves they may constitute the major site of cellular protein synthesis.[1] The machinery for protein synthesis in chloroplasts differs in several respects from that for the synthesis of proteins in the cytoplasm. Chloroplast ribosomes are smaller (70 S) than cytoplasmic ribosomes (80 S), and their RNA components are 16 S and 23 S, whereas the RNA components of cytoplasmic ribosomes are usually 18 S and 25 S.[2] The presence of polyribosomes in chloroplasts[3,4] implicates the existence of messenger RNA in chloroplasts. Distinctive aminoacyl-sRNA synthetases and transfer RNA have been found in mitochondria,[5,6] and it is likely the same is true for chloroplasts. Some evidence for the existence of aminoacyl-sRNA synthetases[7-13] and transfer RNA[12,14] in chloroplasts has been obtained. Although some components of the protein-synthesizing system of chloroplasts may possibly be nuclear or cytoplasmic in origin, chloroplasts after isolation retain the capacity to carry out protein synthesis for a limited period of time. The function of chloroplast DNA in the synthesis of chloroplast proteins is not clear except that is very likely directs the synthesis of the RNA of chloroplast ribosomes.[15-17]

[1] M. Zucker and H. T. Stinson, Jr., *Arch. Biochem. Biophys.* **96**, 637 (1962).

[2] R. M. Smillie and N. S. Scott, *in* "Progress in Molecular and Subcellular Biology" (F. E. Hahn, ed.), Vol. I, p. 136. Springer-Verlag, Berlin, 1969.

[3] M. F. Clark, R. E. F. Matthews, and R. K. Ralph, *Biochim. Biophys. Acta* **91**, 289 (1964).

[4] J. L. Chen and S. G. Wildman, *Science* **155**, 1271 (1967).

[5] W. E. Barnett and D. H. Brown, *Proc. Nat. Acad. Sci. U. S.* **57**, 452 (1967).

[6] W. E. Barnett, D. H. Brown, and J. L. Epler, *Proc. Nat. Acad. Sci. U. S.* **57**, 1775 (1967).

[7] J. M. Clark, Jr., *J. Biol. Chem.* **233**, 421 (1958).

[8] J. Bové and I. D. Raacke, *Arch. Biochem.* **85**, 521 (1959).

[9] A. Marcus, *J. Biol. Chem.* **234**, 1238 (1959).

[10] J. D. Henshall and T. W. Goodwin, *Phytochemistry* **3**, 677 (1964).

[11] R. I. B. Francki, N. K. Boardman, and S. G. Wildman, *Biochemistry* **4**, 865 (1965).

[12] N. M. Sissakian, I. I. Filippovich, E. N. Svetailo, and K. A. Aliyev, *Biochim. Biophys. Acta* **95**, 474 (1965).

[13] D. Spencer, and S. G. Wildman, *Biochemistry* **3**, 954 (1964).

[14] T. A. Dyer and R. M. Leech, *Biochem. J.* **106**, 689 (1968).

[15] N. S. Scott and R. M. Smillie, *Biochem. Biophys. Res. Commun.* **28**, 598 (1967).

[16] K. K. Tewari and S. G. Wildman, *Proc. Nat. Acad. Sci. U. S.* **59**, 569 (1968).

[17] S. P. Gibbs, *Biochem. Biophys. Res. Commun.* **28**, 653 (1967).

Methods for Isolation of Plastids for
Studies of Protein Synthesis

Isolation of Chloroplasts

A buffered solution of sucrose containing $MgCl_2$ and, usually, mercaptoethanol is the most common medium employed for the isolation of chloroplasts for studies of protein synthesis. Various high molecular weight carbohydrates or proteins are frequently included in the isolation medium, as these assist in preserving the integrity of the chloroplasts. The use of the French pressure cell,[18] sonication, and various homogenization techniques, in some cases following partial digestion with proteolytic enzymes, may be used to release chloroplasts from algal cells (see Vol. 23 [20]). Various grinding procedures may be used for leaves, but the method which appears to result in the least damage to chloroplasts is the chopping technique devised by Spencer and Wildman.[13] It is given below[13,19] and has been employed successfully by several different workers to isolate chloroplasts from the leaves of plants including tobacco, spinach, and bean. The extraction medium was developed by Dr. S. Honda and includes Ficoll and Dextran in addition to sucrose.

> *Extraction Medium (Honda Medium)*
> Ficoll, 2.5%
> Dextran-40, 5.0%
> Sucrose, 250 mM
> Tris buffer, 25 mM pH 7.8
> $MgCl_2$, 1 mM
> Mercaptoethanol, 4 mM

Procedure. Tobacco or spinach leaves aré washed, the midrib and large veins are removed, and the laminae are chilled in ice. The chilled leaf material is mixed with Honda medium (2:1, w/v) in a flat-bottomed, shallow polythene dish and chopped to a fine mince with a sharp razor blade. The extract is filtered through several layers of fine cloth or Miracloth, and the filtrate is centrifuged at 1000 *g* for 5 minutes. The chloroplast pellet is suspended in a volume of Honda medium or medium consisting of 100 mM Tris, pH 7.8, 10 mM $MgCl_2$, and 4 mM mercaptoethanol equivalent to half the original weight of leaves. In some studies this fraction has been used directly for studies of amino acid incorporation.[13,19] Alternatively the chloroplasts may be washed

[18]H. W. Milner, N. S. Lawrence, and C. S. French, *Science* 111, 633 (1950).
[19]D. Spencer, *Arch. Biochem. Biophys.* 111, 381 (1965).

several times by centrifugation at 1000 g and resuspension in Tris–Mg–mercaptoethanol medium.

Comment. The chloroplast suspension will contain some remnants of cell structures and a few nuclei. Many of these can be removed by incorporating several centrifugations at very low speeds (40–100 g).[20] Some of the larger mitochondria and possibly peroxisomes are also likely to be present in the chloroplast preparation.

Isolation of Etioplasts

Most studies of protein synthesis by isolated plastids have been done using chloroplasts. Etioplasts (plastids from etiolated leaves) have not been employed partly because of the lack of suitable isolation procedures. Jacobson[21] has now described a procedure which yields maize etioplasts in a good state of preservation. An important component of the extraction medium is bovine serum albumin.

Extraction Medium
  Tris·HCl, 0.5 M, pH 8.0
  Sucrose, 0.5 M
  $MgCl_2$, 1 mM
  Bovine serum albumin, 0.2% (w/v)

Procedure. Leaves of etiolated maize plants (8–10 days old) are chilled, minced, and ground with a chilled pestle and mortar. The homogenate is filtered through two layers of Miracloth, and the filtrate is centrifuged at 40 g for 5 minutes. The resulting supernatant fluid is centrifuged at 370 g for 15 minutes. The pellet of etioplasts is washed once by resuspending it in extraction medium (10 ml) and then is layered onto a discontinuous sucrose gradient. This is prepared by layering 10 ml of 1.3 M sucrose over 25 ml of 2 M sucrose. Both solutions of sucrose also contain 0.5 M Tris·HCl, pH 8.0, 1 mM $MgCl_2$, and 0.2% bovine serum albumin. The gradients and the sample are centrifuged for 30 minutes at 2500 g. The plastids accumulate at the interface between the 1.3 M and 2 M sucrose layers and can be removed with a syringe.

Comment on the Media Used for Isolation of Chloroplasts and Etioplasts and Application of Isolation Procedures to Other Plants

The addition of bovine serum albumin to extraction media containing buffered sucrose has been found to be beneficial for isolating

---

[20] R. M. Smillie, *Aust. J. Biol. Sci.* **9**, 347 (1956).
[21] A. B. Jacobson, *J. Cell Biol.* **38**, 238 (1968).

plastids from both green and dark-grown cells of *Euglena gracilis* and from leaves of pea and wheat seedlings.[22] Partially purified albumin (e.g., fraction V, bacteriological, Nutritional Biochemicals Corporation, Cleveland, Ohio) containing fatty acid impurities should be used. Since the 70 S ribosomes of some plastids break down at concentrations of Mg below 1 mM,[23-25] higher concentrations of Mg (15 mM) and the inclusion of KCl (100 mM) in the extraction medium[26] are recommended provided the high salt concentration does not cause coagulation of the cytoplasmic ribosomes. However, it has been pointed out that the requirement for Mg can vary widely with leaves from different plants,[27] and the optimal Mg concentration should be ascertained for each new plant investigated. The leaves of some plants yield acidic extracts when ground with water and higher concentrations of buffer than is used in the extraction medium for tobacco and spinach leaves may be required. Substitution of HEPES buffer (*N*-2-hydroxyethylpiperazine-*N'*-2-ethanesulfonic acid) for Tris could also be an improvement.

In general, for the isolation of active chloroplasts and for maximum preservation of polyribosomes within chloroplasts, it would seem to be desirable (1) to use Honda media or a medium containing bovine serum albumin; (2) to wash the chloroplasts until they are free of the cell supernatant fluid; (3) to maintain the temperature of the extract until the final wash at very close to 0°; and (4) to keep the time taken between homogenization of the leaves and the final wash to a minimum.

Preparation of Ribosomes from Chloroplasts

Ribosomes isolated from chloroplasts can incorporate amino acids into protein. The following method has been used to isolate chloroplast ribosomes from tobacco leaves.[28] Chloroplasts are first isolated in Honda media as described above, and the unwashed chloroplasts are suspended in 10 mM Tris (pH 7.8), 3 mM magnesium acetate, and 3 mM mercaptoethanol. The suspension is centrifuged at 105,000 g for 60 minutes. The pellet is resuspended in more of the same medium and clarified by centrifugation at 17,000 g for 15 minutes. The ribosomes are washed by repeating the cycle of suspension in buffer, centrifuga-

[22]B. J. Reger, R. M. Smillie, and R. C. Fuller, *Plant Physiol.* In press (1972).
[23]N. K. Boardman, R. I. B. Francki, and S. G. Wildman, *Biochemistry* **4**, 872 (1965).
[24]R. Sager and M. G. Hamilton, *Science* **157**, 709 (1967).
[25]J. W. Lyttleton, *Biochim. Biophys. Acta* **154**, 145 (1968).
[26]J. R. Rawson and E. Stutz, *Plant Physiol.*, Suppl. **43**, S-18, (1968), and private communication.
[27]M. Ranalletti, A. Gnanam, and A. T. Jagendorf, *Biochim. Biophys. Acta* **186**, 192 (1969).
[28]N. K. Boardman, R. I. B. Francki, and S. G. Wildman, *J. Mol. Biol.* **17**, 470 (1966).

tion at 105,000 g for 60 minutes, resuspension, and clarification by centrifugation at 17,000 g for 15 minutes. Similar procedures have been used to prepare chloroplast ribosomes from pea plants[12] and *Euglena gracilis.*[29] The requirements of chloroplast ribosomes for the incorporation of amino acids into protein are similar to those of the isolated chloroplasts, except that a source of aminoacyl-sRNA synthetase and transfer RNA is required.[12,28,29] In the case of tobacco, the 105,000 g supernatant fluid obtained after treating intact chloroplasts with buffer solution lacking sucrose will suffice.[19,28]

Preparation of Polyribosomes from Chloroplasts

Unless special precautions are taken during the isolation of chloroplasts, ribosomes released from chloroplasts will be in the form of the 70 S monomers or even the 30 S and 50 S subunits of the monomer. Clark *et al.*[3] added polyvinyl sulfate (10 mg/ml) to the extraction medium to inhibit ribonuclease, and they succeeded in isolating chloroplast polyribosomes. However, this compound also inhibits the incorporation of amino acids into protein.[28] Chen and Wildman[4] noted that it is important to wash chloroplasts in order to retain significant numbers of polyribosomes. Incubation of washed chloroplasts with a reaction mixture for amino acid incorporation resulted in the release of polyribosomes and 70 S ribosomes from the chloroplasts. After removal of the chloroplasts by centrifugation, polyribosomes in the supernatant fluid could be separated from the 70 S monomers by centrifugation in a density gradient of sucrose. On an RNA basis, polyribosomes were more active than the monomers in incorporating amino acids into protein.

Prolonged contact of preparations of chloroplast ribosomes with cell supernatant fluid is known to inhibit their activity for protein synthesis in the case of peas[12] and tobacco,[28] and partial degradation of the polyribosomal clusters by cytoplasmic ribonuclease may be one of the contributing factors. Where interference by cytoplasmic ribonuclease is suspected, it may be advantageous to include an excess of soluble RNA in the extraction medium.

Measurement of Amino Acid Incorporation by Isolated Chloroplasts

The following procedure is for chloroplasts isolated in an extraction medium consisting of 10 mM HEPES buffer, pH 8.0, 0.33 M sucrose, 1 mM dithiothreitol, 15 mM $MgCl_2$, 100 mM KCl, and 0.2% (w/v)

---

[29]J. M. Eisenstadt and G. Brawerman, *J. Mol. Biol.* **10**, 392 (1964).

bovine serum albumin.[22] The same concentrations of these reagents are retained in the reaction mixture. A similar procedure can be followed for chloroplasts isolated in Honda medium, provided the concentration of $MgCl_2$ in the reaction mixture is increased to 15 mM.

Reagents
    Extraction medium, double strength
    ATP, 40 mM
    GTP, 2.0 mM
    Phosphoenolpyruvate, 250 mM
    Pyruvate kinase, 2 mg/ml
    Mixture of amino acids (omitting the amino acid to be added in the labeled form), each amino acid 0.5 mM.

Procedure. Extraction medium (50 μl, double strength), 25 μl of the mixture of amino acids, and 5 μl of each of the other reagents are combined; 0.4 ml of a suspension of chloroplasts (containing 50–150 μg of chlorophyll) and 5 μl of [^{14}C]-valine or other suitable amino acid (approximately 1 mmole, specific activity, 100–300 mCi/mmole) are added to give a final volume of 0.5 ml. The reaction mixture is incubated at 25°, and samples (e.g., 0.05 ml) are taken at intervals up to 40–60 minutes. The radioactivity incorporated into protein is determined by one of the methods described below. The rate of incorporation should be calculated from the linear portion of the curve. In practice, the difference between the radioactivity of the 5-minute and 15-minute samples usually gives a satisfactory measurement of the incorporation. This method of calculation corrects for nonspecific adsorption of labeled amino acid.

Determination of Activity Incorporated into Protein of Isolated Chloroplasts

Trichloroacetic acid is added to stop the reaction and precipitate the protein. The precipitated protein is extracted with a series of solvents to remove traces of free ^{14}C-labeled amino acid and any activity incorporated into lipid[30] and aminoacyl-sRNA. The extraction can be accomplished by repeatedly centrifuging the protein and resuspending the pellet in fresh medium, but this procedure is tedious and, because of the small amount of protein handled, losses of protein may occur. In the method given below, the washing is carried out using Millipore filters. Where a large number of samples are to be processed, the second

---

[30]B. P. Smirnov and M. A. Rodionova, *Biokhimiya* **29**, 386 (1964).

method given, in which samples of the reaction mixture are acid-precipitated onto paper disks and extractions are carried out by immersing the disks in the extraction medium, is the preferred method. It also incorporates an extraction procedure for the removal of lipids.

Measurement of Incorporation of ^{14}C-Labeled Amino Acid into Protein Employing Extraction on Millipore Filters — Method of Spencer and Wildman[13]

Reagents
[^{12}C]-Valine, 0.1 M
Trichloroacetic acid, 5%
Trichloroacetic acid, 5%, containing 0.05% [^{12}C]-valine

Procedure. The sample of reaction mixture (0.05 or 0.1 ml) is rapidly chilled, and 0.5 ml of 0.1 M [^{12}C]-valine is added followed by 5 ml of 5% trichloroacetic acid. The mixture is allowed to stand at 0° for 15 minutes, then is centrifuged at 12,000 g for 10 minutes. The clear supernatant fluid is withdrawn and discarded. The pellet is frozen in solid CO_2 for 15 minutes and thawed. This step helps the precipitate to coagulate, thus speeding up the subsequent filtrations. The precipitate is suspended in 5 ml of 5% trichloroacetic acid containing 0.05% [^{12}C]-valine and heated to 80° for 15 minutes. The precipitate is transferred quantitatively to a 1.2-μ Millipore filter and washed 5 times on the filter with 5 ml of the 5% trichloroacetic acid containing valine. The filter is dried in a stream of air and cemented to a planchet, and radioactivity is determined.

Measurement of Incorporation of ^{14}C-Labeled Amino Acid into Protein Using a Filter-Paper Disk Procedure — Method of Mans and Novelli[31]

Reagents
Trichloroacetic acid, 10% containing 0.1 M [^{12}C]-valine
Trichloroacetic acid, 5%
Ethyl ether–ethanol, 1:1, v:v
Ethyl ether

Procedure. Whatman filter paper circles (2.3 cm in diameter, No. 3 MM paper) are numbered in pencil and mounted on straight steel pins. (Steel bank pins for heavy-weight material, No. 17 729P, obtainable from Scovill, Oakville Division, Oakville Connecticut.) The paper disks should be used upside down to take advantage of the slight lip. A sample

[31]R. J. Mans and G. D. Novelli, *Arch. Biochem. Biophys.* **94,** 48 (1961).

of reaction mixture (0.05 or 0.1 ml) is spread onto a disk and exposed to a warm stream of air for 15 seconds. This facilitates absorption of the sample into the matrix of the paper. The disk is immersed immediately in 10% trichloroacetic acid containing 0.1 M [^{12}C]-valine at 0° for at least 60 minutes. About 150 ml of solution is convenient for 20 disks. The disks can be stored at this stage in the trichloroacetic acid for at least 4 days. The disks are then washed and extracted as follows. Between each extraction, the disks are collected by filtering the extract through a fluted paper filter, or, more conveniently, they can be kept in a glass beaker containing drainage holes. The beaker is immersed in turn in the various extraction solutions: 5% trichloroacetic acid, 15 minutes at room temperature; 5% trichloroacetic acid, 30 minutes at 90°; 5% trichloroacetic acid, 15 minutes at room temperature; ether–ethanol, at least 30 minutes at 37°; ether–ethanol, 15 minutes at 37°; ethyl ether, 15 minutes at room temperature; ethyl ether, 5–10 minutes at room temperature.

The pins are removed, and the disks are dried in air and placed to lie flat on the bottom of a glass counting vial. Five milliliters of scintillation mixture is added [0.4% 2,5-diphenyloxazole, 0.01% 1,4-bis-2(5-phenyloxazolyl)benzene in toluene], and the radioactivity is counted using a liquid scintillation spectrometer. The method is suitable for measuring the incorporation into protein of both ^{14}C- and ^{3}H-labeled amino acids.

Determination of Aminoacyl-sRNA Formation

The filter paper disk method can also be used to assay for the formation of aminoacyl-sRNA by isolated chloroplasts. Two separate disks are employed, and the procedure described above is followed except that in the case of one of the disks the treatment with 5% trichloroacetic acid at 90° is omitted. The difference between the counts remaining on the two disks indicates the extent of aminoacyl-sRNA formation. For the assay of aminoacyl-sRNA synthetases extracted from chloroplasts, a similar procedure using paper filter disks may be used.[32,33]

Interference by Contaminating Bacteria

Possible contamination of preparations of chloroplasts with bacteria has long been a problem in studies on the incorporation of amino acids into protein by isolated chloroplasts. In a number of published studies adequate precautions against bacterial contamination do not appear to have been taken, and consequently the results are difficult to interpret.

[32]F. J. Bollum, *J. Biol. Chem.* **234**, 2733 (1959).
[33]W. E. Barnett and K. B. Jacobson, *Proc. Nat. Acad. Sci. U. S.* **51**, 642 (1964).

The starting material itself is usually the main source of contamination. *Euglena* and many algae can be grown axenically, although one must still guard against the introduction of bacteria during preparation of the chloroplasts. For instance, in our experiments a persistent bacterial contamination in preparations of chloroplasts from *Euglena* was traced to incomplete washing of the French pressure cell used to break the cells. Although the cell was rinsed with distilled water after each use, small amounts of the isolation medium remained trapped under the O rings and subsequently supported bacterial growth. It became necessary to remove the O rings to ensure thorough washing.

Some plants can be cultivated under aseptic conditions, and this has been done in studies of protein synthesis in chloroplasts isolated from tomato leaves and cotyledons.[34] Leaves harvested from healthy seedlings grown in clean surroundings usually present few problems with bacterial contamination, although it is advisable to surface-sterilize the leaves by immersing them for 10–15 minutes in a solution of hypochlorite or 0.1% cetyltrimethyl ammonium bromide. Plant material obtained from a market, or material stored for several days after harvest, could be heavily contaminated[35] and should be avoided. Some contaminating bacteria are lost during the preparation of chloroplasts—especially if compounds of high molecular weight are included in the extraction medium[36]—and most of the remainder are removed by density gradient centrifugation.[21] Aside from these precautions, the extraction and reaction media should be essentially free of bacteria. Nevertheless, some estimate of the extent of possible bacterial contamination is desirable. The most direct approach is to plate out samples of reaction mixture on agar containing a nutrient medium. The number of bacteria originally present in the reaction mixture can be ascertained and compared with the number of chloroplasts present. The latter value can be determined with the aid of a hemacytometer. With these procedures, it was estimated that chloroplast preparations isolated from young wheat leaves contained as few as one viable bacterium per 10^6 chloroplasts.[22] As another guide, the rapid decrease in rate of incorporation of amino acid observed with chloroplast preparations would not be expected if an appreciable amount of the incorporation was due to the contaminating bacteria. Incorporation by bacteria can also largely be prevented by anaerobiosis or by respiratory inhibitors.[36] Another test of possible bacterial contamination is to treat the reaction mixture at

[34]T. C. Hall and E. C. Cocking, *Biochim. Biophys. Acta* **123**, 163 (1966).
[35]A. A. App and A. T. Jagendorf, *Plant Physiol.* **39**, 772 (1964).
[36]A. Gnanam, A. T. Jagendorf, and M.-L. Ranalletti, *Biochim. Biophys. Acta* **186**, 205 (1969).

the end of the incorporation period with Triton X-100. This reagent solubilizes most of the chloroplast protein, whereas activity incorporated into bacterial protein can still be sedimented by centrifugation.[37,38]

Interference by Nuclei, Mitochondria, and Other Organelles

Nuclei. Preparations of chloroplasts usually contain some nuclei. Again, Triton-X100 has been useful since it does not solubilize the nuclei. By the use of this detergent it was established that the nuclei in preparations of tobacco and spinach chloroplasts contributed little to the amino acid incorporating activity of the preparation.[13,19]

Mitochondria and Other Organelles. Unless sucrose density gradient centrifugation or sucrose flotation procedures are used for purifying chloroplasts, the preparations are invariably contaminated with the larger mitochondria and possibly also with peroxisomes. Although it is known that isolated plant mitochondria can incorporate amino acids into protein,[39] their possible interference in assays of protein synthesis by isolated chloroplasts has been virtually ignored. We have counted the actual number of particles smaller than chloroplasts in preparations of wheat chloroplasts.[22] There was one of these particles (heavy mitochondria, chloroplast fragments, peroxisomes) per 1.3–2.0 intact chloroplasts. However, by comparing the amino acid incorporating activity of the 1000 g fraction with that of a fraction obtained by centrifugation at a higher speed, and by estimating the relative numbers of mitochondria and peroxisomes in the two fractions on the basis of the NAD–malate dehydrogenase activity, it was estimated that these organelles could contribute no more than a few percent to the amino acid incorporating activity of the 1000 g fraction.[22]

Properties of the Protein-Synthesizing System of Isolated Chloroplasts

Ribosomes from tobacco chloroplasts incorporate [^{14}C]-valine into protein at the rate of 200–300 pmoles of valine per milligram of ribosome per 30 minutes.[28] Rates shown by chloroplasts or chloroplast ribosomes from other tissues are usually of this order, but the observed rates are far below the rates of synthesis of chloroplast protein in intact leaves.[40] One characteristic of the incorporating system is that the rate of incorporation decreases with time and saturation is usually reached

---

[37]M. S. Bamji and A. T. Jagendorf, *Plant Physiol.* 41, 764 (1966).
[38]A. Gnanam and J. S. Kahn, *Biochim. Biophys. Acta* 142, 475 (1967).
[39]H. K. Das, K. K. Chatterjee, and S. C. Roy, *Biochim. Biophys. Acta* 87, 478 (1964).
[40]B. Parthier, *Z. Naturforsch.* B20, 1191 (1965).

in 20–40 minutes. Magnesium is essential for activity, and higher concentrations are required by chloroplast ribosomes compared with cytoplasmic ribosomes. Thus in tobacco, the optimal concentration for activity of the 70 S ribosomes of the chloroplast is 11–15 mM, while for the 80 S cytoplasmic ribosomes it is 3–5 mM.[28] The requirements for other components of the reaction mixture varies with the source of chloroplasts. The activity of tobacco,[13] spinach,[19] and wheat[37] chloroplasts decreases to about one-third if ATP is omitted from the reaction mixture. An ATP-generating system such as phosphoenolpyruvate and pyruvate kinase is usually required. However, ATP generated by photophosphorylation is also effective. In this case, the enzyme coupling system is omitted from the reaction mixture, pyocyanine (0.03 mM) is included, and the reaction vessel is illuminated.[19,38] Chloroplasts from *Euglena*,[38] pea,[12] tomato,[34] and *Acetabularia*[41] show only a slight requirement for ATP. GTP stimulates the activity of spinach chloroplasts 2.5-fold,[19] but tobacco chloroplasts are only slightly stimulated.[13] The incorporating activity of most preparations is increased by supplying a complete mixture of the amino acids commonly found in protein.

The activity of chloroplasts, and more so of purified chloroplast ribosomes, is often stimulated by preparations containing aminoacyl-sRNA synthetases and soluble RNA.[12,19] The incorporation of phenylalanine by spinach[19] and tobacco[13] chloroplasts is increased 2- to 3-fold by the addition of polyuridylic acid. The effect of added messenger RNA on the activities of chloroplasts and cytoplasmic ribosomes isolated from the same cells may be quite different.[29]

The activity of spinach and wheat chloroplasts is stimulated 2-fold by NH_4^+ ions[19,37]; for spinach chloroplasts the optimal concentration is 40 mM.

Inhibitors. Chloramphenicol and puromycin both inhibit the incorporation of amino acids into protein by isolated chloroplasts or by chloroplast ribosomes. Some examples are: for chloramphenicol, spinach chloroplasts are inhibited 58% by 20 μg/ml[19] and wheat chloroplasts are inhibited 77% by 40 μg/ml[37]; for puromycin, spinach chloroplasts are inhibited 72% by 10 μg/ml,[19] and tobacco chloroplasts 69% by 10 μg/ml.[13] Only the D-*threo* isomer of chloramphenicol is strongly inhibitory, the L-*threo* and D-*erythro* isomers being comparatively inactive.[42] Similarly in intact leaves, only the D-*threo* isomer inhibits the synthesis of chloroplast protein and the development of photosynthetic capacity.[43] Deoxyribonuclease has little effect on the ability of isolated chloroplasts to incorporate amino acids, but low concentrations of

[41]A. Goffeau and J. Brachet, *Biochim. Biophys. Acta* **95**, 302 (1965).

ribonuclease (< 0.5 μg/ml) inhibit the activities of chloroplasts from bean,[44] spinach,[19] tobacco,[13,42] and wheat.[27,37] On the other hand, amino acid incorporation by chloroplasts from Acetabularia[41] and pea leaves is fairly insensitive to the presence of ribonuclease. Variable results have also been obtained with actinomycin D. The protein-synthesizing activities of spinach[19] and tobacco[13] chloroplasts are only slightly reduced by actinomycin D, but inhibition occurs in the case of chloroplasts from Acetabularia (40–80% inhibition at 50 μg/ml[41]) and Euglena (30% inhibition at 50 μg/ml[38]). Streptomycin (50 μg/ml), while failing to inhibit the incorporation of amino acids by chloroplasts from Acetabularia,[41] is reported to be inhibitory with chloroplasts isolated from Euglena (32% inhibition at 50 μg/ml and 91% inhibition at 250 μg/ml[38]) and from wheat.[27] Penicillin G is noninhibitory with chloroplasts from Acetabularia[41] and Euglena.[38] Cycloheximide at concentrations as high as 200 μg/ml does not appreciably inhibit amino acid incorporation in chloroplasts isolated from Euglena, pea, and wheat[22] or from tobacco.[42]

Distribution. Chloroplasts active in the incorporation of amino acids into protein have been isolated from a wide range of organisms. These include Acetabularia mediterranea,[41,45] Euglena gracilis strain Z[29,38] and, among higher plants, bean,[27,30,44,46] clover,[30] pea,[12,30] spinach,[19,30,40] sunflower,[27] tobacco,[4,11,13,23,28,42,46,47] and wheat.[27,37]

Age of the Plants. The age of a plant is an important consideration when selecting material for studies of chloroplast protein synthesis. Chloroplasts showing high activity for protein synthesis can be obtained only from young expanding tissue.[19,37] In wheat the incorporating activity of chloroplasts isolated from the older leaves is very low.[37] The reason for this became apparent from in vivo studies.[48] Once the expansion of the primary leaf of wheat seedlings had ceased, there was an immediate and large decrease in the ability of the leaf to incorporate uracil into chloroplast ribosomal RNA and amino acid into chloroplast fraction I protein. These changes were followed 1 day later by the onset of a rapid decrease in the amount of chloroplast ribosomal RNA. Thus, aside from problems of bacterial contamination, plant materials ob-

[42] R. J. Ellis, Science 163, 477 (1969).
[43] M. K. Nikolaeva, O. P. Osipova, and Yu, V. K., Dokl. Akad. Nauk SSSR 175, 487 (1967).
[44] F. Parenti and M. M. Margulies, Plant Physiol. 42, 1179 (1967).
[45] A. Goffeau, Biochim. Biophys. Acta 174, 340 (1969).
[46] N. M. Sissakian and I. I. Filippovich, Biokhimiya 22, 375 (1957).
[47] A. Van Kammen, Arch. Biochem. Biophys. 118, 517 (1967).
[48] R. M. Smillie, N. S. Scott, D. Graham, and B. D. Patterson, Proc. I.B.P. Symp. "Productivity of Photosynthetic Systems. Part II: Theoretical Foundations of Optimization of the Photosynthetic Productivity" (A. A. Nichiporovich, ed.). USSR Acad. Sci., Moscow. In press.

tained from markets are generally too old to be suitable for studies of protein synthesis by isolated chloroplasts.

Nature of the Protein Synthesized by Isolated Chloroplasts

About 23% of the activity from [14C]-valine incorporated into spinach chloroplasts is recovered as soluble protein (obtained by subjecting the chloroplasts to osmotic shock followed by centrifugation), and about half of the remaining activity can be solubilized by 3.5% Triton X-100.[19] In similar experiments, the percentage of activity incorporated which could be recovered in the soluble protein fraction was 20%, 34%, 16–56%, and 50% in the case of tobacco chloroplast ribosomes,[13] *Acetabularia* chloroplasts,[41] wheat chloroplasts,[37] and bean chloroplasts,[49] respectively. Some preliminary evidence has been obtained that fraction I protein and photophosphorylation coupling factor are among the proteins which become labeled.[27,49]

The membranous protein fraction of *Acetabularia* chloroplasts has been fractionated further into proteins soluble and insoluble in 0.3% sodium dodecyl sulfate (SDS) and 3% urea.[45] Of the total counts in the membranous fraction, 60% are solubilized by this treatment. "Structural protein" recovered from the SDS-soluble protein as described by Criddle[50] accounted for 14% of the counts in this fraction. The specific activities of the water-soluble protein, water-insoluble protein, and "structural protein" were in the ratio of 1:1.9:0.83.

Thus isolated chloroplasts incorporate amino acids into both soluble and membranous proteins. Two separate protein-synthesizing systems may be involved, since it has been proposed that certain of the lipoproteins of the chloroplast lamellae are synthesized on ribonucleoprotein particles which are bound to the lamellar membranes.[51,52] Chloroplast DNA may be an integral part of this lamella-bound system.[53] In contrast to chloroplasts, isolated mitochondria appear to incorporate amino acids almost exclusively into membranous protein.[54]

Synthesis of Chloroplast Protein *in Vivo*

The experiments with isolated chloroplasts suggest that chloroplasts, in contrast to mitochondria, can synthesize a wide range of their own

[49]M. M. Margulies and F. Parenti, *Plant Physiol.* 43, 504 (1968).
[50]R. S. Criddle, *in* "Biochemistry of Chloroplasts" (T. W. Goodwin, ed.), Vol. I, p. 203. Academic Press, New York, 1966.
[51]E. N. Bezinger, M. I. Molchanov, and N. M. Sissakian, *Biokhimiya* 29, 749 (1964).
[52]E. N. Bezinger, M. I. Molchanov, and N. M. Sissakian, *Dokl. Akad. Nauk SSSR* 166, 738 (1966).
[53]I. I. Filippovich and E. N. Svetailo, personal communication.
[54]L. W. Wheeldon and A. L. Lehninger, *Biochemistry* 5, 3533 (1966).

proteins. This conclusion has been substantiated by studies of the effects of chloramphenicol and cycloheximide on the synthesis of chloroplast proteins *in vivo*. Chloramphenicol inhibits the synthesis of fraction I protein,[2,48,55,56] enzymes of the Calvin cycle,[2,48,55-58] and electron transfer proteins.[2,55,56] Cycloheximide, on the other hand, while inhibiting cell division of *Euglena* in the range of 1-5 μg/ml, actually stimulates the synthesis of fraction I protein.[48,58] The synthesis of chlorophyll and electron transfer proteins (except for ferredoxin, which behaves like fraction I protein[59]) is also stimulated, but only after a lag period which varies with the concentration of cycloheximide. At higher concentrations (10-15 μg/ml) cycloheximide inhibits synthesis of chlorophyll and electron transfer proteins, but the synthesis of fraction I protein continues.[55] These results are consistent with the synthesis of fraction I protein, some enzymes of the Calvin cycle, and electron transfer protein on the 70 S ribosomes of chloroplasts. The differential inhibition shown by cycloheximide in the higher concentration range also suggests that different control mechanisms operate on the synthesis of the soluble proteins (including ferredoxin), on the one hand, and electron transfer proteins of the lamellae on the other hand.

[55]R. M. Smillie, D. Graham, M. R. Dwyer, A. Grieve, and N. F. Tobin, *Biochem. Biophys. Res. Commun.* **28**, 604 (1967).
[56]R. M. Smillie, N. S. Scott, and D. Graham, *in* "Comparative Biochemistry and Biophysics of Photosynthesis" (K. Shibata, A. Takamiya, A. T. Jagendorf, and R. C. Fuller, eds.), p. 332. Univ. of Tokyo Press, Tokyo, 1968.
[57]C. J. Keller and R. C. Huffaker, *Plant Physiol.* **42**, 1277 (1967).
[58]D. Graham, M. D. Hatch, C. R. Slack, and R. M. Smillie, *Phytochemistry* **9**, 521 (1970).
[59]G. C. Gibbons, J. K. Raison, and R. M. Smillie, *Proc. Aust. Biochem. Soc.* p. 43, 1969.

[34] Fatty Acid Synthesis by Spinach Chloroplasts

By P. K. STUMPF

[^{14}C]Acetate is readily incorporated by isolated spinach chloroplasts into long-chain fatty acids, providing the chloroplasts have an intact "mobile phase" or outer envelope, all the necessary cofactors are present, and the reaction is carried out under relatively high light intensity.

Assay

The capacity of chloroplasts to synthesize fatty acids from [14C]acetate is readily measured by incubating [1-14C]acetate with class I chloroplasts under defined conditions, isolating the newly synthesized lipid, and counting the [14C]lipid product by standard procedures.[1]

Reagents
Preparation of chloroplasts with Honda medium[2]:
   Ficoll, 2.5%
   Dextran-40, 5%
   Sucrose, 0.25 M
   Tricine, 0.025 M, pH 7.8
   $MgCl_2$, 0.01 M
Reaction buffer:
   Sorbitol, 0.6 M
   Tricine, 0.1 M, pH 7.8
   Phosphate buffer, 0.1 M, pH 7.8
   Dithiothreitol, 0.001 M
Reaction mixture:
   0.1 Ml containing 25 nmoles of [1-14C] sodium acetate (1 μCi)
   0.5 Ml containing reaction buffer
   0.2 Ml containing 0.5 μmole of CoA, 2 μmoles of ATP, 30 μmoles of $NaHCO_3$
   0.1 Ml of 0.08% Triton X-100 (Rohm and Haas, Philadelphia)
   0.1 Ml of chloroplast, containing 200 μg of chlorophyll

Procedure. Components of the reaction mixture are pipetted into conical 50-ml Quickfit flasks fitted with polyethylene stoppers. The flasks containing the reaction mixture are placed in clamps fitted to a support, which in turn is clamped to the respirometer fittings of a photosynthesis Aminco Warburg heated-refrigerated apparatus. The reaction is run at 3000 ft-c of white light at 20° for 30 minutes. The reaction is terminated by turning off the light source and, as rapidly as possible, removing the flasks from the bath and adding 0.1 ml of 10 N H_2SO_4 and 5 ml of a chloroform:methanol (V/V) mixture. The suspension is transferred to a 17 × 107 mm Pyrex test tube, the flask is washed once with 2.5 ml of the chloroform:methanol solvent, and 4 ml of saturated sodium chloride solution is added. The test tube is vigorously shaken, then centrifuged briefly in a clinical centrifuge to separate

[1]P. K. Stumpf and N. K. Boardman, *J. Biol. Chem.* 245, 2579 (1970).
[2]D. Spencer and S. G. Wildman, *Biochemistry* 3, 954 (1969).

the phases; the top aqueous layer is carefully removed by an aspirator. Anhydrous sodium sulfate is added to the chloroform layer, which is than shaken vigorously. A small wad of defatted absorbent cotton is inserted into the test tube. The blunt needle of a 5-ml Luer-Lok syringe employing a No. 3527 metal pipetting holder (Becton, Dickinson and Company, Rutherford, New Jersey) is used to force the cotton wad to the bottom of the tube; the solvent is then sucked into the syringe. The lipid extract is transferred to a 25 × 150-mm test tube, and the sodium sulfate pellet is extracted once more with 5 ml of chloroform, the same syringe being used as for the reextraction procedure. The lipid extract is then evaporated to dryness over nitrogen in a glycerol bath maintained at 80°; 2 ml of chloroform is carefully pipetted into the cooled tubes, which are shaken to dissolve lipid adhering to the side walls. Aliquots, 0.1 ml, are pipetted into scintillation vials, the chloroform is evaporated under a heat lamp, and the ^{14}C is counted by usual procedures.

Preparation of Chloroplasts

To obtain maximum activity, greenhouse grown spinach leaves which are approximately 4 inches long are harvested in the morning of the preparation, washed, deveined, weighed, and stored in a polyethylene bag at 4° until ready for use. The use of market spinach is discouraged since the history of the tissue is unknown and considerable bacterial contamination may exist on the leaf surfaces.

Chloroplasts containing over 90% of the class I type[3] are quickly prepared by grinding 20 g of leaf tissue in 60 ml of Honda medium in an Omni-mix blender (Sorvall) with a 200-ml stainless steel chamber at 0°. The tissue is blenderized for 5 seconds at 90 V and filtered by squeezing gently through one layer of Miracloth (CalBiochem, Los Angeles); the extract is filtered once more by gravity through a single layer of Miracloth. The green suspension is then centrifuged for 2 minutes at 2000 g in a SS-34 rotor by a RC2-B Sorvall Super speed centrifuge at 2°. The pale green supernatant is discarded, and the green pellet is suspended by gently rubbing it with 1 cm^2 of nylon cloth (nylon stocking) and a sturdy glass stirring rod. The suspension is made up to a suitable volume with fresh Honda medium to yield 200 μg of chlorophyll per 0.1 ml of suspension.

---

[3]Class I chloroplasts are defined [D. Spencer and H. Unt, *Aust. J. Biol. Sci.* **18**, 197 (1965)] as bright, highly refractive organelles of somewhat irregular outline in which the grana are not clearly resolved in any plane of focus. Class II chloroplasts do not possess the outer envelope, and the grana are very distinct.

Properties

Chloroplasts prepared by this technique can be preserved for over 4 hours at 0–2° c with no loss of activity and no loss of class I characteristics. If Honda medium is substituted for the sorbitol medium in the reaction mixture, marked inhibition of acetate incorporation is observed. Prolonged storage of chloroplasts in the sorbitol–tricine–phosphate system results in some loss of activity.

Linearity of acetate incorporation occurs up to 30 minutes but seems to fall off over longer periods of time, presumably since class I chloroplasts convert to class II chloroplasts. Since employing heavier suspensions of chloroplasts leads to loss of light transmission, the concentration recommended in this procedure will allow efficient light transmission leading to 10–20% incorporation of acetate per 30 minutes. Chloroplasts prepared from greenhouse spinach during different periods of the year will show different responses to cofactors. With spinach grown during the winter months, when short days of low light intensity prevail, ATP stimulation is frequently observed whereas with spinach grown in summer, fall, or spring, ATP may have some inhibitory effects. Furthermore, spinach harvested in the early morning tends to yield more intact chloroplast than spinach harvested in the later period of the day, in which starch granules accumulated in the organelles rupture the outer envelope during the preparatory procedures, resulting in low class I yields.

The products of acetate incorporation usually range from 50 to 90% oleic acid and 50 to 20% palmitic acid, less than 5% of stearic, myristic and linolenic acids being synthesized. These acids are found mostly in the free, rather than in the bound, form.

These techniques have been applied to tissues such as green and etiolated pea leaves as well as to other leaf tissues. In general, when leaf tissue is reasonably succulent and free of fibers, the Omni-mix technique is an effective procedure for isolating chloroplasts; however, for more resistant tissue, such as grasses, the Nobel procedure[4] employing small nylon bags and a mortar and pestle is recommended.

[4] P. S. Nobel, *Plant Physiol.* 42, 1389 (1967).

[35] Synthesizing Capability of the Photosynthetic Apparatus: Steroids

By E. I. MERCER and T. W. GOODWIN

Sterols in Chloroplasts

The chloroplast sterol level is likely to vary from plant species to plant species; for instance, the weight ratios of sterol:chlorophyll in French bean (*Phaseolus vulgaris* var. 'Lightning') and maize (*Zea mays* var. 'South African White Horse Tooth') are 1:16.4[1] and 1:50,[2] respectively. The levels quoted in this section refer only to maize chloroplasts.[2]

Assay of Total Sterol and of Esterified and Unesterified Sterols in Chloroplasts

Chloroplasts are probably best prepared for sterol assay by techniques involving aqueous reagents which do not leach out the sterols as do nonaqueous reagents. Chloroplast pellets prepared by the method of Stumpf and James[3] and purified by the density gradient technique of James and Das[4] are very satisfactory, as are those prepared by the method of Kirk.[5]

The sterol extraction and assay scheme is outlined in Fig. 1; by following route 1 the total chloroplast sterol may be analyzed whereas by following route 2 a more complete analysis of esterified and unesterified sterols may be made.

Lipid Extraction

The chloroplast pellet is extracted with acetone until it is colorless; this usually requires four extractions using five times the chloroplast volume of acetone each time. A final extract with dry, peroxide-free diethyl ether ensures the complete removal of the lipid. The extracts are combined, diluted with 4 volumes of water, and extracted 4–5 times with peroxide-free diethyl ether to remove the lipid. The ethereal extract is then washed several times with water to remove the last traces of acetone and dried over anhydrous Na_2SO_4 for 30 minutes in

---

[1] E. I. Mercer and K. J. Treharne, *in* "The Biochemistry of Chloroplasts" (T. W. Goodwin, ed.), Vol. I, p. 181. Academic Press, New York, 1966.
[2] R. J. Kemp and E. I. Mercer, *Biochem. J.* **110**, 119 (1968).
[3] P. K. Stumpf and A. T. James, *Biochim. Biophys. Acta* **70**, 20 (1963).
[4] W. O. James and V. S. R. Das, *New Phytol.* **56**, 325 (1957).
[5] J. T. O. Kirk, *Biochim. Biophys. Acta* **76**, 417 (1963).

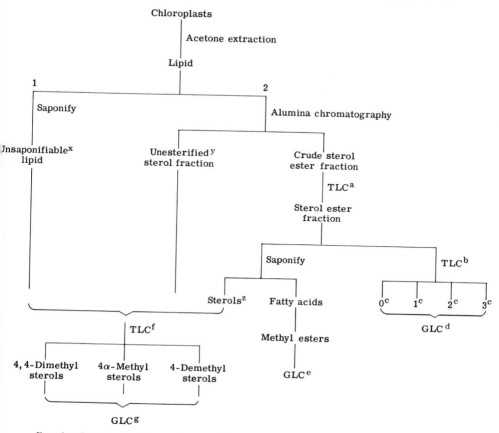

Fig. 1. Sterol extraction and assay scheme. (a) Silica gel G developed with benzene: hexane (2:3, v/v). (b) Silica gel G impregnated with 10% (w/w) AgNO₃ developed with benzene:hexane (2:3, v/v). (c) Sterol esters with 0, 1, 2, and 3 double bonds in their fatty acid moiety. (d) 3% SE-30, temperature programmed between 250° and 275° to give a rise of 1° per minute. (e) 25% DEGS, temperature programmed between 110° and 175° to give a rise of 3° per minute. (f) Silica gel G developed with chloroform. (g) 3% OV-1 at 225°; 1% QF-1 at 225°. (x, y, and z) See text.

the absence of light. Emulsions which may form during the extraction or washing stages can be broken by addition of solid NaCl followed by further shaking. The Na₂SO₄ is filtered off using a sintered funnel, and the filtrate is either distilled or rotary evaporated to a small volume, then transferred to a tared conical flask (e.g., 25 ml) and taken to dryness under nitrogen. The chloroplast pellet obtained from 550 g wet weight of leaf tissue from 21-day-old maize plants yielded 138.4 mg.

Saponification of the Lipid or the Sterol Ester Fraction

The lipid or sterol ester fraction is dissolved in a small quantity of ethanol ($\sim$ 10 ml/100 mg lipid) and a freshly prepared aqueous solution of KOH (60%, w/v) added to the extent of 1 ml/10 ml of ethanolic solution. The mixture is refluxed gently for about 45 minutes. The saponification mixture is then cooled, diluted with 4 volumes of water, and extracted four times with peroxide-free diethyl ether. The bulked ethereal extracts are washed several times with water to remove residual alkali (test the washings with phenolphthalein), dried over anhydrous Na_2SO_4 for 30 minutes, and evaporated to dryness as described previously. The fatty acids liberated by saponification of the sterol ester fraction can be isolated by acidifying the saponification mixture to pH 1 with HCl after extraction of the sterols. The acidified saponification mixture is then extracted four times with peroxide-free diethyl ether. The bulked ethereal extracts are washed with water to remove residual acid, dried over anhydrous Na_2SO_4 for 30 minutes, and evaporated to a small volume under nitrogen.

Column Chromatography of the Lipid

The esterified and unesterified sterol fractions can be separated from one another by chromatography on alumina. The lipid is dissolved in the minimal volume of light petroleum (bp 40°–60°) and added to a column of acid-washed, Brockmann grade III alumina (e.g., M. Woelm Eschwege, Germany). The chromatogram is developed stepwise with successive volumes of light petroleum, 2% E/P[6], and 40% E/P. The sterol esters are eluted in the 2% E/P fraction whereas the free or unesterified sterols are eluted in the 40% E/P fraction. These crude column fractions may than be purified by thin-layer chromatography.

Purification of the Crude Sterol Ester Fraction

The crude sterol ester fraction is purified by chromatography on 0.25-mm layers of silica gel G (E. Merck AG., Darmstadt, Germany) impregnated with the dye Rhodamine 6G[7] using a mixture of benzene and hexane (2:3, v/v) for development. When the developed plates are viewed under ultraviolet light, the sterol esters and other non-ultraviolet-absorbing materials appear as pink zones on a pale-yellow background. The R_f values of a range of sterol esters (sitosterol and campesterol esterified with C_{12}–C_{24} even-numbered and C_{17} saturated

---

[6]Abbreviation: E/P, solution of dry, peroxide-free diethyl ether in light petroleum (bp 40°–60°).
[7]J. Avigan, D. S. Goodman, and D. Steinberg, *J. Lipid Res.* 4, 100 (1963).

fatty acids, palmitoleic acid, linoleic acid, and α-linolenic acid) were found to be between 0.42 and 0.53 on this system.[8] Thus the broad zone chromatographing with R_f values between 0.42 and 0.53 will include the chloroplast sterol esters and can be scraped off the plate and eluted with dry peroxide-free ether.

Separation of the Sterol Esters by Argentation Thin-Layer Chromatography (TLC)

The purified sterol ester fraction is now chromatographed on 0.25-mm layers of silica gel G impregnated with 10% (w/w) $AgNO_3$ using a mixture of benzene and hexane (2:3, v/v) for development. This system separates the sterol esters according to the degree of unsaturation of their fatty acid moieties.[8] The sterol ester zones may be detected by spraying the developed chromatogram with a 0.1% (w/v) Rhodamine 6G solution in acetone and viewing under uv light. On this system, sterol esters of saturated, monounsaturated, diunsaturated, and tri-unsaturated fatty acids chromatograph at R_f 0.65, R_f 0.36, R_f 0.17, and R_f 0.03, respectively. The zones chromatographing at these R_f values may be scraped off the plate and eluted with dry peroxide-free ether. $AgNO_3$-impregnated plates darken rapidly in the light; therefore to obtain reproducible results they should be stored in a dark, desiccated cabinet.

Gas–Liquid Chromatography of Sterol Esters

The four different types of sterol ester separated by argentation TLC may be analyzed by gas–liquid chromatography (GLC) on 3% SE-30 supported upon 60–80-mesh silane-treated Chromosorb W packed into glass columns (18 inches long × 2 mm internal diameter). Dual-column temperature programming is necessary for satisfactory resolution.[9] A starting temperature of 250° rising at a rate of 1° per minute to a final temperature of 275° combined with an argon flow rate of 112 ml/per minute has proved adequate.[7] This system separates sterol esters according to the number of carbon atoms in the molecule regardless of the number of double bonds present (see Fig. 2); hence the need to precede the GLC of the sterol esters by the argentation TLC separation. The carbon number of unknown sterol esters may be determined by comparing their retention times and temperatures with those of a range of authentic samples. The knowledge that a given peak is due to components with a particular number of carbon atoms

---

[8]R. J. Kemp and E. I. Mercer, *Biochem. J.* 110, 111 (1968).
[9]A. Kuksis, *Can. J. Biochem.* 42, 407 (1964).

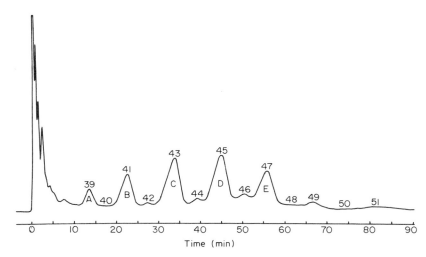

Fig. 2.   Gas–liquid chromatographic separation of the saturated fatty acid steryl esters of maize shoot on 3% SE-30. The number above each peak denotes the number of carbon atoms in the steryl ester molecule.

does not necessarily identify the sterol ester present. For instance peak D in Fig. 2 is given by C_{45} sterol esters of saturated fatty acids and could be a C_{27} sterol esterified to a C_{18} fatty acid (e.g., cholesteryl stearate) or a C_{28} sterol esterified to a C_{17} fatty acid (e.g., campesteryl heptadecanoate), or a C_{29} sterol esterified to a C_{16} fatty acid (e.g., sitosteryl palmitate). It is not yet possible to separate mixtures of sterol esters with the same number of carbon atoms. However, it is possible to get a good idea of what the major sterol esters of such a mixture are by determining the relative abundance of the sterol and fatty acid components in the sterol ester mixture. This can be done by saponifying the sterol ester mixture and isolating the free sterols and free fatty acids as described above in the section on saponification. The composition of the fatty acid mixture can easily be determined by GLC of the methyl esters with, for example, a 25% DEGS column temperature programmed between 110° and 175° to give a rise of 3° per minute.[8]

Separation of the Sterols by Thin-Layer Chromatography

The sterols in the unsaponifiable lipid (x in Fig. 1), or the unesterified sterol fraction (y in Fig. 1) or those derived from the sterol ester fraction (z in Fig. 1) may be separated into three convenient subgroups, the 4,4-dimethyl sterols, the 4α-methyl sterols, and the 4-demethyl sterols by chromatography on 0.25-mm layers of silica gel G impregnated with Rhodamine 6G, chloroform being used for development. The R_f values for the 4,4-dimethyl sterols (e.g., cycloartenol), the 4α-methyl sterols (e.g., 24-methylenelophenol) and the 4-demethyl sterols (e.g., sitosterol) on this TLC system are 0.40, 0.32, and 0.20, respectively. The zones chromatographing at these R_f values can be scraped off the plate, and the sterols can be eluted with peroxide-free diethyl ether.

Gas-Liquid Chromatography of the Sterols

The 4,4-dimethyl, 4α-methyl, and 4-demethyl sterol fractions, separated by TLC (see preceding section), may each be analyzed by GLC on 6-foot × 2-mm i.d. columns of 3% OV-1 and 1% QF-1 supported on 100–120-mesh Gas-Chrom Q developed at 225° with an argon flow rate of 40 ml/min.[8] Cholestane is usually chromatographed with each sample as a marker, and retention times are determined relative to cholestane.

The OV-1 column separates the main chloroplastidic sterols, cholesterol (RRT 1.90), campesterol (RRT 2.51), stigmasterol (RRT 2.74), and sitosterol (RRT 3.16) very well but fails to separate sterols from their corresponding stanols. The QF-1 column separates sterols from their corresponding stanols, but fails to resolve campesterol and stigmasterol (RRT: cholesterol, 2.73; campesterol 3.57; stigmasterol 3.57; campestanol 3.91; stigmastanol 4.67). However, stanols of the main chloroplast sterols were not detected in maize chloroplasts[2] or leaf tissue.[8]

Of the four 4α-methyl sterols detected in chloroplasts, the OV-1 column separates obtusifoliol (RRT 3.06) and 24-ethylidenelophenol (RRT 4.61) from an unresolved mixture of cycloeucalenol and 24-methylenelophenol (RRT 3.49). The QF-1 column, however, will separate cycloeucalenol (RRT 5.11) and 24-ethylidenelophenol (RRT 5.62), but fails to resolve obtusifoliol and 24-methylenelophenol (RRT 4.46).

The main 4,4-dimethyl sterols found in maize chloroplasts are cycloartenol and 24-methylenecycloartanol.[2] These are resolved on the OV-1 column (RRT 3.69 and 4.35, respectively) and are also separated from the pentacyclic triterpenes α-amyrin (RRT 3.47) and β-amyrin (RRT 3.16). The QF-1 column does not separate the components of the 4,4-dimethyl sterol fraction adequately.

Tables 1 and 2 give the composition of the sterols of maize chloroplasts.[2]

TABLE I

COMPOSITION OF THE CHLOROPLAST[a] STEROLS OF MAIZE[b]

Weight of lipid (mg)	Weight of esterified sterol (μg)			Weight of unesterified sterol (μg)		
	4-Demethyl sterol	4α-Methyl sterol	4,4-Dimethyl sterol	4-Demethyl sterol	4α-Methyl sterol	4,4-Dimethyl sterol
138.4	28.0	1.2	9.2	918	20.0	38.5

[a] The chloroplasts were prepared from 550 g wet weight of leaf tissue from 21-day-old maize plants.
[b] R. J. Kemp and E. I. Mercer, *Biochem. J.* 110, 119 (1968).

TABLE II

COMPOSITION OF THE ESTERIFIED AND UNESTERIFIED STEROLS OF MAIZE CHLOROPLASTS[a]

Sterol fraction	Sterol	Percentage of sterols determined after GLC separation	
		Esterified	Unesterified
4-Demethyl sterol	Cholesterol	52	2
	Campesterol	10	13
	Stigmasterol	3	26
	Sitosterol	35	59
4α-Methyl sterol	Obtusifoliol	7	20
	24-Methylenelophenol	64	67
	Cycloeucalenol	Trace	Trace
	24-Ethylidenelophenol	29	13
4,4-Dimethyl sterol	α-Amyrin[b]	29	31
	β-Amyrin[b]	30	10
	Cycloartenol	16	21
	24-Methylenecycloartanol	25	38

[a] R. J. Kemp and E. I. Mercer, *Biochem. J.* 110, 119 (1968).
[b] Pentacyclic triterpenes.

Biosynthesis of Sterols in Chloroplasts

All terpenoids are synthesized from isopentenyl pyrophosphate, the basic isoprenoid precursor, which itself is formed from acetyl-CoA via mevalonic acid (Fig. 3). One molecule of isopentenyl pyrophosphate is isomerized to dimethylallyl pyrophosphate, which acts as starter for the stepwise addition of further molecules of isopentenyl pyrophosphate to form C_{10} (geranyl), C_{15} (farnesyl), and C_{20} (geranylgeranyl) pyrophosphates. This is indicated for geranyl pyrophosphate in Fig. 4. Two molecules of farnesyl pyrophosphate condense to form squalene, the first C_{30} precursor of all triterpenoids, including the sterols, whereas two molecules of geranylgeranyl pyrophosphate condense to form phytoene the first C_{40} precursor of carotenoids.

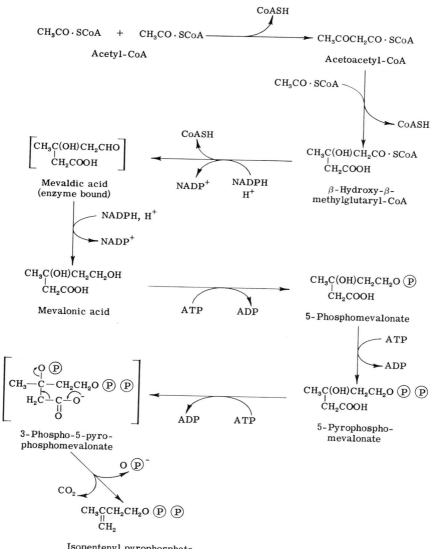

FIG. 3.   Biosynthesis of isopentenyl pyrophosphate.

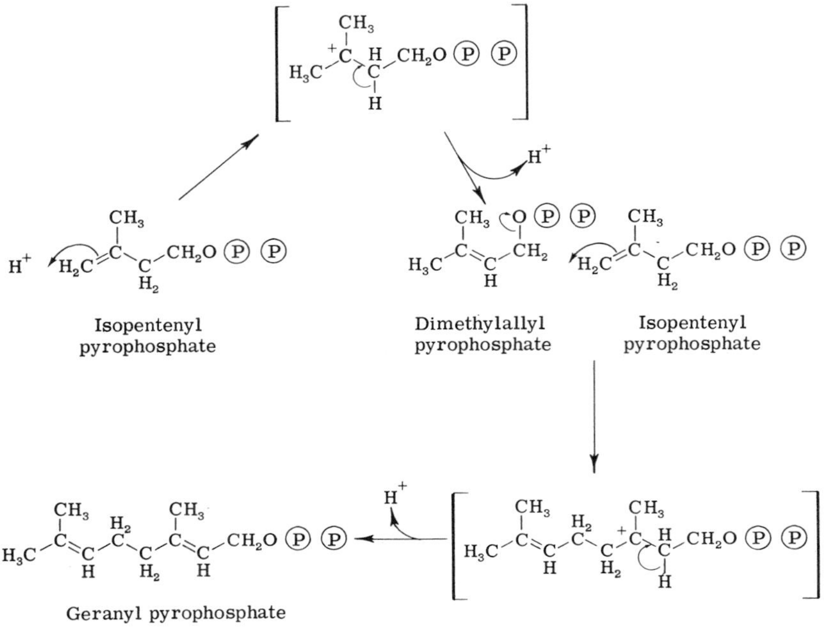

FIG. 4.   Conversion of isopentenyl pyrophosphate into geranyl pyrophosphate.

Chloroplasts formed by the nonaqueous technique will very effectively incorporate mevalonic acid into phytoene whereas the incorporation into squalene is very slight.[10] It is not possible to say whether this synthesis of squalene represents real but limited chloroplast activity or whether it is due to slight contamination with the active extraplastidic synthesizing system, but there is no doubt that the microsomal and supernatant fractions are clearly implicated in sterol synthesis in higher plants.[11] The conclusion at the moment must be that, if the chloroplast can synthesize sterol, its ability is extremely limited and much less than the ability of other regions of the cell.

[10] J. M. Charlton, K. J. Treharne, and T. W. Goodwin. *Biochem. J.* 105, 205 (1967).
[11] F. F. Knapp, R. T. Aexel, and H. J. Nicholas, *Plant Physiol.* 44, 442 (1969).

[36] Use of Mutants of *Rhodopseudomonas spheroides* for Preparation of Bacteriochlorophyll Precursors

By June Lascelles and Thomas P. Hatch

Protoporphyrin is a precursor of hemes and chlorophyll, and most of the enzymatic steps leading to its formation have been established.[1] The enzymatic steps of the magnesium branch of the biosynthetic path are largely unknown, and the sequence shown in Fig. 1 is based mostly upon products accumulated by mutant strains of photosynthetic bacteria and algae blocked in chlorophyll synthesis. The only reaction to be shown so far *in vitro* is the transmethylation of magnesium protoporphyrin.[2]

Mutant strains of *Rhodopseudomonas spheroides* which cannot form bacteriochlorophyll excrete magnesium-containing compounds into the medium when incubated under low aeration in media containing Tween 80[3] and may be exploited as sources of possible substrates for enzymatic work. The following procedures describe the preparation of concentrates of some of these potential intermediates.

Maintenance and Growth of Cultures

Stock cultures of the mutants are maintained on slants of MG medium supplemented with 0.2% yeast extract and 1.5% agar. They are subcultured at intervals of 2–3 weeks, incubated for 24–40 hours at 30°, and stored at 4°.

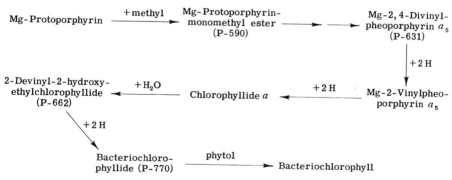

Fig. 1. Presumed sequence of steps in bacteriochlorophyll synthesis.

---

[1]See Vol. V [121], [122].

[2]K. D. Gibson, A. Neuberger, and G. H. Tait, *Biochem. J.* **88**, 325 (1963).

[3]J. Lascelles, *Biochem. J.* **100**, 175 (1966).

MG medium contains per liter: KH_2PO_4, 500 mg; K_2HPO_4, 500 mg; dibasic ammonium phosphate, 800 mg: $MgSO_4\cdot7H_2O$, 200 mg; $MnSO_4\cdot H_2O$, 8.5 mg; $CaCl_2\cdot2H_2O$, 53 mg; iron citrate solution, 2.5 ml (ferrous ammonium sulfate hexahydrate, 784 mg and sodium citrate dihydrate, 588 mg, in 500 ml of water); DL-malic acid, 2.7 g; sodium L-glutamate monohydrate, 1.9 g; nicotinic acid, 1 mg; thiamine hydrochloride, 1 mg; biotin, 0.05 mg. To prevent precipitation, the malic acid and glutamate are first dissolved in about 500 ml of water before the other components are added; the pH is adjusted to 7.0 with 10 N NaOH, and the volume is made to 1 liter.

Cells for the preparation of cell suspensions are grown in MG medium with 0.2% (v/v) Tween 80 under low aeration. The medium is contained in Erlenmeyer flasks filled to 80% of their nominal capacity. The flasks are inoculated with 1% of a broth culture in MG medium with 0.2% yeast extract and are incubated for 18–24 hours at 30° on a shaker operating at 200–220 rpm. The cultures should be pigmented at harvest and should have an absorbance at 680 nm of 1.5–2.

Pigment Formation by Cell Suspensions

The supernatant fluid from the cultures grown as above contain chlorophyll derivatives, but considerably higher yields are obtained by incubation of dense cell suspensions of the mutants in fresh medium. Cells are harvested by centrifuging for 10 minutes at 5000 g. They are resuspended to an absorbance of 4 at 680 nm in MG medium supplemented with 10 mM glycine and sodium succinate, 0.1 mM L-methionine, and 0.2% (v/v) Tween 80. The suspensions are incubated in Erlenmeyer flasks filled to 40–60% of their nominal capacity; these are shaken at 200 rpm at 30°–34° for 12–18 hours. After removal of the cells by centrifugation, the pigments excreted into medium are concentrated as described below. The concentration of pigment in the supernatant fluid is measured spectrophotometrically; the expected yields from the various mutants and absorption maxima are shown in the table.

Isolation of Pigments from Suspension Medium

The methods provide crude concentrates of the magnesium compounds. The magnesium-free derivatives have been purified by chromatography on columns of polyethylene, but this approach has failed with preparations of the magnesium compounds because of the presence of Tween.[4] This detergent is necessary for production of the mag-

---

[4] W. R. Richards and J. Lascelles, *Biochemistry* 8, 3473 (1969).

nesium derivatives by mutants, and methods for its removal from preparations have not yet been found.

Concentrates may be prepared by solvent extraction or by "salting out" with ammonium sulfate. In all procedures, care should be taken to shield preparations from the light to avoid decomposition of the pigments.

Solvent Extraction Methods

Magnesium Protoporphyrin Monomethyester. The supernatant fluid from suspensions of mutant 2-33R contains this pigment, which can be extracted with ether–ethanol.[5] The pH is adjusted to 5.5 by addition of 0.5 ml of glacial acetic acid to 150 ml of supernatant. The pigment is extracted by shaking with 200 ml of diethyl ether–ethanol mixture (1:1) in a separatory funnel. The volume of the top layer is reduced by three extractions with an equal volume of 30% (v/v) ethanol. The material is finally concentrated *in vacuo* to about 5 ml, and 1 M KHCO$_3$ is added to a final concentration of 10 mM. The preparation so obtained contains Tween and has two sharp maxima in the visible spectrum (Fig. 2a): this spectrum is typical of magnesium protoporphyrin or its esterified derivatives. The yield is 80–90%.

Other Pigments. Extraction with ether removes 40–60% of P-631, P-662, P-720, and P-770 from supernatant preparations of the appropriate mutant (see the table). The pH is adjusted to 5.5 by addition of 0.5 ml of glacial acetic acid to 150 ml of supernatant fluid, and the pigments are extracted by shaking with 200 ml of diethyl ether in a separatory funnel. The ether layer is concentrated *in vacuo* to about 2 ml, and 5 ml of 10 mM KHCO$_3$ containing 2% Tween 80 is added. The last traces of ether are removed in a stream of nitrogen. The concentrates of P-631 and P-662 obtained in this manner exhibit sharp maxima (Figs. 2b and 2c). Preparations of P-720 and P-770 are less stable, and the spectra show the presence of decomposition products.

Concentration with Ammonium Sulfate. The supernatant fluid is brought to 30% saturation with ammonium sulfate (17.6 g/100 ml). The mixture is allowed to stand for 30 minutes, then is centrifuged for 10 minutes at 5000 g; the fluid is removed by suction. The precipitated pigment is suspended in about 10 ml of water and again centrifuged. The material is finally dissolved in 10 mM KHCO$_3$ containing 2% (v/v) Tween 80. Yields of up to 80% are obtained by this method, which may be applied to supernatant fluids from all the mutants listed in the table.

[5] R. Cooper, *Biochem. J.* **89,** 100 (1963).

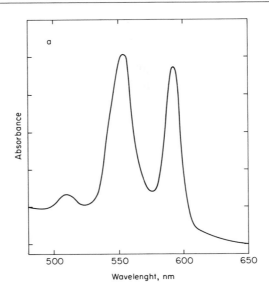

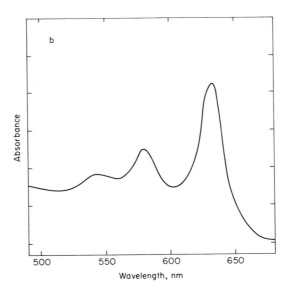

FIG. 2.   Absorption spectra of pigments in 10 mM KHCO$_3$–2% Tween 80. (a) Mg-protoporphyrin monomethylester from mutant 2-33R; (b) Mg-divinylpheoporphyrin a_5 from mutant 8-32; (c) 2-devinyl-2-α-hydroxyethylchlorophyllide a from mutant 8-29.

MAGNESIUM DERIVATIVES ACCUMULATED BY MUTANT STRAINS
OF *Rhodopseudomonas spheroides*

Mutant	Major product	Red adsorption maximum (nm)	Absorbance of supernatant at red maximum
2-33R	Mg-protoporphyrin monomethyl ester (P-590)	590	0.35
8-32	Mg-divinylpheoporphyrin a_5 (P-631)	631	0.1
8-29	2-Devinyl-2-α-hydroxy-ethylchlorophyllide a (P-662)	662	0.93
8-47	2-Deacetyl-2-α-hydroxy-ethylbacteriochlorophyllide (P-720)	720	0.03
8-17	Bacteriochlorophyllide (P-770)	770	0.75

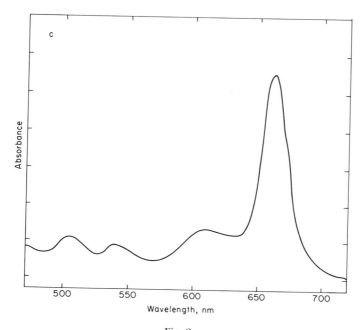

Fig. 2c

Nitrogen Fixation

[37] Nitrogen Fixation—Assay Methods and Techniques

By R. H. Burris

Biological N_2 fixation is a process of great importance to our agricultural economy. In recent years, interest in the process has increased markedly, largely because of the availability of reliable enzymatic preparations and improved and simplified analytical methods.

Although a variety of biological agents are capable of fixing N_2, they apparently share a common enzymatic mechanism and hence can be studied with the same or very similar techniques. The methods described here are generally applicable to all agents.

Determination of N_2 Fixation by an Increase in Total N

Determination of an increase in total fixed N in an organism or enzyme extract furnishes a direct estimate of N fixation. However, in a system initially high in N, e.g., root nodules, the small percentage increases in fixed N observed may be within the experimental error of the method. The method is adequate for cultures of microorganisms capable of vigorous N_2 fixation. Methods for estimating total N have changed little recently, and the method described earlier[1] still is applicable.

Measurement of the Uptake of N_2

Manometric and gasometric analyses of N_2 uptake have been employed occasionally to measure N_2 fixation, but such methods never have been widely accepted. Examples of the application of these methods were cited in an earlier volume.[1] Subsequently, the development of methods for preparing highly active cell-free extracts has made the method more applicable. Mortenson[2] has measured the total $N_2 + H_2$ taken up by preparations using H_2 as the electron donor. As the reaction involves the conversion of $N_2 + 3 H_2 \rightarrow 2 NH_3$, 3 times as much H_2 as N_2 is used, and the micromoles of NH_3 formed are half the total micromoles of $N_2 + H_2$. Reactions are run in conventional respirometers in an atmosphere of 0.5 atm of N_2 and 0.5 atm of H_2; a control vessel with 0.5 atm of argon and 0.5 atm of H_2 always is included. The total micromoles of $H_2 + N_2$ (corrected for control uptake) is divided by 2 to indicate the micromoles of NH_3 produced.

The selection of the pN_2 and pH_2 is somewhat arbitrary, because H_2 not only serves as the electron donor in the system, but also func-

[1] R. H. Burris and P. W. Wilson, Vol. IV, pp. 355–366.
[2] L. E. Mortenson, *Proc. Nat. Acad. Sci. U. S.* **52**, 272 (1964).

tions as a competitive inhibitor of N_2 fixation. Mortenson's observations led him to choose empirically the equal $N_2 + H_2$ mixture for experimental use. His measurements of the time course of N_2 fixation by the manometric method corresponded well with his direct measurements of NH_3 formed.

Use of $^{15}N_2$ to Measure Fixation of N_2

$^{15}N_2$ serves as a sensitive and specific tracer of N_2 fixation. This isotope furnishes a standard method to which other methods should be referred. Recently there has been a strong tendency to use H_2 evolution or acetylene reduction as sole indices of N_2 fixation and to abandon demonstration of NH_3 formation and $^{15}N_2$ fixation. Although the indirect measurements of H_2 evolution and acetylene reduction are sensitive and highly useful, they are indirect methods, and their use to the exclusion of other methods is deplored. They should be compared routinely with NH_3 formation or $^{15}N_2$ fixation to establish a ratio for comparison with a direct method and to assure the investigator that his measurements are valid in terms of N_2 fixation. The use of $^{15}N_2$ as a tracer, its preparation and analysis were covered in some detail earlier,[1] and these methods remain valid.

Purification of $^{15}N_2$. There has been an increased use of commercially prepared $^{15}N_2$, and the user should be warned that certain suppliers are careless about removing oxides of nitrogen from their N_2. It is a desirable practice to pump N_2 from the supplier's vessel into a vessel containing 25 ml (for a liter of gas) of a solution containing 50 g of $KMnO_4$ and 25 g of KOH per liter. If the vessel is shaken, the $KMnO_4$ will oxidize any residual NO to higher oxides, which are absorbed by the KOH. The $^{15}N_2$ then can be pumped (Toepler pump or other pump) to a clean reservoir. If an acid-displacing fluid (e.g., 20% Na_2SO_4 in 5 vol % of H_2SO_4) is used, any residual NH_3 in the gas will be absorbed.

Exposure to $^{15}N_2$. It was suggested earlier[1] that exposures to $^{15}N_2$ be made in respirometer vessels. These are convenient for laboratory work, but for field work small serum bottles are preferable.

Concentrated samples of blue-green algae, bacteria, root nodules, or other N_2-fixing agents are placed in 7-ml (nominal size 5 ml) serum or vaccine bottles (these bottles are available in a variety of sizes), and the bottles are stoppered with a rubber serum or vaccine stopper. From this point, gases and liquids can be added or removed through the stopper with hypodermic syringes and needles. The sample vessels are evacuated with a vacuum pump attached through rubber tubing to a hypodermic needle. Gases are injected through the serum stopper with a hypodermic syringe (Fig. 1 shows a convenient storage vessel for $^{15}N_2$).

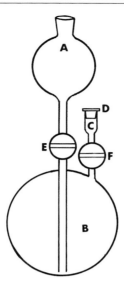

FIG. 1.   Reservoir for $^{15}N_2$. $^{15}N_2$ is stored in B. A contains displacing fluid (20% Na_2SO_4 in 5 volume % H_2SO_4), which can be introduced into B through E. With F closed, C is evacuated through a hypodermic needle piercing vaccine stopper D. After C is evacuated, F and E are opened to fill C with $^{15}N_2$. $^{15}N_2$ is withdrawn through D with a hypodermic syringe and needle.

The pN_2 established in the vessel will be governed by the Michaelis constant for the N_2-fixing system.[3] For most aerobic systems, 0.3 atm of N_2 and 0.1 atm of O_2 is suitable. Experiments should be run to determine whether a higher pN_2 supports a demonstrably higher rate of N_2 fixation; if it does, a suitable conversion factor can be applied. A residual vacuum can be left in the bottles, or they can be brought to atmospheric pressure with an inert gas, such as argon or helium.

After incubation, samples can be inactivated with 0.2 to 1 ml of 5 N H_2SO_4. Whenever possible, incubation should be restricted to an hour. In field studies there has been a tendency to incubate for excessive periods and to omit any measurements of the time course of N_2 fixation.

Analysis of $^{15}N_2$. Mass spectrometric analysis follows the method described earlier.[1,4] However, before sample vessels containing $^{15}N_2$ are opened, the ^{15}N concentration of the N_2 in each vessel should be established. If the vessels were incubated at atmospheric pressure and then were inactivated by addition of H_2SO_4, the gas will be above atmospheric

[3]G. W. Strandberg and P. W. Wilson, *Proc. Nat. Acad. Sci. U. S.* **58**, 1404 (1967).
[4]A. San Pietro, Vol. IV, p. 485.

pressure. If a 1-ml gas sample is to be withdrawn, at least 1 ml of liquid (inactivating 5 N H_2SO_4 or freshly boiled water) should be injected into the serum bottle to produce a pressure above atmospheric. The 1 ml of gas withdrawn from the sample bottle is injected directly into the evacuated mass spectrometer manifold through a gas chromatographic-type septum held in a freezing-trap adaptor, such as is illustrated in Fig. 2. The manifold pressure (a higher than normal pressure must be used to compensate for O_2 and inert gas present) is adjusted, and masses 28, 29, and 30 are measured. In field work, evacuation commonly is incomplete and there may be leakage into the system. Hence, it is necessary to analyze the gas mixture to which the N_2-fixing agent actually was exposed; because any dilution of the $^{15}N_2$ with air will produce a nonequilibrium mixture of masses 28, 29, and 30, all 3 peaks must be measured to permit an accurate determination of the ^{15}N concentration of the gas.

An example will illustrate the calculation of the ^{15}N concentration in a nonequilibrium mixture of the molecular species of N_2. Assume voltage readings were observed with the mass spectrometer as follows: mass 28, 19 V; mass 29, 2 V; mass 30, 0.5 V. The contributions of each of these peaks to the total ^{14}N and ^{15}N are:

Mass	^{14}N (V)	^{15}N (V)
28	19	0
29	1	1
30	0	0.5
	20	1.5
Total voltage from all peaks = 21.5 V		

The atom % ^{15}N then equals $1.5 \times 100/21.5$, or 6.97 atom % ^{15}N.

After the gas from the exposure vessel has been analyzed, the vessel contents are subjected to Kjeldahl digestion, distillation, and conversion of the ammonia to N_2 for mass analysis.[1]

A variation of the original technique used by Rittenberg et al.[5] employs a special conversion tube (commonly called a Rittenberg tube; illustrated in Vol. IV[4]), which is connected directly to the mass spectrometer.[4] The sample of ammonia is placed in one arm of the tube, and alkaline hypobromite in the other arm. After evacuation of the

[5]D. Rittenberg, A. S. Keston, F. Rosebury, and R. Schoenheimer, *J. Biol. Chem.* **127**, 291 (1939).

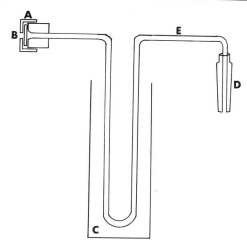

Fig. 2. Adaptor for introducing gas samples into a mass spectrometer. Unit A is threaded to tighten against septum B to provide a gastight seal; C is a Dewar flask to hold dry ice or liquid N_2; E is 1/8-inch or smaller copper tubing which is sealed to 10/30 standard taper joint D with RTV; joint D attaches to the manifold of the mass spectrometer.

tube, the solutions in the two arms are mixed by tilting the tube, and N_2 is generated.

A modification that we have found convenient, especially for class experiments and for samples that must be mailed, employs a conversion vessel shaped like a doorknob (Fig. 3). After digestion, ammonia samples are distilled into 10 ml of 0.01 N H_2SO_4 (this will capture 1.4 mg of N which is adequate for analysis; if appreciably more H_2SO_4 is used,

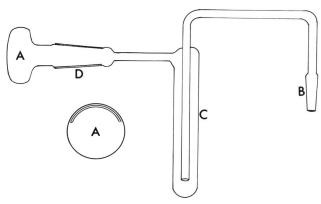

Fig. 3. Apparatus for converting ammonia to N_2 on the mass spectrometer.

the filter papers will be degraded so that they will not hold their position in the doorknob vessels). The distillate is boiled to reduce its volume to 0.2 to 0.5 ml. This concentrated distillate is added to the upper portion of a 0.5 by 4.5-inch piece of filter paper held at the top by a clip; the paper is saturated with distillate (Whatman No. 1 paper supplied in 0.5-inch-wide rolls is suitable; the size is chosen arbitrarily to fit our vessels). The drying of the paper is hastened with a hair dryer, heat lamp, or fan; portions of the ammonia solution are added periodically with a Pasteur pipette until all has been dried on the paper (the transfer need not be quantitative). Glassine weighing papers can be folded around individual sample papers until the samples are analyzed; they may be mailed in this form.

For conversion of the ammonia to N_2, double the paper so that it is 2.25 inches long and then insert it with tweezers into the doorknob vessel. Manipulate the paper carefully with a heavy bent wire until it is held in the position indicated in Fig. 3. Grease the 19/38 inner joint of the doorknob flask A and then pipette 1.5 ml of alkaline hypobromite (to 8 ml of bromine add 40 ml of 13 N NaOH and 60 ml of 0.18% KI; the KI decreases O_2 production[6]) into the flask in a position 180 degrees from the center of the paper. Seat the flask in the 19/38 female joint D. B is attached to the manifold of the mass spectrometer. Open the stopcock between the manifold pump of the mass spectrometer and dry ice freezing trap C; evacuate cautiously to avoid splattering the hypobromite (practice this operation with normal $(NH_4)_2SO_4$ samples on filter paper strips, as one must develop some feel for the evacuation; use a dry ice trap as advised rather than a liquid N_2 freezing trap, because control is easier and volatilization of liquid into the freezing trap is less rapid). Evacuate for 30–60 seconds. Close the stopcock to the vacuum pump. Rotate the doorknob vessel 180 degrees to transfer the alkaline hypobromite to the paper. If the mass spectrometer has a pressure-regulating reservoir, lower the mercury reservoir to provide maximum gas volume. Conversion should be complete within a minute. Close the stopcock to the convertor, and raise the mercury level to provide suitable sample pressure behind the capillary leak of the mass spectrometer. Determine the ^{15}N concentration in the manner appropriate for your specific mass spectrometer.

Barsdate and Dugdale[7] and Desaty et al.[8] have described the use of an automated Dumas apparatus to convert organic compounds to N_2

[6]A. P. Sims and E. C. Cocking, Nature (London) 181, 474 (1958).
[7]R. J. Barsdate and R. C. Dugdale, Anal. Biochem. 13, 1 (1965).
[8]D. Desaty, R. McGrath, and L. C. Vining, Anal. Biochem. 29, 22 (1969).

for mass analysis. The N_2 produced is introduced directly into the mass spectrometer. Modifications of the apparatus have been introduced to reduce memory effects between samples. Compounds with a single nitrogen atom give a random distribution of the molecular species of N_2, but equilibration is incomplete in N_2 derived from certain compounds with N-N, O-N, and C-N bonds.[8] We have not had experience with this method, but it is reported to be more rapid than the conversion dependent upon hypobromite.

Determination of Ammonia

Concern with quantitative analysis for the intermediates in biological N_2 fixation is centered on ammonia because it is the only demonstrated intermediate in the process. Garcia-Rivera and Burris[9] examined other putative intermediates but found no direct evidence for their function. Intact organisms fix N_2 and convert it rapidly to amino acids and proteins; however, extracts from N_2-fixing organisms accumulate ammonia. Crude extracts of C. pasteurianum retain glutamine synthetase and convert sufficient ammonia to glutamine[10] to introduce an appreciable error into estimates of N_2 fixation based upon ammonia formation. In contrast, purified preparations of nitrogenase accumulate ammonia quantitatively, and its increase serves as an excellent index of N_2 fixation.

Mortenson[11] described the quantitative analysis of N_2 fixation by determining ammonia; he employed the Conway microdiffusion technique followed by titration of the ammonia. To avoid confusion from assimilation of ammonia, he added a small amount of an ammonium salt to each control and experimental reaction vessel. Comparison of titration values with tests for assimilation of $^{15}N_2$ established the validity of the method.

Others have preferred to use microdiffusion plus nesslerization to establish the quantities of ammonia produced.[12,13] It is unnecessary to use special Conway microdiffusion vessels, as simple 15 ml (nominal size) vaccine bottles can be substituted (Fig. 4).[14] Round the end of the glass rod by heating, and then etch the flared end with HF or by grinding with a sanding belt or with other abrasive. Mount the rod in a one-hole No. 1 rubber stopper as shown. N_2-fixation reactions are run in the

[9]J. Garcia-Rivera and R. H. Burris, Arch. Biochem. Biophys. 119, 167 (1967).
[10]J. P. Vandecasteele, M. S. Thesis, Univ. of Wisconsin, Madison, Wisconsin, 1965.
[11]L. E. Mortenson, Anal. Biochem. 2, 216 (1961).
[12]M. J. Dilworth, D. Subramanian, T. O. Munson, and R. H. Burris, Biochim. Biophys. Acta 99, 486 (1965).
[13]W. A. Bulen, R. C. Burns, and J. R. LeComte, Proc. Nat. Acad. Sci. U. S. 53, 532 (1965).
[14]S. W. Mayer, F. H. Kelly, and M. E. Morton, Anal. Chem. 27, 837 (1955).

Fig. 4. Microdiffusion apparatus. Ammonia diffuses from the alkaline solution in the 15 ml (nominal size) vaccine bottle to H_2SO_4 held on the etched portion of the central glass rod.

vaccine bottles stoppered with a vaccine stopper, and the reaction mixtures are inactivated by adding 1 ml of saturated K_2CO_3. Dip the end of the glass rod into 1 N H_2SO_4, and then touch the end to a paper towel to remove excess H_2SO_4. Immediately after adding K_2CO_3 to the reaction vessel, insert the stopper carrying the glass rod with H_2SO_4; be careful not to touch the rod to the bottle. Seat the rubber stopper and place a group of bottles on a rotator; this gently agitates the contents of the bottle and speeds diffusion. Rotate for 3 hours at room temperature. Remove the stopper carefully, and use the glass rod as a stirring rod to mix Nessler's reagent (2 ml of H_2O, 2 ml of KI, HgI solution, 3 ml of 2 N NaOH) in a small beaker; this transfers $(NH_4)_2SO_4$ from the glass rod to the Nessler's solution. Allow the color to develop for 15 minutes at room temperature before reading its absorbance in a colorimeter at 490 nm. Include 20-μg N standards, and include as a control reaction vessels which have been incubated in an atmosphere in which argon has been substituted for N_2. The Nessler's reagent is prepared as follows[15]: Dissolve 4 g of KI and 4 g of HgI in 25 ml of water. Select light-colored pieces of gum ghatti and grind 1.75 g in a mortar; add the gum ghatti to 750 ml of water and reflux until it is dissolved. Add the KI, HgI solution to the gum ghatti solution, dilute to 1 liter, and filter. The color intensity follows Beer's law to about 45 μg of NH_3-N under the conditions described; beyond this level turbidity may develop. Smaller amounts of NH_3-N may be measured by reducing the volume of reactants and measuring color intensity in a spectrophotometer equipped with micro cells.

[15]W. W. Umbreit, R. H. Burris, and J. F. Stauffer, "Manometric Techniques," pp. 266–267. Burgess, Minneapolis, Minnesota, 1964.

Ammonia also can be measured at lower concentrations with an indophenol reagent as described by Chaykin.[16] He has specified conditions under which measurements can be made without microdiffusion of ammonia, but we prefer to run a microdiffusion initially as described for analysis with Nessler's reagent. The reagents employed are 0.14 M phenol, 0.25 M nitroprusside, 0.75 M NaOH, and 6% sodium hypochlorite (household bleach). Alkaline hypochlorite is prepared the day of use by mixing 10 ml of NaOH and 0.2 ml of sodium hypochlorite. To a 0.15-ml sample add 3.75 ml of phenol reagent + 0.1 ml of nitroprusside reagent which has been diluted 100-fold (these two reagents may be mixed together immediately before use). Stir with the glass rod carrying ammonia from the microdiffusion. Immediately add 1 ml of alkaline hypochlorite and stir again. Read absorbance at 625 nm 20 minutes after the addition of the alkaline hypochlorite solution. Include a reagent blank and read against a water blank. The response is linear from 0.02 to 1.4 μmoles, so only a single ammonia standard (0.05 ml of 1.0 mM ammonium sulfate) is required.

Levitzki[17] has described a highly sensitive spectrophotometric method for determining ammonia. This depends upon the oxidation of NADH by the reductive amination of α-ketoglutaric acid with ammonia, and measurement of the accompanying rates of change in absorption at 340 nm over a period of 3–5 minutes. The method is reported to be reproducible to ± 5% and to measure 0.02 μmole of ammonia accurately.

Estimation of H$_2$ Evolution, Uptake, and Exchange

H$_2$ Evolution and Uptake. Mortenson[2] employed manometric measurements of H$_2$ and N$_2$ consumption to measure the amount of ammonia formed in reaction mixtures, and he also measured H$_2$ evolution manometrically.[18] Measurements of ATP-dependent, CO-insensitive H$_2$ evolution can be made in an atmosphere of CO; the CO will suppress the activity of hydrogenase. The electron donors most commonly employed to support such evolution of H$_2$ are reduced ferredoxin or Na$_2$S$_2$O$_4$. Standard manometric methods are employed.[15] Kleiner and Burris[19] measured H$_2$ evolution from the hydrogenase of *Clostridium pasteurianum* in 10-ml reaction vessels containing the enzyme, the electron carrier, Tris·HCl buffer in the main vessel, and 0.2 ml 0.2 M Na$_2$S$_2$O$_4$ in the sidearm of the vessel; the contents of the main vessel

[16]S. Chaykin, *Anal. Biochem.* **31**, 375 (1969).
[17]A. Levitzki, *Anal. Biochem.* **33**, 335 (1970).
[18]L. E. Mortenson, *Biochim. Biophys. Acta* **127**, 18 (1966).
[19]D. Kleiner and R. H. Burris, *Biochim. Biophys. Acta* **212**, 417 (1970).

and sidearm were mixed at zero time to give a total liquid volume of 1.0 ml. Measurements were recorded at intervals of 1–5 minutes, depending upon the activity of the preparations. KOH was present in the center well to remove any CO_2.

Reduction of Viologen Dyes. Reduction of methylviologen can be used as a measure of hydrogenase activity.[19] Two milliliters of methylviologen solution (previously saturated with H_2) to give a final concentration of about 10 mM is added to a cuvette (designed to accept a glass stopper) which has been purged with H_2. The cuvette is closed with a rubber serum stopper. The hydrogenase is injected through the serum stopper at zero time (H_2 is released through a second needle piercing the stopper), and the reduction of methylviologen is followed spectrophotometrically at 700 nm. The rate of reduction with H_2 as the electron donor serves as a measure of hydrogenase activity. Care must be taken to establish completely anaerobic conditions for the measurements.

Hydrogen Exchange. H_2/D_2, H_2O/D_2, D_2O/H_2 exchange or ortho/para H_2 exchange can be used as indices of hydrogenase activity. We have employed H_2O/D_2 exchange in 31 ml respirometer-type vessels containing the reaction mixture in about 3 ml, total volume.[19] These vessels are shaken in a 30° water bath and are attached directly to a mass spectrometer manifold system through small-bore flexible tubing. Samples of 0.5–1 ml can be taken at intervals, without drastically altering the conditions in the reaction vessel, and introduced directly through a small-volume freezing trap into the mass spectrometer for analysis. The changes in masses 2, 3, and 4 are followed. The same apparatus can be used for H_2/D_2 and D_2O/H_2 exchange. Ortho/para conversions commonly are measured by changes in thermal conductivity of the H_2.

Utilization of ATP

There is considerable interest in the ratio of ATP used to electrons transported in N_2 fixation and H_2 evolution by nitrogenase.[18,20–22] Experimental results generally have been expressed in terms of ATP molecules used per pair of electrons transported. A pair of electrons is required for H_2 production and 3 pairs for reduction of N_2 to 2 NH_3.

H_2 production can be measured manometrically, and N_2 reduction can be determined by following production of ammonia as described.

[20]H. C. Winter and R. H. Burris, *J. Biol. Chem.* **243**, 940 (1968).
[21]W. A. Bulen and J. R. LeComte, *Proc. Nat. Acad. Sci. U. S.* **56**, 979 (1966).
[22]R. W. F. Hardy and E. Knight, Jr., *Biochim. Biophys. Acta* **122**, 520 (1966).

Measuring the amount of ATP used is complicated by ATPase activity not connected with nitrogenase, and toxicity of ADP to the N_2-fixing reaction. Normal ATPase activity is independent of $Na_2S_2O_4$, whereas reduction of N_2 and utilization of ATP in the process has an absolute requirement for $Na_2S_2O_4$, reduced ferredoxin, or some other strong reductant. Thus, a control run in the absence of $Na_2S_2O_4$ indicates ATP hydrolysis not associated with N_2 reduction; the value determined in this way is subtracted from the total ATP disappearing in the presence of $Na_2S_2O_4$ to yield a corrected value employed in calculating ratios of ATP:electron pair.

Sufficient ATP can be added in stoichiometric amounts to support N_2 fixation for a short time with minimal toxicity. However, to sustain longer periods of N_2 fixation, small amounts of ATP are employed plus creatine phosphate plus creatine phosphokinase to regenerate the ATP. As the ATP level is maintained virtually constant in such a system, a measurement of the change in creatine phosphate level supplies an index of ATP used in the reaction.

As electron flow is involved simultaneously both in H_2 and NH_3 production, H_2 evolution and NH_3 production each must be measured. Reactions are run simultaneously, and phosphocreatine hydrolysis is measured in argon (H_2 evolved by hydrogenase and nitrogenase), in argon + 100 Torr CO, (H_2 evolved by nitrogenase), in N_2 (H_2 evolved by hydrogenase and nitrogenase and N_2 fixed to NH_3), and in N_2 + CO (hydrogenase and nitrogenase blocked by CO, H_2 evolved by nitrogenase). The amount of phosphocreatine hydrolyzed in the absence of $Na_2S_2O_4$ also is measured to estimate ATPase activity.

Winter[23] employed reactants in the following concentrations to study the ATP:2 electron ratios. The respirometer vessels contained 2.0 ml final reaction volume including 0.3 ml of the enzyme preparation in the sidearm and the following in the main vessel: creatine phosphokinase, 0.2 mg; phosphocreatine, 50 μmoles; ATP, 10 μmoles; $MgCl_2$, 20 μmoles; $Na_2S_2O_4$, 20 μmoles added after the vessels were flushed with a stream of argon. The system was gassed with N_2 or other gas mixtures designated.

Creatine arising from hydrolysis of phosphocreatine serves as a measure of ATP used and can be assayed by the method of Eggleton et al.[24] To a neutral solution of creatine in a 10-ml volumetric flask, add 2 ml of 1% α-naphthol solution in a stock alkali containing 30 g of

[23]H. C. Winter, Ph.D. Thesis, Univ. of Wisconsin, Madison, Wisconsin, 1967. *Diss. Abstr. B* **28**, 1352 (1967).
[24]P. Eggleton, S. R. Elsden, and N. Gough, *Biochem. J.* **37**, 526 (1943).

NaOH and 80 g of Na_2CO_3 per 500 ml. Dilute a 1% solution of diacetyl 20-fold, and add 1 ml of this diluted solution to the sample; dilute the contents of the volumetric flask to 10 ml. After allowing 20 minutes for color development, read the absorbancy at 530 nm. A standard curve is linear to 0.6 μmole of creatine, and creatine phosphate and creatinine do not interfere.

Reduction of Alternative Substrates

Until recently there has been little appreciation of the versatility of the N_2-fixing enzyme complex. Nitrous oxide was the first substrate other than N_2 whose reduction by nitrogenase was demonstrated.[25] Later it was shown that azide[26] and acetylene[26,27] were reduced, and this was followed by observations of reduction of cyanide[28] and methylisocyanide.[29] The reduction of acetylene to ethylene and the measurement of ethylene by gas chromatography (flame ionization detector) has become a popular method for measuring the potential of a system for N_2 fixation.[30-32] The method is recommended for its simplicity and sensitivity.

The acetylene reduction method has had its widest application in field studies, where it has made investigations possible on a scale that was impractical when it was necessary to use ^{15}N as a tracer. The acetylene reduction method also is highly useful for enzymatic studies in the laboratory. The high sensitivity of the method permits measurement of reactions over a very short time. The speed of the gas chromatographic analysis for ethylene makes it possible to complete assays in a few minutes. For following enzyme fractionation or for measuring a very dilute enzymatic system, acetylene reduction is the method of choice.

It is generally accepted that the measurement of acetylene reduction serves as a valid measure of N_2 fixation. This opinion is based on considerations that, first, the reduction of acetylene requires ATP and a reducing agent such as dithionite or reduced ferredoxin, as does the reduction of N_2. Second, when the N_2-fixing capacity of the enzyme

[25]M. M. Mozen and R. H. Burris, *Biochim. Biophys. Acta* 14, 577 (1954).
[26]R. Schöllhorn and R. H. Burris, *Fed. Proc., Fed. Amer. Soc. Exp. Biol.* 25, 710 (1966).
[27]M. J. Dilworth, *Biochim. Biophys. Acta* 127, 285 (1966).
[28]R. W. F. Hardy and E. Knight, Jr., *Biochim. Biophys. Acta* 139, 69 (1967).
[29]M. Kelly, J. R. Postgate, and R. L. Richards, *Biochem. J.* 102, 1C (1967).
[30]B. Koch and H. J. Evans, *Plant Physiol.* 41, 1748 (1966).
[31]W. D. P. Stewart, G. P. Fitzgerald, and R. H. Burris, *Proc. Nat. Acad. Sci. U. S.* 58, 2071 (1967).
[32]R. W. F. Hardy, R. D. Holsten, E. K. Jackson, and R. C. Burns, *Plant Physiol.* 43, 1185 (1968).

system is impaired, the capacity for acetylene reduction likewise is impaired. Third, during the purification of the enzyme system for N_2 fixation, there is a parallel purification for acetylene reduction. Fourth, acetylene inhibits N_2 fixation and hence appears to be bound at the same site on the enzyme. Fifth, both the Fe protein and the Mo-Fe protein required in combination for N_2 fixation also are required together to effect reduction of acetylene to ethylene. Although acetylene reduction appears to furnish a valid measurement of N_2 fixation, it must be recognized that the method is indirect and that it should be checked periodically against the capacity of the enzyme system for formation of ammonia from N_2 or for the fixation of $^{15}N_2$. Obviously, if one wishes to translate information from acetylene reduction to rates of N_2 fixation, it is necessary to establish the ratio between acetylene reduction and reduction of $^{15}N_2$ or production of ammonia.

As described earlier, we routinely use vaccine or serum bottles as reaction vessels in following fixation of $^{15}N_2$ or production of ammonia. The same vessels can be utilized for measurement of acetylene reduction. Although in field work investigators often dispense with evacuation of reaction vessels before adding acetylene, in the laboratory it is usually desirable to evacuate and flush the reaction vessels. Serum bottles are stoppered with rubber serum stoppers after the addition of stable reactants. The vessels then are placed on a manifold built by attaching hypodermic needles to a glass manifold. The needles are pushed through the serum stoppers, and all bottles are evacuated in parallel with a mechanical high vacuum pump. Appropriate cylinder gases are added to the vessels separately or as a mixture, the amounts being measured with a mercury manometer. The vessels are removed from the manifold, and labile constituents are injected through the serum stoppers with a hypodermic syringe and needle. To initiate the reaction, 0.1–0.2 atm of acetylene is injected with a hypodermic syringe. As the assembly is gastight, it is immaterial whether the injected acetylene brings the pressure in the bottle above 1 atm. Reaction vessels are placed in a shaken water bath, usually for a period of about 30 minutes; shorter or longer times are entirely feasible. The reaction is terminated by injection of about 0.5 ml of 5 N H_2SO_4.

Alternatively, instead of inactivating the system, one may withdraw a gas sample into a hypodermic syringe[32] and seal the needle of the syringe by plunging it into a small piece of rubber tubing. Conventional syringes with glass plungers are not gastight, but disposable syringes with plungers carrying rubber or plastic rings are reliable for limited periods. Such syringes also are entirely satisfactory for handling acetylene for injection and for removing samples for analysis in the gas chromatographic apparatus. In our experience, the disposable syringes with glass barrels and plastic plungers appear to be the most reliable.

Instead of evacuating reaction vessels, it is suitable to displace the air from them by flushing with a pressurized cylinder of gas. For aerobic systems, a mixture of 80% argon and 20% oxygen is convenient. If photosynthetic organisms are to be tested, it is desirable to add about 0.1% CO_2. Anaerobic systems can be flushed with high purity argon or high purity helium. Cylinder gases are passed through a pressure-reducing valve and into a vaccine or serum bottle through the rubber serum stopper by way of a hypodermic needle; a second hypodermic needle is inserted through the stopper to vent the gas. The venting needle is inserted a shorter distance than the flushing needle to enhance displacement of air. With vigorous passage of gas, adequate displacement is achieved in about 1 minute. As with evacuation, several vessels can be flushed simultaneously through a manifold. However, it may be difficult to judge whether all vessels on the manifold undergo equivalent flushing.

In routine work, it is not obligatory that vessels be flushed or evacuated before the addition of acetylene. The affinity of acetylene for the N_2-fixing enzyme system is sufficiently high so that, when 0.2 atm of acetylene is added, competition by N_2 present is hardly detectable. It is a real convenience in the field to avoid evacuation or flushing, but this simplification presents few advantages in the laboratory. Samples of the particular agent under investigation should be tested to determine whether elimination of evacuation or flushing introduces any appreciable errors.

Commercially available high purity acetylene can be used; however, each cylinder of gas should be analyzed by gas chromatography to determine what impurities are present, as commercial supplies are quite variable in their content of ethylene and methane. Some cylinders are entirely satisfactory and others carry an unacceptable level of ethylene and methane. Acetylene cylinders contain acetone. Acetone apparently has little influence on the N_2-fixing enzyme system, but it may be desirable to remove it by passing the gas through a dry ice trap or through a trap of concentrated sulfuric acid. If heavy gum rubber tubing is placed over the exit from the gas cylinder, is purged and then is plugged, it is possible to remove acetylene conveniently by piercing the tubing with a hypodermic needle and removing a measured amount with the syringe.

A particularly convenient reservoir for field work with acetylene can be made from a rubber football, volleyball, or basketball bladder. The barrel of a 5-ml plastic hypodermic syringe is cemented with epoxy cement to the inflation needle normally used with the bladder. With the inflation needle in place, collapse the bladder. Remove the needle, connect it to an acetylene cylinder, and purge the connecting line. Now insert the needle into the bladder and inflate with acetylene. The inflating needle can be removed for transportation of the bladder. To recover acetylene from the bladder, fill the syringe adaptor with water, stopper the syringe adaptor with a serum stopper, insert the inflation needle into the bladder, pierce the serum stopper with a needle and allow the pressure in the bladder to displace water through the needle to clear the syringe adaptor of water. Acetylene now can be withdrawn through the serum stopper with a hypodermic needle and syringe as needed.

A simple generator for acetylene can be made employing the principle of the Kipp generator. The generator is arranged so that when water comes into contact with calcium carbide, acetylene is generated until the water is displaced from the carbide. The generator is allowed to operate so that air is swept from the system; then acetylene can be produced and used directly from the generator when required. Acetylene is withdrawn through a serum stopper with a hypodermic needle and syringe.

If a sample is to be stored for some time after inactivation with acid, it is advisable to seal the area of the serum stopper which has been punctured by hypodermic needles. Dry the stopper and add a drop of RTV. This is a self-vulcanizing silicone rubber compound, and General Electric RTV 102 or 112 is suitable. Other comparable sealants used as bathtub cement also can be employed. The RTV sets adequately in 60 minutes and provides a barrier which can be penetrated with a hypodermic needle.

Analysis of Ethylene

Samples to be analyzed are removed directly from the serum bottle through the rubber serum stopper with a hypodermic syringe and needle. If the bottle is at atmospheric pressure, 0.5 ml of freshly boiled water should be injected into the bottle before each 0.5-ml sample of gas is removed, so that when the needle is withdrawn through the stopper no air will be pulled into the hypodermic syringe. Disposable 0.5- and 1.0-ml hypodermic syringes with glass barrels and equipped with 25- or 27-gauge needles are suitable for withdrawing gas samples. The 0.5-ml sample withdrawn from the bottle is injected directly into the gas chromatographic apparatus through the rubber septum. Syringes should be flushed with air between samples to preclude cross contamination.

A hydrogen flame ionization detector is particularly effective for analysis of hydrocarbons. It is not necessary that a complex gas chromatographic apparatus be utilized for these analyses. A suitable column for gas chromatographic separation can be prepared from 2 or 3 meters of 1/8-inch diameter aluminum or other metal tubing which is packed with Porapak R or Porapak N of 80–100 mesh. These supports give excellent separation of acetylene and ethylene. The Porapak columns can be operated between room temperature and 80° with a flow rate of about 10 ml of hydrogen per minute at a pressure of about 15 pounds gauge. High purity nitrogen serves as the carrier gas at a pressure of about 40 pounds gauge and a flow rate of 20 ml per minute. Under the conditions specified, separation is achieved in less than 2 minutes. One also may use a 2-foot-long 1/8-inch diameter column of alumina from which particles smaller than 100 mesh have been removed. The alumina column should be operated at about 100°. It is important that flow rates of the gases be kept constant and that the temperature be maintained within close tolerances. If analyses are to be run routinely, it is convenient to keep the instrument running so that constant conditions are maintained.

Gas chromatographic separations on columns as described yield very narrow and sharp peaks. If constant temperature is maintained so that the breadth of the peaks is not subject to temperature alteration, one can quantitate the results adequately by measuring peak heights under standardized conditions. It is customary to include controls which have been inactivated with acid before the addition of acetylene, because such controls furnish a correction for any contaminating ethylene present in the acetylene used. A standard curve is made with the data collected from carefully prepared mixtures of ethylene and air, and the results are plotted as log of concentration vs log of peak height. Ethylene concentration can be derived directly by reference to the standard curve.

When possible, it is desirable to measure samples with more than 50 pmoles of ethylene. Smaller amounts can be detected, but for good quantitation the larger samples are desirable. Usually duplicate samples are measured from a single bottle, and they are expected to give very consistent results. Variations among bottles more commonly result from variations in biological materials and in their handling than from discrepancies in the analytical procedure.

Oxidation of Dithionite

A sensitive method for continuous measurement of the activity of purified nitrogenase or hydrogenase has been developed.[33] It is based

---

[33]T. Ljones and R. H. Burris, *Anal. Biochem.* 45, 448 (1972).

upon spectrophotometric measurement of the oxidation of $Na_2S_2O_4$ (315 nm peak) while it serves as the reductant for nitrogenase or hydrogenase. The method is particularly useful for kinetic studies of these enzyme systems.

[38] Preparation and Properties of Clostridial Ferredoxins

By JESSE RABINOWITZ

Ferredoxin was first isolated from *Clostridium pasteurianum* by Mortenson, Valentine, and Carnahan.[1] The protein functions as an electron transfer factor in the enzymatic formation of acetyl phosphate and hydrogen from pyruvate by extracts of that organism treated with DEAE-cellulose and in a variety of other enzymatic reactions subsequently described.[2] It was characterized as an iron-sulfur protein when the presence of acid-labile sulfur was recognized.[3] Ferredoxin has been found in all clostridial species that have been examined. Procedures for the purification of clostridial ferredoxin have been based on the acidic nature of the protein and its resultant high affinity for DEAE-cellulose,[3] its small size relative to other proteins and resultant solubility in acetone,[4] and the relative ease of crystallization of the protein with ammonium sulfate. Clostridial ferredoxin has a molecular weight of about 6000 and is composed of a single polypeptide chain containing 55 or fewer amino acid residues, 8 of which are cysteine.[5] The protein also contains 8 iron atoms and 8 atoms of acid-labile sulfur.[6] The biological activity of this protein is associated with its unusually low redox potential. Although the amino acid sequence of the protein isolated from several clostridial species is known (Table IV), the molecular structure with respect to the iron and acid-labile sulfur atoms is not known at this time.[6a] The iron and sulfur atoms can be removed by

[1] L. E. Mortenson, R. C. Valentine, and J. E. Carnahan, *Biochem. Biophys. Res. Commun.* **7**, 448 (1962).

[2] R. C. Valentine, *Bacteriol. Rev.* **28**, 497 (1964).

[3] B. B. Buchanan, W. Lovenberg, and J. C. Rabinowitz, *Proc. Nat. Acad. Sci. U.S.* **49**, 345 (1963).

[4] L. E. Mortenson, *Biochim. Biophys. Acta* **81**, 71 (1964).

[5] W. Lovenberg, B. B. Buchanan, and J. C. Rabinowitz, *J. Biol. Chem.* **238**, 3899 (1963).

[6] J.-S. Hong, Ph.D. Thesis, Univ. of California, Berkeley (1969).

[6a] Note added in proof: This is no longer true. The structure of the iron-sulfur complex of the ferredoxin isolated from *Micrococcus aerogenes*, which is a clostridial-type ferredoxin, has been determined: L. C. Sieker, E. Adman, and L. H. Jensen, *Nature* **235**, 40 (1972).

treatment of ferredoxin with mercurial sulfhydryl reagents,[7] by treatment with acid,[8] or by heat.[9] The apoferredoxin formed by these treatments can be isolated, and ferredoxin can be reconstituted from this material prepared in this manner by using ^{59}Fe-labeled salts or [^{35}S]sodium sulfide in the reconstitution reactions.

Growth of *Clostridium pasteurianum*

Potato Tube Cultures. Lyophilized cultures of *C. pasteurianum* (American Type Culture Collection No. 6013) or spore stocks of this organism in soil are most easily revived by growth on "potato medium." This is prepared in the following manner: Solid $CaCO_3$ is added to cover the bottom of a test tube (18 × 150 mm). Fresh potatoes, peeled, washed, and finely diced, are added to the test tube to give a layer about 3 cm high, and 10–12 ml of a 2% sucrose solution in tap water is added to the tubes. The tubes are plugged with nonabsorbent cotton and autoclaved for 20 minutes at 120°. After cooling, the tubes are sealed by adding a plug of absorbent cotton and 2 drops of 2 N K_2CO_3 followed by 2 drops of 4 M pyrogallol. The tube is immediately closed with a rubber stopper and may be stored for 2 weeks. To inoculate, the absorbent cotton is discarded; it is replaced with a fresh seal after inoculation. The inoculated tubes are incubated at 30°. It may take 30–40 hours for a potato-medium culture to start growing when inoculated from soil or lyophilized spore stocks, and about 15 hours when an inoculum grown on the potato-medium culture is used. Since there is a vigorous evolution of hydrogen during growth, tubes should not be stoppered too tightly. The fully grown potato-medium culture may be stored in the refrigerator up to a month and used for subsequent inoculations into liquid media.

Growth on Synthetic Medium. In order to grow a 40-liter culture of the organism, four 500-ml cultures are used as an inoculum. This medium[10] is prepared in the following manner: Solid $CaCO_3$ (1.5 g) and 250 ml of tap water are added to each of four 750-ml Erlenmeyer flasks. The following components are dissolved in 900 ml of tap water:

Sucrose, 40 g
$MgCl_2 \cdot 6H_2O$, 7.4%, 4 ml
NaCl, 20%, 1 ml
$Na_2MoO_4 \cdot 2H_2O$, 10%, 0.2 ml

[7]Richard Malkin and J. C. Rabinowitz, *Biochem. Biophys. Res. Commun.* 23, 822 (1966).
[8]J.-S. Hong and J. C. Rabinowitz, *Biochem. Biophys. Res. Commun.* 29, 246 (1967).
[9]T. Devanathan, J. M. Akagi, R. T. Hersh, and R. H. Himes, *J. Biol. Chem.* 244, 2846 (1969).
[10]J. E. Carnahan and J. E. Castle, *J. Bacteriol.* 75, 121 (1958).

Biotin, 100 μg/ml, 0.1 ml
p-Aminobenzoic acid, 100 μg/ml, 0.1 ml
Na_2SO_4, 7.1%, 2.0 ml
NH_4Cl, 20%, 16 ml

This solution is diluted to 1 liter, and 250 ml is added to each of the 750-ml flasks. The flasks are plugged with nonabsorbent cotton and autoclaved for 15 minutes at 121°. While the flasks are hot, 10 ml of a sterile solution of 5% KH_2PO_4 + 9.2% $K_2HPO_4 \cdot 3H_2O$ is added. When the flasks have cooled to 30°, nitrogen gas that has been sterilized by passage through a cotton filter is bubbled through them. Just before inoculation of each flask, 2 ml of 5% $FeCl_3 \cdot 6H_2O$ (in absolute alcohol) is added with a sterile pipette. Each Erlenmeyer flask is inoculated with 20–25 ml of the culture grown on the potato medium and bubbling with sterile nitrogen is continued. After growth for 16–20 hours at 30°, the A_{660} is 11–12. The doubling time is 1.2–1.8 hours. Three or four of these flasks are then used to inoculate 40 liters of the synthetic medium modified as follows and contained in a large carboy. The calcium carbonate concentration is decreased to 1 g per liter, and the phosphate concentration is doubled. Cells are usually harvested in late log phase when the turbidity as determined by the A_{660} is between 3 and 4. The yield of wet cells from a 40-liter culture is approximately 200–250 g. A portion of the freshly harvested cells is used to prepare the clastic enzymes as described below and the remainder is stored in the deep freeze for use in the preparation of ferredoxin.

Preparation of DEAE-Cellulose

The DEAE-cellulose used in all steps described in this contribution is prepared in the following manner: Newly purchased DEAE-cellulose is suspended in a large volume of 0.5 N NaOH in a large vessel for about 12 hours. The liquid is decanted with suction, and the cellulose derivative is washed with water 3 times. It is then suspended in 0.5 N HCl for 12 hours, and washed with water by decantation 3 times. The wash with 0.5 N NaOH is repeated once again, and the material is washed with water by decantation. The washed DEAE-cellulose is suspended in water, and the suspension is adjusted to pH 6.5 with KH_2PO_4 and diluted to give a final concentration of phosphate of 0.3–1.0 M. This material is used for preparation of the columns for a particlar step, and it is washed with about 10 column volumes of water before use. The DEAE-cellulose recovered from the column after it has been used in a chromatographic step is regenerated by suspending it in 0.5 N NaOH once, washing it with water, and equilibrating it with potassium phosphate at pH 6.5 as described above.

Preparation of Clastic Enzymes

The ferredoxin-free clastic enzyme system is prepared as follows: 100 g of freshly harvested cells of *C. pasteurianum* is suspended in water to give a final volume of 200 ml and broken by sonication of 60-ml aliquots for 1.5 minutes with a Branson Model W185D cell disruptor. The suspension is centrifuged for 1 hour at 40,000 *g*. The supernatant solution is then passed over a column of DEAE-cellulose (2.2 × 20 cm). The pass-through (about 30 mg of protein per milliliter), containing the clastic enzymes, free of ferredoxin, is immediately frozen in 5-ml aliquots and stored in liquid nitrogen.

Assay Methods

Reagents

Potassium phosphate buffer, 0.25 *M*, pH 6.5

Sodium pyruvate, 0.5 *M*

Coenzyme A, 10 mg/ml

Hydroxylamine, 14%, neutralized[11]

Ferric chloride reagent. Mix 100 ml of 10% $FeCl_3$ in 0.2 *N* HCl, 200 ml of 12% trichloroacetic acid, and 200 ml of 3 *N* HCl and dilute to 600 ml with water

Procedure. The following solutions are added to test tubes: 0.10 ml of phosphate buffer, 0.02 ml of pyruvate, 0.005 ml of CoA, 0.3 ml of the clastic enzymes (about 8–10 mg of protein), 0.2–1.0 μg of ferredoxin, and water to make the volume 1 ml. The tubes are incubated at 37°. After 10 minutes, 1.5 ml of the ferric chloride reagent is added. The solutions are filtered through Whatman No. 1 filter paper by gravity (centrifugation is not satisfactory) and the absorbancy of the filtrate at 540 nm is determined using as a blank the control tube incubated in the absence of ferredoxin.

There is an optimum level of clastic enzyme preparation which should be used in the assay so that the amount of reaction is maximal with a particular level of added ferredoxin. This is illustrated in Fig. 1. Since the activities of the clastic enzyme preparations vary, an experiment similar to that shown in Fig. 1 should be done with each enzyme preparation to determine the amount of protein to be used for the ferredoxin assay. A unit of activity in this assay is defined as the amount needed to give a change of 1.0 in absorbance at 540 nm.

---

[11]F. Lipmann and L. C. Tuttle, *J. Biol. Chem.* **159,** 21 (1945).

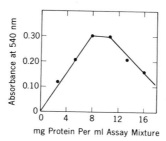

Fig. 1. Dependence of pyruvate "clastic" reaction on enzyme concentration. (See text for details.)

Purification of Ferredoxin from *C. acidi-urici*

Step 1. Extract. Frozen cells of *Clostridium acidi-urici* (Vol. VI [97], 350 g, are thawed overnight at 4° and suspended in 0.05 *M* potassium phosphate buffer, pH 7.8, to a final volume of 900 ml. DNase (0.1 ml of a solution containing 5 mg/ml, Worthington) is added to facilitate suspension of the cells and reduce the viscosity of the mixture. Aliquots (60 ml) are sonicated for 1.5-minute intervals in a Branson Model W185D cell disruptor. The cell debris and unbroken cells are removed by centrifugation at 40,000 *g* for 1 hour. RNase (4 mg) is added to the supernatant solution.

Step 2. DEAE-Cellulose-1. This crude extract (650 ml) is applied to a DEAE-cellulose column (3.0 × 30 cm). The pass-through (650 ml) is collected and used for the purification of formyltetrahydrofolate synthetase (Vol. VI [51]). The column is then washed with 300 ml of water. The ferredoxin is eluted with a linear salt gradient in the absence of buffer obtained by adding 1.8 liters of 0.7 *M* NaCl to 1.8 liter of 0.1 *M* NaCl contained in a mixing chamber, and 10-ml fractions are collected. A typical elution pattern is shown in Fig. 2A. Ferredoxin is recognized from the A_{390} measurement of individual fractions. It is eluted with 1000–1460 ml of eluent. These fractions are pooled and diluted 4-fold with distilled water to give a volume of 2100 ml.

Step 3. DEAE-Cellulose-2. The diluted ferredoxin is reapplied to a second DEAE-cellulose column (3.0 × 23 cm). After application of the sample, the column is washed with 500 ml of 0.15 *M* Tris·chloride buffer, pH 7.4. The ferredoxin is eluted with a salt gradient obtained by adding 1.5 liters of 0.27 *M* NaCl in 0.15 *M* Tris·chloride buffer, pH 7.4, to 1.5 liters of 0.07 *M* NaCl in 0.15 *M* Tris·chloride buffer, pH 7.4, in a mixing chamber, and 10-ml fractions are collected. It is usually possible to allow this chromatographic step to proceed overnight. The ferredoxin is eluted with 900–1600 ml of eluent (Fig. 2B).

Step 4. DEAE-Cellulose-3. To concentrate the ferredoxin the fractions are pooled and diluted 3-fold with distilled water and readsorbed on a third DEAE-cellulose column (3.0 × 8 cm), and eluted with a small volume of 0.15 M Tris·chloride buffer, pH 7.4, containing 0.5 M NaCl.

Step 5. Sephadex G-75. The concentrated solution (26 ml) is then applied to a column of Sephadex G-75 (3.0 × 90 cm) which has been

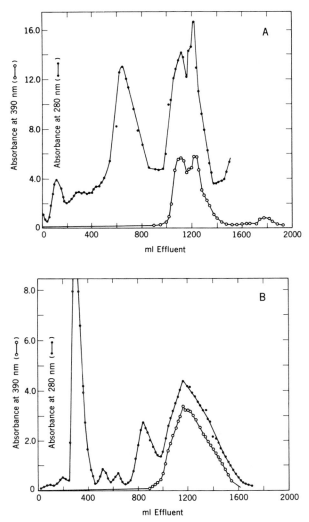

FIG. 2. Purification of ferredoxin from *Clostridium acidi-urici*. (See text for details.)

equilibrated with 0.05 M Tris·chloride buffer, pH 7.4, and 5-ml fractions are collected. It is usual to allow this step to proceed overnight. The ferredoxin is eluted with 250–340 ml of eluent (Fig. 2C).

Step 6. Crystallization. The fractions containing the ferredoxin are combined, and the solution is made 70% in ammonium sulfate. After 1 hour at 0°, the precipitated ferredoxin is collected by centrifugation for 20 minutes at 39,000 g. It is then resuspended in 40 ml of 0.15 M Tris·chloride buffer, pH 7.4, and stored in a tube under vacuum at 4°. The purification of the protein from *C. acidi-urici* is summarized in Table I.

Purification of Ferredoxin from *C. pasteurianum*

The procedure used is very similar to that described for the isolation of ferredoxin from *C. acidi-urici;* however, some differences in results and in the procedure may be noted because the ferredoxin content of *C. pasteurianum* is significantly lower than that of the purine-fermenting organism, and relatively higher concentrations of ammonium sulfate are required for its precipitation.

Step 1. Extract. Frozen cells of *C. pasteurianum* (450 g) grown as described in a previous section, are thawed overnight at 4°. The extract is prepared as described for cells of *C. acidi-urici.*

Step 2. DEAE-Cellulose-1. The extract is applied to a DEAE-cellulose column (4.5 × 30 cm) and is washed with 400 ml of water. The ferre-

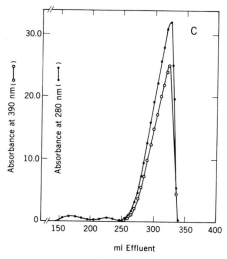

Fig. 2c

TABLE I
PURIFICATION OF FERREDOXIN FROM *Clostridium acidi-urici*

Fraction	Volume (ml)	Protein[a] (mg)	Ferredoxin[b] Units	Ferredoxin[b] Mg	A_{390}/A_{280}
I. Extract[c]	650	11,626	119,600	345	0.046
II. DEAE-cellulose-1	2100	1,196	111,090	321	0.322
III. DEAE-cellulose-2	2040	494	90,168	261	0.726
IV. DEAE-cellulose-3	33	411	88,440	255	0.728
V. Sephadex G-75	93	352	83,700	242	0.775
VI. Crystals	50	354	84,000	243	0.789

[a]Determined by a modification of the Lowry method, with crystalline bovine serum albumin as standard [J. C. Rabinowitz and W. E. Pricer, Jr., *J. Biol. Chem.* **237**, 2898 (1962)].
[b]The ferredoxin was determined by its activity in the enzymatic assay with the *Clostridium pasteurianum* "clastic system." A unit of activity causes a net increase of 1.0 at 540 nm in the assay. A standard solution of ferredoxin is included in each assay, and the unit activity of ferredoxin varies from day to day. The milligrams of ferredoxin recovered are based on this enzymatic test.
[c]Prepared from 350 g of wet cells in the absence of 2-mercaptoethanol.

doxin is eluted with a linear salt gradient in the absence of buffer obtained by adding 1.8 liters of 0.7 M NaCl to 1.8 liter of 0.1 M NaCl, and 10-ml fractions are collected. The elution pattern obtained is shown in Fig. 3A. The ferredoxin is eluted with 1560–1800 ml of eluent.

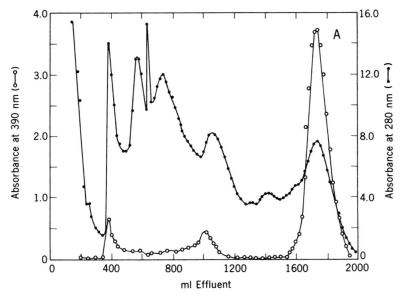

FIG. 3. Purification of ferredoxin from *Clostridium pasteurianum*. (See text for details.)

The fractions containing ferredoxin are combined, diluted to 900 ml, and stored overnight under reduced pressure.

Step 3. DEAE-Cellulose-2. The ferredoxin is readsorbed on a second DEAE-cellulose column (4.5 × 23 cm). Tne column is washed with 500 ml of 0.15 *M* Tris·chloride buffer, pH 7.4, and the ferredoxin is eluted with a linear salt gradient obtained by adding 1.5 liter of 0.27 *M* NaCl in 0.15 *M* Tris·chloride buffer, pH 7.4, to a mixing chamber containing

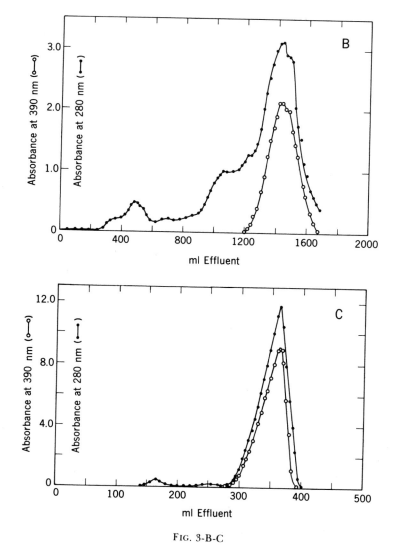

Fig. 3-B-C

1.5 liters of 0.07 M NaCl in 0.15 M Tris·chloride buffer, pH 7.4. Ten-milliliter fractions are collected, and the elution pattern obtained is shown in Fig. 3B. The ferredoxin is eluted with 1300–1600 ml of eluent. The combined fractions are diluted 3-fold and stored overnight in an evacuated flask.

Step 4. DEAE-Cellulose-3. In order to concentrate the ferredoxin, the material is adsorbed on a third DEAE-cellulose column (3.0 × 8 cm) and is eluted with a small volume of 0.5 M NaCl in 0.15 M Tris·chloride buffer, pH 7.4.

Step 5. Sephadex G-75. This material, 20 ml, is applied to a column of Sephadex G-75 equilibrated with 0.05 M Tris·chloride buffer, pH 7.4. The ferredoxin is obtained in the eluate after 310–370 ml of eluate have been collected (Fig. 3C).

Step 6. Crystallization. This solution is brought to 60% saturation with ammonium sulfate, and the small amount of precipitate formed is removed by centrifugation. The supernatant solution is brought to 85% saturation with ammonium sulfate and allowed to stand at 5° overnight. The precipitate is collected by centrifugation and redissolved in 0.15 M Tris·chloride buffer, pH 7.4. The protein is recrystallized by the addition of ammonium sulfate to 80% saturation. Crystals form immediately. The purification of the protein from *C. pasteurianum* is summarized in Table II.

Preparation of Apoferredoxins

Apoferredoxin-I. The iron and acid-labile sulfide are removed from ferredoxin by treatment of the protein with a mercurial.[7] Ferredoxin (15 mg) in 2.7 ml of 0.1 M Tris·chloride buffer, pH 7.4, is treated with 1.8 ml of 0.05 M sodium mersalyl (Sigma Chemical Co.) in the same buffer at room temperature for 2 minutes. The small amount of precipitate

TABLE II
PURIFICATION OF FERREDOXIN FROM *Clostridium pasteurianum*

Fraction	Volume (ml)	Protein[a] (mg)	Ferredoxin[b] (units)	A_{390}/A_{280}
I. Extract[c]	520	4992	79,560	0.0535
II. DEAE-cellulose-1	900	549	73,980	0.366
III. DEAE-cellulose-2	1100	319	62,150	0.625
IV. DEAE-cellulose-3	20	197	56,000	0.635
V. Sephadex G-75	80	142	46,000	0.79
VI. Crystals	17	112	45,220	0.83

[a]See Table I.
[b]See Table I.
[c]Prepared from 400 g of wet cells in the absence of 2-mercaptoethanol.

that forms is removed by centrifugation. The solution is then passed over a column of Chelex 100 (Bio-Rad Laboratories, Berkeley, California) (2 × 17 cm) equilibrated with water. The protein is eluted with water at room temperature and 1.5 ml fractions are collected. The protein is eluted in a single peak with 18–24 ml of eluent. These fractions are pooled, lyophilized, and redissolved in distilled water to give a concentration of 4.0 mg of protein per milliliter. This solution is then passed over a column of Sephadex G-25 equilibrated with water at room temperature. The column is developed with water at room temperature, and 0.5-ml fractions are collected. The absorbancy at 280 nm and the protein content of the fractions are determined. A single protein-containing component is obtained with 2–6 ml of eluent. A component with absorbancy at 280 nm is eluted afterward, but this fraction contains no protein. The fractions containing apoferredoxin-I are lyophilized.

Apoferredoxin-II. The iron and acid-labile sulfide are removed from ferredoxin by treatment of the protein with acid under aerobic conditions.[8] Ferredoxin (14 mg) in 11 ml of 0.15 M Tris·chloride buffer, pH 7.4, is made 4–5% in trichloroacetic acid with a concentrated solution (30%) of the acid. This mixture is allowed to stand for 1 hour at 0°. Evolution of H_2S gas occurs, and the brown color of ferredoxin is completely bleached at the end of the period. The white suspension is centrifuged at 35,000 g for 10 minutes, and the supernatant solution is discarded. The precipitate is washed with either 3 ml of 0.023 M formic acid or 5% trichloroacetic acid, allowed to stand for 30 minutes at 0°, and centrifuged. The precipitate is finally taken up in 4 ml of 0.1 M Tris·chloride buffer, pH 9.0, and the insoluble material is removed by centrifugation. The almost colorless supernatant solution is then passed over a Sephadex G-25 column (1.2 × 38 cm) previously equilibrated with 0.02 M Tris·chloride buffer, pH 8.5, and the protein is eluted with the same buffer. Alternatively, the solution is dialyzed extensively against 0.01 M sodium acetate and then distilled water and lyophilized. The apoprotein is obtained as a white solid.

Apoferredoxin-III. This preparation[8] is similar to that of apoferredoxin-I, but all solutions are 0.05 M with respect to 2-mercaptoethanol. Alternatively, apoferredoxin-III is prepared from apoferredoxin-II as follows: Apoferredoxin-II (54 mg) in 10 ml of 0.1 M Tris·chloride buffer, pH 8.5, containing 8 M urea, 2.5 mM EDTA, and a small amount of antifoam A is flushed at room temperature with prepurified nitrogen for 10 minutes. The mixture is made 0.1 M with respect to 2-mercaptoethanol and flushed with nitrogen for 4 hours. At the end of the period the product, apoferredoxin-III, is reisolated free of urea in 0.02 M Tris·chloride buffer, pH 8.5, containing 0.1 M 2-mercaptoethanol by gel filtration on Sephadex G-25 (1.8×40 cm) which had been equilibrated with the buffer.

Reconstitution of Ferredoxin from Apoferredoxin

Reconstitution from Apoferredoxin-I.[7] Apoferredoxin-I, 1 mg, is dissolved in 1.0 ml of 0.1 M Tris·chloride buffer, pH 7.4, 0.05 M in 2-mercaptoethanol. Then, 30 μl of 0.1 M $Fe_2(NH_4)_2SO_4$ and 30 μl of 0.1 M Na_2S are added and the solution is incubated at 37° for 10 minutes. The protein is reisolated by chromatography on DEAE-cellulose (0.5×8 cm). After applying the protein, the column is washed with water and 0.15 M NaCl to remove excess reagents. The protein is eluted with 0.80 M NaCl.

Reconstitution from Apoferredoxin-II.[8] Apoferredoxin-II, 1.8 mg, is incubated for 4 hours at room temperature in 4 ml of 0.1 M Tris·chloride buffer, pH 8.5, containing 8 M urea, 0.07 M 2-mercaptoethanol and a small amount of antifoam while being flushed with nitrogen gas. Then 0.078 ml of 0.1 M $FeSO_4(NH_4)_2SO_4$, 0.078 ml of 0.1 M Na_2S, and 140 μmoles of 2-mercaptoethanol are added while still flushing with nitrogen gas. The reaction mixture is then diluted 3-fold with deaerated 0.1 M Tris·chloride buffer, pH 8.5, and incubated for 15 minutes at 37°. The reconstituted ferredoxin is reisolated by chromatography on a DEAE-cellulose column (0.8×8 cm). The reaction mixture is applied, then the column is washed successively with 10 ml of 0.15 M Tris·chloride buffer, pH 7.4, and 20 ml of 0.23 M NaCl in 0.005 M Tris·chloride buffer, pH 7.4, to remove excess reagents. The ferredoxin is finally eluted with 0.58 M NaCl in 0.005 M Tris·chloride buffer, pH 7.4.

Reconstitution from Apoferredoxin-III.[8] Apoferredoxin-III, 1.8 mg in 1.6 ml of 0.1 M Tris·chloride buffer, pH 8.5, and 0.05 M 2-mercaptoethanol is incubated for 15 minutes at 37° with 0.078 ml of 0.1 M $Fe_2(NH_4)_2SO_4$ and 0.078 ml of 0.1 M Na_2S. The reconstituted ferredoxin is isolated as described in the previous paragraph.

Radioactive Ferredoxins. The reconstitution procedures have been described here with unlabeled iron and sulfide salts. However, labeled ferredoxins may be obtained by using $^{59}Fe_2SO_4(NH_4)_2SO_4$ or [^{35}S]-sodium sulfide. The labeled sodium sulfide obtained from commercial sources is purified by diffusion. For this purpose, 1.5 ml of 0.05 M [^{35}S]sodium sulfide is acidified with 0.5 ml of 1 N HCl in an all-glass bubbling train and bubbled with prepurified nitrogen. The H_2S evolved is trapped in a receiving tube containing 2 ml of 0.02 N NaOH. The labeled Na_2S solution is mixed with 3 ml of 0.15 M unlabeled Na_2S, and the concentration of sulfide is determined by iodometric titration.

Properties

Ferredoxin from different clostridial species is obtained in different crystalline forms.[3] The ultraviolet spectra of ferredoxin and apoferredoxins prepared from *C. acidi-urici* are shown in Fig. 4. The absorption spectra of other clostridial ferredoxins are similar although not identical to this spectrum. The molar absorption of *C. acidi-urici* ferredoxin at 280 nm is 38,900 and at 390 nm is 30,600.[6] The value $A_{390}:A_{280}$ for *C. acidi-urici* ferredoxin is 0.787. The value for this ratio for *C. pasteurianum* ferredoxin is 0.83. This value is a very useful index of the purity of the protein and can be used to detect deterioration of the crystalline preparation. The molecular weight of *C. acidi-urici* ferredoxin has been determined by sedimentation velocity[5] and differential sedimentation equilibrium,[12] and values of 5600 and 5820, respectively, have been obtained. The molecular weight calculated from the amino acid sequence plus 8 atoms each of iron and sulfur is 6230. The sedimentation coefficient is 1.4, and the partial specific volume of this protein as determined using density gradient columns[5,13] or by the differential sedimen-

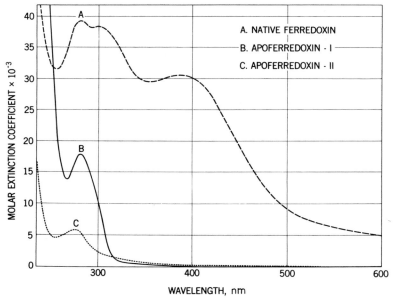

FIG. 4.   The absorption spectrum of *Clostridium acidi-urici* ferredoxin and apoferredoxins.

[12]S. J. Edelstein and H. K. Schachman, *J. Biol. Chem.* **242**, 306 (1967).
[13]A. Hvidt, G. Johansen, K. Linderstrom-Lang, and F. Vaslos, *C. R. Trav. Lab. Carlsberg, Ser. Chim.*, **29**, 129 (1954).

tation equilibrium analysis[12] is 0.63 and 0.61, respectively. *Clostridium acidi-urici* ferredoxin contains 8 atoms of iron per mole of protein as determined by atomic absorption spectroscopy and 8 atoms of acid-labile sulfur determined by the specific activity of ferredoxin reconstituted with [^{35}S]sulfide of known specific activity[6] or through use of the colorimetric assay of Fogo and Popowsky[14] as described previously.[5]

Clostridial ferredoxin accepts two electrons when titrated with sodium dithionite,[15] or on reduction with a crude hydrogenase preparation from *C. pasteurianum*,[16] spinach chloroplasts,[17] or a highly purified pyruvate:ferredoxin oxidoreductase from *C. acidi-urici*.[18]

The behavior of clostridial ferredoxins and apoferredoxins in the Lowry protein assay using the phenol reagent are summarized in Table III.[6] Ferredoxin gives an unusually high color equivalent in this test compared to bovine serum albumin, despite the fact that it contains relatively few aromatic amino acids.

The amino acid sequences of four clostridial-type ferredoxins are given in Table IV. Clostridial ferredoxins contain only a single basic amino acid residue or none. The isoelectric point of this protein is pH 3.7.[5] All the clostridial-type ferredoxins that have been examined contain 8 cysteine residues, and these are in similar positions relative to one another. There is also a significant symmetry in the molecule in that the sequence from residues 30 to 55 is very similar to the sequence that occurs in the first half of the molecule, if two deletions are introduced. A number of amino acids are completely lacking in these materials. For example, *C. acidi-urici* ferredoxin lacks phenylalanine, histidine, trypto-

TABLE III
PHYSICAL CONSTANTS OF *Clostridium acidi-urici*
FERREDOXIN AND DERIVATIVES IN PROTEIN DETERMINATIONS

	Ferredoxin	Apoferredoxin	
		I	II
Molecular weight	6,230	7,390	5,530
A_{390}/mg/ml	5.00	0.0	0.0
ϵ_{390}	30,600	0.0	0.0
Folin protein assay			
$\dfrac{A_{660}/\text{mg ferredoxin}}{A_{660}/\text{mg albumin}}$	1.78	1.10	1.12
$\epsilon_{660} \times 10^{-4}$	19.7	14.8	11.3

[14]J. K. Fogo and M. Popowsky, *Anal. Chem.* 21, 732 (1949).
[15]S. G. Mayhew, D. Petering, G. Palmer, and G. P. Foust, *J. Biol. Chem.* 244, 2830 (1969).
[16]B. E. Sobel and W. Lovenberg, *Biochemistry* 5, 6 (1966).
[17]M. C. W. Evans, D. O. Hall, H. Bothe, and F. R. Whatley, *Biochem. J.* 110, 485 (1968).
[18]K. Uyeda and J. C. Rabinowitz, *J. Biol. Chem.* 246, 3111 (1971).

TABLE IV
Amino Acid Sequence of Clostridial-Type Ferredoxins

Residue No.	Organism				Residue No.	Organism			
	a	*b*	*c*	*d*		*a*	*b*	*c*	*d*
1[e]	Ala	Ala	Ala	Ala	30	Tyr	Phe	Phe	Tyr
2	Tyr	Tyr	Phe	Tyr	31	Val	Val	Val	Ala
3	Val	Lys	Val	Val	32	Ile	Ile	Ile	Ile
4	Ile	Ile	Ile	Ile	33	Asp	Asp	Asp	Asp
5	Asn	Ala	Asn	Asn	34	Ala	Ala	Ala	Ala
6	Glu	Asp	Asp	Asp	35	Asp	Asp	Asp	Asp
7	Ala	Ser	Ser	Ser	36	Thr	Thr	Thr	Ser
8	Cys	Cys	Cys	Cys	37	Cys	Cys	Cys	Cys
9	Ile	Val	Val	Ile	38	Ile	Ile	Ile	Ile
10	Ser	Ser	Ser	Ala	39	Asp	Asp	Asp	Asp
11	Cys	Cys	Cys	Cys	40	Cys	Cys	Cys	Cys
12	Gly	Glu	Gly	Gly	41	Gly	Gly	Gly	Gly
13	Ala	Ala	Ala	Ala	42	Ala	Asn	Asn	Ser
14	Cys	Cys	Cys	Cys	43	Cys	Cys	Cys	Cys
15	Asp	Ala	Ala	Lys	44	Ala	Ala	Ala	Ala
16	Pro	Ser	Gly	Pro	45	Gly	Asn	Asn	Ser
17	Glu	Glu	Glu	Glu	46	Val	Val	Val	Val
18	Cys	Cys	Cys	Cys	47	Cys	Cys	Cys	Cys
19	Pro	Pro	Pro	Pro	48	Pro	Pro	Pro	Pro
20	Val	Val	Val	Val	49	Val	Val	Val	Val
21	Asp	Asn	Ser	Asn	50	Asp	Gly	Gly	Gly
22	Ala	Ala	Ala	—[f]	51	Ala	Ala	Ala	Ala
23	Ile	Ile	Ile	Ser	52	Pro	Pro	Pro	Pro
24	Ser	Ser	Thr	Gln	53	Val	Val	Asn	Asn
25	Glu	Glu	Gln	Gln	54	Gln	Gln	Gln	Pro
26	Gly	Gly	Gly	Gly	55	Ala	Glu	Glu	Glu
27	Asp	Asp	Asp	—[f]					Asp
28	Ser	Ser	Thr	Ser					
29	Arg	Ile	Gln	Ile					

[a]*Clostridium acidi-urici.* S. C. Rall, R. E. Bolinger, and R. D. Cole, *Biochemistry* **8**, 2486 (1969).
[b]*Clostridium pasteurianum.* M. Tanaka, T. Nakashima, A. Benson, H. Mower, and K. T. Yasunobu, *Biochemistry* **5**, 1666 (1966).
[c]*Clostridium butyricum.* A. M. Benson, H. F. Mower, and K. T. Yasunobu, *Arch. Biochem. Biophys.* **121**, 563 (1967).
[d]*Micrococcus aerogenes.* J. N. Tsunoda, K. T. Yasunobu, and H. R. Whiteley, *J. Biol. Chem.* **243**, 6262 (1968).
[e]The sequence is given from the amino-terminal end at position 1 to the carboxy-terminal end at position 55.
[f]The residue number has been shifted to emphasize the similarity of the sequence to that of the other clostridial ferredoxins. This ferredoxin contains a total of 54 amino acid residues.

phan, leucine, methionine, and lysine. Other clostridial ferredoxins also lack several amino acids, although these may differ from those mentioned above.

The reaction of clostridial ferredoxin with iron chelating agents and with sulfhydryl reagents has been investigated.[19] *Clostridium acidi-urici* ferredoxin reacts slowly with the ferrous chelating agent *o*-phenanthroline under anaerobic conditions. After either chemical or enzymatic reduction of the protein, there is a rapid reaction of one of the 8 iron atoms with the chelating agent, but the reaction of the remaining iron is inhibited. In the presence of high concentrations of urea or guanidine hydrochloride and aerobic conditions, all the iron in the protein reacts with *o*-phenanthroline. Only 65–80% of the iron reacts under anaerobic conditions in the presence of these denaturants. Ferredoxin undergoes gradual degradation in the presence of 6.4 *M* urea or 4 *M* guanidine hydrochloride under aerobic conditions, but the protein is relatively stable in these denaturants under anaerobic conditions.

Native, enzymatically reduced, or chemically reduced ferredoxin does not react with the ferric chelating agent Tiron. In 4 *M* guanidine hydrochloride under aerobic conditions, 8 moles of iron per mole of protein react with Tiron. Under anaerobic conditions, however, only approximately half the iron in the protein reacts with Tiron.

Native ferredoxin reacts with the mercurial sulfhydryl reagent CMB in neutral buffered solution.[5] Slightly more than the theoretical value of 24 moles of CMB reacts per mole of protein. However, native ferredoxin does not react with the sulfhydryl reagent DTNB under these conditions, and there is only a slight reaction in the presence of 6.4 *M* urea. In 4 *M* guanidine hydrochloride, under either anaerobic or aerobic conditions, approximately 14 moles of DTNB react per mole of ferredoxin. This reaction is due exclusively to the inorganic sulfide in ferredoxin. The reaction of both the inorganic sulfide and the cysteine residues in ferredoxin with DTNB occurs in the presence of 4 *M* guanidine hydrochloride and EDTA under anaerobic conditions.

[19]R. Malkin and J. C. Rabinowitz, *Biochemistry* 6, 3880 (1967).

[39] Purification of Nitrogenase from *Clostridium pasteurianum*

By LEONARD E. MORTENSON

Nitrogenase from *Clostridium pasteurianum* is an anionic protein complex, and because of this the first successful purification involved adsorption on cationic materials.[1,2] The two most successful steps were selective

[1]L. E. Mortenson, *Biochim. Biophys. Acta* 127, 18 (1966).
[2]L. E. Mortenson, J. A. Morris, and D. Y. Jeng, *Biochim. Biophys. Acta* 141, 516 (1967).

precipitation with protamine sulfate followed by resolubilization with phosphocellulose and adsorption on DEAE-cellulose followed by gradient elution with increasing salt concentration. For removal of other contaminating proteins, these steps were usually followed by gel filtration on Sephadex G-100 or G-200 and by adsorption and elution from calcium phosphate gel. In this way it was shown that clostridial nitrogenase is composed of two *easily* separated protein components which were named molybdoferredoxin (MoFd) or alternatively Mo-Fe protein and azoferredoxin (AzoFd) or alternatively Fe protein.

The isolation from bacteria of the two components catalyzing N_2 reduction is complicated because the components are O_2 sensitive and, in addition, in *C. pasteurianum* one of the components, AzoFd, is cold labile and salt sensitive. All purification steps *must* be accomplished under strict anaerobic conditions and at temperatures of about 10°. Once MoFd was separated from AzoFd, its further purification could be performed at 0°–5° if desired. A second requirement for isolation of either component is that a supply of the other component be available for assay.

Techniques for Maintaining and Monitoring Anaerobic Conditions

Since both MoFd and AzoFd are inactivated by O_2, their purification requires techniques that can easily be performed under anaerobic conditions. All buffers used were deoxygenated on a gassing manifold by evacuating and flushing (with constant mixing) 10 times with deoxygenated H_2 or argon. After removal of O_2, dithionite was added to make the concentration 0.5 mM, and methylviologen was provided as an indicator to assure the continued presence of dithionite. Both the dithionite and methylviologen were introduced by syringe from a solution made by injecting deoxygenated water into a vessel containing the dry dithionite and an argon or nitrogen atmosphere. When dithionite was added to the buffer after this degassing procedure, only small amounts (less than 50 nmole/ml) were oxidized as measured by the amount of reduced methylviologen oxidized. Alternatively the indicator dithionite concentration could be monitored spectrophotometrically at 315 nm. For the preparation of buffer in which no dithionite was to be added a similar procedure was used, but only a sample of the buffer was checked for its inability to oxidize dithionite.

Columns of Sephadex or other materials (such as hydroxyapatite and DEAE-cellulose) were prepared anaerobically by first slurrying the column material in the desired buffer (usually 0.05 M Tris·HCl at pH 8.0) and then degassing the slurry similarly to the buffers. After degassing, up to 1 μmole of dithionite per milliliter was added by syringe to remove final traces of O_2. The slurry was poured into a column (usually 5 cm in diameter by 50 cm high with a 2-liter attached reservoir) containing argon; a volume of anaerobic buffer equivalent to at least two bed volumes of Sephadex was passed through the column, the flow being controlled by a metering pump. The effluent was examined for its dithionite content by addition of methylviologen or spectrophotometrically at 315 nm. Deoxygenated argon or H_2 was continuously bubbled or flushed through the reservoir of the column.

Samples of protein to be introduced into the columns were transferred from glass tubes containing serum stoppers by Hamilton gastight syringes. Before use, the syringes were flushed with degassed buffer containing 1 μmole of dithionite per milliliter. When the sample had entered the column, deoxygenated buffer was added to the reservoir and the separation was initiated by turning on the metering pump.

The separated protein solutions were passed from the metering pump through a syringe needle and a serum cap into thoroughly deoxygenated tubes or bottles. After collection, the tubes were put in an all-glass container under an O_2-free atmosphere. For long-term storage all solutions were frozen in liquid N_2 in the form of pellets and kept in liquid N_2.[3] In this frozen form, both AzoFd and MoFd were stable for months.

Preparation of Crude Extracts from Dry Cells

Techniques for growing *C. pasteurianum*, harvesting the cells, and drying the cell paste are well documented.[4,5] The dry cells from several batches (about 2 kg) were mixed thoroughly and distributed into bottles (100 g); the bottles were filled with H_2 or argon, sealed, and stored at $-15°$. Such cells could be stored for months without noticeable loss in activity.

Crude extracts were prepared by weighing out 100 g of dry cells and 150 mg of lysozyme and adding them with rapid mixing to 1500 ml of anaerobic 0.05 M Tris·HCl buffer at pH 8.0 (different amounts of cell extract can be made with the ratio of cells to buffer approximately the same as above). The O_2 admitted when the dry cells were added was removed rapidly by 10 cycles of evacuation and flushing with deoxygenated H_2. After 1 hour of mixing at 30° (usually on a New Brunswick shaker), the extracted cell mixture was centrifuged at 37,000 g for 15 minutes at 15°. The supernatant solution was decanted from the cellular debris into a 2-liter vacuum flask and immediately evacuated

[3]M. Kelly, R. V. Klucas, and R. H. Burris, *Biochem. J.* **105**, 3c (1967).
[4]J. E. Carnahan, L. E. Mortenson, H. F. Mower, and J. E. Castle, *Biochim. Biophys. Acta* **44**, 520 (1960).
[5]L. E. Mortenson, *Biochim. Biophys. Acta* **81**, 473 (1964).

and filled with H_2 (10 cycles) to assure that no O_2 remained. This preparation with a protein concentration ranging from 18 to 25 mg/ml (total protein, 20–27 g) was assayed for its N_2-fixing and acetylene-reducing ability; the results ranged from 15 to 25 nmoles of N_2 fixed per minute per milligram of protein for N_2 fixation and from 50 to 80 nmoles of acetylene reduced per minute per milligram of protein for acetylene reduction (see [39]).

Separation of Clostridial Nitrogenase into Two Components

The current procedure used to prepare MoFd and AzoFd is outlined in Fig. 1. The crude extract (1 liter or more) is first treated for an hour at 25° with 10 mg of RNase and 4 mg of DNase for each liter of extract. This breaks the contaminating RNA and DNA into pieces so that when protamine sulfate (PS) is added (3% by weight of protein) the contaminating nuclei acids precipitate smoothly rather than in strings, and a cleaner fractionation results. The precipitate is removed easily by centrifugation at 9000 g for 5 minutes in 250-ml bottles. During these steps, transfer of the suspension to the centrifuge bottles and transfer of the supernatant solution from the bottles are accomplished rapidly to minimize aeration. The supernatant solution from the 0–3% PS cut is now treated with PS to make the concentration 6% by weight of original protein, and the centrifugation step is repeated. The supernatant solution from this step is collected and stored at 0°.

Purification of AzoFd[6]

The precipitate from 1 liter of extract (3–6% PS step) is added rapidly to a centrifuge bottle containing a deoxygenated suspension of 3 g of phosphocellulose in 35 ml of 0.05 M Tris·HCl buffer at pH 8.0, and the mixture is again degassed to remove the O_2 introduced during transfer. The suspension of phosphocellulose and the PS precipitate are now mixed at room temperature with the aid of a magnetic stirrer; in less than 10 minutes the PS binds to the phosphocellulose and the protein initially bound to the PS is released. This suspension is centrifuged without transfer since any air contamination now rapidly inactivates the AzoFd.

The supernatant solution containing crude AzoFd freed from phosphocellulose is introduced through a hypodermic needle directly to an anaerobic column containing Sephadex G-100 (5 cm diameter by 50 cm high), and the flow is initiated at 30 ml per hour by starting the metering pump attached to the bottom of the column. In a typical run after 16 hours at room temperature, a well defined AzoFd band can be seen

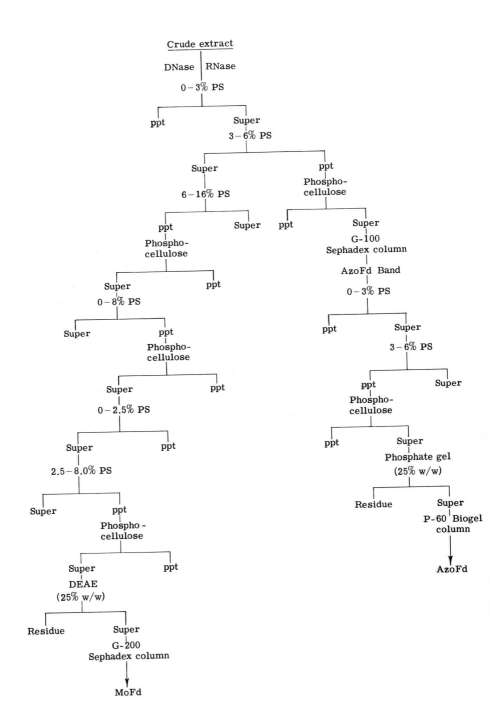

together with 4 less densely colored bands. The bottom band contains small amounts of MoFd whereas the band next to the bottom contains AzoFd. The top band contains traces of Fd. The band next to the top is blue and appears to be a flavoprotein. The band third from the top is red and contains an iron protein of unknown function. After about 24 hours the AzoFd elutes from the column and is collected in a centrifuge bottle containing a magnetic stirring bar (see section on anaerobic conditions). This solution of AzoFd is treated with 3% PS (by weight of protein) by dropwise addition from a syringe through a serum cap on the bottle. Mixing is accomplished by magnetic stirring. The precipitate is separated from the supernatant solution by direct centrifugation of the bottle in which the components were mixed, and the supernatant solution containing the AzoFd is transferred rapidly from the centrifuge bottle through butyl rubber tubing (low O_2 permeability) to a second centrifuge bottle. The whole system is made anaerobic before transfer by the cyclic evacuation and flushing procedure. The precipitate is discarded and the PS concentration in the partially purified AzoFd solution is now increased from 3% to 6% (by weight of original protein). After centrifugation anaerobically as described above, the supernatant solution is discarded, and to the precipitate in the centrifuge bottle is added 3–5 ml of a suspension of phosphocellulose in 0.05 M Tris·HCl buffer at pH 8.0. The amount of phosphocellulose added is equal to 5 times the weight of PS added for the 3–6% fraction. The mixture is magnetically stirred; when complete exchange of the PS from AzoFd to phosphocellulose has occurred, the mixture is centrifuged. After this step AzoFd is better than 90% pure and contains only small traces of two protein contaminants, one of which is hydrogenase. These trace contaminants are removed from the AzoFd by treating the solution anaerobically with 25% (w/w) phosphate gel. As a final step, the AzoFd is passed through an anaerobic column of BioRad P-60 gel (1.5 cm in diameter × 110 cm high). This does not significantly increase the specific activity, but it does remove trace proteins that are known to be present from prior examination of disc gel electrophoresis data. The activity of various fractions and the recovery of units are given in Table I. From 18 g of cell extract protein, we consistently obtain at least 50 mg of pure AzoFd. This represents about a 10% yield.

FIG. 1. Purification scheme. Procedure for the preparation of molybdoferredoxin (MoFd) and azoferredoxin (AzoFd). PS = protamine sulfate; ppt = residue after centrifugation; super = supernatant solution remaining after centrifugation.

TABLE I
PURIFICATION OF AZOFERREDOXIN (AzoFd)

Protein	Purification step	Specific activity[a]	Average[b] (total units)	Percent recovery
18 g	—	60	1.08×10^6	100
0.5–1.0 g	PS I + Sephadex G-100	400–600	3.75×10^5	35
0.2–0.4 g	PS II	900–1100	3.00×10^5	28
60–100 g	Phosphate gel	1700–2000	1.48×10^5	14
50–70 mg	P-60	1700–2200	1.11×10^5	10

[a] Specific activity is expressed as nanomoles of acetylene reduced per minute per milligram of protein in the AzoFd fraction. In the measurement of acetylene reduction, MoFd was added at a molar concentration "equivalent" to the AzoFd. Initial specific activity ranged from 40 to 60, and an "apparent" 30- to 35-fold purification is required before AzoFd is pure (see footnote b).

[b] Note on assay: One can never measure AzoFd maximally since MoFd in excess decreases activity by favoring the formation of an inactive 1:1 MoFd:AzoFd adduct, i.e., the active unit contains at least two AzoFd molecules bound to one MoFd molecule. [D. Y. Jeng and L. E. Mortenson (1970)]

Purification of MoFd (All Steps Anaerobic)[2,7]

The supernatant solution from the 3–6% PS step (Fig. 1) is treated with 6–16% PS (by weight of original protein); after thorough mixing, the precipitate is removed by centrifugation at 0° in 250-ml centrifuge bottles at 9000 g. When kept as a PS precipitate, MoFd is stable in O_2 for several hours. The PS precipitate is treated anaerobically with 5 g of phosphocellulose in 70 ml of 0.05 M Tris·HCl buffer at pH 8.0 and mixed by stirring until all PS exchanged from MoFd to phosphocellulose. This may take up to 60 minutes. The mixture is centrifuged at 0° for 15 minutes to remove the phosphocellulose–PS complex. The dark brown supernatant solution obtained is transferred to a flask, where it is treated anaerobically with 8% PS by weight of protein. The yellow supernatant solution is discarded, and the precipitate is treated with a suspension of phosphocellulose, in 50 ml of 0.05 M Tris·HCl buffer at pH 8.0, containing 5 times the amount of phosphocellulose as the PS added to form the precipitate. The PS treatment is repeated; the first fraction is 0–2.5% (w/w). After centrifugation, the precipitate is discarded and the concentration of PS increased to 8%. After recentrifugation, the supernatant solution is discarded and the precipitate is treated with phosphocellulose (same proportions as before). In contrast to the time required to exchange the MoFd from phosphocellulose in the

[6] E. Moustafa and L. E. Mortenson, *Biochim. Biophys. Acta* **172**, 106 (1969).

[7] I. Kennedy, J. A. Morris, and L. E. Mortenson, *Biochim. Biophys. Acta* **153**, 777 (1968).

earlier PS steps (1 hour), the exchange now occurs completely within 2–5 minutes. A contaminating protein of a very sticky consistency, when combined with PS, was eliminated with the previous supernatant solution. After centrifugation to remove the phosphocellulose complex, the supernatant solution containing MoFd is treated anaerobically with DEAE-cellulose to remove Fd bound to MoFd. This is accomplished by adding anaerobically increments of a slurry of DEAE-cellulose, removing samples of the mixture, centrifuging the sample, and monitoring the supernatant solution for Mo. The DEAE additions are continued until a slight loss of Mo occurs. The DEAE with adsorbed contaminants is removed anaerobically by filtration on a chromatography column. Deoxygenated argon is used to maintain a complete anaerobic atmosphere in the column before and after transfer of the MoFd-DEAE mixture. The MoFd collected from this column is transferred anaerobically to an anaerobic column of Sephadex G-200 (5 cm × 50 cm), and the column flow is initiated at 30 ml per hour by turning on an attached metering pump. On this column the MoFd can be seen to move as a single band that is collected in tubes as previously described for AzoFd. Occasionally after this step MoFd is not completely pure as judged by disc gel electrophoresis. To remove the final traces of contamination, the MoFd is adsorbed at the top of an anaerobic column of DEAE-cellulose (5 cm diameter × 12 cm high) equilibrated with 0.05 M Tris·HCl buffer at pH 8.0. It is then removed with a KCl gradient. A yellow contaminant that appears to be a glycoprotein is removed first and then the MoFd is eluted. This MoFd gives one distinct band when examined by disc gel electrophoresis. It is desalted by passage through a Sephadex G-100 column. The purified MoFd is then pelleted by dropwise addition to liquid N_2 and the pellets are stored in liquid N_2 or at −15°. It is stable for months in this state. From approximately 18 g of protein, 340 mg of pure MoFd are obtained; this represents a recovery of about 35% (Table II).

Variability in Purification Procedure

Extracts from all batches of dry cells do not always respond to this purification procedure in identical manner. We countered this by mixing uniformly several batches of dry cells (2–4 kg) so that each extract made from the cells was uniform. The ideal percentage of protamine sulfate (also from similar large batches) required to remove the AzoFd and MoFd, was determined on the first extract from a batch of dry cells and successive extracts then behaved in a similar way. An example of variability might be that the first step of the purification scheme would require 4.0% PS instead of 3.0% or the AzoFd might be removed

TABLE II
PURIFICATION OF MOLYBDOFERREDOXIN (MoFd)

	Purification step	Protein (g)	Specific activity[a]	Total units	Percent recovery
	—	18.0	60	1.08×10^6	100
PS-I	3.0% PS supernatant	15.6	65	1.01×10^6	94
	6.0% PS supernatant	13.6	70	0.95×10^6	88
	16.0% PS ppt, resolubilized	3.0	280	0.84×10^6	78
PS-II	2.5% PS supernatant	2.0	395	0.79×10^6	73
	8.0% PS ppt, resolubilized	1.0	625	0.63×10^6	58
	DEAE-cellulose (batch)	0.90	650	0.59×10^6	55
	Sephadex G-200	0.34	1120	0.38×10^6	35

[a]Specific activity is expressed as nanomoles of acetylene reduced per minute per milligram of protein in MoFd fractions. Initial specific activity in crude extracts varies from 40 to 60. Usually a 16- to 25-fold purification is required to obtain pure MoFd. The percentage of the total protein as MoFd ranges from 4 to 6%. The assay for maximum MoFd activity requires about a 6-fold molar excess of AzoFd.

NOTE: PS = protamine sulfate (see purification scheme).

from the crude extract by treatment between 3 and 8% rather than 3 and 6%. The MoFd in solution is easily monitored by measuring the Mo concentration by use of an atomic adsorption spectrophotometer. With the latter technique, one can monitor the removal of MoFd from solution by checking small samples for Mo remaining in solution. The degree of separation of AzoFd and MoFd by this PS method is easily seen when either the resolubilized AzoFd or MoFd is passed through the Sephadex G-100 column; there should be only small traces of MoFd in the former and only small traces of AzoFd in the latter.

Properties of Azoferredoxin

The molecular weight of AzoFd was estimated by disc gel electrophoresis,[8] gel filtration[9] and by amino acid analysis to be approximately 27,000 daltons. It occurs in solutions as a dimer of 54,000–58,000 daltons. Estimation of molecular weight by sedimentation in the ultracentrifuge has been extremely variable because of aggregation. Based on 27,000 daltons the chromophore of AzoFd appears to be a complex of 2 iron atoms (which when assayed by α,α'-dipyridyl appear as ferrous iron), 1-2 acid "labile" sulfide groups[10] and 6 free SH groups. Each polypeptide of 27,000 has L-leucine as the carboxyl-terminal amino acid and L-methionine as the N terminal. It contains 6 cysteinyl residues,

[8]P. Andrews, *Biochem. J.* **96**, 595 (1965).
[9]J. Weber and M. Osborn, *J. Biol. Chem.* **244**, 4406 (1969).
[10]J. K. Fogo and M. Popowski, *Anal. Chem.* **21**, 732 (1949).

all of which are present as cysteine. It is extremely O_2 sensitive; complete inactivation occurs after exposure to air for 1–3 minutes. It is cold labile and denatures when the salt concentration is higher than 0.15 M.

AzoFd binds MgATP or ADP or both but does not react with them in the absence of MoFd. AzoFd has an electron spin resonance signal at $g = 2.01$ when it is oxidized for 3–5 minutes and is inactive. Under the latter conditions one sulfide groups, 2 SH groups and 1 iron atom are oxidized. Longer oxidation results in the loss of the $g = 2.01$ signal and oxidation of the remaining sulfide, SH groups, and iron.

On the basis of 27,000 daltons, the amino acid analysis of AzoFd gave the following: Cys-6, Asp(+Asn)-12, Thr-22, Ser-11, Glu(+Gln)-34, Pro-8, Gly-30, Ala-18, Val-16, Met-8, Ile-16, Leu-24, Tyr-8, Phe-6, His-2, Lys-16, Arg-12, Trp-0. The N-terminal end contains Met-Glu (Glu, Val, Ile, Leu, Leu) Asp . . . , while the C-terminus contains . . . (Asp, Gly, Ile) Tyr-Met-Leu-OH (the latter data were obtained in collaboration with Dr. Karl Dus of the University of Illinois, Department of Biochemistry).

AzoFd is brown in the active form, turns red when oxidized, and inactive and colorless when the iron and sulfide are removed. There are no distinct peaks in the visible region, but there is a shoulder in the 300-nm region. Most of the adsorption in the 300–500-nm region is lost when the iron and sulfide are removed.

When combined with molybdoferredoxin the active enzyme (nitrogenase) formed has a ratio of at least two AzoFd to each MoFd. Kinetic studies show that the 1:1 adduct is not active.[11]

Properties of Molybdoferredoxin

The molecular weight of MoFd was estimated by sedimentation in the ultracentrifuge and gel filtration[8] to be approximately 170,000 daltons. The $s_{20,w}$ (zero concentration) is about 9.4 S. Based on 170,000 daltons, MoFd contains 1 Mo atom, 11–14 Fe atoms, and 14–16 acid "labile" sulfide[10] groups. Iron assay indicates the presence of both ferric and ferrous iron.

MoFd is composed of two dissimilar subunits of molecular weights 60,000 and 50,000, respectively. The $s_{20,w}$ of these subunits is in the range of 3.6.

The Mo is bound to MoFd via sulfide and possibly sulfhydryl groups since it and the sulfide are released by treatment with mercury reagents. The iron also is released by this treatment.

[11]D. Y. Jeng and L. E. Mortenson (1972).

MoFd contains all common amino acids. The C-terminal amino acids are alanine and serine.

MoFd when reduced gives ESR signals at g = 4.25, 3.78, and 2.01. When oxidized the signals at g = 4.25, 3.78, and 2.01 decrease and a new large signal at the g = 2 region appears.

Structure of Nitrogenase

Present data suggest that the nitrogenase complex of *Clostridium pasteurianum* consists of two 60,000, one 50,000, and three to four 27,000 molecular weight subunits. The molecular weight of this complex would be 251,000. When MoFd and AzoFd are mixed, the resulting complex could be the complex suggested above or a larger multiple of it.

[40] Nitrogenase Complex and Its Components

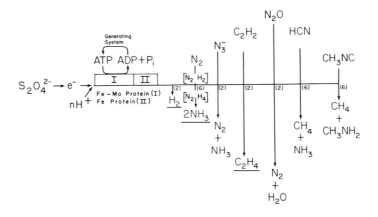

by WILLIAM A. BULEN and JACK R. LeCOMTE

Assay Methods

Principle. Three principal methods are available for the measurement of nitrogenase activity: the colorimetric measurement of ammonia formed from N_2, the manometric measurement of H_2 formation in the absence of reducible substrates, and the measurement of ethylene from acetylene reduction by gas chromatography. The H_2 evolution assay is convenient for kinetic studies and is the only known reaction in which all reducing electrons appear in a single product. The acetylene reduction assay is the most sensitive and is preferred when low levels of activity are to be detected. All are based on the $Na_2S_2O_4$ assay procedure of Bulen et al.[1] Since ATP and ADP above approximately 5 mM are inhibitory, a system for regenerating ATP is generally used.

Reagents

$Na_2S_2O_4$, 0.2 M. The reagent is prepared just before use by dissolving the salt in H_2-saturated water and adjusting the pH to 7.2. A 20-ml portion is conveniently prepared in a 30-ml test tube with a sidearm attached to a hydrogen tank equipped with a Deoxo purifier (Engelhard Industries, Inc.). A slight H_2 pressure is maintained over the solution, and a syringe is used to transfer the solution to reaction vessels.

ATP, 0.1 M. An aqueous solution is adjusted to pH 7 and stored frozen when not in use.

Creatine phosphate, 0.5 M; the solution is adjusted to pH 7.2. We have found the results more consistent using a sodium salt prepared in our own laboratory.[2]

Creatine kinase, 2.5 mg (40 units/mg) per milliliter prepared fresh daily

$MgCl_2$, 0.1 M

N-Tris(hydroxymethyl)methyl-2-aminoethanesulfonic acid (TES)– KOH buffer, pH 7.25, 0.5 M

Procedure. The assay system contains, in a final volume of 2 ml: 0.2 ml of TES buffer (100 μmoles), 0.1 ml of $MgCl_2$ (10 μmoles), 0.2 ml of creatine phosphate (100 μmoles), 0.1 ml of creatine kinase (0.25 mg), 0.1 ml of ATP (10 μmoles), 0.2 ml of $Na_2S_2O_4$ solution (40 μmoles), and enzyme. Water is added as required to give the 2-ml volume.

N_2 reduction is conducted in Warburg flasks attached to a gassing manifold constructed of capillary tubing and ground-glass joints to accommodate 4 or 6 flasks, a stopcock, and a ball joint. Manifolds are mounted on a support permitting attachment to a Warburg bath for incubation with agitation (120 oscillations per minute) at controlled temperatures. Reactions are conducted by adding buffer, water, $MgCl_2$, creatine phosphate, and creatine kinase to the main chambers and ATP to the sidearms of the flasks. Control reactions generally contain all reagents except the ATP generating system. The manifold is then

[1]W. A. Bulen, R. C. Burns, and J. R. LeComte, *Proc. Nat. Acad. Sci. U.S.* **53**, 532 (1965).
[2]This compound was synthesized by Dr. J. Corbin by a modification of a synthesis described by J. Anatal, F.P.A. 75327; *Chem. Abstr.* **58**, 576g (1963).

attached via the ball joint to a gassing apparatus consisting of a vacuum line, a mercury manometer, and a tank of N_2 (Matheson prepurified) or other gases if desired.[3] To establish anaerobic conditions, the flasks are evacuated and filled with N_2 (3×), then preincubated at 30° in the Warburg bath for 5 minutes. After a second flushing (2×), sidearms are opened individually under a slight pressure of N_2. $Na_2S_2O_4$ is added to the sidearms, and enzyme is added under the surface of the solution in the main chamber with a hypodermic syringe equipped with a short piece of polyethylene tubing attached to a 22-gauge cannula. The flasks are flushed (2×) and filled to 1 atm with N_2, then the stopcock is closed. The manifold is removed from the gassing apparatus, and the reactions are started by tipping the sidearms. The flasks are incubated at 30° in the Warburg bath, the usual reaction time for standard assays being 20 minutes. This procedure was developed to provide uniform conditions of gassing procedure, atmosphere, and temperature of incubation for a series of flasks. Other useful procedures[4] have been reported from other laboratories.

The ammonia formed is separated by microdiffusion and determined colorimetrically with Nessler's reagent. A 1-ml portion of the reaction mixture and 1 ml of a saturated K_2CO_3 solution is added to a 15-ml serum bottle. The bottle is quickly closed with a rubber stopper holding an etched-glass rod previously dipped in 1 N H_2SO_4.[5] Distillation is allowed to proceed for 90 minutes with slow reciprocal shaking (or rotation) at 30°. The etched tip of the rod is then placed in a 10-ml beaker containing 4 ml of Nessler's reagent (prepared as described by Johnson,[6] clarified by centrifugation, and diluted 1:1 with water). Three milliliters of 2 N NaOH are added, the mixture is stirred, and the glass rod is removed. After 25 minutes at room temperature, absorptivity is measured at 490 nm, and ammonia values are obtained from a standard curve. This procedure is suitable for determining up to approximately 60 μg of N.

ATP-dependent H_2 evolution is measured with Warburg respirometers attached with flexible tubing to a central manifold providing

---

[3]When gas mixtures are used, a Toepler pump is attached to the gassing apparatus, and, after the final flushing, the gas mixture is added and withdrawn (3×) into the pump reservoir to assure proper mixing. When time course data are desired, individual flasks are separated from the manifold by stopcocks permitting the removal of a single flask without disturbing the atmosphere in the manifold.

[4]For example, see, G. W. Strandberg and P. W. Wilson, *Proc. Nat. Acad. Sci. U.S.* **58**, 1404 (1967); *Can. J. Microbiol.* **14**, 25 (1967).

[5]This equipment is available from Scientific Industries, Inc., New York.

[6]M. J. Johnson, *J. Biol. Chem.* **137**, 375 (1941).

argon at low pressure. All reagents except $Na_2S_2O_4$ and enzyme are added as described for N_2 reduction, and the flasks are flushed with argon for 10 minutes with shaking. $Na_2S_2O_4$ and enzyme are added, and the flasks are flushed an additional 2 minutes; the gassing arms are closed, and the flasks are equilibrated for 5 minutes before tipping. The first reading is taken 3–5 minutes after tipping, and the first measured increment is corrected to zero time. This technique is used also for determining the amount of H_2 evolved during N_2 reduction. In this case the amount of gas evolved is corrected for the N_2 uptake calculated from a simultaneous measurement of the ammonia formed.

Acetylene reduction is conducted in Warburg flasks equipped with serum stoppers in the sidearms. Each flask is connected to a central gassing manifold via an adaptor containing a stopcock. Reagents are added and anerobic conditions established as described for N_2 reduction using argon as the flushing gas. $Na_2S_2O_4$ and enzyme are added through the serum stoppers with hypodermic syringes. The final gas mixture contains 0.1 atm of C_2H_2 and argon to 1 atm premixed with a Toepler pump. After incubation as described above, 0.5-ml samples of the gas phase are removed via the serum stopper with a gastight syringe, and the C_2H_2 is measured in a gas chromatograph equipped with an alumina column[7] and a hydrogen flame detector using He as the carrier gas. This method has been successfully adapted as an index of N_2 fixation in field studies.[8]

Inorganic phosphate is assayed by an adaptation of the method of Furchgott and de Gubareff,[9] especially designed to measure P_i in the presence of the acid-labile creatine phosphate.

Definition of Specific Activity. Specific activity is defined as the nanomoles of substrate (N_2, C_2H_2) reduced or H_2 evolved per minute per milligram of protein. As described below, the enzymatically active complex consists of two protein components, and there is kinetic evidence suggesting a dissociation of the components under assay conditions. When describing isolation of the complex, the total protein added is used for the calculations. When the component proteins are added separately, specific activities have been reported in terms of the amounts of only one component. The success of this method depends on the demonstra-

---

[7] B. Koch and H. J. Evans, *Plant Physiol.* 41, 1748 (1966).
[8] W. D. P. Stewart, G. P. Fitzgerald, and R. H. Burris, *Proc. Nat. Acad. Sci. U.S.* 58, 2071 (1967); R. W. F. Hardy, R. D. Holsten, E. K. Jackson, and R. C. Burns, *Plant Physiol.* 43, 1185 (1968).
[9] R. F. Furchgott and T. de Gubareff, *J. Biol. Chem.* 223, 377 (1956); W. A. Bulen and J. R. LeComte, *Proc. Nat. Acad. Sci. U.S.* 56, 979 (1966).

tion that one component is actually in excess and that no inhibition occurs.[10]

Purification Procedure

Nitrogenase from *Azotobacter vinelandii* can be isolated as a purified complex (PNC) providing a more stable and easily stored preparation useful for many enzyme studies. PNC can then be separated into its component proteins for individual study. This procedure reduces the complications accompanying separate isolation procedures for each component. In general, after the initial purification step, the preparations are unstable in air and the entire isolation is conducted under anaerobic conditions. Enzyme preparations are handled under an atmosphere of tank H_2 passed through a Deoxo purifier and stored as required under argon. Unless otherwise noted, all enzyme solutions are maintained at ice-bath temperature and column separations are performed at room temperature. Reagents are made anaerobic by evacuation and storage under H_2. All columns are coated with dichlorodimethylsilane and all eluting solvents are sparged for 1 hour per liter with purified H_2 passed through coarse fritted cylinders.[11] When dithiothreitol is used, the solvents are sparged for 30 minutes before it is added. Gel suspensions are prepared by heating to boiling followed by evacuation and flushing with H_2. The columns are packed under a slow stream of H_2. The final anaerobic state of all columns is attained by elution, a minimum of 5 fluid volumes of the H_2-saturated solvent being used. Samples are collected in graduated tubes fitted with serum stoppers. The tubes are evacuated in a desiccator and filled with argon (2×) through hypodermic needles which are rapidly removed when the desiccator is opened. Columns are continuously monitored with a 280-nm photometric detector unit. Eluates are collected in the sample tubes via a hypodermic needle (16 gauge), a second needle (23 gauge) being used to vent the system.

Growth of Cells. A. vinelandii (strain OP)[12] are maintained for regular use in liquid culture by daily transfer of a 5% inoculum to 150 ml of

---

[10]R. H. Burris, *Proc. Roy. Soc. Ser. B* **172,** 339 (1969).

[11]Adaptations of apparatus for anaerobic chromatography described by W. Sakami [*Anal. Biochem.* **3,** 358 (1962)] and by T. O. Munson, M. J. Dilworth, and R. H. Burris [*Biochim. Biophys. Acta* **104,** 278 (1965)] are used in our laboratory, but other designs easily constructed from equipment available in most laboratories may prove to be just as satisfactory.

[12]Obtained through the courtesy of Dr. P. W. Wilson, University of Wisconsin, Madison, Wisconsin.

Burk's medium[13] and incubation at 30° on a rotary shaker. Cells for enzyme isolation are obtained either from 6-liter cultures incubated in 10-liter bottles or from 360-liter cultures incubated in a 100-gallon tank both equipped with Carborundum spargers to provide high aeration. Cells are harvested by continuous-flow centrifugation (Sharples) after approximately 16 hours. Tank cultures yield approximately 1.5 kg of cell paste, which provides active extracts up to several months when stored frozen at −20°.

Step 1. Extraction. In a typical preparation, 1 kg of frozen cell paste is thawed in 3 liters of H_2-saturated 25 mM potassium phosphate, pH 7.5, and stirred until evenly suspended. This suspension is centrifuged at 2000 g for 20 minutes. Any slime layer present is slurried into the supernatant liquid by gentle rotation, and the liquid phase is removed by suction. The cells are resuspended in fresh buffer at ice-bath temperature with a weight-to-volume ratio of 1:2 based on the wet weight of the washed cells. The cells are ruptured by the French pressure cell technique; in fact, small amounts of cell suspension are conveniently handled with commercially available pressure cells of this type. For larger preparations, cell suspensions are passed (2×) through a Manton-Gaulin homogenizer (Model 15M-8TA) containing a ball-valve cylinder with a tapered Kennametal seat. Before each operation, the homogenizer is precooled with ice water. Cold, H_2-saturated buffer (500 ml) both precedes and follows addition of the cell suspension. The homogenate is collected anaerobically and centrifuged at 205,000 g for 30 minutes.[14] The supernatant solution is carefully decanted (leaving any loosely packed material with the sediment), diluted, when required, with buffer to a protein content of approximately 35 mg/ml, and adjusted to pH 7 with 0.5 N KOH, yielding the fraction designated S-205 (see the table).

Step 2. Protamine Sulfate-Cellulose Phosphate Fractionation. In this step, protamine sulfate is used both to remove nucleic acids and to precipitate the nitrogenase complex, cellulose phosphate being used to resolubilize the nitrogenase.[15] Approximately 50 g of S-205 protein (approximately 1.5 liters) is treated with 4.6 ml of a 2% protamine sulfate

---

[13]D. Burk and H. Lineweaver, *J. Bacteriol.* **19**, 389 (1930). The medium contains per liter: K_2HPO_4, 0.8 g; KH_2PO_4, 0.2 g; $MgSO_4 \cdot 7H_2O$, 0.2 g; NaCl, 0.2 g; $CaSO_4 \cdot 2H_2O$, 0.05 g and sucrose, 20 g. Mo (1 ppm) and Fe (3 ppm) are added, the latter as a citrate complex.
[14]With the amounts of material described here, this and all subsequent centrifugations are made using 90-ml capped tubes in the No. 42 rotor of a Beckman L2-65B centrifuge.
[15]W. A. Bulen, J. R. LeComte, R. C. Burns, and J. Hinkson, *in* "Non-Heme Iron Proteins: Role in Energy Conversion" (A. San Pietro, ed.), p. 261. Antioch Press, Yellow Springs, Ohio, 1965.

PURIFICATION OF NITROGENASE COMPLEX FROM *Azotobacter vinelandii*

Step	Fraction	Volume (ml)	Units	Protein[a]	Specific activity
1	Crude extract	1785	548,710	100.8 g	5.4
	S-205	1435	430,560	43.5 g	8.3
2	PSC	540	392,742	10.1 g	38.8
3	C-42I	180	268,074	4.3 g	62.3
	C-42II[c]	108	212,380	2.7 g	77.1[a]
					99.9[b]
3	C-42II[d]	29	46,197	610 mg[b]	75.7[b]
4	G-150 (PNC)	86	37,305	420 mg[b]	88.8[b]

[a] Protein calculated using $A_{540} = 0.092$ per milligram.
[b] Protein calculated using $A_{540} = 0.119$ per milligram (see text footnote 20).
[c,d] This table also illustrates the general magnitude of the difference in specific activities observed when: [c]C-42II is prepared as a continuous operation; [d]C-42II is obtained when fractions are stored between execution of different steps (see Properties).

solution[16] per gram of protein and stirred slowly for 30 minutes; the nucleic acids are separated by centrifugation at 205,000 g for 10 minutes. The supernatant solution is carefully decanted, leaving all loosely packed material with the sediment, and $MgCl_2$ (2 M in 10 mM TES, pH 7.1) is added to give a final concentration of 10 mM. The pH is adjusted to 6.5 with 0.5 N acetic acid, and 1.4 ml[17] of the protamine solution is added per gram of S-205 protein. After 15 minutes the protamine complex is collected in six tubes by centrifugation at 11,000 g for 20 minutes, the supernatant solution is removed by suction, and the pellets are each covered with 20 ml of 20 mM TES, pH 6.8. Purified cellulose phosphate[18] equal to one-tenth the weight of the starting S-205 protein is suspended in 20 mM TES, pH 6.8, diluted to 90 ml, saturated with H_2, and centrifuged at 11,000 g for 20 minutes. The sediment is

[16] This solution is freshly prepared in water, adjusted to pH 6 with NaOH, cleared of any insoluble material by centrifugation, and held at room temperature. Deaeration is affected by stirring under a slow flow of Argon (2 hours) followed by H_2 (30 minutes) just before use.

[17] Minor variations in the amounts required to give maximal recovery without precipitating unwanted proteins may be required when different batches of reagents or different cell cultures are used.

[18] Commercial cellulose phosphate (Sigma Chemical Co.) is cleared of fine particles by suspension (5×) of 10 g in 500 ml of water and removal of suspended material by suction after most of the material has settled (approximately 1 hour). Ammonia is removed by transferring the sedimented portion to a column and washing with 1 liter of 10 mM potassium phosphate, pH 7, followed by 1.5 liters of demineralized water. The material is dried and, after the clumps are disintegrated, stored dry in closed containers.

resuspended in 20 ml of the TES buffer, combined with the pellets of the protamine complex, and the mixture is evenly suspended with a magnetic stirrer. This suspension is transferred to a 4 × 40-cm column equipped with a fritted-glass disk. The pass-through solution and 50 ml of wash buffer are collected anaerobically, diluted to 540 ml with buffer, and centrifuged at 205,000 g for 30 minutes. The clear supernatant fraction is decanted and designated as fraction PSC.

Step 3. MgCl₂ Precipitation. In this step, remaining small contaminating proteins are removed. Sufficient 2 M MgCl₂ solution is added to the PSC fraction to give a final concentration of 20 mM. The solution is immediately adjusted to pH 7 with 0.5 N KOH, stirred intermittently for 30 minutes, and centrifuged at 30,000 g for 15 minutes. The supernatant solution is decanted and centrifuged at 205,000 g for 2 hours to sediment the complex. Suction is used to remove and discard the supernatant material, and the pellets are suspended in 20 mM TES, pH 6.8, containing 0.1 mg of dithiothreitol (DTT) per milliliter. The pellets are disintegrated with a stirring rod, and resuspension is facilitated by brief intermittent treatments with a 20-kHz ultrasonic probe. The solution is diluted to 180 ml and centrifuged at 60,000 g for 10 minutes. The clear supernatant fraction contains about 20 mg of protein per milliliter and is designated C-42I. Sufficient 2 M MgCl₂ is again added to give a final concentration of 10 mM, and the solution is stirred intermittently for 30 minutes. Centrifugation at 20,000 g for 15 minutes followed by centrifugation of the supernatant solution at 205,000 g for 2 hours again provides pellets of the complex which are overlayered with 20 mi of the TES–DTT buffer. The pellets are combined in one centrifuge tube with 5 ml of rinse buffer and fully suspended by slow magnetic stirring.[19] Protein concentration is controlled at this point by dilution with the buffer to approximately 25 mg/ml. After centrifugation at 60,000 g for 10 minutes, the supernatant solution (approximately 90 ml) is decanted and designated C-42II.

Step 4. Sephadex Column Chromatography. Remaining large molecular weight contaminants (some of which contribute cytochrome-type spectra) are removed by Sephadex G-150 column chromatography. A 4 × 30-cm glass column (Fischer-Porter Co.) is closed with a rubber stopper through which pass two pieces of glass tubing that can be connected via rubber tubing to (1) a Y connector leading to two 2-liter reservoir bottles with side tubes and (2) an H₂ tank.[11] One reservoir contains water plus 0.1 mg of DDT per milliliter, and the other contains

---

[19]The use of the sonic oscillator probe for this suspension will frequently cause unwanted gel formation.

20 mM TES and 10 mM MgCl$_2$ at pH 7.2 with 0.1 mg of DTT per milliliter. The column is packed to a length of 20 cm, a sample applicator is inserted, and the reservoirs are saturated with H$_2$. The packed column is equilibrated from the water reservoir, and approximately 30 ml of C-42II (approximately 600 mg of protein) is added from a hypodermic syringe through the rubber tubing into the sample applicator. The sample, followed by an equal volume of water, is passed into the column. The flow rate is maintained at 30 ml per hour until all sample has entered the column and then is increased to 60 ml per hour. Elution is completed with the TES buffer solution, and the purified nitrogenase complex is collected after approximately 140 ml has eluted. A typical elution pattern is shown in Fig. 1. The amount of material eluted in fraction (1) will vary somewhat with different preparations, but fractions of the nitrogenase complex freed of this material are routinely obtained. A check for cytochrome-type spectra is useful here. This procedure yields 400–500 mg of protein[20] in fraction (2), which is designated G-150 and subsequently referred to as the purified nitrogenase complex (PNC). For prolonged storage, the G-150 fraction is

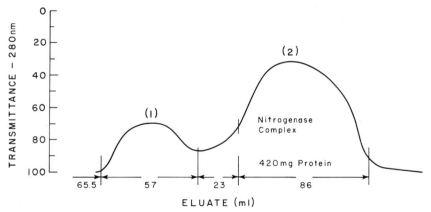

Fig. 1. Elution of nitrogenase complex from Sephadex G-150 column. Fraction (1) is not retained on the column and contains material giving cytochrome-type spectra. Fraction (2), the G-150 fraction of the table, is referred to as purified nitrogenase complex (PNC) and by this procedure generally contains 400–500 mg of protein.

[20]Protein is determined with the biuret reagent of A. G. Gornall, C. J. Bardawill, and M. M. David, *J. Biol. Chem.* **177**, 751 (1949). With crude or partially purified preparations, bovine serum albumin giving an absorbance at 540 nm of 0.092/mg was used as the standard. With the purified nitrogenase complex, amino acid analysis and dry weight determinations made by Dr. S. Klein in our laboratory established that $A_{540} = 0.119$ per milligram of protein was a more accurate value.

concentrated by ultrafiltration (Amicon Diaflo, Model 50 with an XM50 membrane) at room temperature to approximately 20 mg of protein per milliliter, enough of the 2 M MgCl$_2$ solution is added to give 5 mM MgCl$_2$, and the pH is adjusted to 7 with 0.5 M KOH if required.

Step 5. Separation of Components. This step is a modification of the original DEAE-cellulose column procedure of Bulen and LeComte[21]; when it is to be applied, the G-150 column (Step 4) is eluted with 20 mM Tris·HCl, pH 7.2, containing 0.1 mg of DTT per milliliter (rather than the TES buffer solution previously described) and concentrated by ultrafiltration to 15–20 mg of protein per milliliter. A 2.5 × 9-cm column is prepared from Whatman DE-52 (H. Reeve Angel Inc., Clifton, New Jersey) from which fine particles have been removed, as recommended by the manufacturer, by repeated suspension in 25 mM Tris·HCl, pH 7.2. The 9-cm length refers to the column packed in air under gravity in a 15-cm column, washed successively with 100 ml of 0.5 N KOH, 50 ml of water, and 300 ml of 25 mM Tris·HCl, pH 7.2 (allow 1.5 cm for shrinkage), and topped with a sample applicator. This same washing procedure is used to regenerate the column for subsequent separations. Through a stopper closing the column pass two glass tubes connected by rubber tubing to an H$_2$ tank providing a slight pressure and to a 3-way stopcock, one arm of which leads to a 1-liter reservoir containing 800 ml of 25 mM Tris·HCl, pH 7.2. The third arm of the stopcock leads to a 4-way glass connector. Three arms of the connector lead to 500-ml reservoirs cntaining 200 ml of 0.15, 0.27, and 0.4 M NaCl in 25 mM Tris·HCl, pH 7.2, respectively. The 0.4 M solution also contains 5 mM MgCl$_2$. The solvents are saturated with H$_2$ and supplied with DTT by gassing simultaneously with sparging systems of the type previously cited.[11] Reservoirs are connected in series so as to permit the addition of one reservoir at a time to the sequence, with accompanying pressure increases as required. The column is equilibrated and rendered anaerobic by elution with 400 ml of the Tris-buffer solution, and 300–350 mg of the concentrated complex is injected into the sample applicator. The sample is allowed to enter the column at a flow rate of 18 ml per hour and followed by 5 ml of the 25 mM Tris buffer solution. The column is then sequentially eluted with 20 ml of the 0.15 M, 70 ml of the 0.27 M, and, finally, the 0.4 M NaCl solutions, the flow rate being increased to 36 ml per hour when the 0.27 M solution is started. All samples are collected anaerobically as previously described. A typical elution pattern is shown in Fig. 2. Recovery of protein from this column is approximately 95% of that added. Only

[21]W. A. Bulen and J. R. LeComte, *Proc. Nat. Acad. Sci. U.S.* **56**, 979 (1966).

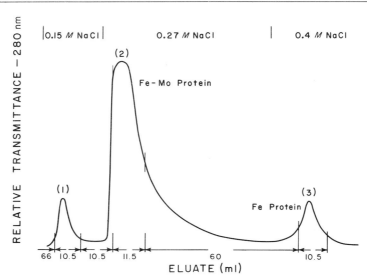

Fig. 2. Elution of components of the nitrogenase complex from a DEAE-cellulose column. Fraction (1) eluted by the 0.15 M NaCl solution has been tentatively identified from spectra and electron spin resonance data as the pink, Azotobacter II, protein of Shethna *et al.* [Y. I. Shethna, D. V. Der Vartanian, and H. Beinert, *Biochem. Biophys. Res. Commun.* 31, 862 (1968)]. Fraction (2) is eluted with the 0.27 M and fraction (3) is eluted with the 0.4 M NaCl solutions, respectively. The vertical lines indicate eluates collected for analysis. Fractions 1–3 contained 15, 210, and 22.5 mg of protein ($A_{540} = 0.119$ per milligram; see text footnote 20), respectively.

the Fe-Mo protein and the Fe protein are required for catalysis of the nitrogenase reactions by the $Na_2S_2O_4$ assay. This requirement is illustrated by the results of the combination of 0.91 mg of the Fe-Mo protein with 0.42 mg of the Fe protein (obtained from the fractions collected in the separation depicted in Fig. 2). The combined fractions yielded 71.4 units of activity, and the activity of the individual fractions was negligible. Adaptations or modifications of this procedure have been used by other workers to separate nitrogenase preparations from *Azotobacter* sp. and other sources at various stages of purity into component fractions requiring recombination for activity.[22-26]

[22]R. V. Klucas, B. Koch, S. A. Russell, and H. J. Evans, *Plant Physiol.* 43, 1906 (1968).

[23]R. W. F. Hardy, R. D. Holsten, E. K. Jackson, and R. C. Burns, *Plant Physiol.* 43, 1185 (1968).

[24]R. W. Detroy, D. F. Witz, R. A. Parejko, and P. W. Wilson, *Proc. Nat. Acad. Sci. U.S.* 61, 537 (1968); *J. Bacteriol.* 98, 325 (1969).

[25]M. Kelly, *Biochim. Biophys. Acta.* 171, 9 (1969).

[26]M. Kelly, R. V. Klucas, and R. H. Burris, *Biochem. J.* 105, 3C (1967).

Properties

In defining properties it is to be emphasized that one is dealing with a system subject to some inactivation during purification and examination, even under the most carefully controlled conditions. Reaction with molecular oxygen is perhaps the major contributing factor. We view the presentation of absolute number values for some of the properties with reservation pending confirmation from other laboratories.

Stability. The nitrogenase complex is considerably more stable than either the Fe-Mo protein or the Fe protein. Susceptibility to molecular oxygen increases with purification but is retarded by the presence of $MgCl_2$. When the preparation of PNC is to be interrupted, temporary storage of the S-205 fraction, of the pellets from the protamine separation after covering with buffer, of the PSC fraction, or of the C-42I or C-42II fractions is recommended. C-42II can be stored anaerobically for up to 6 weeks at ice-bath temperature with little or no loss of activity. The PNC is stable for up to 30 days after concentration when anaerobic conditions and ice-bath temperatures are maintained. The Fe-Mo protein is subject to oxidation but can be stored anaerobically at ice-bath temperatures. The Fe protein is cold labile and has a half-life of approximately 18 hours when dilute solutions are stored at room temperature. Both proteins can be stored in liquid nitrogen for up to 14 days without loss of activity.[26]

Purity. Purity is followed routinely by examination under anaerobic conditions in the analytical ultracentrifuge and evaluated by the symmetry of the observed refractive index gradients and the area under the curve. The successful evaluation of purity requires that the samples be examined in solutions of low ionic strength and essentially free of $MgCl_2$. For example, 0.1 M NaCl will cause the PNC to give multiple gradients all of which sediment faster than the desalted sample. Approximately 10 mg protein per milliliter is routinely analyzed after passage through a polyacrylamide P-2 gel column equilibrated with 10 mM TES, pH 7.2. The calculated $s_{20,w}^0$ values for the PNC and the Fe-Mo protein obtained to date average 11.9 S and 11.0 S, respectively.[27] Additional peaks representing the presence of contaminants, inactivation, or (as suspected) dimerization, when present, do not account for more than 5% of the total area under the curves. The Fe protein fraction has not been extensively examined by this technique, and $s_{20,w}^0$ values are not available. Evidence for the integrity of the complex

[27]We are indebted to R. R. Knotts with Dr. R. Darrow for numerous ultracentrifuge examinations.

is both physical and chemical. For example, it can be sedimented by centrifugation and it behaves as a single species both on the G-150 column and during ultrafiltration with membranes designed to retain only molecules larger than 50,000 molecular weight, even though it contains two components of lower molecular weight. The constancy of amino acid analysis performed on different preparations and the relative amounts of the components obtained on separation also supports this concept.

pH and Buffers. Phosphate and cacodylate buffers used in early experiments produced inhibition and the formation of toxic arsines, respectively; therefore TES buffer[28] (pK_a, 7.3 at 30°) is recommended. Azotobacter nitrogenase is active over a broad pH range with an optimum at about 7.25.[29]

Substrate Specificity. Azotobacter nitrogenase reactions exhibit a specificity for ATP (K_m, 0.3 mM[30]). In the absence of reducible substrate, all reducing electrons formed are evolved as H_2 in what has been designated as the ATP-dependent hydrogen evolution reaction.[1,31] When products other than P_i are not measured, this reaction has also been referred to as a reductant-dependent ATPase.[32] In addition to N_2 (K_m, 0.1 mM[33,34]), the following compounds have been shown to undergo reduction by extracts or partially purified preparations from *Azotobacter* sp: N_3^- (K_m, 1.3 mM[35,36]), C_2H_2 (K_m, 0.3 mM[37]), N_2O (K_m, 1.2 mM[36]), HCN (K_m, approximately 4 mM[36]), and CH_3NC.[38] In addition to methane, small amounts of ethylene and ethane were also found upon reduction of HCN or CH_3NC. The reduction of various analogs of nitriles and acetylenes has also been reported.[39] Nitrogenase from *Azotobacter* sp. also catalyzes the formation of HD when D_2 is present in the atmosphere or when the reaction is conducted in

[28]N. E. Good, G. D. Winget, W. Winter, T. N. Connolly, S. Izawa, and R. M. M. Singh, *Biochemistry* 5, 467 (1966).

[29]K. L. Hadfield and W. A. Bulen, *Biochemistry* 8, 5103 (1969).

[30]R. C. Burns, *Biochim. Biophys. Acta* 171, 253 (1969).

[31]R. C. Burns and W. A. Bulen, *Biochim. Biophys. Acta* 105, 437 (1965).

[32]R. W. F. Hardy and E. Knight, Jr., *Biochim. Biophys. Acta* 122, 520 (1966).

[33]G. W. Strandberg and P. W. Wilson, *Proc. Nat. Acad. Sci. U.S.* 58, 1404 (1967).

[34]Determined by K. L. Hadfield in our laboratory.

[35]R. Schöllhorn and R. H. Burris, *Proc. Nat. Acad. Sci. U.S.* 57, 1317 (1967).

[36]R. W. F. Hardy and E. Knight, Jr., *Biochim. Biophys. Acta* 139, 69 (1967).

[37]R. Schöllhorn and R. H. Burris, *Proc. Nat. Acad. Sci. U.S.* 58, 213 (1967).

[38]M. Kelly, J. R. Postgate, and R. L. Richards, *Biochem. J.* 102, 1C (1967).

[39]R. W. F. Hardy and E. K. Jackson, *Proc. Fed. Amer. Soc. Exp. Biol.* 26, 725 (1967).

D_2O with H_2 in the atmosphere. Opposing views regarding the requirement for N_2 in this reaction have been presented.[40,41]

Activators and Inhibitors. A divalent cation, preferably Mg^{2+}, is required for nitrogenase activities, as demonstrated using substrate levels of ATP.[31] Mn^{2+}, Co^{2+}, or Fe^{2+} replace Mg^{2+} but are less effective, and Ca^{2+} is inhibitory.[30] CO and H_2 are reversible inhibitors of N_2 reduction. H_2 inhibits competitively[33,40] (K_i, 0.19 atm), but CO inhibition appears to be noncompetitive.[42] CO inhibits the reduction of all other substrates listed[36,38,39] and the formation of HD,[40,41] but does not inhibit the ATP-dependent H_2 evolution reaction.[15] H_2 appears not to inhibit the reduction of substrates other than N_2,[42] nor does it inhibit the ATP-dependent H_2 evolution reaction. Metal binding agents, e.g., *o*-phenanthroline, 2,2'-dipyridyl, and Tiron[15] and conditions of high ionic strength, e.g., NaCl above 50 mM,[21] inhibit. P_i at concentrations above 30 mM and ATP or ADP at levels above 5 mM[1] produce marked inhibition. Hydrazine and some of its derivatives are not reduced and act as weak inhibitors.[35]

Metals and Labile Sulfur. The PNC contains Mo, Fe, and labile sulfur. From nine preparations, the Mo content was 6.52 ± 0.35 and the Fe content 159.1 ± 8.8 nanomoles per milligram of protein, giving an Fe:Mo ratio of 24.4. The Fe-to-labile sulfur ratios are in the range 1.8:1 but, since the incorporation of the $MgCl_2$ purification steps, a sufficient number of labile sulfur analyses are not available on which to base a reliable standard deviation.

Among the components, the pink (Azotobacter II) Fe protein, fraction (1) of Fig. 2, was previously characterized by Shethna *et al.*[43] Both the Fe-Mo protein and the Fe protein contain labile sulfur in addition to the metals. Sufficient analysis of preparations of these components is not yet available for statistical evaluation.

Spectra and Other Properties. Spectra of the PNC preparations have a shoulder in the 420 nm region (appearing as a peak only if all cytochrome-containing material has not been removed) but show only a general reduction in absorbance at higher wavelengths. Reduction with either $Na_2S_2O_4$ or formamidine sulfinic acid produces a decrease both in the 420-nm region and in the general absorbance through the visible region. Difference spectra (protein as isolated vs reduced) show

[40] E. K. Jackson, G. W. Parshall, and R. W. F. Hardy, *J. Biol. Chem.* 243, 4952 (1968).
[41] M. Kelly, *Biochem. J.* 109, 322 (1968).
[42] J. C. Hwang and R. H. Burris, *Fed. Proc. Fed. Amer. Soc. Exp. Biol.* 27, 639 (1968).
[43] Y. I. Shethna, D. V. Der Vartanian, and H. Beinert, *Biochem. Biophys. Res. Commun.* 31, 862 (1968).

shoulders in the 420- and 460-nm regions; some of the former and most of the latter can be ascribed to the reduction of the 418 and 460 nm absorbance of the pink (Azotobacter II) protein. Upon exposure to air or O_2, the complex exhibits a general decrease in absorbance over the entire visible spectra. A farther decrease rather than an increase accompanies reduction of the oxygen-reacted protein with $Na_2S_2O_4$. An ESR signal at $g = 2.0$ forms during exposure to air. It reaches a maximum and then decreases on continued exposure. Both the Fe-Mo protein and the Fe protein display shoulders in the 416–420-nm region showing only a nondescript, general reduction in absorbance at higher wavelengths. Reduction with $Na_2S_2O_4$ produces a general decrease in absorbance, with the shoulder still discernible. With 416 nm used as an arbitrary wavelength, absorptivity (a) values ($gl^{-1} cm^{-1}$) calculated for what is considered a typical isolation are: PNC—as isolated 0.35, reduced 0.25; Fe-Mo protein—as isolated 0.39, reduced 0.35; Fe protein—as isolated 0.27, reduced 0.19. These values are considered to represent ranges rather than specific numbers and, pending determination of exact molecular weight values and an assessment of the degree of oxidation of the isolated material, molar absorptivity values can only be approximated.

The presentation of exact molecular weight data is prevented by abnormal behavior on gel columns and the demonstration that (at the low concentrations used) the meniscus depletion, sedimentation equilibrium method of ultracentrifuge analysis shows the solutes to become polydisperse. Based on the Mo content, a minimum molecular weight of 120,000 has been calculated for the Fe-Mo protein.

[41] Preparation of Nitrogenase from Nodules and Separation into Components

By HAROLD J. EVANS, BURTON KOCH, and ROBERT KLUCAS

For several years, cell-free preparations of nitrogenase from *Clostridium pasteurianum* have been utilized in the elucidation of the biochemistry of nitrogen fixation in these organisms.[1] Until recently, all efforts to prepare cell-free nitrogenase from the nodules of leguminous plants had met with failure. After the realization that nitrogenase was sensitive to oxygen, Bergerson[2] utilized a specially designed an-

[1]R. W. F. Hardy and R. C. Burns, *Annu. Rev. Biochem.* 37, 331 (1968).
[2]F. J. Bergersen, *Biochim. Biophys. Acta* 130, 304 (1966).

aerobic press for the preparation of nodule brei which exhibited nitrogen-fixing capacity. Cell-free extracts of nodule bacteroids could not be prepared until it was realized that the phenolic compounds that are prevalent in legume nodules react with, and denature, the nitrogenase when nodules are macerated in air.[3,4] In the method to be described, nodules are macerated and extracts of bacteroids are prepared under conditions where contact with oxygen is minimized. Ascorbate is used in the preparative medium to prevent the oxidation of phenols, and phenolic compounds in nodule breis are removed as an insoluble complex with polyvinylpolypyrrolidone (PVP).[5] By use of these procedures, crude extracts of nodule bacteroids with specific activities comparable to those of extracts of free-living nitrogen-fixing bacteria may be prepared consistently. The preparations have been purified and fractionated into two components, both of which are essential for nitrogenase activity.

Preparation Methods

Reagents and Materials

Nutrient Solution (see table)

COMPOSITION OF NITROGEN-FREE NUTRIENT SOLUTION

Salt	Concentration
	(g/liter)
K_2SO_4	0.279
$MgSO_4 \cdot 7H_2O$	0.493
KH_2PO_4	0.023
K_2HPO_4	0.145
$CaSO_4 \cdot 2H_2O$	1.033
$CaCl_2$	0.056
	(mg/liter)
H_3BO_3	1.43
$MnSO_4 \cdot 4H_2O$	1.02
$ZnSO_4 \cdot 7H_2O$	0.22
$CuSO_4 \cdot 5H_2O$	0.08
$Na_2MoO_4 \cdot 2H_2O$	0.05
$CoCl_2 \cdot 4H_2O$	0.10
Ferric ethylenediamine di-O-hydroxyphenyl acetate[a]	16.67

[a] This material contains 6% Fe and is obtained from the Geigy Chemical Co., P.O. Box 430, Yonkers, New York.

[3] B. Koch, H. J. Evans, and S. Russell, *Plant Physiol.* **42**, 466 (1967).
[4] B. Koch, H. J. Evans, and S. Russell, *Proc. Nat. Acad. Sci. U. S.* **58**, 1343 (1967).
[5] W. D. Loomis and J. Battaile, *Phytochemistry* **5**, 423 (1966).

Reagents for Preparation of Extracts

Potassium phosphate-ascorbate buffer, 0.02 *M* solution of potassium phosphate buffer to which is added 0.2 *M* sodium ascorbate. The pH is adjusted to 7.4.

Insoluble polyvinylpolypyrrolidone (Polyclar AT, General Aniline Corporation, New York, New York) Suspend 1 weight of PVP in 5 weights of 3 *N* HCl and allow to stand for 12 hours. Remove the acid by filtration in a Büchner funnel and wash thoroughly with distilled water. Follow this by washing with 0.02 *M* potassium phosphate buffer, pH 7.4, until the pH is 7.0 or above.

Tris(hydroxymethyl)aminomethane (Tris buffer), 0.05 *M* solution, pH 8.0

Tris buffer, 0.1 *M* solution, pH 8.5

Protamine sulfate (Eli Lilly and Co., Indianapolis, Indiana), 2% solution in water adjusted to pH 6.0

DEAE-cellulose (Whatman DE32, H. Reeve Angel Inc., Clifton, New Jersey)

NaCl, 0.15 *M* solution in 0.025 *M* Tris buffer, pH 7.5

$MgCl_2$, 0.035 *M*, 0.06 *M*, and 0.10 *M* solutions in 0.025 *M* Tris buffer, pH 7.5

Cellulose phosphate (Sigma Chemical Co., St. Louis, Missouri)

Reagents for the Nitrogenase Assay

Adenosine 5-triphosphate, disodium salt (Sigma Chemical Co., St. Louis, Missouri), 0.05 *M* solution in 0.05 *M* Tris buffer, pH 8.0

Phosphocreatine, disodium salt (Sigma Chemical Co., St. Louis, Missouri), 0.5 *M* solution in 0.05 *M* Tris buffer, pH adjusted to 7.5

Creatine phosphokinase (a preparation from Sigma Chemical Co., St. Louis, Missouri, that will transfer 35 μmoles of phosphate from phosphocreatine to ADP per minute per milligram of enzyme), 2 mg/ml in 0.05 *M* Tris buffer, pH 7.5

$MgCl_2$, 0.1 *M* solution in distilled H_2O

Sodium dithionite (J. T. Baker Chemical Co., Phillipsburg, New Jersey). The solid dithionite is maintained in a desiccator under vacuum. A 0.2 *M* solution is freshly prepared in oxygen-free 0.05 *M* Tris buffer, pH 8.0, and maintained under anaerobic conditions.

Nitrogen gas, prepurified (National Cylinder Gas Co., Chicago, Illinois), containing less than 8 ppm of oxygen

Argon (National Cylinder Gas Co., Chicago, Illinois) containing less than 1 ppm of oxygen

Calcium carbide (Allied Chemical Co., Morristown, New Jersey)

Procedure

Source of Nodules. Soybean plants (*Glycine max* Merr.) have been used routinely as a source of nodules for the preparation of nitrogenase. This legume grows relatively rapidly in the greenhouse and produces large nodules that are removed easily from the roots. In some experiments, nodules of serradella (*Ornithopus sativus*) and blue lupines (*Lupinus angustifolius*) have been utilized, and active extracts have been prepared from the nodule bacteroids of these species.

Soybean seeds inoculated with a commercial strain of *Rhizobium japonicum* (The Nitragen Company, Milwaukee, Wisconsin) are planted in plastic pots (8 inches in diameter) containing either perlite or vermiculite. A supply of 300–400 g of nodules may be obtained from 150 pot cultures, each containing eight plants. Each pot is flushed daily with nitrogen-free nutrient solution and every fourth day with water to prevent the accumulation of salts in the growth medium. An excess of nutrient solution is added to each pot, and the solution not held by perlite or vermiculite escapes through drain holes in the bottom of each pot. During the winter months, plants are supplemented with fluorescent light of about 400 ft-c (3 feet from the pot) to prevent premature flowering. When the plants are 30–35 days old, they are removed from the perlite, root systems are flushed with tap water, and nodules are removed from the plants. After the nodules have been washed with distilled water to remove perlite and other debris, the nodules are either used directly for the preparation of nitrogenase or frozen in liquid nitrogen and stored at −80° until they are used for the preparation of extracts.

Preparation of Bacteroids. All steps in the preparation of bacteroids and the subsequent preparation of cell-free extracts are carried out under conditions that minimize contact with oxygen. This is accomplished as follows: by sparging reagent solutions with purified nitrogen or argon and maintaining them in bottles sealed with rubber serum caps; by transferring solutions by use of plastic syringes; and by use of a glove box filled with nitrogen gas for certain operations. In the preparation of bacteroids and cell-free extracts, all operations are carried out at 0–4°.

In a typical experiment, 150 g of fresh or frozen nodules, 50 g of PVP, 300 mg of dithionite, and 400 ml of ascorbate–phosphate buffer are placed in the vessel of an Omni-mixer (Ivan Sorvall and Co.) fitted with gassing vents. The contents of the vessel are gassed for 10 minutes while being shaken with nitrogen. The nodules are then macerated by blending for 5 minutes at a voltage setting of 50 on a variable transformer. The gas vents are closed, and the vessel is transferred to a glove box. The macerate is squeezed through four layers of cheesecloth, and the liquid portion is transferred to 50-ml centrifuge tubes, which are then sealed. After centrifugation of the macerate for 10 minutes at 5000 g, the tubes are returned to the glove box, the supernatant solution is discarded, and the bacteroids are resuspended in 0.05 M Tris buffer at pH 8.0. Again, bacteroids are collected by centrifugation at 5000 g for 10 minutes, about 20 mg of dithionite are added to the bacteroids, and the supernatant fluid is discarded.

Preparation of Cell-Free Extracts. The washed bacteroid cells and a precooled Aminco French press are placed in a nitrogen filled glove box. Then the bacteroids are suspended in 20 ml of 0.1 M Tris buffer at pH 8.5 and the suspension is transferred to the press. The press is closed and removed from the glove box, then the cells ruptured by being forced through the orfice of the press at a pressure of 16,000 psi. The ruptured cells are collected in centrifuge tubes that are continuously gassed with a stream of nitrogen. The centrifuge tubes are closed with caps fitted with rubber seals and are centrifuged at 40,000 g for 1 hour; the crude cell-free supernatant liquid is collected. The crude extract is either used immediately or transferred to polyethylene bottles filled with liquid nitrogen and then stored in liquid nitrogen.

The cell-free extract from three different 150-g preparations of nodules are combined for further purification procedures. About 60 ml of crude extract is analyzed for protein,[6] and sufficient 2% protamine sulfate is added slowly with stirring to obtain 80 mg of protamine sulfate per gram of protein. The extract is stirred gently for 10 minutes and then centrifuged at 40,000 g for 10 minutes. To the supernatant solution, additional 2% protamine sulfate is added to obtain a final protamine sulfate concentration of 160 mg per gram of protein. The preparation is stirred again for 10 minutes, then the precipitate is collected by centrifugation at 15,000 g for 10 minutes. The pellet is suspended in 24 ml of 0.025 M Tris buffer at pH 7.5, and 100 mg of cellulose phosphate is added for each gram of protein in the original crude extract.[7] The preparation is stirred slowly for 10 minutes, the cellulose phosphate is removed by centrifugation at 15,000 g for 10 minutes, and the supernatant liquid is collected.

Some variation has been observed in the quantity of protamine sulfate required to precipitate most of the nitrogenase. The amount utilized, therefore, must be determined by the individual researcher.

Chromatography. The protamine sulfate eluate is further purified on a column of DEAE-cellulose using the anaerobic technique described

[6]A. G. Gornall, C. J. Bardawill, and M. M. David, *J. Biol. Chem.* 177, 751 (1949).
[7]W. A. Bulen and J. R. LeComte, *Proc. Nat. Acad. Sci. U. S.* 56, 979 (1966).

by Munson *et al.*[8] DEAE-cellulose is suspended in 0.5 M HCl, washed with water, then suspended in 0.5 M KOH and washed again with water as recommended by the manufacturers. The washed DEAE in a Büchner funnel is then leached with 0.025 M Tris buffer at pH 7.5 until the pH of the eluate is 7.0 or above. A column of DEAE-cellulose (2.5 × 13 cm) is prepared and attached to the anaerobic elution apparatus.[9] The column is leached overnight with about 2-liters of 0.025 M Tris buffer at pH 7.5 to which has been added 10 mg of sodium hydrosulfite. The temperature of the column is maintained at 9° by an external water jacket connected to a temperature-controlled water bath. The nitrogenase (about 24 ml of protamine sulfate eluate) is layered on the top of the DEAE-cellulose with a hypodermic syringe to which is attached a section of polyethylene tubing 50 cm in length and of 1.5 mm internal diameter. When the meniscus of the extract reaches the top surface of the DEAE column, elution of proteins is initiated with a series of salt solutions.

Before the buffered salt solutions are added to the column, each solution is thoroughly sparged with high-purity nitrogen and is then treated with about 5 mg of $Na_2S_2O_4$ per 50 ml of solution. Fractions (10 ml each) are collected through a hypodermic needle attached to the end of the column. For the collection of each fraction, the needle is injected through the rubber serum cap of a 21-ml bottle that previously was thoroughly gassed with high-purity nitrogen. A series of bottles sealed with rubber serum caps and gassed with nitrogen are utilized for the manual collection of the fractions. When the eluate from the column begins to drip into the bottle, a second needle is injected into the serum cap to allow gas to escape.

The column is eluted first with 60 ml of 0.15 M NaCl, then successively with 50 ml of 0.035 M $MgCl_2$, 50 ml of 0.06 M $MgCl_2$, and finally with 100 ml of 0.1 M $MgCl_2$. All the salt solutions are prepared in 0.025 M Tris at pH 7.5. A flow rate of about 60 ml per hour is maintained.

In a typical chromatographic experiment, proteins without nitrogenase activity are eluted with the NaCl solution and a greenish-brown component is eluted with the solution of 0.035 M $MgCl_2$. This component shows a high concentration of both nonheme iron and molybdenum and is designated as fraction I. The solution of 0.06 M $MgCl_2$ removes all the remaining fraction I from the column, and then a second colored component is eluted with the solution of 0.1 M $MgCl_2$. This component, designated as fraction II, contains appreciable nonheme

[8]T. O. Munson, M. J. Dilworth, and R. H. Burris. *Biochim. Biophys. Acta* **104**, 278 (1965).
[9]R. V. Klucas, B. Koch, S. A. Russell, and H. J. Evans, *Plant. Physiol.* **43**, 1906 (1968).

iron, but no greater concentration of Mo than the crude extract. In some cases, fraction II exhibits some nitrogenase activity when assayed alone; however, in many cases the two components are separated completely and neither component alone shows activity in the ammonia synthesis assay. As shown by Klucas et al.,[9] vigorous N_2 reduction takes place in assay mixtures containing both fractions. As pointed out by Klucas et al.[9] if one wishes to assay the activity of one component, it is essential to add the other component in excess and to establish that the response to the fraction varied is approximately linear.

Assay

The synthesis of ammonia by the nitrogenase reaction is measured by a procedure similar to that described by Dilworth et al.[10] The reaction is carried out in a 21-ml glass vial. The standard assay mixture in a final volume of 1.5 ml contains 75 μmoles of Tris·HCl at pH 7.5, 7.5 μmoles of Na_2 ATP, 0.2 mg of creatine phosphokinase, 50 μmoles of creatine phosphate, and 10 μmoles of $MgCl_2$. The bottle is sealed with a rubber serum cap (No. 2330, Arthur H. Thomas and Co.) and is evacuated and flushed five times with high-purity nitrogen gas or argon. Then 20 μmoles of $Na_2S_2O_4$ from a 200 mM solution in 0.05 M Tris buffer at 8.0 is injected into the bottle with a hypodermic syringe and the reaction is initiated by injection of an appropriate amount of nitrogenase. The reaction is incubated with shaking for 20 minutes in a water bath at 30° and is terminated by injecting 2 ml of saturated K_2CO_3 solution into the vial.

A glass rod with an etched tip is fitted into a rubber stopper of appropriate size to seal the reaction vessel. The etched surface of the glass rod is dipped into 1 N H_2SO_4 and the rubber stopper is inserted into the reaction vessel immediately after the addition of saturated K_2CO_3 to terminate the reaction. During a 2-hour period of shaking, ammonia is distilled into the acid and then measured by nesslerization.[11] This procedure has been described by Dilworth et al.[10]

Instead of measuring ammonia synthesis, reaction mixtures may be gassed out with argon, 0.1 atm of acetylene is added, and the ethylene production is measured by the procedure of Kelly et al.[12] Acetylene for use in this procedure may be produced conveniently from calcium carbide or obtained from a commercial source.

[10]M. J. Dilworth, D. Subramanian, T. O. Munson, and R. H. Burris, Biochim. Biophys. Acta 99, 486 (1965).
[11]W. M. Umbreit, R. H. Burris, and J. F. Stauffer, "Manometric Techniques," p. 274. Burgess, Minneapolis, Minnesota, 1957.
[12]M. Kelly, R. V. Klucas, and R. H. Burris, Biochem. J. 105, 1c (1967).

[42] Clostridial Rubredoxin

By WALTER LOVENBERG

Rubredoxin is an iron-containing protein which serves as an electron carrier in bacterial enzyme systems. This red protein was isolated first from *Clostridium pasteurianum*[1] and later from several other anaerobic organisms.[2-5] No specific function is known for this protein in anaerobic organisms. A similar protein has been isolated from *Pseudomonas oleovorans*.[6] In this organism, rubredoxin functions as an electron carrier in the ω-hydroxylation system.

Assay Method

There is no common and specific biochemical assay for the rubredoxins, and the rubredoxin content of extracts is generally monitored during isolation by absorbancy at 490 nm. A biochemical procedure which has been used[7] is the reduction of cytochrome *c* by NADPH, spinach ferredoxin-NADP reductase, and limiting amounts of rubredoxin. In this reaction rubredoxin transfers an electron from the reduced flavin enzyme to the cytochrome *c*. The procedure of Peterson *et al.*[7] for this spectrophotometric assay is as follows: Reaction mixtures containing 100 μmoles of Tris·HCl, pH 7.5, 1 mg of bovine serum albumin, about 20 μg of purified spinach ferredoxin-NADP reductase,[8] 0.05 μmole of horse heart cytochrome *c*, and 0.01–0.04 mμmole of rubredoxin are placed in cuvettes. After a preincubation of several minutes to allow equilibration with experimental temperature (usually 30°), 0.3 μmole of NADPH is added, bringing the final volume to 1.0 ml. The rate of cytochrome *c* reduction (Δ OD 550 nm) is approximately proportional to the rubredoxin concentration under these conditions. This assay is extremely sensitive. One mμmole of *C. pasteurianum* rubredoxin catalyzes an increase in absorbancy at 550 nm of about 2 absorbance units/minute. The blank used is a cuvette from which NADPH has been omitted, and the control is a cuvette which contains no rubre-

[1] W. Lovenberg and B. E. Sobel, *Proc. Nat. Acad. Sci. U.S.* **54,** 193 (1965).

[2] T. C. Stadtman, *in* "Non-heme Iron Proteins Role in Energy Conversion" (A. San Pietro, ed.), p. 439. Antioch, Yellow Springs, Ohio, 1965.

[3] J. LeGall and N. Dragoni, *Biochem. Biophys. Res. Commun.* **23,** 145 (1966).

[4] S. G. Mayhew and J. L. Peel, *Biochem. J.* **100,** 80 (1966).

[5] H. Bachmayer, K. T. Yasunobu, and H. R. Whiteley, *Biochem. Biophys. Res. Commun.* **26,** 435 (1967).

[6] E. T. Lode and M. J. Coon, *J. Biol. Chem.* **246,** 791 (1971).

[7] J. A. Peterson, M. Kusunose, E. Kusunose, and M. J. Coon, *J. Biol. Chem.* **242,** 4334 (1967).

[8] M. Shin, K. Tagawa, and D. I. Arnon, *Biochem. Z.* **338,** 84 (1963).

doxin. Although the assay is extremely sensitive, it is not specific; clostridial ferredoxin also catalyzes the reduction of cytochrome c, but at about 20% the rate of rubredoxin. The assay therefore is only valid when the samples contain relatively small contaminating amounts of ferredoxin.

Purification Procedure

The following procedure is described primarily for the purification of the protein from clostridial organisms. The steps for the isolation of the more complex rubredoxin from *P. oleovorans* have recently been detailed.[6]

The acidity of this small protein permits relatively easy purification from *Clostridium pasteurianum* on anion exchange celluloses. Steps 1–3 are done at 0°, subsequent steps may be done at room temperature.

Step 1. Preparation of Acetone Extract. Eight hundred to 1000 g of frozen cells of *C. pasteurianum* are suspended in enough water to make about 1700 ml; one hundred ml of 1 M Tris·HCl, pH 7.3, is added. This suspension is chilled to 0° and an equal volume (1800 ml) of acetone precooled to 0° is added; the suspension is stirred in an ice–salt bath to return the temperature to 0° as quickly as possible. After 15 minutes the material is centrifuged at 5000 rpm for 10 minutes using a large refrigerated centrifuge.

Step 2. First DEAE-Cellulose Chromatography. A 2 × 10 cm column of DEAE-cellulose is prepared. The DEAE-cellulose obtained from Biorad Corporation is soaked in 1 M KPO_4 buffer, pH 7.0, prior to use, and the column is washed before use with 10 volumes of deionized water. The supernatant fraction obtained in Step 1 is applied to the column. The rubredoxin and ferredoxin which are in the 50% acetone supernatant fraction adhere as a dark band at the top of the column. The column is next washed with 5 volumes of water followed by sufficient buffer (0.15 M Tris·HCl, pH 7.3, containing 0.05 M NaCl) to stretch the dark band on the column to within about 1 cm of the bottom of the column. The rubredoxin and ferredoxin are eluted with high chloride buffer (0.15 M Tris·HCl containing 0.65 M NaCl). This eluate is desalted using Sephadex G-25 which has been equilibrated with 0.05 M Tris·HCl, pH 7.3.

Step 3. First Ammonium Sulfate Fractionation. The above eluate contains the majority of the ferredoxin and rubredoxin. Crystalline ammonium sulfate is added to bring this solution to 90% saturation. Under these conditions the majority of the ferredoxin precipitates, whereas a considerable portion of the rubredoxin remains in the supernatant fraction. After centrifugation the precipitate is again dissolved in 100 ml of 0.05 M Tris·HCl and crystalline ammonium sulfate is added to bring again the concentration to 90% saturation. After centrifugation, the second supernatant fraction is combined with the first. The precipitate can be used as a source of ferredoxin.

Step 4. Second DEAE-Cellulose Chromatography. The combined 90% ammonium sulfate saturated supernatant fractions are applied to a small DEAE-cellulose column (1 × 10 cm) prepared in a manner similar to that described in Step 2. Under these conditions both ferredoxin and rubredoxin adhere tightly to the top of the column. The column is next eluted with a 50% saturated ammonium sulfate solution. This procedure selectively elutes rubredoxin. The red-colored fractions are collected, brought to 90% saturation with respect to ammonium sulfate, and readsorbed on another 1 × 10-cm DEAE-cellulose column. This column is eluted with the 0.8 *M* HCl buffer described in Step 2. The rubredoxin at this point is essentially pure, having a 280 nm/490 nm absorbancy ratio of 2.4 or better. This is the ratio which has been observed with crystalline material.

Step 5. Crystallization. The rubredoxin eluate obtained is desalted on a Sephadex G-25 column which has been equilibrated with water. The salt-free rubredoxin is lyophilized and redissolved in water to yield a solution containing 1–2 mg/ml. This solution is adjusted to pH 3.0, and any precipitate formed is removed by centrifugation. The solution is adjusted to pH 4.0 and made about 75% saturated with respect to ammonium sulfate at 23°. Crystals appear as rhombs about 0.2 mm along an edge within a week.

Properties

Chemistry. Rubredoxin from *C. pasteurianum* has a molecular weight of about 6000 and contains a single iron atom. The protein can undergo a reversible 1-electron reduction and has a redox potential of about −0.057 V. Rubredoxin is bright red in the oxidized state and has absorption maxima at 750, 490, 380, and 280 nm with molar extinction coefficients of 340, 8.8×10^3, 10.7×10^3, and 21.1×10^3, respectively.[1,9] When rubredoxin is reduced by a hydrogen–hydrogenase system, it becomes colorless, with absorption maxima at 333, 311, and 275 nm, and with molar extinction coefficients of 6.1×10^3, 10.8×10^3 and 24.8×10^3, respectively. The above electronic absorption bands are optically active, with the strongest Cotton effects associated with the peak at 490 nm and a shoulder at 570 nm.

The single iron atom appears to be tetrahedrally coordinated in the center of the four sulfur atoms of the four cysteine residues of the pro-

[9] W. A. Eaton and W. Lovenberg, unpublished observation.

tein. This conclusion is based on chemical observations[10,11] and 2.5-Å resolution X-ray crystallography data.[12]

Catalytic Activity. Rubredoxin interacts with several systems. It is readily reduced by H_2 in the presence of hydrogenase from *C. pasteurianum* and by NADPH and extracts of that organism. Rubredoxin will function as a carrier in the reduction of NADP by molecular hydrogen in this organism, but is not very effective because of its relatively high redox potential. Pyruvate and pyruvate dehydrogenase also reduce rubredoxin, as will NADPH and spinach ferredoxin-NADP reductase. Reduced rubredoxin is extremely autooxidizable by atmospheric oxygen and can easily give up its electron to oxidized cytochrome *c*.

[10]W. Lovenberg and W. M. Williams, *Biochemistry* **8**, 141 (1969).
[11]H. Bachmayer, K. T. Yasunobu, and H. R. Whiteley, *Proc. Nat. Acad. Sci. U.S.* **59**, 1273 (1968).
[12]J. R. Herriott, L. C. Sieker, L. H. Jensen, and W. Lovenberg, *J. Mol. Biol.* **50**, 391 (1970).

[43] Purification of Nitrogenase and Crystallization of Its Mo-Fe Protein

By R. C. BURNS and R. W. F. HARDY

Nitrogenase,[1,2] the enzyme of nitrogen fixation, is a complex of Mo-Fe protein and Fe protein. Enzyme preparations from numerous sources catalyze the following reaction:

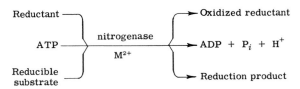

In addition to N_2, reducible substrates include C_2H_2, azide, cyanide, N_2O, nitriles, and isonitriles. Hydrogen ion also functions as a reducible substrate to yield H_2. This is the sole reduction product in the absence of added reducible substrates and is produced to a lesser extent in their presence. Functional reductants include ferredoxins, flavodoxin, azotoflavin, viologen dyes, and dithionite.

[1]R. W. F. Hardy and R. C. Burns, *Annu. Rev. Biochem.* **37**, 331 (1968).
[2]R. W. F. Hardy and E. Knight, Jr., *in* "Phytochemistry" (L. Reinhold and Y. Liwschitz, eds.), p. 407. Wiley (Interscience), New York, 1968.

The procedures described here enable the preparation of sufficient purified *Azotobacter* nitrogenase for over a thousand individual assays and for the crystallization of the Mo-Fe protein in amounts of several hundred milligrams with excellent overall recovery of activity. The complete sequence from rupture of bacterial cells through third crystallization of the Mo-Fe protein requires little specialized apparatus and less than 1 week of work for one technician.

Assay Methods

Principle. Early work on the biochemistry of N_2 fixation depended heavily on the ^{15}N analysis method,[3] but advances in N_2 fixation research have led to simpler and more sensitive assays. These include assays based on the measurement of NH_3 formed from N_2,[4] ethylene formed from acetylene,[5-7] and H_2 evolved in the absence of added reducible substrate.[8] These new assays are utilized now almost exclusively, although the ^{15}N assay remains the most definitive method. These assays have been most useful with the *Azotobacter* preparations, but are not restricted to these preparations; C_2H_2 reduction methods in particular have been used extensively with preparations from other microorganisms as well as whole organism systems both in the laboratory and *in situ.*[9,10]

The ammonia formation assay utilizes the physiological substrate of nitrogenase and may be presumed to reflect maximum integrity of the enzyme. Hydrogen evolution assay directly measures nitrogenase-catalyzed electron transfer and enables easy rate measurements by manometry. Acetylene reduction assay is much more sensitive than all other methods, surpassing the ^{15}N method by a factor of 10^3. With a given enzyme preparation, the ratio of specific activities observed for N_2 reduction, H_2 evolution, and C_2H_2 reduction is 4:1:1 (Table I).[11] This ratio reflects an equivalency in the rates of electron transfer which appear to be independent of the nature of electron acceptor. The

[3] R. H. Burris and P. W. Wilson, Vol. IV [16].
[4] L. E. Mortenson, *Anal. Biochem.* **2**, 216 (1961).
[5] M. J. Dilworth, *Biochim. Biophys. Acta* **127**, 285 (1966).
[6] R. Schöllhorn and R. H. Burris, *Proc. Nat. Acad. Sci. U. S.* **58**, 213 (1967).
[7] R. W. F. Hardy and E. Knight, Jr., *Biochim. Biophys. Acta* **139**, 69 (1967).
[8] R. C. Burns and W. A. Bulen, *Biochim. Biophys. Acta* **105**, 437 (1965).
[9] W. D. P. Stewart, G. P. Fitzgerald, and R. H. Burris, *Proc. Nat. Acad. Sci. U. S.* **58**, 2071 (1967).
[10] R. W. F. Hardy, R. D. Holsten, E. K. Jackson, and R. C. Burns, *Plant Physiol.* **43**, 1185 (1968).
[11] R. C. Burns and R. W. F. Hardy, unpublished results.

TABLE I

COMPARISON OF SPECIFIC ACTIVITIES OF NITROGENASE PREPARATIONS
UTILIZING DIFFERENT ASSAYS

	Assay		
Preparation	N_2 reduction (units[a]/mg)	H_2 evolution (units[b]/mg)	C_2H_2 reduction (units[c]/mg)
Crude extract	11	40	45
PS-1	20	80	79
Δ60	42	169	169
PS-2	91	348	381

[a] One unit = 1 nmole of N_2 reduced per minute.
[b] One unit = 1 nmole of H_2 evolved per minute.
[c] One unit = 1 nmole of C_2H_2 reduced per minute.

transfer of four pairs of electrons is associated with N_2 reduction because concomitant with N_2 reduction (6 electron transfer), H_2 is evolved (2 electron transfer). Little or no H_2 is evolved when C_2H_2 is the electron acceptor, and thus the 2-electron reduction of C_2H_2 to C_2H_4 provides the same specific activity value as observed in the H_2 evolution assay.

The N_2 reduction assay, the H_2 evolution assay, and the C_2H_2 reduction assay will be described. Since these are variations of the same procedure, only the C_2H_2 assay will be described in detail, and the N_2 reduction and H_2 evolution assays will be described as modifications of it.

A sigmoidal relationship between activity and protein concentration is frequently observed, presumably because nitrogenase is a dissociating system and/or because of partial loss of the more labile component in handling.[10] Valid activity measurements thus often necessitate the assay of two or more enzyme concentrations; activity is then calculated from the linear part of an activity-concentration curve. A unit of activity is defined as the amount of enzyme required to catalyze the reduction of 1 nmole of substrate per minute or the formation of 1 nmole of reduced product per minute. Because of the wide range of substrates reducible by nitrogenase, any reference to activity units should indicate the appropriate substrate or product. Specific activity is defined as units per milligram of protein.

Reagents

$MgCl_2$, 0.1 M
ATP or ADP, 0.1 M, pH 7.2
$Na_2S_2O_4$, 0.2 M, pH 7.2, prepared anaerobically

Creatine phosphate, 0.3 M, pH 7.2
Creatine kinase solution, 200 enzyme units/ml in water
C_2H_2 gas[12]
Enzyme solution in dilute phosphate or Tris·HCl buffer, pH 7–8

Procedure for C_2H_2 Reduction Assay. Reaction vessels of 20–40-ml volume similar to respirometer flasks with a sidearm and a sidearm port fitted with a serum bottle stopper are used. To the main compartment are added 0.5 ml of 0.1 M $MgCl_2$, 0.05 ml of 0.1 M ATP, 0.1 ml of 0.3 M creatine phosphate, and enough water to make the final reaction volume 1.0 ml; 0.05 ml of creatine kinase solution is added to the sidearm. The vessel is attached to a manifold connected to a vacuum pump and sources of argon and C_2H_2; it is evacuated and filled with argon. With syringes 0.1 ml of 0.2 M $Na_2S_2O_4$ is added to the main compartment, and 0.1 to 300 units (nmoles of C_2H_2 reduced per minute) of nitrogenase is added to the side arm by injection through the serum bottle cap. Twenty percent of the atmosphere is pumped out of the flask and replaced with C_2H_2 scrubbed with concentrated H_2SO_4 to trap impurities. The flask is equilibrated at 30°, and the reaction is initiated by tipping the sidearm contents; the flask is shaken mechanically at 30° for 20 minutes. Incubation is terminated by the addition of 0.2 ml of 1 N H_2SO_4 to the reaction mixture by injection through the serum bottle cap. A 200-μl aliquot of gas is withdrawn from the flask for gas chromatographic separation and flame ionization analysis of C_2H_4 and C_2H_2. The amount of C_2H_4 formed is determined by comparison with gas mixtures of known pC_2H_4. A correction for the small amount of C_2H_4 present as an impurity in the C_2H_2 is required. Suitable gas chromatographic columns include (1) a $\frac{1}{8}$-inch × 10-foot column containing 20% ethyl, N,N'-dimethyl oxalamide on 100–120-mesh acid-washed firebrick at 0° with a He flow rate of 30 ml per minute,[10] and (2) a 4-mm × 4-foot Porapak R column at 45° or 75° with a N_2 flow rate of 75 ml per hour.[13] Alumina columns, used earlier,[7,14] are not recommended.

Procedure for H_2 Evolution Assay. The above procedure is followed except that respirometers are used, the reaction is incubated under an inert atmosphere, and standard manometric techniques are followed. Between 15 and 300 units (nmoles of H_2 evolved per minute) of nitrogenase should be used.

[12]Acetylene is highly explosive, and adequate precaution must be observed.
[13]M. Kelly, *Biochem. J.* **107,** 1 (1968).
[14]B. Koch and H. J. Evans, *Plant Physiol.* 41, 1748 (1966).

Procedure for N₂ Reduction Assay. Additional reagents and equipment include saturated K_2CO_3, Nessler reagent,[15] 2 N NaOH, 1 N H_2SO_4, and 15-ml serum bottles with a one-hole stopper fitted with a length of 8-mm glass rod which extends half way to the bottom of the bottle and terminates in a slightly enlarged and roughened tip. The above procedure is followed, using N_2 as the gas phase. At the end of the incubation an aliquot of reaction mixture containing 0.3–6 μmoles of NH_3 (as NH_4^+) is transferred to a serum bottle containing at least an equal volume of saturated K_2CO_3. A stopper assembly containing a drop of 1 N H_2SO_4 on the tip is immediately inserted into the bottle. The bottle is agitated gently at 30° for 90 minutes to allow distillation of the NH_3 into the acid drop.

The stopper assembly is transferred to a 10-ml beaker containing 2 ml of Nessler reagent and 2 ml of H_2O. Three milliliters of 2 N NaOH is added and stirred in with the stopper assembly. The absorbancy of the solution is determined at 490 nm 20 minutes after addition of the NaOH; NH_4^+ content is determined from a standard curve.

Purification and Crystallization Procedure

The Mo-Fe protein crystallization procedure (step 8) has been described by Burns, Holsten, and Hardy.[16] Preliminary steps (1–7) involving partial purification and fractionation are based on procedures in the literature[8,17,18] and are used routinely in our laboratory to provide large quantities in high yields of highly active preparations for nitrogenase assays. Other procedures for partial purification and/or fraction-action of *Azotobacter* nitrogenase have been described.[10,19–21]

Step 1. Growth of Cells. A. vinelandii OP (ATCC 13705) is grown with vigorous aeration at 30° on a modified Burk's medium containing, in grams per liter: sucrose, 20; $MgSO_4 \cdot 7H_2O$, 0.2; $CaCl_2$ (anhydrous), 0.06; $FeCl_3 \cdot 6H_2O$, 0.05; $Na_2MoO_4 \cdot 2H_2O$, 0.0025; KH_2PO_4, 0.2; K_2HPO_4, 0.8; the pH is 7.0–7.2. The culture is subcultured daily in 500-ml shake flasks containing 100 ml of medium and 3.0 ml of inocu-

[15]W. W. Umbreit, R. H. Burris, and J. F. Stauffer, "Manometric Techniques," p, 238. Burgess, Minneapolis, Minnesota, 1957.

[16]R. C. Burns, R. D. Holsten, and R. W. F. Hardy, *Biochem. Biophys. Res. Commun.* **39**, 90 (1970).

[17]W. A. Bulen, J. R. LeComte, R. C. Burns, and J. Hinkson, *in* "Non-Heme Iron Proteins: Role in Energy Conversion" (A. San Pietro, ed.), p. 261. Antioch Press, Yellow Springs, Ohio, 1965.

[18]W. A. Bulen and J. R. LeComte, *Proc. Nat. Acad. Sci. U. S.* **56**, 979 (1966).

[19]M. Kelly, R. V. Klucas, and R. H. Burris, *Biochem. J.* **102**, 1c (1967).

[20]M. Kelly, *Proc. Int. Congr. Microbiol. 9th*, p. 277 (1966).

[21]K. L. Hadfield and W. A. Bulen, *Biochemistry* **8**, 5103 (1969).

lum. Large cultures may be grown in any suitable chamber with provision for vigorous aeration, agitation, and temperature control: 20-liter cultures initiated by a 5% inoculum of a 24-hour culture yield 140–160 g of packed cell paste when harvested with an unrefrigerated Sharples supercentrifuge after 16 hours while the cells are still in logarithmic growth. The cell paste is either used immediately or frozen in liquid N_2 and stored at −70°.

Step 2. Preparation of Crude Extract. Cell paste is suspended in 0.025 M potassium phosphate buffer, pH 7.5, at 0° at the ratio of 1.5 ml buffer per gram of cell paste. Cell paste stored at −70° for as long as 14 months provides preparations equivalent to those obtained with freshly harvested cells; when frozen paste is used, it is thawed in the suspending buffer. The pH of the cell suspension is adjusted to 7.0, and the cells are ruptured in a French pressure cell operated between 16,000 and 20,000 psi; the broken cell preparation is passed through the French cell a second time to assure maximum cell breakage. The preparation is kept in an ice bath before and after cell rupture, but no precautions are taken to keep the French cell cold during operation. The broken cell suspension is adjusted to pH 7.0, and the preparation is centrifuged 16 hours at 20,000 g (e.g., at 11,000 rpm in a Sorvall GSA rotor). The colorless layer which comprises the top 2–10% of the supernatant liquid is aspirated off and discarded; the remaining dark brown supernatant liquid designated crude extract is decanted into a suction flask held at 0°.

Step 3. Removal of Nucleic Acid with Protamine Sulfate. A solution of 2% protamine sulfate (Eli Lilly Co.), pH 6.0, is added to the crude extract to obtain a maximum 280:260 nm ratio of 0.9–1.0. This requires approximately 3 ml per gram of crude extract protein but is determined exactly for each crude extract by trials on small aliquots. The protamine sulfate solution (at room temperature) is added from a burette to magnetically stirred crude extract (at 0°) at 1.5–2.0 ml per minute. Nitrogen is flushed through the flask containing the crude extract, and the preparation is maintained under N_2 from this point on. The protamine sulfate-treated extract is stirred for 20 minutes at 0° and is then centrifuged for 50 minutes at 20,000 g. The combined supernatant liquids are adjusted to pH 7.0 and designated the PS-1 preparation; the pellets are discarded.

Step 4. Heat Treatment. The PS-1 preparation is heated to 60° by vigorous stirring in a boiling water bath, then immediately transferred to a 60° bath and incubated for 7 minutes with occasional swirling by hand. Heating anaerobically is conveniently accomplished using 500-ml suction flasks containing 250–300 ml of preparation and plugged with a

2-hole stopper, one hole for a thermometer and the other an outlet port for N_2 which is flushed through the sidearm. The heated extract is immediately cooled in an ice bath to $< 20°$ while being stirred magnetically; denatured protein is removed by centrifugation for 30 minutes at 27,000 g in a Sorvall SS-34 rotor. The supernatant liquid, designated $\Delta 60$ preparation, is decanted and placed in an ice bath.

Step 5. Precipitation of Nitrogenase with Protamine Sulfate. The $\Delta 60$ preparation is adjusted to pH 6.50 with HCl and is then treated with half the volume of 2% protamine sulfate solution used to remove nucleic acids. The procedure is similar to step 3 except that the protamine sulfate solution is first made anaerobic (by evacuating and filling the protamine sulfate flask 5× with N_2) and is added from a syringe. The insoluble protamine–nitrogenase complex is accumulated as a pellet in several 50-ml centrifuge tubes which are used repeatedly to centrifuge successive portions of the preparation for 1 minute at 27,000 g; the supernatant liquids are discarded. The gray-black pellets are overlaid with a small volume of 0.01 M Tris·HCl buffer, pH 7.2, and stored at room temperature overnight. All subsequent operations are performed at room temperature unless indicated otherwise.

Step 6. Solubilization of Nitrogenase. The pellets are combined and suspended in a minimum amount of 0.01 M Tris·HCl buffer. To this suspension is added 0.25 g of dry cellulose phosphate[22] (Cellex-P, Bio-Rad Laboratories) per milliliter of protamine sulfate solution used in step 5. This mixture is adjusted to pH 7.2, and it is stirred, Tris·HCl buffer, pH 7.2, being added as needed to assure a uniform consistency. The mixture is then filtered through sintered glass. The brown filtrate is collected. The retained solids are washed free of brown solution by the addition of Tris·HCl buffer, pH 7.2, and this wash is added to the filtrate to give PS-2 preparation which is stored under N_2 at room temperature. Representative data for purification through this step are shown in Table II.

Step 7. Fractionation of PS-2 Preparation by DEAE-Cellulose Chromatography. A sample of freshly prepared PS-2 preparation containing 2–4 g of protein in 60–120 ml is treated with approximately 5 mg of $Na_2S_2O_4$ per gram of protein. The preparation is adjusted to pH 7.2, stirred for 5 minutes, and then placed on a DEAE-cellulose (Bio-Rad Laboratories)

---

[22]A stock quantity of cellulose phosphate is prepared by suspending 20 g in 500 ml of 0.01 M potassium phosphate buffer, pH 7.0, packing this slurry in a column, washing it with 1000 ml of phosphate buffer followed by 1000 ml of H_2O, and then transferring to a crystallizing dish to dry at 80°. Prior to use, the required amount is weighed into a suction flask, which is then evacuated and filled 5× with N_2.

TABLE II
PURIFICATION OF *Azotobacter* NITROGENASE

| Prepa-ration | Volume (ml) | Protein (g) | Activity | | Percent | | Purifi-cation (fold) |
			Units[a] ×10⁻³	Units/mg	recovery of protein	activity	
Crude ex-tract	599	41.50	2224	54	100	100	1.00
PS-1	609	30.66	2144	70	74	96	1.30
Δ60	504	12.70	1830	144	31	82	2.74
PS-2	147	5.44	1772	326	13	79	6.10

[a]One unit = 1 nmole of H_2 evolved per minute.

column (10 cm × 5 cm in diameter) which was prepared from a slurry of 20 g of DEAE-cellulose in 0.02 M Tris·HCl, pH 7.2, washed with 100 ml of 0.6 N KOH and 500 ml of N_2-saturated 0.02 M Tris·HCl, pH 7.2, and maintained under N_2. All solutions added to the column are first saturated with N_2. The sample, which enters the column at approximately 1 ml per minute and occupies the top one-third to one-half of the column as a dark brown band, is washed in with 25 ml of 0.02 M Tris·HCl, pH 7.2. Elution with 0.15 M NaCl in 0.02 M Tris·HCl, pH 7.2, is begun and is continued until no further protein is eluted; this requires 400–600 ml at an elution rate of 1–1.5 ml per minute, the rate used throughout the procedure. The 0.15 M NaCl fraction is discarded.

The brown Mo-Fe protein of nitrogenase is eluted with 0.25 M NaCl in 0.02 M Tris·HCl, pH 7.2, and is collected in four to eight 50-ml fractions in 125-ml suction flasks continuously flushed with N_2.

The column is then washed with 0.25 M NaCl in 0.02 M Tris·HCl to remove remaining Mo-Fe protein, which would otherwise contaminate the Fe protein fraction. The last traces of Mo-Fe protein are removed only with difficulty and may require up to 2000 ml of salt solution. The Fe protein is eluted with 0.35 M NaCl in 0.02 M Tris·HCl and is collected in two to four 50-ml fractions under N_2.

All operations are conducted at room temperature. Development of the column may be interrupted at any time; e.g., the Fe protein fraction is conveniently eluted after the column has stood over the weekend. An anaerobic column similar to that described by Munson et al.[23] may be used, although less elaborate apparatus is satisfactory. Both proteins are highly colored, and their movement on the column is conveniently monitored visually.

[23]T. O. Munson, M. J. Dilworth, and R. H. Burris, *Biochim. Biophys. Acta* **104,** 278 (1965).

Step 8. Crystallization of Mo-Fe Protein. The brown Mo-Fe protein fractions obtained in the DEAE-cellulose fractionation are concentrated to at least 30 mg of protein per milliliter using an Amicon ultrafiltration cell fitted with an XM-50 filter (50,000 MW cutoff). The dark brown concentrated solution is then added dropwise to 6–10 volumes of 0.01 M Tris·HCl, pH 7.2. A heterogeneous mixture of white crystalline material and brown amorphous material forms immediately or after a few minutes of gentle stirring; stirring is continued for 20 minutes, and the insoluble material is collected as a brown to gray pellet by centrifugation for 5 minutes at 10,000 g and 5°; the almost colorless supernatant liquid is discarded. The pellet is washed by suspending it in 0.01 M Tris·HCl and centrifuging as before; the clear supernatant liquid is discarded. The pellet is dissolved in a minimum volume of cold 0.25 M NaCl in 0.02 M Tris·HCl, pH 7.2; any insoluble material is removed by centrifugation as before and discarded. Dilution of this solution at room temperature with 3 volumes of 0.01 M Tris·HCl, pH 7.2, initiates immediate formation of a dense mass of crystals. The protein may be recrystallized repeatedly with little or no loss of activity. Data from a typical preparation are given in Table III.

With some preparations where the yield of crystalline material from the diluted column fractions is low, precipitation of essentially all activity is obtained by first reducing the pH of the concentrated fraction to 5.8 with 0.2N acetic acid at 0°, then stirring 5 minutes, diluting in

TABLE III
CRYSTALLIZATION OF MO-FE PROTEIN

| | | | Activity | |
Fraction	Volume (ml)	Protein (mg)	Units[a] $\times 10^{-3}$	Units/ (mg)
DEAE-column fraction[b] (0.25 M NaCl)	88	1282	754	587
Concentrated column fraction	18	1159	730	622
First crystallization[c]	15	460	638	1388
Second crystallization[d]	12	407	590	1450
Third crystallization[d]	12	378	545	1444

[a]One unit = 1 nmole of H_2 evolved per minute. Specific activity is based on Mo-Fe protein only.
[b]Product from the fractionation of 2.82 g of PS-2 protein, containing 866,000 units.
[c]Prepared by adding 126 ml of 0.01 M Tris·HCl, pH 7.2, to concentrated column fraction, centrifuging, washing pellet in 30 ml of Tris·HCl and dissolving pellet in 0.01 M Tris·HCl containing 0.25 M NaCl.
[d]Prepared by adding 105 ml of Tris·HCl to previous preparation, centrifuging, and dissolving pellet in 0.25 M NaCl–0.01 M Tris·HCl.

water instead of buffer, stirring 15 minutes, and centrifuging as above.[11] With these modifications, the resulting pellet is dissolved in 0.20M NaCl in 0.05M N-tris(hydroxymethyl)methyl-2-aminoethane sulfonic acid (TES), pH 7.2.

Properties

Properties of Nitrogenase Purified through Step 6. Enzyme activity is stable for several weeks when the preparations are maintained at room temperature, under an O_2-free atmosphere and at pH 7–8. Complete and irreversible inactivation follows exposure to air for 6 hours at 22° or overnight anaerobic storage at −15°; 80% of activity is lost on overnight anaerobic storage at 5°. Even under optimal storage conditions, a black sludge forms after several days and an increasingly intense stench characteristic of volatile sulfur compounds develops; however, these changes are not well correlated with activity loss.

The following compounds, in addition to H^+,[8] are reduced by nitrogenase and are mutually inhibitory; products and K_m values are indicated where known.

$N_2 \rightarrow NH_3$, 0.16 atm (0.1 mM)[7,13]

$C_2H_2 \rightarrow C_2H_4$, 0.01–0.03 atm (0.4–1.2 mM)[6,10,13]

$N_2O \rightarrow N_2 + H_2O$, 0.05 atm (1.2 m$M$)[24]

Azide $\rightarrow N_2 + NH_3$, 1.3 mM[7,25]

Cyanide $\rightarrow CH_4 + NH_3 + CH_3NH_2 +$ traces of C_2H_4 and C_2H_6, 0.5–1.4 mM[7,13]

$CH_3NC \rightarrow CH_4 + C_2H_6 + C_2H_4 + C_3H_6 + C_3H_8 + CH_3NH_2$[13,26]

$C_2H_5NC \rightarrow CH_4 + C_2H_6 + C_2H_4$[13,26]

Acrylonitrile $\rightarrow C_3H_6 + C_3H_8$, 10 m$M$[27]

$RCN \rightarrow RCH_3 + NH_3$[28]

1-Propyne, 1-butyne, and allene are reduced to the corresponding alkenes.[28]

Carbon monoxide and H_2 are competitive inhibitors. A K_i of 4.6 × 10^{-4} atm (4 × 10^{-7} M) is observed for CO inhibition of the reduction of N_2 and C_2H_2.[10,13] The K_i for H_2 inhibition of N_2 reduction is 0.14–0.22

[24]R. W. F. Hardy and E. Knight, Jr., *Biochem. Biophys. Res. Commun.* 23, 409 (1966).
[25]R. Schöllhorn and R. H. Burris, *Proc. Nat. Acad. Sci. U. S.* 57, 1317 (1967).
[26]R. W. F. Hardy and G. W. Parshall, *Abstr. 158th Nat. Meeting Amer. Chem. Soc.*, Abstr. No. 226 (1969).
[27]W. H. Fuchsman and R. W. F. Hardy, *Bacteriol. Proc.* p. 148 (1970).
[28]R. W. F. Hardy and E. K. Jackson, *Fed. Proc. Fed. Amer. Soc. Exp. Biol.* 26, 725 (1967).

atm $(1.1-1.7 \times 10^{-4} M)^{21,29,30}$; H_2 shows no inhibition of C_2H_2 reduction.[31] Other inhibitors, types of inhibition unknown, include metal-binding reagents, such as o-phenanthroline, 2,2′-dipyridyl, Tiron (1,2-dihydroxybenzene 3,5-disulfonate), and 2,3-dimercaptopropanol, which inhibit N_2 reduction and H_2 evolution with K_i values between 10^{-3} and $10^{-4} M$[17]; O_2, Cu^{2+}, Zn^{2+},[32] ADP,[33] and conditions of high ionic strength[8] also inhibit. Nitric oxide, a potent inhibitor of clostridial nitrogenase,[34] has not been tested with *Azotobacter* nitrogenase. Conventional uncouplers of oxidative and photosynthetic phosphorylation and inhibitors of electron transport are without effect except at unusually high levels.[35]

The enzyme is absolutely specific for ATP; CTP, UTP, and GTP do not support the reaction.[10] Because of inhibition by ADP, ATP is best supplied by an ATP generating system. The K_m for ATP is $3 \times 10^{-4} M$.[32] Hydrosulfite is the preferred reductant. The natural electron donor is unidentified, but azotoflavin[36] and *Azotobacter* ferredoxin[37] have been implicated, and clostridial ferredoxin[38] is effective. The K_m for hydrosulfite is $9 \times 10^{-3} M$. The divalent metal cation requirement is satisfied by Mg^{2+}, Mn^{2+}, Co^{2+}, Fe^{2+}, or Ni^{2+} in that order of effectiveness; when substrate levels of ATP are used, the reaction is best supported by a concentration of divalent cation one-half the concentration of ATP, but when an ATP generating system is used the ATP-to-cation ratio is not critical.[32] Nitrogenase catalyzes the exchange reaction, $D_2 + H_2O \rightleftharpoons HD + HDO$, but only while actively catalyzing the reduction of N_2.[30] At temperatures above about 20°, the activation energy of nitrogenase for all its reactions is 12.0–14.6 kcal/mole, but it is 39–54 kcal/mole at lower temperatures.[10,32]

Properties of Crystallized Mo-Fe Protein.[11,16,39] The needlelike crystalline protein appears as shown in Fig. 1; the crystals are white, or possibly slightly yellow, and are typically 30–60 μ long and 1–4 μ wide. They

[29]G. W. Strandberg and P. W. Wilson, *Proc. Nat. Acad. Sci. U. S.* **58**, 1404 (1967).

[30]E. K. Jackson, G. W. Parshall and R. W. F. Hardy, *J. Biol. Chem.* **243**, 4952 (1968).

[31]J. C. Hwang and R. H. Burris, *Fed. Proc. Fed. Amer. Soc. Exp. Biol.* **27**, 639 (1968).

[32]R. C. Burns, *Biochim. Biophys. Acta* **171**, 253 (1969).

[33]W. A. Bulen, R. C. Burns, and J. R. LeComte, *Proc. Nat. Acad. Sci. U. S.* **53**, 532 (1965).

[34]A. Lockshin and R. H. Burris, *Biochim. Biophys. Acta* **111**, 1 (1965).

[35]R. W. F. Hardy and E. Knight, Jr., *Biochim. Biophys. Acta* **122**, 520 (1966).

[36]J. R. Benemann, D. C. Yoch, R. C. Valentine, and D. I. Arnon, *Proc. Nat. Acad. Sci. U. S.* **64**, 1079 (1969).

[37]D. C. Yoch, J. R. Benemann, R. C. Valentine, and D. I. Arnon, *Proc. Nat. Acad. Sci. U. S.* **64**, 1404 (1969).

[38]W. A. Bulen, R. C. Burns, and J. R. LeComte, *Biochem. Biophys. Res. Commun.* **17**, 265 (1964).

[39]K. T. Fry, R. C. Burns, and R. W. F. Hardy, unpublished results.

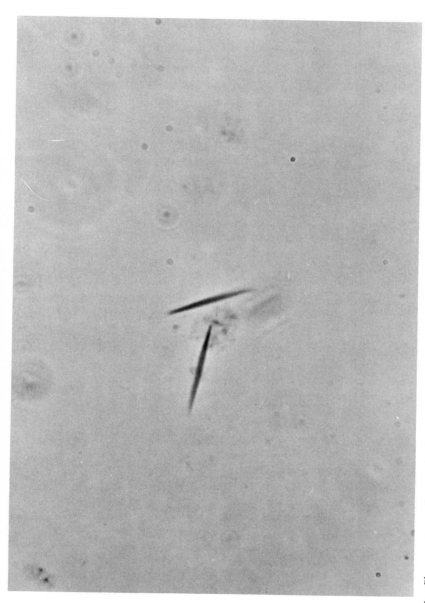

Fig. 1.   Phase contrast micrograph of crystals of Mo-Fe protein. From R. C. Burns, R. D. Holsten, and R. W. F. Hardy, *Biochem. Biophys. Res. Commun.* **39**, 90 (1970).

convert readily to an amorphorus state when compressed and are not conveniently collected in the crystalline state by centrifugation or filtration. Visible spectra of the native and reduced protein (Fig. 2) show on reduction a red shift in the absorption at 412–420 nm and the development of absorption at 525 and 557 nm. Millimolar extinction coefficients of the native protein are 470 at 280 nm and 85 at 412 nm, based on a molecular weight of 270,000. In 0.25 M NaCl–0.01 M Tris·HCl, pH

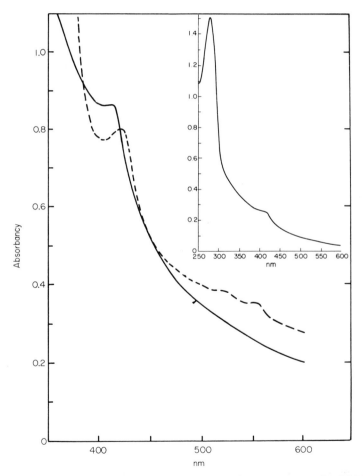

FIG. 2.   Visible UV spectra of crystallized Mo-Fe protein. Cuvettes contained 2.73 mg or (inset) 0.87 mg of protein in 2 ml of 0.25 M NaCl, 0.01 M Tris·HCl, pH 7.4, under N_2. Native protein (—) plus $Na_2S_2O_4$ (---). From R. C. Burns, R. D. Holsten, and R. W. F. Hardy, *Biochem. Biophys. Res. Commun.* **39,** 90 (1970).

TABLE IV
Amino Acid[39] Mo, Fe, and Acid-Labile Sulfide Analysis
of Crystallized Mo-Fe Protein

Aspartic	249[a]	Leucine	190
Threonine	115	Tyrosine	79
Serine	134	Phenylalanine	102
Glutamic	250	Lysine	177
Proline	101	Histidine	55
Glycine	206	Arginine	108
Alanine	169	Tryptophan	50
Valine	173	Half-cystine	34
Methionine	86	Mo	2
Isoleucine	134	Fe	32–36
		Acid-labile sulfide	28

[a] Residues or atoms per 270,000 molecular weight.

7.2, ultracentrifugal analyses by Archibald approach to equilibrium and by Yphantis depletion of meniscus methods show molecular weight values of 270,000–290,000. Amino acid, Fe, Mo, and acid-labile sulfide contents are given in Table IV. The protein is composed of at least two types of subunits.

The Mo-Fe protein has no known intrinsic catalytic activity, but combines with the Fe protein of nitrogenase to form the active enzyme with the catalytic properties described above. Titration of the Fe protein fraction with Mo-Fe protein produces the activity curve shown in Fig. 3. Specific activity, in terms of the Mo-Fe protein only, is 1400–1500 nmoles of H_2 evolved or C_2H_2 reduced per minute per milligram of protein based on the initial linear portion of the curve; assuming one active site per enzyme molecule, these data indicate a turnover number of 380–420. The protein is irreversibly inactivated by O_2 but is stable under anaerobic conditions for days at 0° and indefinitely at −15°.

Magnetic susceptibility measurements by both nuclear magnetic resonance (NMR) and Faraday cage techniques indicate 3 Bohr magnetons per Fe. The electron paramagnetic resonance (EPR) spectra at 4.2°K (Fig. 4a,b) are qualitatively identical to EPR spectra of purified, unfractionated nitrogenase (Fig. 4c,d); resonances are observed at g = 2.01, 3.67, and 4.30 in the native protein; these signals are intensified on reduction, and an additional signal develops at g = 1.94. The Mössbauer spectrum (Fig. 5) is unchanged between 20° and 200°K and shows a large doublet with an additional signal on the high field shoulder; the latter signal increases at low pH and on treatment with dithionite;[40] the

[40] G. V. Novikov, L. A. Syrtsova, G. I. Likhtenshtein, V. A. Trukhtanov, V. F. Rachek, and V. L. Gol'danskii, Dokl. Phys. Chem. USSR 181, 590 (1968).

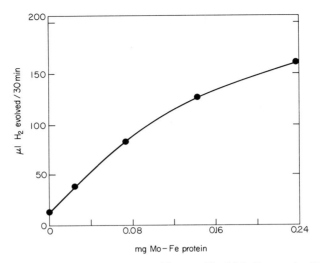

FIG. 3. Titration of Fe protein fraction with crystallized Mo-Fe protein. H_2 evolution was assayed as described in the section on assay methods; vessels contained 0.34 mg of Fe protein fraction and indicated amounts of 3× crystallized Mo-Fe protein.

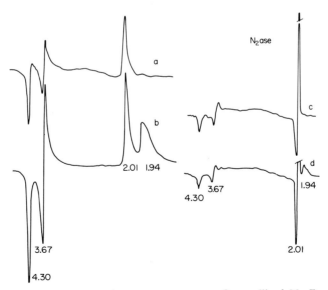

FIG. 4. Electron paramagnetic resonance spectra of crystallized Mo-Fe protein and purified nitrogenase. Mo-Fe protein (65 mg/ml): a, as prepared; b, plus $Na_2S_2O_4$. Purified nitrogenase (48 mg of PS-2 protein per milliliter): c, as prepared; d, plus $Na_2S_2O_4$. g values are indicated; temperature 4.2°K.[11]

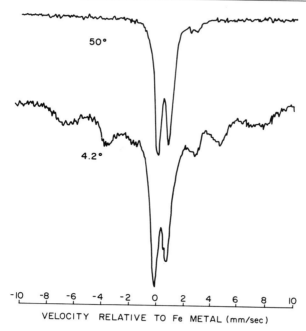

50°

4.2°

-10 -8 -6 -4 -2 0 2 4 6 8 10
VELOCITY RELATIVE TO Fe METAL (mm/sec)

Fig. 5   Mössbauer spectra of Mo-Fe protein crystallized from preparations of *Azotobacter vinelandii* grown in 92.8% [57]Fe-enriched medium. Temperature 50°K and 4.2°K.[11]

doublet has a quadrupole split of 0.84 mm/second and an isomer shift of 0.31 mm/second. At temperatures below 20°K the doublet begins to collapse, giving rise to outer lines in the spectrum.

Protein solubility is highly responsive to ionic strength, as indicated by Fig. 6; heat of solution is −5.4 kcal/mole. Results qualitatively similar to those shown with NaCl are observed with KCl, KNO₃, Na₂SO₄, MgCl₂, NH₄Cl, sodium formate, sodium acetate, ammonium formate, Tris·HCl, and sodium phosphate.

Properties of the Fe Protein. This protein, contained in the 0.35 *M* NaCl DEAE column fraction, is highly sensitive to air and to cold (0°),[18] which lead to irreversible denaturation. In dilute solutions (4–6 mg of protein per milliliter) at room temperature under N₂ the protein loses all activity in about 1 week, but it retains activity longer at higher protein concentrations; almost no loss of activity is incurred in preparations stored in liquid N₂ at high protein concentration (40–80 mg/ml) for more than a month, and such preparations can be thawed rapidly to room temperature and refrozen in liquid N₂ repeatedly without ill effects. Activity is here understood to mean ability to recombine with Mo-Fe protein to

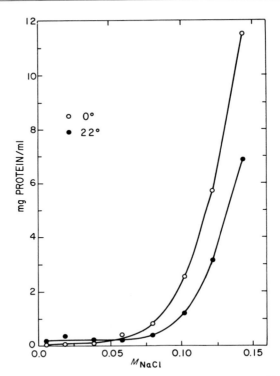

FIG. 6.   Solubility of Mo-Fe protein as a function of M_{NaCl} at $0°$ and $22°$. At each temperature Mo-Fe protein was added to salt solutions in the indicated range, undissolved protein was removed by centrifugation, and the protein concentration of the supernatant solution was determined; all solutions were $0.015\ M$ in Tris·HCl, pH 7.4, and contained NaCl as indicated; all operations were performed anaerobically. From R. C. Burns, R. D. Holsten, and R. W. F. Hardy, *Biochem. Biophys. Res. Commun.* **39,** 90 (1970).

form enzymatically active nitrogenase, since the Fe protein alone has no known intrinsic catalytic activity. Molecular weight is approximately 40,000, and the protein is reported to contain 2–3 equivalents of Fe[18,41] and presumably labile sulfur.

[41]R. H. Burris, *Proc. Roy. Soc. Ser. B* **172,** 339 (1969).

[44] N₂-Fixing Plant-Bacterial Symbiosis in Tissue Culture

By R. D. HOLSTEN and R. W. F. HARDY

The infection of specific plants of the family Leguminosae by selective bacteria of the genus *Rhizobium* results in the formation of the vital N₂-fixing plant–bacterial symbiosis. The chemistry and biology of the development of this symbiosis are only poorly defined, largely because of the complexity and diversity of experimental systems utilizing whole plants[1] or more recently root organ cultures.[2, 3] The tissue culture symbiotic system combining plant tissue and bacterial culture techniques provides the most defined system available as yet for the study of the symbiotic process. This experimental system developed by Holsten *et al.*[4,5] will be described.

Growth of Soybean Root Tissue Culture

Soybean seeds (*Glycine max* var. 'Acme'), surface sterilized by a 15-minute immersion in a commercial hypochlorite solution (Zonite, Chemway Corporation, Wayne, New Jersey) and thoroughly rinsed with sterile distilled water, are germinated on the surface of 1% (w/v) agar-solidified medium based on a modification of that described by Murashige and Skoog[6] here designated as MS and composed as follows, in milligrams per liter: KNO₃, 1900; NH₄NO₃, 1650; CaCl₂ · 2H₂O, 440; MgSO₄ · 7H₂O, 370; KH₂PO₄, 170; Na₂EDTA, 37.3; FeSO₄ · 7H₂O, 27.8; MnSO₄ · 4H₂O, 22.3; ZnSO₄ · 4H₂O, 8.6; H₃BO₃, 6.2; KI, 0.83; Na₂MoO₄ · 2H₂O, 0.25; CuSO₄ · 5H₂O, 0.025; CoCl₂ · 6H₂O, 0.025; thiamine · HCl, 0.4; inositol, 100; sucrose 30,000. Media are routinely adjusted to pH 6.4 and sterilized by autoclaving at 15 psi for 15 minutes. All incubations are at 26° ± 1° and in the dark unless otherwise noted.

A 5-mm explant aseptically excised at 2.5 cm from the growing tip of a 4-cm primary root is incubated on the surface of 50 ml of agar-

[1] E. K. Allen and O. N. Allen, *in* "Handbuch der Pflanzenphysiologie – Encyclopedia of Plant Physiology" (W. Ruhland, ed.), Vol. 8, p. 48. Springer-Verlag, Berlin, 1958.

[2] M. Raggio, N. Raggio, and J. G. Torrey, *Amer. J. Bot.* **44**, 325 (1957).

[3] N. Raggio, M. Raggio, and R. H. Burris, *Biochim. Biophys. Acta* **32**, 274 (1959).

[4] R. D. Holsten, R. R. Hebert, R. C. Burns, and R. W. F. Hardy, *Bacteriol. Proc.*, p. 149 (1970).

[5] R. D. Holsten, R. R. Hebert, R. C. Burns, and R. W. F. Hardy, *Nature* **232**, 173–176 (1971).

[6] T. Murashige and F. Skoog, *Physiol. Plant.* **15**, 473 (1962).

solidified MS medium supplemented with 15% (v/v) deproteinized coconut milk (CM) (Grand Island Biological Co., Grand Island, New York) and 2 mg of 2,4-dichlorophenoxyacetic acid (2,4-D) per liter contained in a cotton-stoppered 250-ml Erlenmeyer flask. After 20–25 days, 25–40-mg portions of the actively growing callus produced on the root explant are aseptically removed and transferred to 225 ml of liquid media of similar composition, i.e., MS + CM + 2,4-D, in a multisidearm 1000-ml flask (Blaessig Glass Co., Rochester, New York) and rotated at 1 rpm on a clinostat (Fig. 1). An actively growing cell and callus popu-

Fig. 1. Clinostat and multisidearm flasks used for growth of large-volume plant tissue culture and for initiation of symbiosis.

lation of about 20 g fresh weight results after incubation for 14–21 days. For maintenance of an active cell line, the liquid system is subcultured by pipetting 1 ml of the liquid media containing cells and tissue pieces into fresh MS + CM + 2,4-D media every 20–30 days. Viable populations of soybean root cell tissue have been maintained for several years by this technique. It is important that a rapidly growing plant cell and callus culture with a doubling time of 18 hours or less be established; otherwise, the bacterial culture will rapidly outgrow the plant culture during the period required to initiate a functional symbiosis.

Growth of Bacterial Cultures

Rhizobium japonicum strain 61A76 (Dr. J. C. Burton, The Nitragin Co., Milwaukee, Wisconsin) is cultured in the light at $30 \pm 1°$ in liquid medium[7] composed as follows, in grams per liter: K_2HPO_4, 1:0; KH_2PO_4, 1.0; $FeCl_3 \cdot 6H_2O$, 0.005; $MgSO_4 \cdot 7H_2O$, 0.36; $CaSO_4 \cdot 2H_2O$, 0.17; KNO_3, 0.7; yeast extract, 1.0; sucrose or mannitol, 3.0, at pH 6.8. Subcultures are made every 7 days. Strain 61A76, a comparatively slow-growing *Rhizobium*, is used in an effort to balance the rate of plant and bacterial growth. Nodulating effectiveness of the bacteria is checked by application to soybean seeds and observation of nodule development and N_2-fixing activity by C_2H_2–C_2H_4 assay[8] of the root systems.

Establishment of Symbiosis

Approximately 20 g, fresh weight, of rapidly growing cell and callus culture of root origin in 225 ml of MS + CM + 2,4-D liquid medium is aseptically inoculated with about 0.1 ml, or 25×10^6, cells of an actively growing liquid culture of *R. japonicum*. The plant cell–bacterial culture is incubated for about 5 days in the multisidearm 1000-ml flask rotated at 1 rpm on the clinostat. Although all prior and later subcultures of individual plant or plant–bacterial tissue may be made using semisolid agar surfaces or liquid media in Erlenmeyer flasks on rotary shakers, the gentle shearing forces and continuous aeration produced by the slowly rotating multisidearm flasks facilitate the initial establishment of the symbiosis. After about 5 days an effective symbiosis, referred to as stage 1, is usually established; it is characterized by a rough-surfaced callus mottled with both light and dark areas. The stage 1 incubation mixture is filtered through sterile cheese-cloth, and the retained plant callus tissue is washed with 250 ml of MS

---

[7] S. Ahmed and H. J. Evans, *Soil Sci.* **90**, 205 (1960).
[8] R. W. F. Hardy, R. D. Holsten, E. K. Jackson, and R. C. Burns, *Plant Physiol.* **43**, 1185 (1968).

medium to remove extracellular bacteria. The washed callus may be assayed or may be transferred to either liquid or agar-solidified media for an additional incubation period. Variation of the media with respect to auxins or cytokinins at this stage will alter the further development of the symbiosis (stage 2); thus MS medium unsupplemented with either cytokinins or auxin is preferred during this period.

Characteristics of Symbiosis

The cytology of the callus symbiosis at the light and electron microscope level shows similarities to the nodule system formed by bacteria

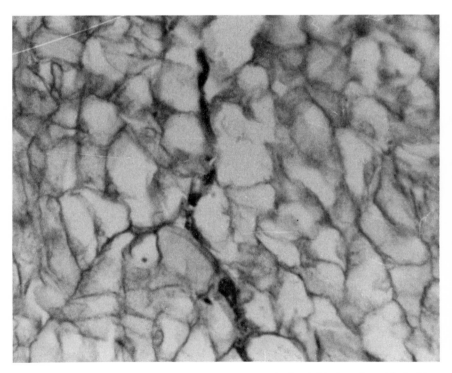

Fig. 2. Light micrograph of a soybean root callus during initiation of symbiosis. Note infection thread containing bacteria. From R. D. Holsten, R. R. Hebert, R. C. Burns, and R. W. F. Hardy, *Nature* **232**, 173–176 (1971). Callus was killed and fixed in formalin:acetic acid:ethanol (5:5:90; v/v), dehydrated through the ethanol series, and embedded in Paraplast of melting point 56–57°. Sections 8–10 μ thick were cut, stained with safranin-fast green, and photographed in a Zeiss microscope (D. A. Johansen, "Plant Microtechnique." McGraw-Hill, New York, 1940).

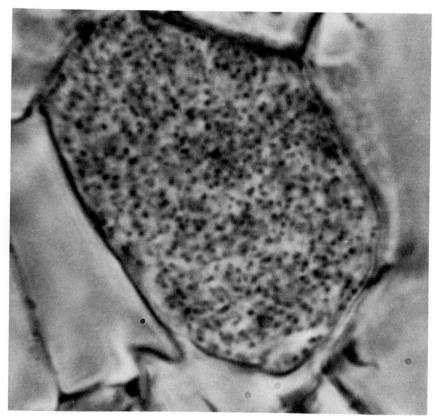

FIG. 3.   Light micrograph of soybean callus cell at the end of stage 1 filled with *Rhizobium japonicum*. ×2100. From R. D. Holsten, R. R. Hebert, R. C. Burns, and R. W. F. Hardy, *Nature* **232**, 173–176 (1971).

interacting with the whole plants.[9,10] Infection threadlike structures are observed in the callus during stage 1. The threads penetrate some distance into the interior of the callus before releasing the infecting bacteria (Fig. 2). At the end of stage 1, some cells are densely packed with bacteria and the cytoplasmic contents are displaced to the periphery of the cell (Fig. 3). During stage 2 there is further bacterial involvement of the callus tissue with some suggestion of the development of enclosing vesicles (Figs. 4 and 5) within the infected cells. An inclusion

[9]D. J. Goodchild and F. J. Bergersen, *J. Bacteriol.* **92**, 204 (1966).
[10]R. D. Holsten, E. K. Jackson, R. R. Hebert, R. C. Burns, and R. W. F. Hardy, *Bacteriol. Proc.*, p. 149 (1969).

MS

MS + CM

MS + CM + 2,4-D

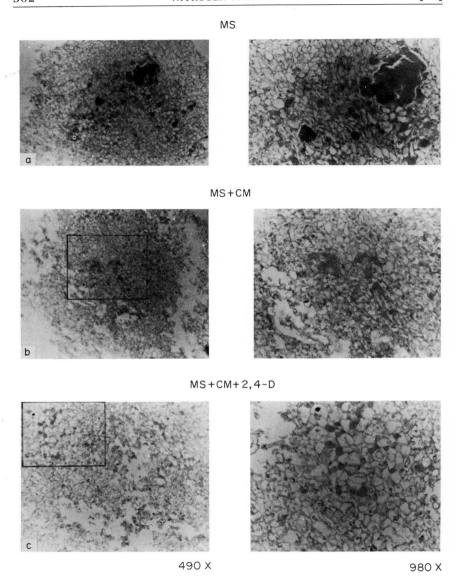

490 X   980 X

FIG. 4. Light micrograph of stage 2 soybean callus cultured on (a) MS, (b) MS + CM, and (c) MS + CM + 2,4-D media. Note increasing bacterial invasion with elimination of exogenous growth factors. From R. D. Holsten, R. R. Hebert, R. C. Burns, and R. W. F. Hardy, *Nature* **232**, 173–176 (1971).

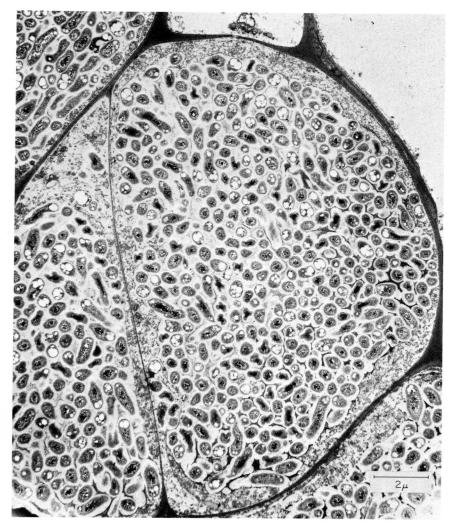

FIG. 5. Electron micrograph of stage 2 soybean callus cell. Callus tissue was overlaid with 5% glutaraldehyde in Millonig's phosphate buffer,[11] pH 7.2–7.4, fixed for 60 minutes at room temperature, rinsed with buffer for 20 minutes, treated with 1% OsO_4 at room temperature for 60 minutes, rinsed in buffer for 20 minutes, and dehydrated through an ethanol series up to 100% propylene oxide. Samples were embedded in Epon 812 (Miller-Stevenson, Danbury, Connecticut), polymerized at 13° for 24 hours, sectioned at 500 Å with a diamond knife, stained with uranyl acetate followed by Karnovsky's lead, and examined in an RCA EMU-36 electron microscope. From R. D. Holsten, R. R. Hebert, R. C. Burns, and R. W. F. Hardy, *Nature* **232**, 173–176 (1971).

[11]G. Millonig, *J. Appl. Phys.* **32**, 1637 (1961).

product, possibly β-hydroxybutyric acid, becomes prominent, as has been observed in intact nodules.

The callus symbiosis at stage 1 or later displays N_2-fixing activity as measured by the C_2H_2-C_2H_4 assay (see the table). The activity is about 1% of that of intact nodules and is markedly affected by exogenous hormones. Dark conditions are recommended for all developmental stages of the symbiotic system.

N₂-FIXING ACTIVITY OF SOYBEAN ROOT CALLUS WITH AN ESTABLISHED SYMBIOSIS

Tissue	Media	N_2-fixing activity (nmoles C_2H_4/mg fresh wt·24 hours)
Infected callus[a]	MS	0.32
Infected callus[a]	MS + 10% CM	0.13
Infected callus[a]	MS + 10% CM + 2 mg/l 2,4-D	0.04
Uninfected callus	MS	0.005

[a]Callus tissue was assayed after incubation for 14 days on semisolid medium in the dark. Washed stage 1 or later callus is assayed for N_2-fixing activity utilizing the C_2H_2-C_2H_4 method as described by Hardy et al. [R. W. F. Hardy, R. D. Holsten, E. K. Jackson, and R. C. Burns, Plant Physiol. 43, 1185 (1968)]. Approximately 200 mg fresh weight of callus is placed in a 50-ml disposable syringe and repeatedly flushed with Ar:O_2 (80:20) to remove air. The syringe is sealed containing 30 ml of gas, and incubation is initiated by the addition of 20 ml of Ar:O_2 containing 20% C_2H_2. Syringes are rotated at approximately 10 rpm for 1 hour of incubation at room temperature. Product gas is removed and analyzed for C_2H_4 by gas chromatography.

Author Index

Numbers in parentheses are reference numbers and indicate that an author's work is referred to, although his name is not cited in the text.

H

Hadfield, K. L., 468, 484, 490(21)
Halbach, K., 83
Hall, D. O., 304, 368, 444
Hall, T. C., 389, 391(34)
Hallier, U. W., 193, 261
Hamilton, M. G., 384
Hammond, E. C., 34, 36(18)
Hardy, R. W. F., 424, 426, 427(32), 459, 466, 468, 469, 470, 480, 481, 483(7, 10), 484, 489, 490, 492, 493(39), 494(11), 495(11), 496, 497, 499, 500, 501, 502, 503, 504
Harriman, G. R., 288
Harris, E. J., 371
Harris, S. E., 282
Hart, R. W., 116
Harth, E., 159
Haskins, R. H., 279
Hasselberger, F. X., 89, 91(13)
Hatakeyama, H., 100
Hatch, M. D., 394
Hatchard, C. G., 142
Haxo, F. T., 114, 115, 123
Hayashi, T., 203
Hayt, W. H., Jr., 299
Heath, O. V. S., 249
Heath, R. L., 358, 359(4)
Heber, U., 122, 190, 192, 194, 205, 261
Hebert, R. R., 497, 500, 501, 502, 503
Heidt, L. J., 288
Hellwarth, R. W., 30
Henderson, M., 87
Henderson, P. J. F., 373
Hendrich, W., 244
Henshall, J. D., 381
Herriott, D. R., 280
Herriott, J. R., 308, 480
Hersh, R. T., 432
Hesketh, J., 258
Hess, B., 330, 331(21), 332(21)
Hew, C. S., 259
Heytler, P. G., 368
Hildreth, W. W., 32, 221
Hilgenberg, W., 247
Hilgenheger, H., 205
Hill, K. L., 360
Hill, R., 146, 156, 157, 219, 221(6)
Hilton, J. L., 360
Himes, R. H., 432

Hind, G., 103, 107, 357, 358, 359(4), 360, 361, 365, 368(8)
Hinkson, J., 461, 469(15), 484, 490(7)
Hirt, R. C., 284, 285
Hiyama, T., 221
Ho, C., 311
Hoch, G. E., 93, 95, 167, 169, 170(6), 293, 366
Hodgson, B. W., 333
Höfer, M. P., 371
Hoering, T. C., 166
Hofnung, M., 118, 128, 129(4)
Hohl, M. C., 366
Hohorst, H. J., 264, 265
Holsten, R. D., 426, 427(32), 459, 466, 481, 483(10), 484, 489(10), 490(10, 16), 492, 496, 497, 499, 500, 501, 502, 503, 504
Holt, A. S., 288
Homann, P., 359
Honda, S. I., 203
Hong, J.-S., 431, 432, 441(8), 442(8), 443(6), 444(6)
Hongladarom, T., 203
Hopkins, D. W., 22, 23
Horio, T., 96, 97, 100, 363, 364, 366(45), 367, 370, 371(45)
Horiuti, Y., 97, 367
Houssier, C., 211, 212(27)
Huber, J. R., 40
Hudson, M. A., 261
Huffaker, R. C., 394
Hunt, R. E., 291
Hurwitz, J., 262
Huth, B. G., 36
Hvidt, A., 443
Hwang, J. C., 469, 490

I

Ilgenfritz, G., 337
Inesi, G., 351
Inhoffen, H. H., 212
Interrante, L. V., 114
Isemura, T., 202
Ito, T., 71
Itoh, M., 179, 182, 189, 198, 199, 202, 205
Izawa, S., 55, 148, 151, 179, 182, 189, 198, 199, 202, 205, 218, 358, 359, 360, 361(27), 365, 366, 368, 371, 375, 376, 468

J

Jackson, E. K., 426, 427(32), 459, 466, 468, 469, 481, 483(10), 484(10), 489, 490, 499, 501, 504
Jackson, J. B., 336, 337(2), 339, 342, 373, 374
Jackson, W. A., 169, 170, 247, 258(7)
Jacobson, A. B., 383, 389(21)
Jacobson, K. B., 388
Jagendorf, A. T., 59, 68, 69(1), 70, 103, 107, 109, 111, 112, 113, 319, 365, 366, 368, 369, 370, 374, 384, 389, 390, 391(37), 392(27, 37), 393(27, 37)
Jagger, W. S., 371
James, A. T., 360, 398
James, W. O., 398
Jardetzky, O., 314
Javan, A., 280
Jeng, D. Y., 446, 452(2), 455
Jenkins, F. A., 289
Jenner, E. L., 368
Jensen, L. H., 308, 480
Jensen, R. G., 59, 60(4), 62(4)
Jeumann, J., 377
Jobsis, F. F., 330, 343
Johansen, G., 443
Johnson, J. H., 371
Johnson, M. J., 458
Johnston, J. A., 167, 168
Joliot, A., 118, 129, 133, 134, 223, 224, 227, 238 358
Joliot, P., 118, 128, 129, 130, 131(8), 132, 133, 134, 223, 224, 226, 227, 234, 238
Jones, D. W., 306
Josse, J., 89

K

Kadota, K., 337
Kagan, M. R., 36
Kahn, J. S., 377, 390, 391(38), 392(38)
Kaiser, W., 279
Kakuno, T., 97, 367
Kamen, M. D., 89, 90(9), 96, 106, 364
Kanazawa, T., 176
Kanwisher, J. W., 115, 122
Kaplan, J. H., 112, 113, 319, 366
Kapphahn, J. I., 91, 92
Karlish, S. J. D., 373, 374
Karu, A. E., 366

Kasha, M., 291
Katoh, S., 151, 159, 164, 221, 359, 361
Katsumata, M., 96, 97(3), 367, 370
Katz, J. J., 207, 212
Kay, I. T., 207
Ke, B., 32, 39, 46, 52(14), 208, 214, 242
Keene, J. P., 333
Keister, D. L., 162, 245, 362, 364(38), 367, 371, 373, 374
Keller, C. J., 394
Kelly, F. H., 421
Kelly, J., 193
Kelly, M., 426, 448, 466, 467(26), 468, 469, 476, 483, 484, 489(13)
Kemmerly, J. E., 299
Kemp, R. J., 398, 401, 402(8), 403(2, 8), 404
Kennedy, I., 452
Kessler, E., 359
Keston, A. S., 418
Kihara, R., 221
Kikuchi, G., 363
Kinmonth, R., 284, 285
Kirk, J. T. O., 398
Kirkwood, J. G., 208
Kleiner, D., 423, 424(19)
Kleinhaus, H., 330, 331(21), 332(21)
Klemme, J.-H., 363, 364(43)
Klenert, M., 28, 29
Klucas, R. V., 448, 466, 467(26), 475, 476, 484
Knapp, F. F., 406
Knight, E., Jr., 424, 426, 468, 469(36), 480, 481, 483(7), 489, 490
Koch, B., 426, 459, 466, 471, 475, 476(9), 483
Kok, B., 85, 133, 134, 144, 154, 156(15), 160, 167, 169, 170(6), 220, 221, 226, 227(27), 229, 230, 235, 237(29), 238, 357
Koller, D., 246, 248, 258
Koshii, K., 202
Krasnovskii, A. A., 243
Kraut, J., 313
Kreutz, W., 53, 235
Krippahl, G., 64
Kroes, H. H., 207
Krogmann, D. W., 365, 366, 371
Krotkov, G., 253, 258(26), 259
Krueger, W. C., 207
Kuhn, W., 208

Subject Index

F

Fatty acid synthesis, 394–397
  assay, 395–396
  chloroplast preparation, 396–397
Ferredoxin, 221, 229
  assay, 434–435
  Hill reaction, 154–155
  magnetic susceptibility, 307, 309–311
  preparation, 434, 435–440
  properties, 443–446
  reconstruction from apoferredoxin, 442
  reduction, 155
Ferricyanide, reduction, 158, 159–160, 296
Flash kinetic spectrophotometry
  excitation flash, 27–37
  double-flash excitation, 52–53
  light modulation, 41–42
  measuring geometry, 39–40
  measuring light, 37–39
  monochromatic light, 42–43
  optical interference, 43–44, 51–52
  photomultipliers, 44
  signal averaging, 47–50
  signal-noise ratio, 45–46
Flash lamps
  organic dye laser, 33–36
  Q-switched ruby laser, 29–32
  stimulated Raman emission, 32–33
  xenon, 27–28
  "Z-pinch," 28–29
Fluorescence
  chlorophyll *in vivo*, 135–138
Fluorescence leakage, 51–52
Fluorometric assay
  glucose-6-phosphate, 91–92
Folin protein assay, 60, 61, 62
Fructose-1,6-diphosphate assay, 264, 265–266
Fructose-6-phosphate assay, 265–266

G

Gas-liquid chromatography
  sterol esters, 401–402
  sterols, 403–404
Glucose-1-phosphate, assay, 265–266

Glucose-6-phosphate assay
  enzymatic-optical, 265–266
  fluorometric, 91–92
  radiochemical, 90–91
Glucose transport, 375
Glyceraldehyde-3-phosphate, assay, 264
Glycerate-3-phosphate, assay, 266
Glycine buffer, 64
Glycine max
  root tissue culture, 497–499
  symbiosis, 499–504
Glycolate assay, 267–268
Glycylglycine buffer, 64
Green safelight, 321–322

H

HEPES buffer, 62, 67–68
HEPPS buffer, 62
High-potential iron protein, 316–317
Hill reaction
  ferredoxin, 154–155
  ferredoxin independent, 156–159
  general conditions, 150–154
  inhibition, 361
  modulated electrode, 134
  photosystem I donors, 160–163
  photosystem II, 159–160, 165
  photosystem II donors, 163–165
  reaction type I, 154–155
  reaction type II, 156–159
  reaction type III, 159–160
  reaction type IV, 160–163
  reaction type V, 163–165
  reaction type VI, 165
Honda medium, 382
Hydrazobenzene, 240
Hydrogenase activity, 423–424
Hydrogen exchange phenomena, 68–74
Hydrogen ion buffers, 53–68
Hydrogen ion concentration, indicators, 336–342
Hydroquinone, electron donor, 164
N-2-Hydroxyethylpiperazine-N'-ethane-sulfonic acid, 62
N-2-Hydroxyethylpiperazine-N'-propane-sulfonic acid, 62
Hydroxylamine, 358

DATE DUE			
APR 2 2 1985			
DEC 3 1 1987			
FEB 1 5 1995			
DEC 0 9 1996			
NOV 1 5 1998			
GAYLORD			PRINTED IN U.S.A